SIXTH EDITION

Juvenile JUSTICE

SIXTH EDITION

Juvenile JUSTICE

A Guide to Theory, Policy, and Practice

Steven M. Cox
Professor Emeritus, Western Illinois University

Jennifer M. Allen
Western Illinois University

Robert D. Hanser
University of Louisiana at Monroe

John J. Conrad
Western Illinois University

SAGE Publications
Los Angeles • London • New Delhi • Singapore

For information:

Sage Publications, Inc.
2455 Teller Road
Thousand Oaks, California 91320
E-mail: order@sagepub.com

Sage Publications Ltd.
1 Oliver's Yard
55 City Road
London EC1Y 1SP
United Kingdom

Sage Publications India Pvt. Ltd.
B 1/I 1 Mohan Cooperative Industrial Area
Mathura Road, New Delhi 110 044
India

Sage Publications Asia-Pacific Pte. Ltd.
33 Pekin Street #02-01
Far East Square
Singapore 048763

Printed in the United States of America

Library of Congress Cataloging-in-Publication Data

Juvenile justice : a guide to theory, policy, and practice/Steven Cox . . . [et al.].—6th ed.
p. cm.
Includes bibliographical references and index.
ISBN 978-1-4129-5133-3 (pbk.)
1. Juvenile justice, Administration of—United States. I. Cox, Steven M.

HV9104C63 2008
364.360973—dc222007020910

This book is printed on acid-free paper.

07 08 09 10 11 10 9 8 7 6 5 4 3 2 1

Acquisitions Editor: Jerry Westby
Associate Editor: Deya Saoud
Editorial Assistant: Melissa Spor
Production Editor: Denise Santoyo/Karen Wiley
Copy Editor: D. J. Peck
Typesetter: C&M Digitals (P) Ltd.
Proofreader: Caryne Brown
Indexer: Rick Hurd
Cover Designer: Candice Harman
Senior Marketing Manager: Jennifer Reed

Contents

Preface xi
Approach xi
The Sixth Edition xii
Pedagogical Aids xiii
Instructor Supplements xiii

Acknowledgments xv
Photo Credits xvi

Chapter 1: Juvenile Justice in Historical Perspective 1
In Practice 1.1 2
Juvenile Justice Historically 4
Continuing Dilemmas in Juvenile Justice 11
In Practice 1.2 13
Rethinking Juvenile Justice 16
Career Opportunities in Juvenile Justice 17
Summary 17
Critical Thinking Questions 18
Suggested Readings 18

Chapter 2: Defining and Measuring Offenses by and Against Juveniles 19
Legal Definitions 20
Changing Definitions 20
Age Ambiguity 20
In Practice 2.1 21
Inaccurate Images of Offenders and Victims 22
Behavioral Definitions 24
Official Statistics: Sources and Problems 25
Official Delinquency Statistics 25
Official Statistics on Abuse and Neglect 27
The National Crime Victimization Survey 27
In Practice 2.2 28
Sources of Error in Official Statistics 30
Unofficial Sources of Data 31
Self-Report Studies 32
Police Observation Studies 34

Summary 35
Critical Thinking Questions 37
Suggested Readings 37

Chapter 3: Characteristics of Juvenile Offenders 39
In Practice 3.1 41
Social Factors 43
Family 43
In Practice 3.2 46
Education 50
In Practice 3.3 52
Social Class 56
Gangs 58
Drugs 58
Physical Factors 61
Age 61
Gender 67
Race 73
Summary 78
Critical Thinking Questions 80
Suggested Readings 80

Chapter 4: Theories of Causation 81
Scientific Theory 83
Some Early Theories 84
Demonology 84
Classical Theory 85
Rational Choice Theory 85
Deterrence Theory 85
Routine Activities Theory 86
The Positivist School 87
Biological Theories 88
Cesare Lombroso's "Born Criminal" Theory 88
Other Biological Theories 88
In Practice 4.1 91
Psychological Theories 92
Sigmund Freud's Psychoanalytic Approach 93
Psychopathology 94
Behaviorism and Learning Theory 95
In Practice 4.2 96
Sociological Theories 98
Anomie and Strain Theory 98
The Ecological/Social Disorganization Approach 99
Edwin Sutherland's Differential Association Theory 101
Labeling Theory 103
Conflict/Radical/Critical/Marxist Theories 104
Feminism 105
Control Theories 105

Integrated Theories 106
Summary 107
Critical Thinking Questions 109
Suggested Readings 109

Chapter 5: Purpose and Scope of Juvenile Court Acts 111
Purpose 112
Comparison of Adult Criminal Justice and Juvenile Justice Systems 113
Protecting the Juvenile From Stigmatization 114
In Practice 5.1 115
Maintaining the Family Unit 116
Preserving Constitutional Rights in Juvenile Court Proceedings 116
In Practice 5.2 117
Scope 120
Age 120
Delinquent Acts 122
Unruly Children 124
Deprived, Neglected, or Dependent Children 125
Jurisdiction 127
Concurrent or Exclusive Jurisdiction 128
Waiver 129
Double Jeopardy 135
Summary 136
Critical Thinking Questions 137
Suggested Readings 137

Chapter 6: Juvenile Justice Procedures 139
Rights of Juveniles 141
Bail 142
Taking Into Custody 143
Interrogation 144
In Practice 6.1 145
The Detention Hearing 147
In Practice 6.2 148
Detention/Shelter Care 152
The Preliminary Conference 153
The Petition 154
Notification 155
The Adjudicatory Hearing 156
The Social Background Investigation 160
The Dispositional Hearing 161
Summary 165
Critical Thinking Questions 166
Suggested Readings 167

Chapter 7: Juveniles and the Police 169
Police Discretion in Encounters With Juveniles 170
In Practice 7.1 174

Unofficial Procedures 175
Official Procedures 179
Training and Competence of Juvenile Officers 180
Police–School Consultant and Liaison Programs 181
In Practice 7.2 183
Community-Oriented Policing and Juveniles 185
Police and Juvenile Court 185
Summary 186
Critical Thinking Questions 187
Suggested Readings 188

Chapter 8: Key Figures in Juvenile Court Proceedings 189
The Prosecutor 190
Defense Counsel 192
Relationship Between the Prosecutor and Defense Counsel: Adversarial or Cooperative? 194
The Juvenile Court Judge 196
In Practice 8.1 197
In Practice 8.2 198
In Practice 8.3 200
The Juvenile Probation Officer 201
Children and Family Services Personnel 203
Court-Appointed Special Advocates 204
Training and Competence of Juvenile Court Personnel 204
Summary 206
Critical Thinking Questions 207
Suggested Readings 207

Chapter 9: Prevention and Diversion Programs 209
Prevention 211
Diversion Programs 214
In Practice 9.1 215
Some Examples of Prevention and Diversion Programs 216
School Programs 216
Wilderness Programs 219
Restorative Justice Programs 220
In Practice 9.2 221
Children and Family Services 224
Federal Programs 225
Other Diversion and Prevention Programs 228
Some Criticisms 230
Child Abuse and Neglect Prevention Programs 231
Summary 233
Critical Thinking Questions 234
Suggested Readings 234

Chapter 10: Dispositional Alternatives 237
Probation 239
Foster Homes 245
In Practice 10.1 246
Treatment Centers 247
In Practice 10.2 248
Juvenile Corrections 251
Capital Punishment and Youthful Offenders 254
Some Possible Solutions 255
Summary 259
Critical Thinking Questions 259
Suggested Readings 260

Chapter 11: Child Abuse and Neglect 261
Physical Abuse 263
In Practice 11.1 264
Child Neglect 267
Emotional Abuse of Children 268
Sexual Abuse of Children 268
In Practice 11.2 270
Internet Exploitation 271
Intervention 272
Summary 274
Critical Thinking Questions 275
Suggested Readings 275

Chapter 12: Violent Juveniles and Gangs 277
In Practice 12.1 279
Violent Juveniles 279
Firearms and Juvenile Violence 281
Gangs 284
A Brief History of Gangs 285
The Nature of Street Gangs 287
Delinquent and Criminal Gang Activities 288
In Practice 12.2 289
Gang Membership 293
Characteristics 294
Age 294
Gender 295
Monikers 296
Graffiti 296
Jargon 296
Recruitment 296
Response of Justice Network to Gangs 297
Public, Legislative, and Judicial Reaction 298

Alternatives to Incarceration for Violent Juveniles 298
Establishing a Juvenile Gang Exit Program 301
Summary 302
Critical Thinking Questions 304
Suggested Readings 304

Chapter 13: The Future of Juvenile Justice 307
In Practice 13.1 310
In Practice 13.2 318
Suggested Readings 320

Appendix 321

Glossary 347

References 379

Index 397

About the Authors 417

Preface

Since we wrote the first version of this text some 30 years ago, the juvenile justice network has undergone dramatic and nearly constant change. The pace of this change has been rapid, and the changes have sometimes been confusing. Those who believe that the network "coddles" juveniles have been successful in convincing legislators in a variety of jurisdictions that juveniles who commit serious offenses should be treated as adults. At the same time, those who believe that treatment and education are better alternatives for most juveniles with problems have established restorative justice programs and other intermediate sanctions as alternatives, or additions, to official processing. Violent crime committed by juveniles, which had declined for a decade beginning in the mid-1990s, currently appears to be increasing in some areas, and concern with juvenile gangs seems to be proliferating. In addition, increased concerns in the development of school-based programs, victimization on school property, and bullying have been pushed to the forefront of delinquency prevention. New programs aimed at delinquency and promising to be more effective and efficient have been initiated, while older programs have largely disappeared from the scene. What sense, if any, can we make of these changes, and what are their implications for policy and practice in juvenile justice?

As both practitioners in the juvenile justice network and instructors in criminology, criminal justice, and sociology courses, we have time and again heard, "That's great in theory, but what about in practice?" We remain convinced that a basic understanding of the interrelationships among notions of causation, procedural requirements, and professional practices is a must if one is to understand, let alone practice in, the juvenile justice network.

With these concerns in mind, we have attempted to write a text that is reader friendly and comprehensive yet concise. As we revised the text for this new edition, these concerns remain. We have expanded discussions to keep them contemporary and have updated reference and legal materials throughout the text. In addition, we have made greater use of materials and resources available through the Internet.

Approach

In this text, we integrate juvenile law, theories of causation, and procedural requirements while examining their interrelationships. We have attempted to make our treatment of these issues both relevant and comprehensible to those who are actively employed in the juvenile justice network, to those who desire to become so employed, and to those whose interest in juvenile justice is more or less academic. We address the juvenile justice network as a composite of interacting individuals whose everyday decisions have very real consequences for others involved in the network. The day-to-day practical aspects of the network are discussed in terms of theoretical considerations and procedural requirements:

- The network approach allows us to examine the interrelationships among practitioners, offenders, victims, witnesses, and others involved with delinquency, abuse, neglect, and other varieties of behavior under the jurisdiction of the juvenile court.
- The roles of practitioners in the network are discussed in relationship to one another and with respect to discretion, politics, and societal concerns. Thus, the police, juvenile probation officers, and social service agents all have roles to play in providing services for juveniles with problems. Unless each contributes, the network is likely to be ineffective in dealing with these problems.
- The law, of course, plays a key role in juvenile justice, and we have attempted to present the most recent and important changes in juvenile law based on an overview of a number of states.
- What we know about theories of behavior should dictate the procedures and treatments employed in dealing with juveniles. To ignore theory is to ignore possible explanations for behavior, and treatment is likely to be ineffective if explanations of behavior are lacking. Thus, we spend time discussing theories of behavior and their importance in juvenile justice.

In the following pages, we define technical terms clearly where they are presented, and we have included numerous practical examples—which we call "In Practice" boxes—in an attempt to present readers with a basic understanding of both the theoretical and practical aspects of the juvenile justice network. These real-world "In Practice" boxes are designed to help students connect theory and practice and to focus on a number of critical issues.

The Sixth Edition

With this edition, a change of authors occurs. John Conrad, one of the original authors of the text, has retired and is actively involved in new pursuits. Because his insights and contributions remain an integral part of the text, we have continued to list him as one of the coauthors for this edition. Rob Hanser has graciously agreed to lend his expertise to this edition, and we welcome him aboard as the newest member of the team. His experience in youth counseling makes him a vital component to the textbook.

We have also made numerous substantive changes to this edition:

- Updated references
- Coverage of current concerns and recent trends in juvenile justice
- Coverage of restorative justice programs
- Expanded discussion of theory
- Discussion of recent changes in juvenile codes from a variety of states
- Expanded discussion of gangs
- A new look at the future of juvenile justice

Pedagogical Aids

To enhance learning, we have included the following devices in every chapter:

- "In Practice" boxes to help students see the practical applications of what they are reading
- "Career Opportunity" boxes
- Lists of key terms and end-of-chapter summaries to help students prepare for exams
- Suggested readings lists for students who are interested in reading more information on the topics discussed in the respective chapters
- End-of-chapter "Critical Thinking Questions" to encourage students to go beyond memorization of terms and concepts in their learning
- A glossary of terms commonly used in juvenile justice, as well as in this textbook, to assist students in learning the "language" of the system
- A Web-based study site with Internet links and Internet activities to encourage students to use the Internet as a research and learning tool and to provide additional information on juvenile justice and other study aids for students. (To visit the site, go to www.sagepub.com/juvenilejustice6study.)

Instructor Supplements

- The Instructor's Resource CD-Rom provides instructors with PowerPoint presentations, suggested course projects and classroom activities, suggested scholarly, film, and Web resources, sample syllabi, as well as a Computerized Test Bank, which allows for easy test creation using the provided multiple-choice, true/false, and short answer questions. This material is provided in an electronic format which makes it easy for instructors to pick-and-choose what they wish to utilize and to easily modify and incorporate the material into their own teaching resources for this course.

Acknowledgments

A number of people have helped in the preparation of this book. For their encouragement and assistance, we thank William P. McCamey, Gene Scaramella, Michael H. Hazlett, Giri Raj Gupta, Dennis C. Bliss, Terry Campbell, and Courtney Cox.

We also thank the following reviewers of the manuscript for their many helpful suggestions.

From 5th Edition:

Stephanie R. Bush, Baskette, Florida State University

Alejandro del Carmen, University of Texas at Arlington

John E. Holman, University of North Texas

Stephan D. Kaftan, Hawkeye Community College

William Kelly, Auburn University

Donna Massey, University of Tennessee at Martin

Janet McClellan, Southwestern Oregon Community College

Greg Scott, DePaul University

Anne T. Sulton, New Jersey City University

For 6th Edition:

Dennis Bliss, Western Illinois University

George E. Buzzy, Liberty University

Dorinda Dowis, Columbus State University

Kurt D. Siedschlaw, University of Nebraska at Kearney

We welcome your comments concerning the text:

Steven Cox (SM-Cox1@wiu.edu)

Jennifer M. Allen (JM-Allen4@wiu.edu)

Rob Hanser (Hanser@ulm.edu)

Photo Credits

Chapter 1, p. 3: Reprinted with permission of the Chicago History Museum; p. 7: © CORBIS; p. 9: © CORBIS.

Chapter 2, p. 22: © CORBIS; p. 24: © Tony Savino/Corbis; p. 33: © Getty Images.

Chapter 3, p. 40: © Jupiter Images/William Fritsch; p. 44: © Paul S. Howell/Getty Images; p. 49: © Jupiter Images/Bananastock; p. 54: © SW Productions/Getty Images; p. 56: © David Young-Wolff/Getty Images; p. 59: © Jérome Sessini/In Visu/CORBIS.

Chapter 4, p. 82: © Will & Deni McIntyre/Getty Images; p. 87: © Bettmann/CORBIS; p. 97: © Bettmann/CORBIS; p. 100: © Brenda Ann Kenneally/CORBIS.

Chapter 5, p. 129: © Sal Dimarco Jr./Getty Images.

Chapter 6, p. 140: Jupiter Images/William Fritsch; p. 143: © THE Oregonian/CORBIS SYGMA.

Chapter 7, p. 170: © Gabe Palmer/CORBIS; p. 176: © Davis Turner-Pool/Getty Images; p. 184: © Mark Perlstein/Getty Images.

Chapter 8, p. 190: © Andrew Lichtenstein/Getty Images; p. 193: © CORBIS.

Chapter 9, p. 210: © Andrew Lichtenstein/Getty Images; p. 218: © Bob Sciarrino/Star Ledger/CORBIS; p. 225: © Ed Kashi/CORBIS.

Chapter 10, p. 238: Jupiter Images/Curtis Johnson; p. 251: © Robert King/Getty Images.

Chapter 11, p. 263: © Robert A. Sabo/Getty Images; p. 267: Photo reproduced courtesy of The George Sim Johnston Archives of The New York Society for the Prevention of Cruelty to Children.

Chapter 12, p. 281: © MARK LEFFINGWELL/Getty Images; p. 284: © STAN HONDA/Getty Images; p. 291: © Christian Poveda/CORBIS.

Chapter 13, p. 309: © JERRY/Express News/CORBIS SYGMA.

JUVENILE JUSTICE IN HISTORICAL PERSPECTIVE

CHAPTER LEARNING OBJECTIVES

On completion of this chapter, students should be able to

- *Understand the history of juvenile justice in the United States*
- *Understand contemporary challenges to the juvenile justice network*
- *Discuss the controversy between due process and informality in juvenile justice*
- *Recognize discrepancies between the ideal and real juvenile justice networks*

KEY TERMS

Age of responsibility

Common law

Mens rea

Chancery courts

Parens patriae

In loco parentis

Houses of refuge

Reform schools

Era of socialized juvenile justice

Holmes case

Kent case

Gault case

Winship case

Breed v. Jones

McKeiver v. Pennsylvania

Roper v. Simmons

Therapeutic approach

Legalistic approach

The juvenile justice network in the United States grew out of, and remains embroiled in, controversy (see In Practice 1.1). More than a century after the creation of the first family court in Illinois (1899), the debate continues as to the goals to be pursued and the procedures to be employed within the network, and a considerable gap between theory and practice remains. Meanwhile, concern over delinquency in general, and violent delinquents in particular, continues to grow while confidence in the juvenile justice network continues to erode. As Bilchik (1999a) indicated, "The reduction of juvenile crime, violence, and victimization constitutes one of the most crucial challenges of the new millennium" (p. 1). Ironically, however, as Johnson (2006b) noted, "Police across the nation are linking the recent jump in the nation's violent-crime rate to an increasing number of juveniles involved in armed robberies, assaults, and other incidents" (p. 1A).

In Practice 1.1

Cops Zero in on City's "Wolf Packs"

There are predators on Jersey City streets that run in packs to bring down their prey. Police say these "wolf packs" of young people are responsible for up to 20 criminal incidents a week.

They are youth gangs, and there is nothing new under the sun about them since the 1950s except they have more firepower, have better transportation, and have expanded their sphere of influence beyond the traditional "neighborhood turf."

In Jersey City, police will have to determine if these wolf packs are just a small group of loosely structured, like-minded young toughs or a more sophisticated organization in the tradition of MS-13, the notorious gang founded in Los Angeles by Salvadorans.

In any case, street gangs are nearly 10 times more likely to use a gun in a crime than is the regular juvenile offender, according to the National Youth Gang Center (NYGC), a national law enforcement research and training organization.

Gangs are usually a factor in crime being on the rise in a city, and organizations like the NYGC provide surveys that suggest that gang crimes are cyclical and that a flare-up of gang violence in a year is typically local and not a national trend. This is bad news for Jersey City, where law enforcement officials suggest as much as 40 percent of the city crime is committed by juveniles.

One interesting study made available by the NYGC shows that gangs exist because they are accepted in some communities as a way of life. Gangs require a "safe haven" and an ability to draw members from a "recruitment pool." Gangs also have a psychological tie, a bond, with a community because its members may be children of residents or a neighborhood can relate to the economic conditions that help gangs to flourish. This is not to say that these communities may be in fear of gangs and want more police protection.

There are a dizzying number of state and federally funded programs out there addressing prevention and ways of dealing with youth gangs. There are many service providers competing for this money.

Jersey City police have their work cut out for them. It does not help that many of the city's youth are without summer jobs, have few recreational opportunities, and have too much time on their hands. It may take more than just one city department to battle the influence of gangs in poorer sections of the city that boasts an area called the Gold Coast.

The juvenile court was supposed to have provided due process protections along with care, treatment, and rehabilitation for juveniles while protecting society. Yet there is increasing doubt as to whether the juvenile justice network can meet any of these goals. Violence committed by juveniles, which some suggest occurs in cycles (Johnson, 2006b), has again attracted nationwide attention and raised a host of questions concerning the juvenile court, even though such violence had actually declined significantly during the prior decade. Can a court designed to protect and care for juveniles deal successfully with those who, seemingly without reason, kill their peers and parents? Is the juvenile justice network too "soft" in its dealings with such juveniles? Isn't the "get tough" approach adopted over the past two decades what is needed to deal with violent adolescents? Was the juvenile court really designed to deal with the types of offenders we see today?

▲ The Juvenile Court Building, at Ewing and Halsted in Chicago in 1907, is shown. As noted in this chapter, the first family court in the United States was in Cook County, Illinois.

While due process for juveniles (discussed in detail later, but consisting of things such as the right to counsel and the right to remain silent), protection of society, and rehabilitation of youthful offenders remain elusive goals, frustration and dissatisfaction among those who work in the juvenile justice network, as well as among those who assess its effectiveness, remain the reality. Some observers have called for an end to juvenile justice as a separate system in the United States. Others maintain that the juvenile court and associated agencies and programs have a good deal to offer juveniles in trouble. For example, evidence from a survey of Tennessee residents indicates that, in that state at least, the public believes that rehabilitation should still be an integral goal of the juvenile justice network (Moon, Sundt, Cullen, & Wright, 2000). In a similar vein, in the 1997 legislative session in Maryland, the legislature revised the Juvenile Causes Act with a focus on balanced, restorative, and victim-centered justice. The revised act emphasizes prevention through development of programs for at-risk juveniles and also focuses on improving the network's response to offenders by providing a continuum of sanctions and treatment alternatives (Simms, 1997, p. 94). Yet during the 1990s, public fear of juvenile crime led the public to demand that legislators enact increasingly severe penalties for young offenders. Fanton (2006), in discussing the juvenile justice network in Illinois, concluded that "by the end of the 20th century the line between the Illinois juvenile justice and criminal justice systems was hopelessly blurred, reflecting a national trend" (p. A5). As Snyder and Sickmund (2006) pointed out, however, "America's youth are facing an ever-changing set of problems and barriers to successful lives. As a result, we are constantly challenged to develop enlightened policies and programs to address the needs and risks of those youth who enter our juvenile justice system. The policies and programs we create must be based on facts, not fears."

Can the reality and the ideal of the juvenile justice network be made more consistent? What can be done to bring about such consistency? What are the consequences of lack of consistency? A brief look at the history of juvenile justice and a detailed look at the network as it currently operates should help us answer these questions.

Juvenile Justice Historically

The distinction between youthful and adult offenders coincides with the beginning of recorded history. Some 4,000 years ago, the Code of Hammurabi (2270 BC) discussed runaways, children who disowned their parents, and sons who cursed their fathers. Approximately 2,000 years ago, both Roman civil law and later canon (church) law made distinctions between juveniles and adults based on the notion of **age of responsibility.** In ancient Jewish law, the Talmud specified conditions under which immaturity was to be considered in imposing punishment. There was no corporal punishment prior to puberty, which was considered to be the age of 12 years for females and 13 years for males. No capital punishment was to be imposed for those under 20 years of age. Similar leniency was found among Muslims, where children under the age of 17 years were typically exempt from the death penalty (Bernard, 1992).

By the 5th century BC, codification of Roman law resulted in the Twelve Tables, which made it clear that children were criminally responsible for violations of law and were to be dealt with by the criminal justice system (Nyquist, 1960). Punishment for some offenses, however, was less severe for children than for adults. For example, theft of crops by night was a capital offense for adults, but offenders under the age of puberty were only to be flogged. Adults caught in the act of theft were subject to flogging and enslavement to the victims, but children received only corporal punishment at the discretion of a magistrate and were required to make restitution (Ludwig, 1955). Originally, only those children who were incapable of speech were spared under Roman law, but eventually immunity was afforded to all children under the age of 7 years as the law came to reflect an increasing recognition of the stages of life. Children came to be classified as *infans, proximus infantia, and proximus pubertati.* In general, infans were not held criminally responsible, but those approaching puberty who knew the difference between right and wrong were held accountable. In the 5th century AD, the age of infantia was fixed at 7 years and children under that age were exempt from criminal liability. The legal age of puberty was fixed at 14 years for boys and 12 years for girls, and older children were held criminally liable. For children between the ages of 7 years and puberty, liability was based on capacity to understand the difference between right and wrong (Bernard, 1992).

Roman and canon law undoubtedly influenced early Anglo-Saxon **common law** (law based on custom or use), which emerged in England during the 11th and 12th centuries. For our purposes, the distinctions made between adult and juvenile offenders in England at this time are most significant. Under common law, children under the age of 7 years were presumed to be incapable of forming criminal intent and, therefore, were not subject to criminal sanctions. Children between the ages of 7 and 14 years were not subject to criminal sanctions unless it could be demonstrated that they had formed criminal intent, understood the consequences of their actions, and could distinguish right from wrong (Blackstone, 1803, pp. 22–24). Children over the age of 14 years were treated much the same as adults.

The question of when and under what circumstances children are capable of forming criminal intent (***mens rea*** or "guilty mind") remains a point of contention in juvenile justice proceedings today. For an adult to commit criminal homicide, for instance, it must be shown not only that the adult took the life of another human without justification but also that he or she *intended* to take the life of that individual. One may take the life of another accidentally (without intending to do so), and such an act is not regarded as criminal homicide. In other words, it takes more than the commission of an illegal act to produce a crime. Intent is also required (and, in fact, in some cases it is assumed as a result of the seriousness of the act, e.g., felony murder statutes).

But at what age is a child capable of understanding the differences between right and wrong or of comprehending the consequences of his or her acts before they occur? For example, most of us would not regard a 4-year-old who pocketed some money found at a neighbor's house as a criminal act because we are confident that the child cannot understand the consequences of this act. But what about an 8- or 9- or 12-year-old?

Another important step in the history of juvenile justice occurred during the 15th century when chancery or equity courts were created by the King of England. **Chancery courts,** under the guidance of the king's chancellor, were created to consider petitions of those who were in need of special aid or intervention, such as women and children left in need of protection and aid by reason of divorce, death of a spouse, or abandonment, and to grant relief to such persons. Through the chancery courts, the king exercised the right of ***parens patriae*** ("parent of the country") by enabling these courts to act ***in loco parentis*** ("in the place of parents") to provide necessary services for the benefit of women and children (Bynum & Thompson, 1992). In other words, the king, as ruler of his country, was to assume responsibility for all of those under his rule, to provide parental care for children who had no parents, and to assist women who required aid for any of the reasons just mentioned. Although chancery courts did not normally deal with youthful offenders, they did deal with dependent or neglected children, as do juvenile courts in the United States today. The principle of *parens patriae* later became central to the development of the juvenile court in America and today generally refers to the fact that the state (government) has ultimate parental authority over juveniles in need of protection or guidance. In certain cases, then, the state may act *in loco parentis* and make decisions concerning the best interests of children. This includes removing children from the home of their parents when circumstances warrant.

In 1562, Parliament passed the Statute of Artificers, which stated that children of paupers could be involuntarily separated from their parents and apprenticed to others (Rendleman,

1974, p. 77). Similarly, the Poor Law Act of 1601 provided for involuntary separation of children from impoverished parents, and these children were then placed in bondage to local residents as apprentices. Both statutes were based on the belief that the state has a primary interest in the welfare of children and the right to ensure such welfare. At the same time, a system known as the "City Custom of Apprentices" operated in London. The system was established to settle disputes involving apprentices who were unruly or abused by their masters in an attempt to punish the appropriate parties. When an apprentice was found to be at fault and required confinement, he or she was segregated from adult offenders. Those in charge of the City Custom of Apprentices attempted to settle disputes in a confidential fashion so that the juveniles involved were not subjected to public shame or stigma (Sanders, 1974, pp. 46–47).

Throughout the 1600s and most of the 1700s, juvenile offenders in England were sent to adult prisons, although they were at times kept separate from adult offenders. The Hospital of St. Michael's, the first institution for the treatment of juvenile offenders, was established in Rome in 1704 by Pope Clement XI. The stated purpose of the hospital was to correct and instruct unruly juveniles so that they might become useful citizens (Griffin & Griffin, 1978, p. 7).

The first private separate institution for youthful offenders in England was established by Robert Young in 1788. The goal of this institution was "to educate and instruct in some useful trade or occupation the children of convicts or such other infant poor as [were] engaged in a vagrant and criminal course of life" (Sanders, 1974, p. 48).

During the early 1800s, changes in the criminal code that would have allowed English magistrates to hear cases of youthful offenders without the necessity of long delays were recommended. In addition, dependent or neglected children were to be appointed legal guardians who were to aid the children through care and education (Sanders, 1974, p. 49). These changes were rejected by the House of Lords due to the opposition to the magistrate's becoming "judges, juries, and executioners" and due to suspicion concerning the recommended confidentiality of the proceedings, which would have excluded the public and the press (pp. 50–51).

Meanwhile in the United States, dissatisfaction with the way young offenders were being handled was increasing. As early as 1825, the Society for the Prevention of Juvenile Delinquency advocated separating juvenile and adult offenders (Snyder & Sickmund, 1999). Up to this point in time, youthful offenders had been generally subjected to the same penalties as adults, and little or no attempt was made to separate juveniles from adults in jails or prisons. This caused a good deal of concern among reformers who feared that criminal attitudes and knowledge would be passed from the adults to the juveniles. Another concern centered on the possibility of brutality directed by the adults toward juveniles. Although many juveniles were being imprisoned, few appeared to benefit from the experience. Others simply appealed to the sympathy of jurors to escape the consequences of their acts entirely. With no alternative to imprisonment, juries and juvenile justice officials were inclined to respond emotionally and sympathetically to the plight of children, often causing them to overlook juvenile misdeeds or render lenient verdicts (Dorne & Gewerth, 1998, p. 4).

In 1818, a New York City committee on pauperism gave the term *juvenile delinquency* its first public recognition by referring to it as a major cause of pauperism (Drowns & Hess, 1990, p. 9). As a result of this increasing recognition of the problem of delinquency, several institutions for juveniles were established between 1824 and 1828. These institutions were oriented toward education and treatment rather than punishment, although whippings, long periods of silence, and loss of rewards were used to punish the uncooperative. In addition, strict regimentation and a strong work ethic philosophy were common.

Under the concept of *in loco parentis,* institutional custodians acted as parental substitutes with far-reaching powers over their charges. For example, the staff members of the New York House of Refuge, established in 1825, were able to bind out wards as apprentices, although the consent of the child involved was required. Whether or not such consent was voluntary is questionable given that the alternatives were likely unpleasant. The New York House of Refuge was soon followed by others in Boston and Philadelphia (Abadinsky & Winfree, 1992).

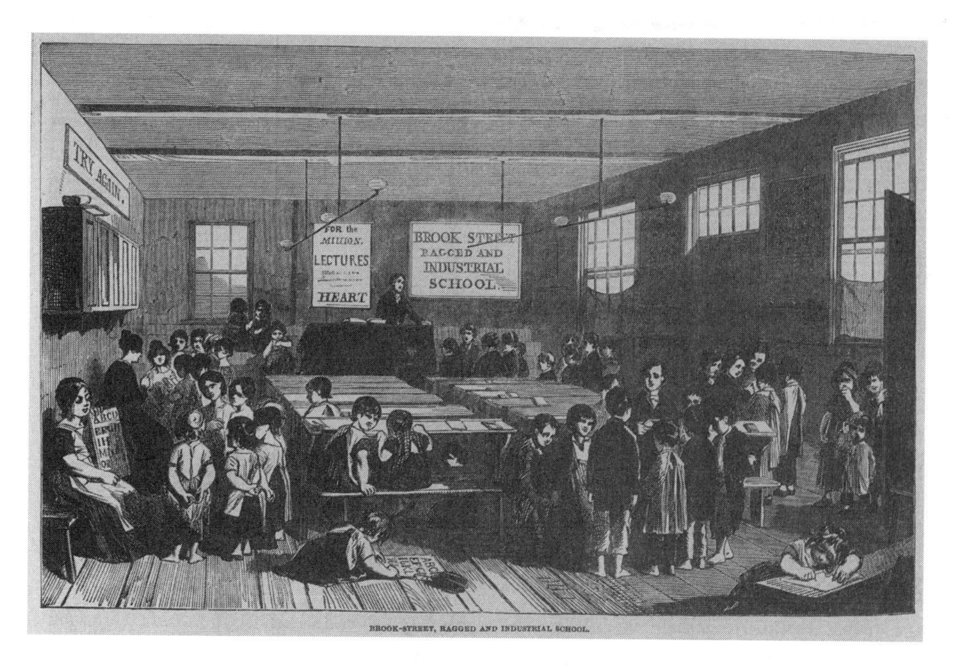

▲ Founded in 1843 in Hampstead Road, Birmingham, and known as the Brook-Street Ragged and Industrial school, this was an early reform school.

"By the mid-1800s, **houses of refuge** were enthusiastically declared a great success. Managers even advertised their houses in magazines for youth. Managers took great pride in seemingly turning total misfits into productive, hard-working members of society" (Simonsen & Gordon, 1982, p. 23, boldface added). However, these claims of success were not undisputed, and by 1850 it was widely recognized that houses of refuge were largely failures when it came to rehabilitating delinquents and had become much like prisons. As Simonsen and Gordon (1982) stated, "In 1849 the New York City police chief publicly warned that the numbers of vicious and vagrant youth were increasing and that something must be done. And done it was. America moved from a time of houses of refuge into a time of preventive agencies and reform schools" (p. 23).

In Illinois, the Chicago Reform School Act was passed in 1855, followed in 1879 by the establishment of Industrial Schools for dependent children. These schools were not unanimously approved, as indicated by the fact that in 1870 the Illinois Supreme Court declared unconstitutional the commitment of a child to the Chicago Reform School as a restraint on liberty without proof of crime and without conviction for an offense (*People ex rel. O'Connell v. Turner,* 1870). In 1888, the provisions of the Illinois Industrial School Act were also held to be unconstitutional, although the courts had ruled previously (1882) that the state had the right, under *parens patriae,* to "divest a child of liberty" by sending him or her to an industrial school if no other "lawful protector" could be found (*Petition of Ferrier,* 1882). In spite of good intentions, the new **reform schools,** existing in both England and the United States by the 1850s, were not effective in reducing the incidence of delinquency. Despite early enthusiasm among reformers, there was little evidence that rehabilitation was being accomplished. Piscotta's (1982) investigation of the effects of the 19th-century *parens patriae* doctrine led him to conclude that although inmates sometimes benefited from their incarceration and reformatories were not complete failures in achieving their objectives (whatever those were), the available evidence showed that the state was not a benevolent parent. In short, there was significant disparity between the promise and practice of *parens patriae.*

> Discipline was seldom "parental" in nature; inmate workers were exploited under the contract labor system, religious instruction was often disguised proselytization, and the indenture system generally failed to provide inmates with a home in the country. The frequency of escapes, assaults, incendiary incidents, and homosexual relations suggests that the children were not separated from the corrupting influence of improper associates. (Piscotta, 1982, pp. 424–425)

The failures of reform schools increased interest in the legality of the proceedings that allowed juveniles to be placed in such institutions. During the last half of the 19th century, there were a number of court challenges concerning the legality of failure to provide due process for youthful offenders. Some indicated that due process was required before incarceration (imprisonment) could occur, and others argued that due process was unnecessary because the intent of the proceedings was not punishment but rather treatment. In other words, juveniles were presumably being processed by the courts in their own "best interests."

During the post–Civil War period, an era of humanitarian concern emerged, focusing on children laboring in sweat shops, coal mines, and factories. These children, and others who were abandoned, orphaned, or viewed as criminally responsible, were a cause of alarm to reformist "child savers." The child savers movement included philanthropists, middle-class reformers, and professionals who exhibited a genuine concern for the welfare of children.

One of the outcomes of the Zeitgeist ("spirit of the times") of the late 19th century was the development of the first juvenile court in the United States. During the 1870s, several states (Massachusetts in 1874 and New York in 1892) had passed laws providing for separate trials for juveniles, but the first juvenile or family court did not appear until 1899 in Cook County, Illinois. "The delinquent child had ceased to be a criminal and had the status of a child in need of care, protection, and discipline directed toward rehabilitation" (Cavan, 1969, p. 362).

By incorporating the doctrine of *parens patriae,* the juvenile court was to act in the best interests of children through the use of noncriminal proceedings. The basic philosophy contained in the first juvenile court act reinforced the right of the state to act *in loco parentis* in

▲ Life in the reform schools of the 19th century was not easy.

cases involving children who had violated the law or were neglected, dependent, or otherwise in need of intervention or supervision. This philosophy changed the nature of the relationship between juveniles and the state by recognizing that juveniles were not simply miniature adults but rather children who could perhaps be served best through education and treatment. By 1917, juvenile court legislation had been passed in all but three states, and by 1932 there were more than 600 independent juvenile courts in the United States. By 1945, all states had passed legislation creating separate juvenile courts.

The period between 1899 and 1967 has been referred to as the **era of socialized juvenile justice** in the United States (Faust & Brantingham, 1974). During this era, children were considered not as miniature adults but rather as persons with less than fully developed morality and cognition (Snyder & Sickmund, 1999). Emphasis on the legal rights of the juvenile declined, and emphasis on determining how and why the juvenile came to the attention of the authorities and how best to treat and rehabilitate the juvenile became primary. The focus was clearly on offenders rather than the offenses they committed. Prevention and removal of the juvenile from undesirable social situations were the major concerns of the court. As Faust and Brantingham (1974) noted, "The blindfold was, therefore, purposefully removed from the eyes of 'justice' so that the total picture of the child's past experiences and existing circumstances could be judicially perceived and weighed against the projected outcomes of alternative courses of legal intervention" (p. 145).

It seems likely that the developers of the juvenile justice network in the United States intended legal intervention to be provided under the rules of civil law rather than criminal law. Clearly, they intended legal proceedings to be as informal as possible given that only through suspending the prohibition against hearsay and relying on the preponderance of evidence could the "total picture" of the juvenile be developed. The juvenile court exercised considerable discretion in dealing with the problems of youth and moved further and further from the ideas of legality, corrections, and punishment toward the ideas of prevention, treatment, and rehabilitation. This movement was, however, not unopposed. There were those who felt that the notion of informality was greatly abused and that any semblance of legality had been lost. The trial-and-error methods often employed during this era made guinea pigs out of juveniles who were placed in rehabilitation programs, which were often based on inadequately tested sociological and psychological theories (Faust & Brantingham, 1974, p. 149).

Nonetheless, in 1955, the U.S. Supreme Court reaffirmed the desirability of the informal procedures employed in juvenile courts. In deciding not to hear the ***Holmes*** **case,** the Court stated that because juvenile courts are not criminal courts, the constitutional rights guaranteed to accused adults do not apply to juveniles (*in re Holmes,* 1955).

Then, in the ***Kent*** **case** of 1961, 16-year-old Morris Kent, Jr., was charged with rape and robbery. Kent confessed, and the judge waived his case to criminal court based on what he verbally described as a "full investigation." Kent was found guilty and sentenced to 30 to 90 years in prison. His lawyer argued that the waiver was invalid, but appellate courts rejected the argument. He then appealed to the U.S. Supreme Court, arguing that the judge had not made a complete investigation and that Kent was denied his constitutional rights because he was a juvenile. The Court ruled that the waiver was invalid and that Kent was entitled to a hearing that included the essentials of due process or fair treatment required by the Fourteenth Amendment. In other words, Kent or his counsel should have had access to all records involved in making the decision to waive the case, and the judge should have provided written reasons for the waiver. Although the decision involved only District of Columbia courts, its implications were far-reaching by referring to the fact that juveniles might be receiving the worst of both worlds—less legal protection than adults and less treatment and rehabilitation than that promised by the juvenile courts (*Kent v. United States,* 1966).

In 1967, forces opposing the extreme informality of the juvenile court won a major victory when the U.S. Supreme Court handed down a decision in the case of Gerald Gault, a juvenile from Arizona. The extreme license taken by members of the juvenile justice network became abundantly clear in the ***Gault*** **case.** Gault, while a 15-year-old in 1964, was accused of making an obscene phone call to a neighbor who identified him. The neighbor did not appear at the adjudicatory hearing, and it was never demonstrated that Gault had, in fact, made the obscene comments. Still, Gault was sentenced to spend the remainder of his minority in a training school. Neither Gault nor his parents were notified properly of the charges against the juvenile. They were not made aware of their right to counsel, their right to confront and cross-examine witnesses, their right to remain silent, their right to a transcript of the proceedings, or their right to appeal. The Court ruled that in hearings that may result in institutional commitment, juveniles have all of these rights (*in re Gault,* 1967). The Supreme Court's decision in this case left little doubt that juvenile offenders are as entitled to the protection of constitutional guarantees as their adult counterparts, with the exception of participation in a public jury trial. In this case and in the *Kent* case, the Court raised serious questions about the concept of *parens patriae* or the right of the state to informally determine the best interests of juveniles. In addition, the Court noted that the handling of both Gault and Kent raised serious issues of Fourteenth Amendment (due process) violations. The free reign of socialized juvenile justice had come to an end, at least in theory.

During the years that followed, the U.S. Supreme Court continued the trend toward requiring due process rights for juveniles. In 1970, in the ***Winship* case,** the Court decided that in juvenile court proceedings involving delinquency, the standard of proof for conviction should be the same as that for adults in criminal court—proof beyond a reasonable doubt (*in re Winship*, 1970). In the case of ***Breed v. Jones*** (1975), the Court decided that trying a juvenile who had previously been adjudicated delinquent in juvenile court for the same crime as an adult in criminal court violates the double jeopardy clause of the Fifth Amendment when the adjudication involves violation of a criminal statute. The Court did not, however, go so far as to guarantee juveniles all of the same rights as adults. In 1971, in the case of ***McKeiver v. Pennsylvania,*** the Court held that the due process clause of the Fourteenth Amendment did not require jury trials in juvenile court. Nonetheless, some states have extended this right to juveniles through state law.

In March 2005, in the case of ***Roper v. Simmons,*** the U.S. Supreme Court reversed a 1989 precedent and struck down the death penalty for crimes committed by people under the age of 18 years. Christopher Simmons started talking about wanting to murder someone when he was 17 years old. On more than one occasion, he discussed with friends a plan to commit a burglary, tie up the victim, and push him or her from a bridge. Based on the specified plan, he and a younger friend broke into the home of Shirley Crook. They bound and blindfolded her and then drove her to a state park, where they tied her hands and feet with electrical wire, covered her whole face with duct tape, walked her to a railroad trestle, and threw her into the river. Crook drowned as a result of the juveniles' actions. Simmons later bragged about the murder, and the crime was not difficult to solve. On being taken into custody, he confessed, and the guilt phase of the trial in Missouri state court was uncontested (Bradley, 2006). The U.S. Supreme Court held that "evolving standards of decency" govern the prohibition of cruel and unusual punishment and found that "capital punishment must be limited to those offenders who commit a narrow category of the most serious crimes and whose extreme culpability makes them the most deserving of execution" (Death Penalty Information Center, n.d.). The Court further found that there is a scientific consensus that teenagers have "an underdeveloped sense of responsibility" and that, therefore, it is unreasonable to classify them among the most culpable offenders: "From a moral standpoint, it would be misguided to equate the failings of a minor with those of an adult, for a greater possibility exists that a minor's character deficiencies will be reformed" (Death Penalty Information Center, n.d.). In addition, the Court concluded that it would be extremely difficult for jurors to distinguish between juveniles whose crimes reflect immaturity and those whose crimes reflect "irreparable corruption." (Bradley, 2006). Finally, the Court pointed out that only seven countries in the world have executed juveniles since 1990, and even those countries now disallow the juvenile death penalty. Thus, the United States was the only country to still permit it. The pros and cons of this decision are discussed in Chapter 10, but suffice it to say now that the decision in this case furthered the considerable controversy that has characterized the juvenile justice network since its inception.

Continuing Dilemmas in Juvenile Justice

Several important points need to be made concerning the contemporary juvenile justice network. First, most of the issues that led to the debates over juvenile justice were evident by the 1850s, although the violent nature of some juvenile crimes over the past quarter century has raised serious questions about the juvenile court's ability to handle such cases. The issue of protection and treatment rather than punishment had been clearly raised under the 15th-century chancery court system in England. The issues of criminal responsibility and separate facilities for youthful offenders were apparent in the City Custom of Apprentices in 17th-century England and again in the development of reform schools in England and the United States during the 19th century.

Second, attempts were made to develop and reform the juvenile justice network along with other changes that occurred during the 18th, 19th, and early 20th centuries. Immigration, industrialization, and urbanization had changed the face of American society. Parents working long hours left children with little supervision, child labor was an important part of economic life, and child labor laws were routinely disregarded. At the same time, however, treatment of the mentally ill was undergoing humanitarian reforms as the result of efforts by Phillipe Pinel in France and Dorothea Dix and others in the United States. The Poor Law Amendment Act had been passed in England in 1834, providing relief and medical services for the poor and needy. Later in the same century, Jane Addams sought reform for the poor in the United States. Thus, the latter part of the 18th century and all of the 19th century may be viewed as a period of transition toward humanitarianism in many areas of social life, including the reform of the juvenile justice network.

Third, the bases for most of the accepted attempts at explaining causes of delinquency and treating delinquents were apparent by the end of the 19th century. We discuss these attempts at explanation and treatment later in the book. At this point, it is important to note that those concerned with juvenile offenders had, by the early part of the 20th century, clearly indicated the potentially harmful effects of public exposure and were aware that association with adult offenders in prisons and jails could lead to careers in crime.

Fourth, the *Gault* decision obviated the existence of two major, and more or less competing, groups of juvenile justice practitioners and scholars. One group favors the informal, unofficial, treatment-oriented approach, referred to as a casework or **therapeutic approach;** the other group favors a more formal, more official, more constitutional approach, referred to as a formalistic or **legalistic approach.** The *Gault* decision made it clear that the legalists were on firm ground, but it did not deny the legitimacy of the casework approach. Rather, it indicated that the casework approach may be employed, but only within a constitutional framework. For example, a child might be adjudicated delinquent (by proving his or her guilt beyond a reasonable doubt) but ordered to participate in psychological counseling (as a result of a presentence investigation that disclosed psychological problems).

All of these issues are very much alive today. Caseworkers continue to argue that more formal proceedings result in greater stigmatization of juveniles, possibly resulting in more negative self-concepts and eventually in careers as adult offenders. Legalists contend that innocent juveniles may be found to be delinquent if formal procedures are not followed and that ensuring constitutional rights does not necessarily result in greater stigmatization, even if juveniles are found to be delinquent.

Similarly, the debate over treatment versus punishment continues. On the one hand, status offenders (those committing acts that would not be violations if they were committed by adults) have been removed from the category of delinquency, in part as a result of the passage of the Juvenile Justice and Delinquency Act of 1974 (Snyder & Sickmund, 1999). On the other hand, beginning in the 1980s and continuing to the present, more severe punishments for certain violent offenses have been legislated, and waiver to adult court for such offenses has been made easier. At the same time, the U.S. Supreme Court's decision in *Roper v. Simmons* has denied the possibility of the ultimate punishment—death. The perceived increase in the number of violent offenses perpetrated by juveniles has led many to ponder whether the juvenile court, originally established to protect and treat juveniles, is adequate to the task of dealing with modern-day offenders. Simultaneously, the concept of restorative justice, which involves an attempt to make victims whole through interaction with and restitution by their offenders, has become popular in juvenile justice (see Chapter 10). This approach emphasizes a treatment philosophy as opposed to the "get tough" philosophy so popular during recent years. Both of these approaches lead observers to believe that if the juvenile court survives, major changes in its underlying philosophy (see In Practice 1.2) are likely to occur (Cohn, 2004b; Ellis & Sowers, 2001; Schwartz, Weiner, & Enosh, 1998).

In Practice 1.2

Cutting Crime's Roots: New Initiatives Aim to Reach Young Offenders Before They Become Violent—and to Stay on Their Cases if They Do

The caravan of five vehicles rolled up to the northeast Minneapolis house quietly. Out jumped police, probation officers, and U.S. marshals wearing bulletproof vests, some with guns drawn.

"Police, open up!" Sgt. Ron Stenerson shouted through the front door as officers surrounded the house. They had come looking for a 17-year-old with a long rap sheet who had violated probation.

Stenerson banged on the door with a flashlight. No answer. Finally, the teen's grandmother, Rosie McKnight, appeared and said he had not been there for weeks. But as officers returned to their cars, she burst back into view. Her grandson was on the phone, ready to turn himself in.

Minutes later, he was in the middle of an empty street, hands in the air. "Why am I being arrested?" he muttered.

The answer, in part, is because Minneapolis and Hennepin County are making extensive, unusual changes in how they deal with juvenile criminals, who figure prominently in the city's 19 percent increase in violent crime this year.

McKnight's grandson is among 155 chronic, sometimes violent, teen offenders who have been arrested in the seven months since Minneapolis police formed a juvenile criminal apprehension unit, one of the first of its kind in the nation.

Police also have arrested 2,559 juveniles since May over truancy, petty crimes, and curfew violations—more than double the tally last year.

Mike Freeman, just elected Hennepin County attorney, said the issue will be a priority when he takes office in January. "We must focus on truancy and curfew violations if we want to keep juvenile crime down," he said.

Even before crime spiked, local officials were searching for new ways to deal with troubled teens. Some are studying juvenile justice makeovers in Portland, Chicago, and elsewhere—places that now lock up fewer kids by figuring out which ones pose risks to society.

(Continued)

(Continued)

"It is bigger than the Police Department," said Minneapolis Police Chief Tim Dolan.

On Friday, the Minneapolis City Council declared youth violence a public health crisis requiring an ambitious community effort to steer kids away from gangs, guns, and crime.

In another recent change, Hennepin County's Juvenile Detention Center has begun housing violent offenders apart from low-level offenders. Previously, the youths were assigned to living units based on age. That practice created "almost a little mini-criminal school," said Fred LaFleur, director of the county's community corrections.

Also, youths who commit petty crimes such as vandalism soon will report to caseworkers at youth programs instead of to probation officers, who had little time to see them.

"Everyone would agree, let's get those really crazy violent offenders off the street," said Rebecca Saito, a youth development consultant working to steer kids into youth programs. "But the clear majority of young people are open to a whole bunch of positive experiences and opportunities if they are presented in a way that is attractive to them."

Some Youth Crime Surging

Robberies by juvenile suspects jumped nearly 80 percent this summer—with 311 reported from June through September compared with 174 last year, police say. Arrests of juveniles in assault cases also rose and followed increases in violent crime last year in Minneapolis and many other cities nationwide.

Teens have been charged in fatal holdups of a man leaving an Uptown restaurant and a 15-year-old boy robbed of his football jersey on the North Side.

Yet not all crime is up, and court and police data suggest a late-summer ebb in serious juvenile cases. It's unclear whether a decade-long decline in juvenile crime is headed toward the alarming levels of the early 1990s.

"We won't really know for several years," said Melissa Sickmund, co-author of reports on juvenile crime for the U.S. Justice Department. "It is so frustrating. It's like driving and looking in your rear-view mirror."

In St. Paul, police believe juvenile crime is down but lack data to confirm that trend. Crime reports from the past year are being recounted by hand because computer counts had "anomalies," said department spokesman Tom Walsh.

Pete Cahill, chief deputy of the Hennepin County Attorney's Office, said prosecutors urged former Minneapolis Police Chief Bill McManus to restore the department's juvenile crime investigations unit in 2004. The unit was cut during the 2003 state budget crisis. McManus resigned to become San Antonio police chief in April.

"We dropped the ball," said Dolan, McManus's second-in-command at the time.

After Dolan was named interim chief, he assigned nine detectives to the new unit to focus on armed robberies and aggravated assaults. Lt. Bryan Schafer, who leads the unit, said it is tracking new criminal trends, including a rise in purse snatchings and armed robberies near light-rail stations.

Wake Up, You're Under Arrest

It's 5 a.m. on a chilly October day, and the new unit is hoping to catch some kids—in bed.

Sgt. Stenerson, flanked by unit members in a north Minneapolis house used as a probation office, pages through a clipboard with more than a dozen arrest warrants for juvenile felons. Some of the juveniles are violent; most violated probation.

(Continued)

(Continued)

"It takes a concerted effort," he said. "You have to do research on the juvenile's associates, addresses, where they might go to school. Even when we do hit the right address, they may not be there. Juveniles are very mobile."

Today, the top priority is a teenager who ran away months ago from a group home in Shakopee. Other teens are wanted because they tested positive for marijuana, ignored their probation officer, or skipped required antiviolence classes.

Knocking on doors for four hours, officers kept striking out. They talked to the 13-year-old sister of one suspect and the grandmother of another; neither knew where the kids were. At another stop, Stenerson greeted a mother he has visited repeatedly in search of her 16-year-old son, who's wanted on a gun conviction. She told him he ran away to Las Vegas.

But the team's luck changed after it approached a mother who couldn't speak English. She showed officers a certificate indicating her son attended Roosevelt High School. They arrested him there an hour later. The intensive effort has prompted some parents to get their children to turn themselves in, Stenerson said.

"We need to send a message that they will be held accountable," he said.

Alternatives to Lockup

As police strive to round up juveniles who commit violence, some corrections officials are rethinking how they deal with teen offenders who have not—yet.

Low-level offenders, including some picked up for missing a court appearance, often end up in juvenile detention centers, sometimes filling them. Corrections officials say that is costly and unnecessary and that home monitoring, secure shelters, and community-based programs can better steer low-risk kids away from crime.

That's how it's done now in juvenile centers in Cook County, Ill., and Multnomah County, Ore., whose new intervention efforts are admired by juvenile justice officials nationwide.

Corrections officials, police, and judges from Hennepin, Ramsey, and Dakota counties visited those places this year in the hope of developing similar plans. The Annie E. Casey Foundation, which funds research into detention alternatives, paid for the travel.

"To keep locking them up doesn't cut crime," said Robert Hargesheimer, head of youth investigations for the Chicago Police Department. He and other officers now team up with social and mental health workers, probation officers, school officials, and faith-based mentors at a neighborhood center for low-level offenders. "You have to intervene when they are younger and help the families as well," Hargesheimer added.

In Minnesota, officials also hope such efforts might reduce racial disparities in Hennepin, Ramsey, and Dakota counties' juvenile detention centers. Hennepin detains black juveniles at 15 times the rate of white kids, a March 2005 state Department of Public Safety report found.

Focusing Attention and Money

By declaring juvenile violence a public health crisis, Minneapolis officials are hoping to focus attention and money on what some supporters call an effort to "outrecruit the gangs." A 15-member panel is being formed to oversee the effort.

"The low-level offenders who are living in the community where the negative influences are 24/7, we want to get them involved," said Jan Fondell, a city youth specialist working on the effort.

(Continued)

(Continued)

On a recent weeknight, two teens sat behind a video camera in a North Side church basement as another adjusted a spotlight. A few minutes later, they began filming an interview for their seventh documentary, this one about a peace rally. It is one of the youth-led "positive peer group" efforts overseen by Edwin Irwin, executive director of the Kwanzaa Community Church's Nia Imani Youth and Family Development Center.

"Most of these youth are A students, but some have made bad choices," he said.

Fondell and others would like to replicate such efforts for troubled kids ages 10 to 15 years. In a small step, the city is proposing to spend $120,000 to focus on 60 to 80 at-risk children next year, she said. A separate city effort is mapping existing youth programs by neighborhood, hoping to attract more kids.

It is hard and can be risky to approach troubled youth on the street. New Age Urban Development, run out of the north Minneapolis house of Craig Bell, offers a boxing program, jobs with the group's construction company, and résumé and homework help. Bell and others walk high-crime neighborhoods to find teens in need.

"Sometimes I have to pull drugs and guns out of their hands," he said.

SOURCE: David Chanen and David Shaffer, November 19, 2006, p. 1A. Copyright 2006. *Star Tribune: Newspaper of the Twin Cities.*

Rethinking Juvenile Justice

Finally, the issue of responsibility for delinquent acts continues to surface. The trend has been to hold younger and younger juveniles accountable for their offenses, to exclude certain offenses from the jurisdiction of the juvenile court, and to establish mandatory or automatic waiver provisions for certain offenses.

There are a number of practical implications of the various dilemmas that characterize the juvenile justice network. Juvenile codes in many states were changed during the 1990s to reflect expanded eligibility for criminal court processing and adult correctional sanctions. All states now allow juveniles to be tried as adults under certain circumstances. Because the juvenile justice network does not exist in a vacuum, laws dealing with juveniles change with changing political climates, whether or not such changes are logical or supported by evidence. Thus the cycle of juvenile justice is constantly in motion. Disputes between those who represent the two competing camps are common and difficult to resolve. Finally, the discrepancy between the ideal (theory) and practice (reality) remains considerable. What should be done to, with, and for juveniles, and what is possible based on the available resources and political climate, may be quite different things. Bilchik (1999b) asked,

> As a society that strives to raise productive, healthy, and safe children, how can we be certain that our responses to juvenile crime are effective? Do we know if our efforts at delinquency prevention and intervention are really making a difference in the lives of youth and their families and in their communities? How can we strengthen and better target our delinquency and crime prevention strategies? Can we modify these strategies as needed to respond to the ever-changing needs of our nation's youth? (p. iii)

As we enter the 21st century, these are among the questions that remain unanswered in the field of juvenile justice.

Career Opportunities in Juvenile Justice

In each of the following chapters, look for the Career Opportunities Box, which provides you with information concerning specific occupations, typical duties, and job requirements within or related to the juvenile justice network. Keep in mind that different jurisdictions have different requirements, so we are presenting you with information that is typical of the occupations discussed. We encourage you to discuss career options with faculty and advisers and to contact the placement office at your university or college for further information. You might also seek out individuals currently practicing in the juvenile justice field to discuss your interest and concerns. Good hunting!

SUMMARY

Although the belief that juveniles should be dealt with in a justice system different from that of adults is not new, serious questions are now being raised about the ability of the juvenile justice network to deal with contemporary offenders, particularly those who engage in violent conduct. The debate rages concerning whether to get increasingly tough on youthful offenders or to retain the more treatment/rehabilitation-centered approach of the traditional juvenile court. The belief that the state has both the right and responsibility to act on behalf of juveniles was the key element of juvenile justice in 12th-century England and remains central to the juvenile justice network in the United States today.

Age of responsibility and the ability to form criminal intent have also been, and remain, important issues in juvenile justice. The concepts of *parens patriae* and *in loco parentis* remain cornerstones of contemporary juvenile justice, although not without challenge. Those who favor a more formal approach to juvenile justice continue to debate those who are oriented toward more informal procedures, although decisions in the *Kent, Gault,* and *Winship* cases made it clear, in theory at least, that juveniles charged with delinquency have most of the same rights as adults.

Although some (e.g., Hirschi & Gottfredson, 1993) have argued that the juvenile court rests on faulty assumptions, it appears that the goals of the original juvenile court (1899) are still worth pursuing. It is becoming increasingly apparent that the political climate of the time is extremely influential in dictating changing, and sometimes contradictory, responses to juvenile delinquency.

Note: Please see the Companion Study Site for Internet exercises and Web resources. Go to www.sagepub.com/juvenilejustice6study.

Critical Thinking Questions

1. What do the terms *parens patriae* and *in loco parentis* mean? Why are these terms important in understanding the current juvenile justice network?
2. List and discuss three of the major issues confronting the juvenile justice network in the United States today. Are these new issues, or do they have historical roots? If the latter, can you trace these roots?
3. What is the significance of each of the following court decisions?
   a. *Kent*
   b. *Winship*
   c. *Gault*
   d. *Roper v. Simmons*
4. Discuss some of the historical events that have had an impact on the contemporary juvenile justice network in the United States. What do you think the long-term effects of these events will be on the juvenile justice network?

Suggested Readings

Bazemore, G., & Feder, L. (1997). Rehabilitation in the new juvenile court: Do judges support the treatment ethic? *American Journal of Criminal Justice, 21,* 181–212.

Bilchik, S. (1999, December). *Juvenile justice: A century of change* (Juvenile Justice Bulletin). Washington, DC: U.S. Department of Justice.

Bishop, D. (2004). Injustice and irrationality in contemporary youth policy. *Criminology and Public Policy, 3,* 633–644.

Cohn. A. W. (2004). Planning for the future of juvenile justice. *Federal Probation, 68*(3), 39–44.

Lindner, C. (2004, Spring). A century of revolutionary changes in the United States court systems. *Perspectives,* pp. 24–29.

Moon, M. M., Sundt, J. L., Cullen, F. T., & Wright, J. P. (2000). Is child saving dead? Public support for juvenile rehabilitation. *Crime & Delinquency, 46,* 38–60.

Reno, J. (1998). Taking America back for our children. *Crime & Delinquency, 44,* 75–82.

Rosenheim, M. K., Zimring, F. E., Tanenhaus, D. S., & Dohrn, B. (Eds.). (2002). *A century of juvenile justice.* Chicago: University of Chicago Press.

Sprott, J. B. (1998). Understanding public opposition to a separate youth justice system. *Crime & Delinquency, 44,* 399–411.

DEFINING AND MEASURING OFFENSES BY AND AGAINST JUVENILES

CHAPTER LEARNING OBJECTIVES

On completion of this chapter, students should be able to

- *Understand and discuss the importance of accurately defining and measuring delinquency*
- *Understand the impact of differences in definitions of delinquency*
- *Discuss legal and behavioral definitions of delinquency*
- *Discuss official and unofficial sources of data on delinquency and abuse and the problems associated with each*

KEY TERMS

Legal definitions
Behavioral definitions
Age ambiguity
Uniform Crime Reports (UCRs)
National Incident-Based Reporting System
Offenses known to the police
Victim survey research
National Center for Juvenile Justice
Office of Juvenile Justice and Delinquency Prevention
National Center on Child Abuse and Neglect
National Children's Advocacy Center
National Crime Victimization Survey
Unofficial sources of data
Self-report studies
Police observational studies

One of the major problems confronting those interested in learning more about offenses by and against juveniles involves defining the phenomena. Without specific definitions, accurate measurement is impossible, making development of programs to prevent and control delinquency and offenses against juveniles extremely difficult.

There are two major types of definitions associated with delinquency. Strict **legal definitions** hold that only those who have been officially labeled by the courts are offenders. **Behavioral definitions** hold that those whose behavior violates statutes applicable to them are offenders whether or not they are officially labeled. Each of these definitions has its own problems and implications for practitioners and leads to different conclusions about the nature and extent of offenses. For example, using the legal definition, a juvenile who committed a relatively serious offense but was not apprehended would not be classified as delinquent, whereas another juvenile who committed a less serious offense and was caught would be so classified.

Legal Definitions

Changing Definitions

A basic difficulty with legal definitions is that they differ from time to time and from place to place. An act that is delinquent at one time and in one place might not be delinquent at another time or in another place. For example, wearing gang colors or using gang signs may be a violation of city ordinances in some places but not in others. Or the law may change so that an act that was considered delinquent yesterday is not considered delinquent today. For instance, the Illinois Juvenile Court Act of 1899 defined as delinquent any juvenile under the age of 16 years who violated a state law or city or village ordinance. By 1907, the definition of delinquency had changed considerably to include incorrigibility, knowingly associating with vicious or immoral companions, absenting oneself from the home without just cause, patronizing poolrooms, wandering about the streets at night, wandering in railroad yards, and engaging in indecent conduct. The current Illinois Juvenile Court Act (*Illinois Compiled Statutes* [*ILCS*], ch. 705, art. 2, sec. 405, 2006) more closely resembles the 1899 version except that the maximum age for delinquency has been changed to 17 years and attempts to violate laws are also included. Legal definitions are limited in their applicability to a given time and place because of these inconsistencies. You will note as we proceed through the text that many of the examples provided are from the Illinois Juvenile Court Act (*ILCS*, ch. 705, 2006). There are several reasons for the use of these examples. First, Illinois has been and remains a national leader in the field of juvenile justice (Fanton, 2006; see In Practice 2.1). Second, because most of the authors teach and/or practice in Illinois, these are the statutes with which we are most familiar. Third, it is impossible to cite all of the statutes from the 50 states in the confines of the text. We strongly encourage you to access online the statutes of the state in which you reside to compare and contrast them with the sample statutes cited in the text.

Age Ambiguity

Another problem with legal definitions has been the ambiguity reflected with respect to age (**age ambiguity**). What is the lower age limit for a juvenile to be considered delinquent? At what age are children entitled to the protection of the juvenile court? Although custom has

In Practice 2.1

Illinois, a National Leader in Juvenile Justice Reforms

In 1899, Illinois established the nation's first juvenile court. The court was created with the enlightened goal of providing individual attention to young people in trouble with the law. The Illinois system quickly became an international model.

But by the end of the 20th century, the line between Illinois' juvenile justice and criminal justice systems was hopelessly blurred, reflecting a national trend. The 1990s marked the peak of the system's breakdown. In reaction to the public's fear of juvenile crime, harsher and more punitive measures for young offenders were enacted. As a result, a growing number of young people were tried as adults, and many were jailed in facilities with adult criminals.

Both inside and outside the system, there were limited resources available to help these young people repair their lives, such as treatment for mental illnesses or substance abuse. Consequently, recidivism among juvenile offenders increased during the 1990s, reducing educational and employment prospects and exacerbating already-troubling racial disparities in arrest and detention rates. In short, the system failed too many kids, and the prospects for reform were bleak.

In the past few years, however, real progress has been made in Illinois. The state is now reemerging as a national leader in juvenile justice reform. Through a combination of public and private efforts, Illinois is enacting policies and programs not only to protect public safety and hold young people accountable for their actions but also to provide for their rehabilitation.

Illinois is the only state in the nation to curtail the automatic transfer of young people arrested on drug offenses to the adult criminal system, giving greater discretion to judges in these cases. Illinois is a model for community-based alternatives to detention, such as night reporting centers and electronic monitoring.

These are all steps in the right direction. How they are implemented—and how adequately they are funded—will determine ultimate success or failure.

Illinois recognizes that its juvenile justice system must be fair and rational. Adolescents are simply not as developmentally mature as adults. Although adolescents should be held responsible for their actions, they should not be held accountable in the same way as adults.

The MacArthur Foundation recently selected Illinois as one of only four states nationwide to participate in our juvenile justice reform initiative, "Models for Change."

Over the next five years, the foundation will provide up to $7.5 million to government and nonprofit entities to build upon reform efforts already under way. We want to help create a model for change here that can be replicated elsewhere and to attract more resources, in Illinois and across the nation, for juvenile justice reform efforts.

In this second century of juvenile justice, Illinois is once again taking the lead in demanding justice, fairness, and accountability in the treatment of young people in trouble with the law. We have great faith that through these efforts, Illinois will reclaim its role as a model for change around the nation.

SOURCE: *Peoria Journal Star,* May 14, 2006, page A5.

established a lower limit for petitions of delinquency at roughly 7 years of age, some states set the limit higher. For example, the youngest age for juvenile court jurisdiction for delinquency in several states is 10 years or under (Office of Juvenile Justice and Delinquency Prevention, 2006, p. 103). Our thinking with respect to the minimum age at which children should be afforded court protection changed with the emergence of crack cocaine and methamphetamines, both of which may have serious prenatal effects (Wells, 2006). According to Illinois statutes, for example, any infant whose blood, urine, or meconium contains any amount of a controlled substance is defined as neglected (*ILCS*, ch. 705, art. 2, sec. 405, 2006).

There is also considerable diversity with respect to the upper age limit in delinquency cases. As of 2002, 35 states defined juveniles as those who had not yet reached their 18th birthday, 2 states set the upper age limit as the 17th birthday, 10 states set the upper age limit as the 16th birthday, and 3 states set the upper age limit as the 15th birthday (Mitchell & Kropf, 2002). Some states set higher upper age limits for juveniles who are abused, neglected, dependent, or in need of intervention than for delinquents in an attempt to provide protection for juveniles who are still minors even though they are no longer subject to findings of delinquency. And in most states, juvenile court authority over a juvenile may extend beyond the upper age of original jurisdiction (frequently to the age of 21 years).

▲ Are race and ethnicity really major factors in delinquency?

An example of the confusion resulting from all of these considerations is the Illinois Juvenile Court Act (*ILCS*, ch. 705, 2006). This act establishes no lower age limit, establishes the 17th birthday as the upper limit at which an adjudication of delinquency may be made, makes it possible to automatically transfer juveniles over the age of 15 years to adult court for certain types of violent offenses, and sets the 18th birthday as the upper age limit for findings of abuse, dependency, neglect, and minors requiring intervention. Adding to the confusion is the distinction made in the Illinois Juvenile Court Act between minors (those under 21 years of age) and adults (those 21 years of age and over). This raises questions about the status of persons over the age of 18 but under 21 years. For example, a 19-year-old in Illinois is still a minor (although he or she may vote) but cannot be found delinquent, dependent, neglected, abused, or in need of intervention. Such ambiguities with respect to age make comparisons across jurisdictions difficult.

Inaccurate Images of Offenders and Victims

Yet another difficulty with legal definitions is that they may lead to a highly unrealistic picture of the nature and extent of delinquency, abuse, neglect, and dependency. Because these definitions depend on official adjudication, they lead us to concentrate on only a small portion of those actually involved as offenders and victims.

Similar problems arise when considering abuse and neglect because only a small portion of such cases are reported and result in official adjudication. In short, most juvenile offenders and victims never come to the attention of the juvenile court, and a strict legal definition is of little value if we are interested in the actual size of offender and victim populations. It may well be, for example, that females are more involved in delinquent activities than official statistics would lead us to believe. It may be that they are not as likely to be arrested by the police as their male counterparts. Not infrequently, we have seen police officers search male gang members for drugs and/or weapons while failing to search females who are with the gang members. It does not take long for the males involved to decide who should carry drugs and weapons. Similarly, blacks and other minority group members may be overrepresented in official statistics simply because they live in high-crime areas that are heavily policed and, therefore, are more likely to be arrested than those living in less heavily policed areas. For example, in 2004, blacks represented roughly 16% of the overall juvenile population, but as Chart 2.1 indicates, 46% of juvenile arrests for violent crime and 28% of property crime arrests involved black juveniles.

Chart 2.1 Black Proportions of Juvenile Arrests in 2004

Most Serious Offense	Black Proportion of Juvenile Arrests in 2004 (%)
Murder/nonnegligent manslaughter	50
Forcible rape	34
Robbery	63
Aggravated assault	39
Burglary	27
Larceny theft	27
Motor vehicle theft	40
Weapons	33
Drug abuse violations	27
Curfew and loitering	32
Runaway	21

SOURCE: Federal Bureau of Investigation. (2005). *Crime in the United States,* Table 43b. Available: www.fbi.gov/ucr/cius_04

A final difficulty with legal definitions also characterizes behavioral definitions and results from the broad scope of behaviors potentially included. Does striking a child on the buttocks with an open hand constitute child abuse? What does the phrase "beyond the control of parents" mean? How is "incorrigible" to be defined? What does a "minor requiring authoritative intervention" look like? Although all of these questions may be answered by referring to definitions contained in state statutes, in practice they are certainly open to interpretation by parents, practitioners, and juveniles themselves. The broader the interpretation, the greater the number of victims and offenders.

Behavioral Definitions

In contrast to legal definitions, behavioral definitions focus on juveniles who offend or are victimized even if they are not officially adjudicated. Using a behavioral definition, a juvenile who shoplifts but is not apprehended is still considered delinquent, whereas that juvenile would not be considered delinquent using a legal definition. The same is true of a child who is abused but not officially labeled as abused. If we concentrate on juveniles who are officially labeled, we get a far different picture from that if we include all of those who offend or are victimized. Estimates of the extent of delinquency and abuse based on a legal definition are far lower than those based on a behavioral definition. In addition, the nature of delinquency and abuse appears to be different depending on the definition employed.

▲ Police prepare to search juveniles for drugs and weapons.

We might assume, for example, that the more serious the case, the greater the likelihood of official labeling. If this assumption is correct, relying on official statistics would lead us to believe that the proportion of serious offenses by and against juveniles is much higher than it actually is (using the behavioral definition). Finally, relying on legal definitions (and the official statistics based on such definitions) would lead us to overestimate the proportion of lower-social-class children involved in delinquency and abuse. The reasons for this overestimation are discussed later in this chapter.

In general, we prefer a behavioral definition because it provides a more realistic picture of the extent and nature of offenders and victims. It may be applied across time and jurisdictions because it is broad enough to encompass the age and behavioral categories of different jurisdictions and statutes. In addition, the broader perspective provided may help in the development of more realistic programs for preventing or controlling delinquency. In spite of its advantages, however, there is one major difficulty with the behavioral definition. Because it includes many juveniles who do not become part of official statistics, we need to rely on unofficial, and sometimes questionable, methods of assessing the extent and nature of unofficial or "hidden" delinquency and abuse.

Official Statistics: Sources and Problems

Official Delinquency Statistics

What do current official statistics on delinquency and abuse indicate? Despite growth in the juvenile population over the past decade, crime and violence by juveniles have declined. For example, the arrest total for juveniles in 2004 decreased 1.7% from that in 2003. Over the same 2-year period, arrests of juveniles for violent crimes declined 1.0% and for property crimes dropped 2.9% (Federal Bureau of Investigation [FBI], 2005). In fact, children are at a much greater risk of being the victims of violent crime than of being the perpetrators of violent crime, with an estimated 803,000 being victims of child abuse and neglect in 2003, a rate of 12.4 maltreatment victims for every 1,000 children in the United States. This is a 5.0% increase in the number of children investigated or assessed, and a 1.1% increase in the number of substantiated child victims, compared with 2002 figures (U.S. Department of Health and Human Services, 2005). Some 2,827 children and teens died as a result of gun violence in 2003 (Children's Defense Fund, 2007).

The number of black and Hispanic juveniles in custody remains high, with black juveniles making up 38% of those in residential placement and Hispanic juveniles accounting for 19%. From 1997 to 2003, the minority population of juvenile offenders in custody increased for females (51% to 55%) and decreased for males (64% to 62%). From 1991 to 2003, the number of juvenile female offenders in custody increased 52% (Office of Juvenile Justice and Delinquency Prevention, 2006). Where do such varied statistics come from, and how accurate are they likely to be?

Official statistics on delinquency are currently available at the national level in *Crime in the United States,* published annually by the FBI based on **Uniform Crime Reports (UCRs).** Since 1964, these reports have contained information on arrests of persons under 18 years of age. In addition, since 1974, the reports have included information on police dispositions of juvenile offenders taken into custody as well as urban, suburban, and rural arrest rates. For the year 2004, the FBI claimed that UCRs covered roughly 94% of the total national population, with the most complete reporting from urban areas and the least complete reporting from rural areas (FBI, 2005, p. 1). Although the FBI statistics are the most comprehensive official statistics available, they are not totally accurate for several reasons.

First, because UCRs are based on reports from law enforcement agencies throughout the nation, errors in reporting made by each separate agency become part of national statistics. Sources of error include mistakes in calculating percentages and in placing offenders in appropriate categories. Statistics reported to the FBI are based on "offenses cleared by arrest" and, therefore, say nothing about whether the offenders were actually adjudicated delinquent for the offenses in question.

Assuming that more serious offenses are more likely to lead to arrests (however defined) than are less serious and more typically juvenile offenses, arrest statistics would show a disproportionate number of serious juvenile offenses. These types of cases actually account for only a very small proportion of all delinquent acts. Black and Reiss (1970) found that in urban areas, only about 5% of police encounters with juveniles involved alleged felonies. Lundman, Sykes, and Clark (1978) replicated the Black and Reiss study and also found a 5% felony rate, noting that only approximately 15% of all police–juvenile encounters result in arrests, leaving 85% of these encounters that cannot become a part of official police statistics. Empey, Stafford, and Hay (1999) concluded, "We have seen that the police traditionally have been inclined to avoid arresting juveniles. Because they have been granted considerable discretion, however, the police continue to counsel and release many of those whom they have arrested, albeit less frequently than in the past" (p. 331).

There are a variety of other difficulties with UCR data. If one wants to know the number of juveniles arrested for specific serious offenses during a given period of time in specific types of locations, UCR data are useful. But if one wants to know something about the actual extent and distribution of delinquency, or about police handling of juveniles involved in less serious offenses, UCR data are of little value because "many juveniles who commit crimes (even serious crimes) never enter the juvenile justice system. Consequently, developing a portrait of juvenile law-violating behavior from official records gives only a partial picture" (Office of Juvenile Justice and Delinquency Prevention, 2006).

In an attempt to combat some of the reporting problems found in UCR data, since 1987 the FBI has been implementing an incident-based reporting system, a modification of the original UCR reporting system, throughout the United States. As of 2003, 23 states and some 5,200 agencies had been certified through the new system, and a number of other states were working toward certification or developing the system (Bureau of Justice Statistics, 2005b). The new system, called the **National Incident-Based Reporting System** (NIBRS), was developed to collect information on each crime occurrence. Under this reporting system, policing agencies report data on **offenses known to the police** (offenses reported to or observed by the police) instead of only those offenses cleared by arrest, as was done in the original UCR crime reporting process. Of all official statistics, offenses known to the police probably provide the most complete picture of the extent and nature of illegal activity, although there is considerable evidence from **victim survey research** (discussed later in this chapter) that even these statistics include information on fewer than 50% of the offenses actually committed (Hart & Rennison, 2003, p. 1). Criminal justice agencies are also allowed to customize the NIBRS to meet agency statistical needs while still meeting the requirements of the UCRs without biasing the data. In addition, crimes that were not discussed in UCRs originally are included in the new reporting system, including terrorism, white-collar crime, children missing due to criminal behaviors, hate crimes, juvenile gang crimes, parental kidnapping, child and adult pornography, driving under the influence, and alcohol-related offenses.

Data at the national level are also available from the **National Center for Juvenile Justice,** which collects and publishes information on the number of delinquency, neglect, and dependency cases processed by juvenile courts nationwide. In addition, the **Office of Juvenile Justice and Delinquency Prevention** in the U.S. Department of Justice maintains and publishes statistics on juveniles. Unfortunately, much of the information available from these two agencies is out of date by the time it is published (2- to 4-year time lags are not uncommon).

There are a variety of sources of official statistics available at local, county, and state levels as well. Many social service agencies, such as police departments, children and family services departments, and juvenile and adult court systems, maintain statistics on cases in which they are involved. These statistics are often focused on agency needs and are used to secure funding from local or private sources, the county, the state, and/or the federal government. The statistics may also be used to justify to the community or the media certain dispositions employed by the agencies and to alert the community to specific needs of the agencies.

Official Statistics on Abuse and Neglect

Official statistics on abused and neglected children are available from a number of sources but are probably even more inaccurate than other crime statistics because of underreporting, as In Practice 2.2 indicates. Part II of the UCRs contains data on "offenses against family and children." The **National Center on Child Abuse and Neglect,** the **National Children's Advocacy Center,** and the National Resource Center on Child Sexual Abuse (all under the auspices of the U.S. Department of Health and Human Services, the American Humane Association, and the National Committee for the Prevention of Cruelty to Children), as well as the Office of Juvenile Justice and Delinquency Prevention, publish data on abuse and neglect of children. Data are also kept and periodically published by departments of children and family services of each state. Of the 50,000 cases of child maltreatment referred weekly, roughly 18% are substantiated (Office of Juvenile Justice and Delinquency Prevention, 2006). This does not mean that all cases of maltreatment are reported (in fact, according to the Office of Juvenile Justice and Delinquency Prevention, because parents are the perpetrators of maltreatment in approximately 80% of substantiated cases, and because most substantiated maltreatment occurs in private settings, it is likely that the majority of such cases are not reported) or that the 82% that are not substantiated do not involve maltreatment (they simply cannot be substantiated at the time).

The National Crime Victimization Survey

The U.S. Department of Justice (Bureau of Justice Statistics) and the U.S. Bureau of the Census annually provide us with official data on crime from the perspective of victims. The **National Crime Victimization Survey** (NCVS) involves interviews every 6 months with roughly 135,000 individuals in more than 75,000 households. These interviews allow us to estimate the amount of crime and the likelihood of victimization and to gain information about the characteristics of victims and their perceptions of offenders. When the data collected by the NCVS are compared with the data from the UCRs, we can make some rough estimates of the extent to which certain types of crime occur but are not reported. In general, these surveys indicate that fewer than half of the crimes occurring are reported to the police (Hart & Rennison, 2003, p. 1). As noted in Table 2.1, the reasons for not reporting are diverse.

In addition to the NCVS, the Bureau of Justice Statistics has worked with the Office of Community Oriented Policing Services (COPS) to develop a statistical software program measuring victimization and citizen attitudes on crime. Local policing municipalities participating in community policing programs use the software program in conjunction with telephone

(Text continues on page 30)

In Practice 2.2

Three Missed Chances to Help Child: Lawsuit in Tot's Death Claims Reluctance to Report Possible Abuse

One doctor suspected child abuse when he examined 10-month-old Chance Chilton, whose head was so swollen that one of his tiny ears was nearly obscured.

A second doctor didn't see results of a medical exam that likely would have triggered concerns of abuse and rebuffed requests from Chance's family to look into the possibility his head injury was the result of abuse. And a child protection caseworker who visited the boy's home could not confirm allegations of abuse "as long as doctors had no concern."

After the series of missed opportunities, Chance was returned to his mother's home, where her boyfriend beat him to death less than a month later.

Now the boy's father, Riley L. Chilton, is suing three doctors and Methodist Hospital in Marion Superior Court. He alleges they did not adequately investigate the family's concerns and failed to photograph the injury or report the suspected abuse to police or child protection officials as required under Indiana law.

Chilton filed the lawsuit after reviews of the case were completed by state and other agencies. A malpractice medical review panel completed its review last week. The outcome of the review has not been made public.

The doctors named in the suit are Gary D. Thompson, Michael S. Turner, and Mary E. Wermuth. Their attorneys declined to comment on the case this week. A Methodist spokesman said the care received by Chance was "completely appropriate" based on information available at the time.

"People we trusted to help my grandson let him down," said Chance's grandmother, Pat Chilton of Indianapolis. "They could and should have reported that Chance was abused. And they didn't."

The lawsuit was filed as Gov. Mitch Daniels and the Department of Child Services, Prevent Child Abuse Indiana, and local volunteer groups like Champions for Children in Marion County are waging campaigns to promote awareness of the reporting law. The case offers a glimpse at the potentially tragic outcome when warning signs of abuse are missed or the law is ignored. That law requires anyone—including a medical professional—who suspects a child is being abused or neglected to file a report with police or child protection officials.

Failing to report suspicions about abuse and neglect is a common theme in the majority of child maltreatment deaths in Indiana, which has the country's highest rate of abuse and neglect fatalities.

The doctors' responses to a malpractice complaint and testimony at the criminal trial of John Beauchamp, who fatally battered Chance, paint a picture of communication breakdowns and a reluctance to report suspected abuse.

Doctor Was "Concerned"

Chance's mother, Carolyn Tolbert, took the boy to the office of their family physician, Dr. Thompson, on Aug. 12, 1998. She said Chance had fallen from a crib and hit his head on a wooden box.

Thompson examined the boy and sent him to Methodist's emergency room, where Wermuth also examined Chance. "I am very concerned about the possibility of abuse," Thompson wrote in his exam notes.

(Continued)

(Continued)

Thompson contends he notified Methodist officials the case should be referred to the hospital's social services staff but acknowledges he didn't notify police or child protection officials.

Court records reveal conflicting accounts of what happened next.

Turner, the neurosurgeon who treated Chance at Methodist, contends Thompson never made him aware of his concerns about abuse.

Medical records show hospital social workers met with Tolbert and her live-in boyfriend Beauchamp. Those contacts focused on parenting and supervision skills.

"That was not a medical investigation," said Indianapolis attorney Michael J. Woody, who is representing Chance's father in the lawsuit. "It had nothing to do with investigating child abuse."

At Methodist, a resident who saw Chance in the emergency room ordered a long-bone scan, a type of X-ray commonly used to check for signs of abuse. Results—which revealed a possible spinal injury—should have triggered concerns about abuse, several doctors testified. But Turner said he was not aware the exam had been ordered and did not see the results before releasing Chance on Aug. 14.

Turner determined that Tolbert's explanation of Chance's injuries sounded plausible.

When Pat Chilton begged Turner to reconsider, he reportedly declined. Turner finally gave the persistent grandmother a number to call and report abuse. "If you want to contact child protective services, you can call them yourself," Pat Chilton said the doctor told her.

In court filings, Turner said that the only indication he had of abuse was second-hand reports from Pat and Riley Chilton and that he had no responsibility to file a report with authorities based on what he knew.

Pat Chilton called the Johnson County Office of Family and Children, which sent a caseworker Aug. 15 to visit Tolbert and Beauchamp. The investigator could not confirm the allegations, noting "(abuse) would be unsubstantiated as long as doctors had no concern."

Less than a month later, Chance was back in a hospital emergency room with head and back injuries. Chance died Sept. 9, 1998, at Riley Hospital for Children. The coroner ruled the death a homicide by child abuse. Beauchamp pleaded guilty to battery and served about six years of a 20-year sentence. He is now out of prison.

Abuse Often Unreported

Dr. Antoinette Laskey, a forensic pediatrician and child abuse expert who heads the state's Child Fatality Review Team, said she could not comment on specifics of the lawsuit or about the doctors involved in the case. But Laskey, who practices at Methodist and Riley, said too many instances of abuse go unreported in Indiana.

The fatality review team's examination last year of 19 abuse and neglect deaths found they all could have been prevented if parents, caregivers, relatives, and medical professionals had acted on concerns they had about the children's safety. In six of the cases, action on the part of medical providers could have prevented the deaths.

"It is essential that everyone involved in a child's life needs to take responsibility, make a report, and let the system work."

SOURCE: Evans, T. (2006, April 20). Three missed chances to help child: Lawsuit in tot's death claims reluctance to report possible abuse. *The Indianapolis Star,* A01. Reproduced with the permission of Gannett Co., Inc., by NewsBank, Inc.

Table 2.1 Estimated Percentage Distributions of Reasons for Not Reporting Personal and Property Victimizations to Police by Type of Crime, 2002

Reasons for Not Reporting	Total (%)	Violent Crimes (%)	Property Crimes (%)
Reported to another official	17.0	17.1	9.7
Private or personal matter	21.2	21.7	5.2
Object recovered; offender unsuccessful	15.7	15.8	25.7
Not important enough	4.9	5.1	3.4
Too inconvenient or time-consuming	2.6	2.5	3.6
Fear of reprisal	4.5	4.6	0.7
Police inefficient, ineffective, or biased	3.0	3.0	3.3
Police would not want to be bothered	5.0	4.8	8.3
Lack of proof	3.8	3.1	11.8
Unable to recover property; no ID number	0.8	0.3	7.4
Not aware crime occurred until later	0.6	0.4	5.8
Insurance would not cover	0.1	0.1	2.4
Other reasons	20.9	21.3	12.7

SOURCE: Adapted from U.S. Department of Justice, Bureau of Justice Statistics. (2004). *Sourcebook of criminal justice statistics 2003*, p. 208, Table 102. Available: www.ojp.usdoj.gov/bjs/pub/pdf/cvus02.pdf

NOTE: Percentages might not add to 100 because of rounding.

(Text continued from page 27)

surveys of local residents to collect data on crime victimization, attitudes toward the police, and other community issues. The results are used to identify which community programs are needed and where those programs should be located in the community.

Although victimization surveys would appear to be a better overall indicator of the extent and nature of crime, delinquency, and abuse, they also have their limitations. As is the case with all self-report measures (see the following section), there are serious questions about the accuracy and specificity of reports by victims. In addition, the surveys do not include interviews with children under the age of 12 years and do not include questions about all types of crime (the NCVS focuses primarily on violent offenses).

Sources of Error in Official Statistics

Official statistics are collected at several different levels in the juvenile justice network, and each level includes possible sources of error. Table 2.2 indicates some sources of error that may

Table 2.2 Some Sources of Error at Specified Levels in the Juvenile Justice System

Data May Be Collected	Sources of Error in Official Statistics
Offenses known to the police	All offenses not detected All offenses not reported to or recorded by the police
Offenses cleared by arrests	Errors from Level 1 All reported offenses that do not lead to arrests
Offenses leading to prosecution	Errors from Levels 1 and 2 All offenses that result in arrests but do not lead to prosecution
Offenses leading to adjudication of delinquency	Errors from Levels 1, 2, and 3 All offenses prosecuted that do not lead to adjudication of delinquency
Offenses leading to incarceration	Errors from Levels 1, 2, 3, and 4 All offenses leading to adjudication of delinquency but not to incarceration

affect official statistics collected at various levels. Each official source has its uses, but generally the sources of error increase as we move up each level in the network.

There are two additional sources of error that may affect all official statistics. First, those who are least able to afford the luxury of private counsel and middle-class standards of living are probably overrepresented throughout all levels. Thus, official statistics might not represent actual differences in delinquency and abuse by social class but rather might represent the ability of middle- and upper-class members to avoid being labeled (for a more thorough discussion, see Empey & Stafford, 1991, pp. 315–317; Garrett & Short, 1975; Knudsen, 1992, p. 31). Second, it is important to remember that agencies collect and publish statistics for a variety of administrative purposes (e.g., to justify more personnel and more money). This does not mean that all or even most agencies deliberately manipulate statistics for their own purposes. All statistics are open to interpretation and may be presented in a variety of ways, depending on the intent of the presenters.

Unofficial Sources of Data

It is clear that relying on official statistics on delinquency and abuse is like looking at the tip of an iceberg; that is, a substantial proportion of these offenses remain hidden beneath the surface. Although it is certain that much delinquency and maltreatment is not reported to or recorded by officials (**unofficial sources of data**), there is no perfect method for determining just how many of these behaviors remain hidden.

Self-Report Studies

Recognizing that official statistics provide a "false dichotomy" between those who are officially labeled and those who are not, a number of researchers have focused on comparing the extent and nature of delinquency among institutionalized (labeled) delinquents and noninstitutionalized (nonlabeled) juveniles. Short and Nye (1958) used self-reports of delinquent behavior obtained by distributing questionnaires to both labeled and nonlabeled juveniles. These questionnaires called on respondents to indicate what types of delinquent acts they had committed and the frequency with which such acts had been committed. Short and Nye concluded that delinquency among noninstitutionalized juveniles is extensive and that there is little difference between the extent and nature of delinquent acts committed by noninstitutionalized juveniles and those committed by institutionalized juveniles. In addition, the researchers indicated that official statistics lead us to misbelieve that delinquency is largely a lower-class phenomenon given that few significant differences exist in the incidence of delinquency among upper-, middle-, and lower-class juveniles. Conclusions reached in similar **self-report studies** by Porterfield (1946), Akers (1964), Voss (1966), and Bynum and Thompson (1992, pp. 78–79) generally agree with those of Short and Nye (1958). Based on these self-report studies, it is apparent that the vast majority of delinquent acts never become a part of official statistics (Conklin, 1998, p. 67). This, of course, parallels information from victim survey research at the adult level.

More recent studies of self-reported delinquency have been conducted by Taylor, McGue, and Iacono (2000), Pagani, Boulerice, and Vitaro (1999), Williams and Dunlop (1999), and Farrington and colleagues (2003), indicating that the technique is still in use. Self-report studies, however, are subject to criticism on the basis that respondents may under- or overreport delinquency or abuse as a result of either poor recall or deliberate deception. To some extent, this criticism applies to victimization surveys as well even though victims are not asked to incriminate themselves. Mistakes in recalling the date of an incident, the exact nature of the incident, or the characteristics of the parties involved may occur. Or for reasons of their own, victims may choose not to report particular incidents. NCVS interviewers attempt to minimize these problems by asking only about crimes during the prior 6 months and by avoiding questions requiring personal admissions of offenses, but there are still no guarantees of accuracy, and this is certainly the case when asking juveniles to report their own crimes or abuse. Hindelang, Hirschi, and Weis (1981, p. 22), for example, contended that illegal behaviors of seriously delinquent juveniles are underestimated in self-report studies because such juveniles are less likely to answer questions truthfully. Farrington and colleagues (2003), for example, concluded that criminal career research based on self-reports sometimes yields different conclusions compared with research based on official records.

Some researchers have included "trap questions" to detect these deceptions. In 1966, Clark and Tifft used follow-up interviews and a polygraph to assess the accuracy of self-report inventories. They administered a 35-item self-report questionnaire to a group of 45 male college students. The respondents were to report the frequency of each delinquent behavior they had engaged in since entering high school. At a later date, each respondent was asked to reexamine his questionnaire, and to correct any mistakes, after being told he would be asked to take a polygraph test to determine the accuracy of his responses. Clark and Tifft (1966) found that all respondents made corrections on their original questionnaires (58% at the first opportunity and 42% during the polygraph examination). Three fourths of all changes increased the frequency of admitted deviancy, all respondents underreported the frequency of their misconduct on at least one item, and 50% overreported on at least one item. With respect to self-reported delinquency, Clark and Tifft concluded that "those items most frequently used on delinquency scales were found to be rather inaccurate" (p. 523).

There are ways of attempting to improve the accuracy of self-reports. In a study of convicted child molesters, official records concerning the sexually abusive activity of the inmates could be compared with their self-reports of behavior. In some cases, it was also possible to confirm through official records the inmates' claims that they themselves had been abused as children (Rinehart, 1991). Without some corroboration, however, the use of self-reports to determine the extent and nature of either delinquency or child abuse is, at best, risky. Still, self-report studies based on either community or school samples have increased in number during recent decades. Empey and colleagues (1999) concluded, "In short, self-report surveys, like other ways of estimating delinquent behavior, have their limitations. Nonetheless, they are probably the single most accurate source of information on the actual illegal acts of young people" (p. 87). As noted earlier, self-reports of delinquency are more comprehensive than official reports because the former include behaviors not reported, or not otherwise known, to the authorities. At least some research indicates that juveniles are willing to report accurate information about their delinquent acts (Farrington, Loeber, Stouthamer-Loeber, Van Kammen, & Schmidt, 1996). Based on a review of self-reported delinquency studies, Espiritu, Huizinga, Crawford, and Loeber (2001) found that the vast majority of juveniles age 12 years or under reported involvement in some form of aggression or violence, but only roughly 5% reported being involved in violence serious enough to be considered a delinquent/criminal offense. Furthermore, the authors noted that self-report rates for major forms of delinquency were nearly the same in 1976 and 1998.

▲ What role do drugs play in delinquency?

Police Observation Studies

Another method for determining the extent and nature of offenses by and against juveniles is observation of police encounters related to juveniles (**police observation studies**). Several studies over the years have found that most delinquent acts, even when they become known to the police, do not lead to official action and, thus, do not become a part of official statistics (Black & Reiss, 1970; Piliavin & Briar, 1964; Terry, 1967; Werthman & Piliavin, 1967). These studies indicate that 70% to 85% of encounters between police and juveniles do not lead to arrests and inclusion in official delinquency statistics. The reasons given by the police for dealing informally with juvenile offenders are both numerous and critical to a complete understanding of the juvenile justice network. These reasons are discussed in some detail in Chapter 7. The point is that the number of juveniles who commit delinquent acts but do not become part of official statistics seems to be considerably larger than the number of juveniles who do become part of official statistics. Relying only on official statistics to estimate the extent and nature of delinquency, thus, can be very misleading. Morash (1984) concluded,

> Youths of certain racial groups and in gang-like peer groups were more often investigated and arrested than other youths. Evidence of the independent influence of subject's race and gang-likeness of peers was not provided by the multivariate analysis, however. Thus, there is some question about whether race and gang qualities have an independent influence on police actions, or whether they are related to police actions because they are correlated with other explanatory variables. The multivariate analysis did provide evidence that the police are prone to arrest males who break the law with peers and who have delinquent peers. Alternatively, they are prone not to investigate females in all-female groups. These tendencies cannot be attributed to the delinquency of the youths or to correlations with other independent variables. There is, then, a convincing demonstration of regular tendencies of the police to investigate and arrest males who have delinquent peers regardless of these youths' involvement in delinquency. (pp. 108–109)

Furthermore, Frazier, Bishop, and Henretta (1992) found that black juveniles were more likely to receive harsher dispositions in areas where the proportion of whites was high, thereby introducing another possible source of bias (relative proportion of whites and blacks in the community) in police statistics. Engel, Sobol, and Worden (2000) found that police action was affected by state of intoxication when combined with displays of disrespect on the part of the suspect. Overall, however, they concluded, "It appears that police officers expect their authority to be observed equally by all suspects, and do not make distinctions based on race, sex, location, and the seriousness of the situation" (pp. 255–256).

Observation of police behavior with respect to abused children reflects a number of concerns. According to Peters (1991),

> As a result of insufficient investigation and unsophisticated prosecution, some innocent people have been wrongly charged [in child abuse cases]. More frequently, however, valid cases have not been charged—or were dismissed or lost at trial—because evidence was overlooked. While the police are mandated to report suspected cases of child abuse, they are frequently faced with determining where discipline ends and abuse begins. (p. 22)

Bell and Bell (1991) found that the police often fail to take official action, preferring instead to handle incidents of domestic violence (involving child as well as adult victims) by referring the parties to another agency.

Career Opportunity—Chief Juvenile Probation Officer

Job Description: Supervise juvenile probation officers as they supervise probationers, conduct presentence investigations, and hold preliminary conferences. Coordinate with police, judges, and other juvenile justice practitioners. Supervise probationers if dictated by caseloads.

Employment Requirements: A master's degree in social work, criminal justice, corrections, or a related field. Ten years of experience in juvenile justice, with at least 5 years of direct service and casework experience.

Beginning Salary: $30,000 to $50,000. Typically good retirement and benefits packages.

SUMMARY

Clearly, there are several potential problems arising from definitional difficulties. First, we need to keep in mind the fact that defining a juvenile as a delinquent is often interpreted as meaning a "young criminal." Although some juveniles who commit serious offenses are certainly young criminals, it is important to note that others who commit acts that are offenses solely because of their age, or who are one-time offenders, may also be labeled as young criminals. Yet these offenses (e.g., underage drinking, illegal possession of alcohol, curfew violations) would not have been considered criminal if the juveniles had been adults.

Second, rehabilitation and treatment programs are almost certainly doomed to failure if they are based solely on information obtained from officially labeled abused children and delinquents. Recognition of the wide variety of motives and behaviors that may be involved is essential if such programs are to be successful, particularly with respect to prevention.

Third, labels (e.g., delinquent, abused child, minor requiring authoritative intervention) tell practitioners very little about any particular juvenile. All parties involved would benefit far more from focusing on the specific behaviors that led to the labels.

There is no doubt that a good deal more delinquency and abuse occur than are reported, although the exact amount is very difficult to determine. There are scores of delinquent acts and abused children never reported. Although it is tempting to divide the world into those who have committed delinquent acts and those who have not, or those who have been abuse victims and those who have not, this polarizes the categories and overlooks the fact that there are many in the official nondelinquent, nonabused category who actually are delinquent or abused.

It is easy to perceive those who are delinquent or abused as abnormal when, in fact, the only abnormal characteristic of many of these juveniles may be that they were detected and labeled. In most other respects, except for extreme cases, these juveniles may differ little from their cohorts. With respect to delinquency at least, there are reasons to be both optimistic and pessimistic based on this view. If most juveniles engage in behavior similar to that which causes some to be labeled as delinquent, there is reason to believe there is no serious underlying pathology in most delinquents. Some types of delinquency occur as a "normal" part of adolescence. Activities such as underage drinking, curfew violation, and experimentation with sex and marijuana seem to be widespread among adolescents. Although these activities may be undesirable when engaged in by juveniles, they are not abnormal or atypical. Thus, reintegration or maintenance within the community should be facilitated.

Those viewing activities that are widespread among juveniles as atypical or abnormal are faced with essentially two choices. Either they can define the majority of juveniles as delinquent, thereby increasing official delinquency rates, or they can reevaluate the legal codes that make these activities violations and remove such behaviors from the category of delinquent. Clearly, many prefer to ignore the latter option and instead continue to polarize "good" and "bad" juveniles.

To some extent, the same argument holds for abused and neglected juveniles. Although those who are labeled are victims instead of perpetrators (as is the case with delinquents), in many cases they are not so terribly different from their peers either. If, as we suspect, the vast majority of abuse and neglect cases go unreported, many juveniles experience a lot of the same behaviors as do those labeled as abused or neglected. Thus the way we treat those who are labeled may be crucial in determining the extent of psychological damage done. If we recognize them as victims but also recognize that they are not abnormal, our efforts at reintegration and rehabilitation may be more effective.

Practitioners in the juvenile justice network, particularly juvenile court judges and those involved in prevention and corrections, may have an inaccurate image of delinquents and maltreated juveniles. Discussions with numerous practitioners at these levels indicate that many view the lower-social-class black male as the typical delinquent and the lower-social-class female as the typical victim of maltreatment. Some social science research perpetuates these mistaken impressions by focusing on labeled juveniles, but other research indicates that such juveniles are typical only of those who have been detected and labeled. Prevention programs and dispositional decisions based on erroneous beliefs about the nature and extent of delinquency and maltreatment can hardly be expected to produce positive results.

Both legal and behavioral definitions of delinquency and child maltreatment present problems. Legal definitions assess, more or less accurately, numbers and characteristics of juveniles who become officially labeled. However, use of legal definitions can be misleading with respect to the actual extent and nature of offenses by and against juveniles. Behavioral definitions assess the extent and nature of such activities more accurately but raise serious problems in the area of data collection. How do we identify those juveniles who commit delinquent acts or who are mistreated but not officially detected?

Official statistics reflect only the tip of the iceberg with respect to delinquency and mistreatment and are subject to errors in compilation and reporting. The use of self-report techniques, victim survey research, and police observational studies helps us to better assess the extent of unofficial or hidden delinquency, abuse, and neglect, although each of these methods has weaknesses. Success in preventing and correcting offenses by and against juveniles depends on understanding not only the differences but also the similarities between labeled and nonlabeled juveniles.

Note: Please see the Companion Study Site for Internet exercises and Web resources. Go to www.sagepub.com/juvenilejustice6study.

Critical Thinking Questions

1. What are the two major types of definitions of delinquency and child maltreatment? Discuss the strengths and weaknesses of each. How might legal definitions lead to mistaken impressions of delinquents and abused juveniles on behalf of juvenile court personnel?
2. What are the national sources of official statistics on delinquency? On child abuse? Discuss the limitations of these statistics.
3. What is the value of self-report studies? Of victim survey research? What are the weaknesses of these two types of data collection?
4. Compare and contrast the nature and extent of delinquency and child abuse as seen through official statistics on the one hand, and self-report, victim survey, and police observational studies on the other.

Suggested Readings

Forum on Child and Family Statistics. (2006). *America's children in brief.* Available at http://childstats.gov

Maxfield, M. G. (1999). The National Incident-Based Reporting System: Research and policy implications. *Journal of Quantitative Criminology, 15,* 119–149.

National Criminal Justice Reference Service. (2006). *Women and girls in the criminal justice system.* Available at www.ncjrs.gov/spotlight/wgcjs/Summary.html

Puzzanchera, C. M. (2000, February). *Self-reported delinquency by 12-year-olds, 1997* (OJJDP Fact Sheet). Washington, DC: Office of Juvenile Justice and Delinquency Prevention.

Trahan, J. (2006, April 22). Dallas again leads nation in crime rate, but skeptics abound. *Dallas Morning News.* Available at http://web.ebscohost.com.ezproxy.wiu.edu/ehost/detail?vid=4&hid=121&sid=cbbe2454-9818-4e4c-891a-be7ff63cc0fe%40sessionmgr104

Wyatt, G. E., Loeb, T. B., & Solis, B. (1999). The prevalence and circumstances of child sexual abuse: Changes across a decade. *Child Abuse and Neglect, 23,* 45–60.

CHARACTERISTICS OF JUVENILE OFFENDERS

CHAPTER LEARNING OBJECTIVES

On completion of this chapter, students should be able to

- *Recognize differences between delinquency profiles based on official statistics and behavioral profiles*
- *Recognize and discuss the multitude of factors related to delinquency*
- *Discuss the impact of social factors (e.g., family, schools, social class) on delinquency*
- *Discuss the effects of physical factors (e.g., gender, age, race) on delinquency*

KEY TERMS

Social factors

Socialization process

Broken homes

Latchkey children

Socioeconomic status

Learning disabled

Youth culture

Criminal subculture

Underclass

Methamphetamine

Dropouts

Crack

◀ Are juveniles who hang out on street corners to be feared?

In any discussion of the general characteristics of juvenile offenders, we must be aware of possible errors in the data and must be cautious concerning the impression presented. In general, profiles of juvenile offenders are drawn from official files based on police contacts, arrests, and/or incarceration. Although these profiles may accurately reflect the characteristics of juveniles who are or will be incarcerated or who have a good chance for an encounter with the justice network, they might not accurately reflect the characteristics of all juveniles who commit offenses.

Studies have established that the number of youthful offenders who formally enter the justice network is small in comparison with the total number of violations committed by juveniles. Hidden offender surveys, in which juveniles are asked to anonymously indicate the offenses they have committed, have indicated repeatedly that far more offenses are committed than are reported in official agency reports. In addition, even those juveniles who commit offenses resulting in official encounters are infrequently officially processed through the entire network. The determination of who will officially enter the justice network depends on many variables that are considered by law enforcement and other juvenile justice personnel. It is important to remember that official profiles of youthful offenders might not actually represent those who commit youthful offenses but rather represent only those who enter the system.

It is common practice to use official profiles of juveniles as a basis for development of delinquency prevention programs. Based on the characteristics of known offenders, prevention programs that ignore the characteristics of the hidden and/or unofficial delinquent have been initiated. For example, there is official statistical evidence indicating that the major proportion of delinquents comes from lower socioeconomic families and neighborhoods. The correlates of poverty and low social status include substandard housing, poor sanitation, poor medical care, unemployment, and so forth. It has been suggested that if these conditions were altered, delinquency might be reduced. However, as Harcourt and Ludwig (2006) found out in their study of broken windows policing, changing the disorder does not necessarily reduce or eliminate criminal behavior. (Recall our comments on middle-class delinquency in Chapter 2.)

Even with the trend toward development of large mixed-income housing projects, Holzman (1996) suggested that big-city public housing will continue to suffer from crime. Venkatesh (1997) suggested that public housing projects are characterized by their own type of social support networks and that the move to scattered site housing destroys such networks; thus mixed-income developments, even large ones, may have something to offer.

The factors causing delinquency seem to be numerous and interwoven in complex ways. Multiple factors must be considered if we are to improve our understanding of delinquency. Thornberry, Huizinga, and Loeber (2004) found that drug, school, and mental health problems are strong risk factors for male adolescent involvement in persistent and serious delinquency, although more than half of persistent serious offenders do not have such problems. Still, more than half of the males studied who did have persistent problems with drugs, school, or mental health were also persistent and serious delinquents. Fewer than half of persistent and serious female delinquents studied had drug, school, or mental health problems, but these problems alone or in combination were not strong risk factors for serious delinquency.

Most criminologists contend that a number of different factors combine to produce delinquency (see In Practice 3.1). Unfortunately, simplistic explanations are often appealing and sometimes lead to prevention and rehabilitation efforts that prove to be of very little value. With this in mind, let us now turn our attention to some of the factors that are viewed as important determinants of delinquent behavior. It must be emphasized once again that most of the information we have concerning these factors is based on official statistics. For a more accurate portrait of the characteristics of actual juvenile offenders, we must also concentrate on the vast majority of juveniles who commit delinquent acts but are never officially labeled as delinquent.

In Practice 3.1

Various Factors Led to Rise in Juvenile Crime: Arrests of Youths Shot Up by Nearly 40 Percent in January—A Rise in Adult Arrests May Have Contributed

Carlos Petty, the 16-year-old accused of shooting a Tastykake deliveryman this month, was a ninth grader at West Philadelphia High School. He had 96 unexcused absences this year. Last year, his first try at ninth grade, he missed 57 days.

His juvenile record includes arrests in the theft of two automobiles, police said. He also was arrested when a teacher was assaulted, police said, though the school district says it has no record of that.

Investigators believe that Petty—who has declined to cooperate—may have robbed the deliveryman to satisfy his taste for designer clothes, such as the Rocawear sweatshirt and the street-hip AND1 sneakers he was wearing when he was caught last Monday after a brief chase near his home in Southwest Philadelphia. Police still are seeking a teenage accomplice.

(Continued)

(Continued)

Incidences of juvenile violence, such as the videotaped beating of a Drexel University student by local high school students earlier this year, have shocked the city. Who are these youths who are so quick to pull the trigger, to act brutally without hesitation, and are their attacks part of a rising trend?

Philadelphia police data show a rise in juvenile arrests in connection with violent crimes over the last four years, from 1,795 in 2002 to 1,956 last year, a 9 percent increase. But in January, juvenile arrests in violent incidents spiked by nearly 40 percent—from 126 last year to 176.

"That's a big jump," said Jeffrey A. Roth, associate director of research at the Jerry Lee Center of Criminology at the University of Pennsylvania.

Roth said that police targeting young offenders may account for some of the increase but that the larger number seems to indicate more juvenile crime was being committed.

Nationally, juvenile crime peaked in the early to mid-1990s and has generally been declining. A report to be released tomorrow by the U.S. Department of Justice says the arrest rate for juvenile violent crime in 2003 was below the levels of the 1980s.

"The data suggest an ebb and flow in this," said Lawrence W. Sherman, director of the Jerry Lee Center.

And there's always the perception that the current crop of youth is behaving worse than earlier generations.

"The people who come out of Graterford [Prison] say the younger people are wilder, but they always say that," Sherman said.

Various factors contribute to the rise and fall of juvenile violence and delinquency.

Sherman noted that high school dropout rates declined in the 1990s, which some experts believe helped reduce juvenile crime.

Roth, however, stressed that dropout rates may be a "good leading indicator but not necessarily a cause" of juvenile crime.

Philadelphia has recorded a slight decline in secondary school dropouts in recent years, according to state records. But each year, more than 5,000 students give up on their education—often with nothing to turn to except the streets. Petty, though he basically stopped going to school this year, was not yet considered a dropout but rather, in district jargon, a "non-attender."

Experts also are examining the impact of arrests and the removal of adults from communities as a factor in juvenile crime.

Ralph B. Taylor, a Temple University criminologist, has conducted preliminary research of Philadelphia police districts that shows a rise in serious juvenile delinquency in specific districts after an increase in adult arrests.

"As the arrest rates go higher and higher, you are taking out adults who are performing important supervisory functions," Taylor said.

Petty, who police say nearly killed Tastykake driver Kyle Winkfield, 20, with a large-caliber revolver March 16 in West Philadelphia, lived at home with his mother, Cyrena Bowman, 33, several siblings, and other relatives in the city's Kingsessing section.

"He is a good boy," said his mother, who declined to comment further.

Marion Bowman, Petty's great-grandmother, said he was "in the house all day long" the day of the shooting except when he went out to pick up one or two younger siblings from school.

"He takes care of me and fixes me breakfast," she said before Petty's mother arrived and angrily ordered a reporter to leave the house.

Virginia Coverson, 79, a retired Philadelphia public school teacher who lives next door to Petty's family, said that the teen had gotten into trouble at school last year and that his mother had asked her to write a letter to the school in support of him.

(Continued)

(Continued)

Coverson, who briefly served as a substitute teacher at West Philadelphia High after she retired, said she wrote that "he was a good boy around here as far as I know."

Coverson said Petty was quiet, polite, and would offer to shovel snow in front of her rowhouse. "I'd give him five dollars" for his work, she said.

Gwen Morris, the district's interim assistant superintendent for alternative education, said Petty, with his excessive truancy, was put in a program last year for troubled students called Opportunity for Success.

But Petty's truancy worsened. In January, he and his mother met with school officials, and Petty signed a contract in which he vowed to show improvement, Morris said.

The last day he attended school was Feb. 13.

Marvin Daughtry, 18, another neighbor, said Petty was "not the type to pick up a gun to solve his problems."

Daughtry said Petty was usually on the block hanging out or at the Kingsessing Recreation Center playing basketball.

Both neighbors said they were not aware of Petty's brushes with the law.

He is now charged as an adult with attempted murder and is being held on $1 million bail.

SOURCE: Moran, Robert. (2006, March 27). Various Factors Led to Rise in Juvenile Crime—Arrests of Youth Shot Up by Nearly 40 Percent in January. *The Philadelphia Inquirer,* page B1.

Social Factors

As they grow up, children are exposed to a number of **social factors** that may increase their risk for problems such as abusing drugs and engaging in delinquent behavior.

> Risk factors function in a cumulative fashion; that is, the greater the number of risk factors, the greater the likelihood that youth will engage in delinquent or other risky behavior. There is also evidence that problem behaviors associated with risk factors tend to cluster. For example, delinquency and violence cluster with other problems, such as drug abuse, teen pregnancy, and school misbehavior. (Helping America's Youth, n.d.)

Shown in Chart 3.1 are a number of factors experienced by juveniles as individuals, as family members, in school, among their peers, and in their communities. For further information concerning the indicators of these risks and data sources associated with such indicators, visit the Web site from which the chart was adapted.

Family

One of the most important factors influencing delinquent behavior is the family setting. It is within the family that the child internalizes those basic beliefs, values, attitudes, and general patterns of behavior that give direction to subsequent behaviors. Because the family is the initial transmitter of the culture (through the **socialization process**) and greatly shapes the personality characteristics of the child, considerable emphasis has been given to family structure, functions, and processes in delinquency research (Smith & Stern, 1997). Although it is not

◀ Juveniles relax in front of buildings in New Orleans. Drug use, high crime rates, and dilapidated housing persist in the wake of Hurricane Katrina, but not all juveniles from areas characterized by these problems are delinquents.

Chart 3.1 Risk Factors for Health and Behavior Problems

Individual

Antisocial behavior and alienation/delinquent beliefs/general delinquency involvement/drug dealing
Gun possession/illegal gun ownership/carrying
Teen parenthood
Favorable attitudes toward drug use/early onset of alcohol/other drug (AOD) use
Early onset of aggression/violence
Intellectual and/or developmental disabilities
Victimization and exposure to violence
Poor refusal skills
Life stressors
Early sexual involvement
Mental disorder/mental health problem

Family

Family history of problem behavior/parent criminality
Family management problems/poor parental supervision and/or monitoring
Poor family attachment/bonding
Child victimization and maltreatment
Pattern of high family conflict
Family violence
Having a young mother
Broken home
Sibling antisocial behavior
Family transitions
Parental use of physical punishment/harsh and/or erratic discipline practices

(Continued)

(Continued)

Low parent education level/illiteracy
Maternal depression

School

Low academic achievement
Negative attitude toward school/low bonding/low school attachment/commitment to school
Truancy/Frequent absences
Suspension
Dropping out of school
Inadequate school climate/poorly organized and functioning schools/negative labeling by teachers
Identified as learning disabled
Frequent school transitions

Peer

Gang involvement/gang membership
Peer alcohol/tobacco/other drug (ATOD) use
Association with delinquent/aggressive peers
Peer rejection

Community

Availability/use of alcohol, tobacco, and other drugs in neighborhood
Availability of firearms
High-crime neighborhood
Community instability
Low community attachment
Economic deprivation/poverty/residence in a disadvantaged neighborhood
Neighborhood youth in trouble
Feeling unsafe in the neighborhood
Social and physical disorder/disorganized neighborhood

SOURCE: Adapted from Helping America's Youth. (n.d.). *Site map.* Available: http://guide.helpingamericasyouth.gov/sitemap.htm

possible to review all such research here, we concentrate on several areas that have been the focus of attention.

A great deal of research focuses on the crucial influence of the family in the formation of behavioral patterns and personality. Contemporary theories attach great importance to the parental role in determining the personality characteristics of children. More than half a century ago, Glueck and Glueck (1950) focused attention on the relationship between family and delinquency, a relationship that has remained in the spotlight ever since (see In Practice 3.2).

To young children, home and family are the basic sources of information about life. Thus many researchers and theorists have focused on the types of values, attitudes, and beliefs maintained and passed on by the family over generations. Interest has focused on the types of behavior and attitudes transmitted to children through the socialization process resulting in a predisposition toward delinquent behavior.

In Practice 3.2

Families Can Open Their Homes to Troubled Teens

Kendall homeowner Raysa Rodriguez recently took in a 15-year-old girl with a shaky past. She did not adopt the girl, who has bounced in and out of juvenile detention on battery charges and probation violations, and she is not her foster parent.

Rather, Rodriguez will house the girl for a year as part of a new program that aims to save troubled youths from lives of delinquency by placing them in stable homes.

As a counselor for juvenile delinquents at the alternative Bay Point Schools in Cutler Bay, Rodriguez said it was only natural for her to do something to help.

"I wanted to impact a youth's life," she said.

Using a proven behavioral system, host parents and a group of clinical therapists encourage teenagers with serious delinquency or behavioral problems to develop academic skills and positive work habits, helping them become model citizens.

A teenager's legal guardian is simultaneously taught more effective parenting skills.

Jonelle K. Dougery, a clinical program supervisor at Liberty Resources, which started the Community-Based Residential Alternative Program, said there are two aims: "to create opportunities for youth to live successfully in a family while preparing their parent to provide them with effective parenting," Dougery said, adding that the program is modeled after the Oregon-based Multidimensional Treatment Foster Care Program.

Liberty Resources needs more volunteer parents to host children for up to a year at an $18,000 yearly stipend. They just opened the first office in West Kendall this spring at 13016 SW 120th St.

This is the only program of its kind in the state.

Gerard Bouwman, president of Oregon-based TFC Consultants, whose purpose is to help implement the model in other cities, said research has proven the program is effective.

"We see significant reductions in contact with authorities subsequent to treatment, a lot less delinquency, a lot less behavior problems, and also [the ability] to function in family settings," Bouwman said.

The program is voluntary, but the Florida Department of Juvenile Justice must first refer the teenager. Children must be between 14 and 17 years old, have prior treatment or placement, serious and chronic delinquent behavior, and family problems.

"They're coming to us if they get a new misdemeanor charge or don't abide by conditional release by missing curfew or skipping school and are at risk of being committed again," Dougery said.

How does the program work? Children earn points as they exhibit appropriate behavior, such as getting up on time, doing extra chores, and going to each class.

"It's very encouraging as opposed to only pointing out that they did something wrong," Dougery said.

The parents keep tabs on the daily activities of the teen by giving [him or her] a card—like a progress report—which [the teen has] to take daily to school or any after-school job in which teachers and employers have to sign off on the time the teen gets there and leaves.

Once a week, the child sees an individual therapist.

"They'll role-play situations like, say, over the week the teenager got into a fight with the professional parent, the therapist would say, 'How could you have handled that differently? Let's talk about it,'" Dougery said.

(Continued)

(Continued)

A skills trainer is also assigned to go out with the children into the community and help them get involved, whether through sports or just interacting with others.

"If we're out in the community and the youth doesn't want to order food, we do it, and through that they learn to model appropriate behavior," Dougery said.

A family therapist is assigned to the after-care guardian or the person who will care for the child after the treatment period is over.

"They go into the home and work with the parents on what were the challenges, how can you do that differently, and teaching them more practical parenting skills because whatever they were doing obviously wasn't working," Dougery said.

Host families receive a daily phone call, a two-day training and certification course in the behavioral program, and 24-hour, seven-days-a-week on-call support.

The program has room for 10 kids at the time, there are already six referrals, and about six families are undergoing the process to get certified with the state.

Rodriguez is the only licensed host at the time.

SOURCE: Pineiro, Yudy. (2006, Sept. 14). Families Can Open Their Homes to Troubled Teens. *The Miami Herald,* page 15SD.

For example, research indicates a relationship between delinquency and the marital happiness of the children's parents. Official delinquency seems to occur disproportionately among juveniles in unhappy homes marked by marital discord, lack of family communication, unaffectionate parents, high stress and tension, and a general lack of parental cohesiveness and solidarity (Davidson 1990; Fleener 1999; Gorman-Smith, Tolan, & Loeber, 1998; Wallerstein & Kelly, 1980; Wright & Cullen, 2001). In unhappy familial environments, it is not unusual to find that parents derive little sense of satisfaction from their childrearing experience. Genuine concern and interest is seldom expressed except on an erratic and convenient basis at the whim of the parents. Also typical of this familial climate are inconsistent guidance and discipline marked by laxity and a tendency to use children against the other parent (Simons, Simons, Burt, Brody, & Cutrona, 2005). It is not surprising to find poor self-images, personality problems, and conduct problems in children of such families. If there is any validity to the adage "chip off the old block," it should not be surprising to find children in unpleasant family circumstances internalizing the types of attitudes, values, beliefs, and modes of behavior demonstrated by their parents.

It seems that in contemporary society, the family "home" has in many cases been replaced by a house where a related group of individuals reside, change clothes, and occasionally eat. It is somewhat ironic that we often continue to focus on **broken homes** (homes disrupted through divorce, separation, or desertion) as a major cause of delinquency rather than on nonbroken homes where relationships are marked by familial disharmony and disorganization. There is no doubt that the stability and continuity of a family may be shaken when the home is broken by the loss of a parent through death, desertion, long separation, or divorce. At a minimum, one half of the potential socializing and control team is separated from the family. The belief that one-parent families produce more delinquents is supported both by official statistics and by numerous studies. Canter (1982), for example, indicated that "youths from broken homes reported significantly more delinquent behavior than youths from intact homes. The general finding of greater male involvement in delinquency was unchanged when the focus was restricted to children from broken homes. Boys from broken homes reported more delinquent behavior than did girls from broken homes" (p. 164). Canter concluded, "This finding gives credence to the proposition that broken homes reduce parental supervision, which in turn may increase involvement in delinquency, particularly among males" (p. 164). In the Pittsburgh Youth Study, Browning and Loeber (1999) found that the demographic variable most strongly related to delinquency was having a broken family. According to the Forum on Child and Family Statistics (2006), "Living with two parents who are married to each other is associated with more favorable outcomes for children." However, the proportion of children under the age of 18 years living with two married parents fell from 77% in 1980 to 67% in 2005.

There is also, however, some evidence that there may be more social organization and cohesion, guidance, and control in happy one-parent families than in two-parent families marked by discord. It may be that the broken family is not as important a determinant of delinquency as are the events leading to the broken home. Disruption, disorganization, and tension, which may lead to a broken family or may prevail in a family staying intact "for the children's sake," may be more important causative factors of delinquency than the actual breakup (Browning & Loeber, 1999; Emery, 1982; Stern, 1964). According to Rebellon (2002), broken homes are strongly associated with a range of delinquent behaviors, including minor status offenses and more severe property/violent offenses. However, several factors, including divorce/separation, recent remarriage, and the long-term presence of a stepparent, appear to be related to different types of delinquency.

Not all authorities agree that broken homes have a major influence on delinquency. Wells and Rankin (1991), reviewing the relationship between broken homes and delinquency, concluded that there is some impact of broken homes on delinquency, although it appears to be moderately weak, especially for serious crime. Bumphus and Anderson (1999) concluded that traditional measures of family structure relate more to criminal patterns of Caucasians than to those of African Americans. Rebellon (2002) found that single-parenthood per se does not appear to be associated with delinquency; rather, certain types of changes in family composition appear to be related to delinquency. Demuth and Brown (2004), using data from the 1995 National Longitudinal Survey of Adolescent Health, extended prior research investigating the effects of growing up in two-parent versus single-mother families by also examining delinquency in single-father families. The results indicate that juveniles in single-parent families are significantly more delinquent than their counterparts residing with two biological married parents. However, the authors found that family processes fully account for the higher levels of delinquency exhibited by adolescents from single-father versus single-mother families. In 2005, among children under the age of 18 years, 23% lived with only their mother, 5% lived with only their father, and 4% lived with neither of their parents (Forum on Child and Family Statistics, 2006).

The American family unit has changed considerably during the past 50 years. Large and extended families, composed of various relatives living close together, at one time provided mutual aid, comfort, and protection. Today, the family is smaller and has relinquished many of its socialization functions to specialized organizations and agencies that exert a great amount of influence in the education, training, care, guidance, and protection of children. This often results in normative conflict for children who find their attitudes differing from the views and standards of their parents. These changes have brought more economic wealth to the family, but they may have made it more difficult for parents to give constructive guidance and protection to their children. In addition, the rise of "mixed families," in which each parent

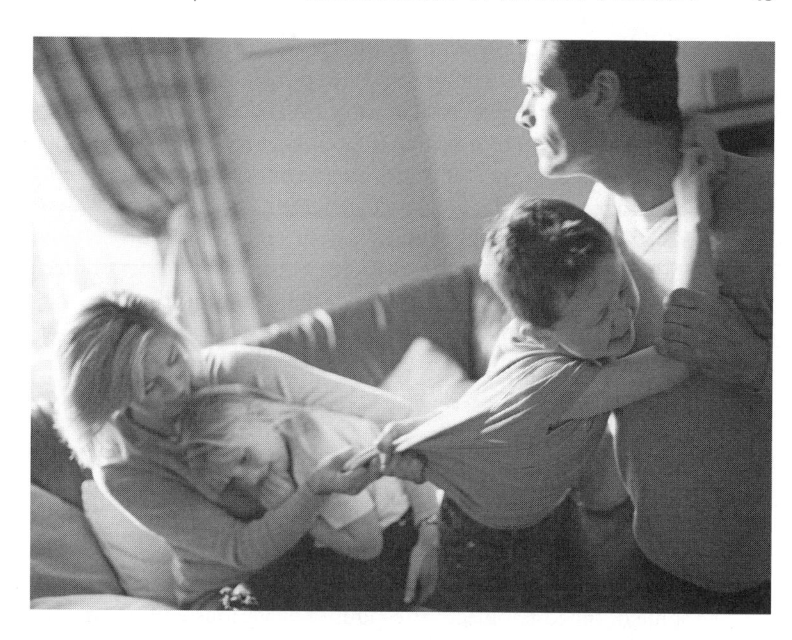

▶ Problems with children occur in families of all races and social classes.

brings children of his or her own into the family setting, may result in conflicts among the children or between one parent and the children of the other parent.

Over the years, there has been considerable interest in children with working parents who have come to be known as **latchkey children.** This term generally describes school-age children who return home from school to an empty house. Estimates indicate that there may be as many as 10 million children left unsupervised after school (Willwerth, 1993). These children are often left to fend for themselves before going to school in the morning, after school in the afternoon, and on school holidays when parents are working or otherwise occupied. This has resulted in older (but still rather young) children being required to care for younger siblings during these periods and is also a factor in the increasing number of children found in video arcades, in shopping malls, on the Internet, and in other areas without adult supervision at a relatively young age. Although the vast majority of latchkey children appear to survive relatively unscathed, some become involved in illegal or marginally legal activity without their parents' knowledge (Coohey, 1998; Flannery, Williams, & Vazsonyi, 1999; Vander Ven, Cullen, Carrozza, & Wright, 2001; Vandivere, Tout, Capizzano, & Zaslow, 2003).

There is little doubt that family structure is related to delinquency in a variety of ways. However, relying on official statistics to assess the extent of that relationship may be misleading. It may be that the police, probation officers, and judges are more likely to deal officially with juveniles from broken homes than to deal officially with juveniles from more "ideal" family backgrounds. Several authorities, including Fenwick (1982) and Simonsen (1991), have concluded that the decision to drop charges against a juvenile depends, first, on the seriousness of the offense and the juvenile's prior record and, second, on the juvenile's family ties. "Youths are likely to be released if they are affiliated with a conventional domestic network" (Fenwick, 1982, p. 450). "When parents can be easily contacted by the police and show an active interest in their children and an apparent willingness to cooperate with the police, the likelihood is much greater (especially in the case of minor offenses) that a juvenile will be warned and released to parental custody" (Bynum & Thompson, 1999, p. 364). Fader, Harris, Jones, and Poulin (2001) concluded that, in Philadelphia at least, juvenile court decision makers appear to give extra weight to child and family functioning factors in deciding on dispositions for first-time offenders.

It often appears that the difference between placing juveniles in institutions and allowing them to remain in the family setting depends more on whether the family is intact than on the quality of life within the family. Concentrating on the broken family as the major or only cause of delinquency fails to take into account the vast number of juveniles from broken homes who do not become delinquent as well as the vast number of juveniles from intact families who do become delinquent (Krisberg, 2005, p. 73).

Education

Schools, education, and families are very much interdependent and play a major role in shaping the future of children. In our society, education is recognized as one of the most important paths to success. The educational system occupies an important position and has taken over many functions formerly performed by the family. The total social well-being of children, including health, recreation, morality, and academic advancement, is a concern of educators. Some of the lofty objectives espoused by various educational commissions were summarized by Schafer and Polk (1967) more than a quarter century ago:

> All children and youth must be given those skills, attitudes, and values that will enable them to perform adult activities and meet adult obligations. Public education must ensure the maximum development of general knowledge, intellectual competence, psychological stability, social skills, and social awareness so that each new generation will be enlightened, individually strong, yet socially and civically responsible. (p. 224)

The child is expected by his or her parents, and by society, to succeed in life, but the child from a poor family, where values and opportunities differ from those of white middle-class America, encounters many difficulties early in school. Studies indicate that students from middle-class family backgrounds are more likely to have internalized the values of competitiveness, politeness, and deferred gratification that are likely to lead to success in the public schools (Braun, 1976). Braun (1976) also found that teachers' expectations were influenced by physical attractiveness, **socioeconomic status,** race, gender, name, and older siblings. Lower expectations existed for children who came from lower socioeconomic backgrounds, belonged to minority groups, and had older siblings who had been unsuccessful in school. Alwin and Thornton (1984) found that the socioeconomic status of the family was related to academic success both during early childhood and during adolescence. Blair, Blair, and Madamba (1999) found that social class–based characteristics were the best predictors of educational performance among minority students.

Numerous studies show that although some difficulties may be partially attributable to early experience in the family and neighborhood, others are created by the educational system itself (see In Practice 3.3). The label of low achiever, slow learner, or **learning disabled** may be attached shortly after, and sometimes even before, entering the first grade based on the performance of other family members who preceded the child in school. Teachers may expect little academic success as a result. Identification as a slow learner often sets into motion a series of reactions by the student, his or her peers, and the school itself that may lead to negative attitudes, frustrations, and eventually a climate where school becomes a highly unsatisfactory and bitter experience. Kelley (1977) found that early labeling in the school setting had a lasting impact on children's educational careers and that such labeling occurred with respect to children with both very great and very limited academic potential.

Kvaraceus (1945) believed that although school might not directly cause delinquency, it might present conditions that foster delinquent behavior. When aspirations for success in the educational system are blocked, the student's self-assessment, assessment of the value of education, and assessment of the school's role in his or her life may progressively deteriorate. Hawkins and Listiner (1987) indicated that low cognitive ability, poor early academic performance, low attachment to school, low commitment to academic pursuits, and association with delinquent peers appear to contribute to delinquency. Unless the student is old enough to drop out of this highly frustrating experience, the only recourse may be to seek others within the school who find themselves in the same circumstances.

Thornberry, Moore, and Christenson (1985) noted that dropping out of school was positively related to delinquency and later crime over both the long and short terms. Although the presence of others who share the frustrating experience of the educational system may be a satisfactory alternative to dropping out of school, the collective alienation may lead to delinquent behavior. Rodney and Mupier (1999) found that being suspended from school, being expelled from school, and being held back in school increased the likelihood of being in juvenile detention among adolescent African American males. Lotz and Lee (1999) found that negative school experiences are significant predictors of delinquent behavior among white teenagers. Jarjoura (1996) found that dropping out of school is more likely to be associated with greater involvement in delinquency for middle-class youth than for lower-class youth.

Most theorists agree that negative experiences in school act as powerful forces that help to project juveniles into delinquency. Achievement and self-esteem will be satisfied in the peer group or gang. In many ways, the school contributes to delinquency by failing to provide a meaningful curriculum to lower-class youth in terms of future employment opportunities. There is a growing recognition by many juveniles of the fact that satisfying educational requirements is no guarantee of occupational success (Monk-Turner, 1990). More than a quarter century ago, Polk and Schafer (1972) noted that the role of the school was rarely acknowledged as producing these unfavorable conditions. Instead of recognizing and attacking deficiencies in the learning structure of the schools, educational authorities place the blame on "delinquent youth" and thus further alienate them from school. In summarizing, Polk and Schafer listed the following as unfavorable experiences:

> (1) Lower socioeconomic–class children enter the formal educational process with a competitive disadvantage due to their social backgrounds; (2) The physical condition and educational climate of a school located in working class areas may not be conducive for the learning process; (3) Youths may be labeled early and placed in ability groups where expectations have been reduced; and (4) Curriculum and recognition of achievement revolve around the "college bound youth" and not the youth who intends to culminate his educational pursuit by graduating from high school. (p. 189)

Yablonsky and Haskell (1988), Battistich and Hom (1997), Yogan (2000), and Kowaleski-Jones (2000) all have discussed how school experiences may be related to delinquency. First, if

In Practice 3.3

Forum Tackles Discipline, At-Risk Youths

A group of Miami–Dade educators, politicians, and juvenile justice officials held a forum on Friday to discuss preventive measures to decrease racial disparities in school discipline.

At the center of the debate, held inside the Miami–Dade School Board auditorium, was what most leaders called a poorly defined and overused zero-tolerance policy in the district and how it has resulted in a staggering number of arrests, expulsions, and suspensions due to minor offenses.

"We're throwing away children at an early age by funneling them into the juvenile justice system," said Carlos Martinez, chief assistant Miami–Dade public defender. "I would have been considered delinquent for many of the things I did as a child."

Martinez gave the opening remarks to the forum titled "Improving Educational Outcomes and Reducing Disparities in Arrests and Discipline by Doing What Works."

The session was the third in a series that explored the linkages between school failure, zero-tolerance polices, race, and delinquency and their relevance to the Supreme Court's decision in *Brown v. Board of Education.*

Last year, when Miami–Dade schools police arrested about 2,500 students, black students accounted for more than half of those arrests, though they only make up 28 percent of county enrollment, district records show. Blacks also accounted for more than half of the 29,000 students sent home on outdoor suspension.

"Our primary mission is to chart a course to reduce racial disparities in suspensions and arrests, so all our children can share in the American dream," said event organizer Bennett H. Brummer, Miami–Dade public defender.

Throughout the four-hour-long dialogue, officials stressed the importance of prevention before punishment and the role that early intervention plays in making that philosophy a reality.

"We're aggressively implementing measures to address what's going on in the lives of kids who are exhibiting bad behavior," said schools Police Chief Gerald Darling. "Using us as the bullies and bad guys does not fix the problem."

Darling has implemented a civil citation program, which officials say will cut down on most of the arrests. Under the program, first- and second-time offenders will get citations for minor offenses such as disorderly conduct and trespassing.

The district has also proposed a new kind of suspension starting next school year. Suspended students who commit certain violations of the code of conduct will be required to go to alternative centers where they'll be assigned conflict resolution and other forms of anger management intervention.

But some officials present suggested that the zero-tolerance policy mandate sending thousands into handcuffs should be restructured as well as the student code of conduct guidelines.

"We pay for programs that are reactive," argued state NAACP President Adora Obi Nweze. "We need to change the zero-tolerance statute."

Sharon Frazier-Stephens said the policy has taken a personal toll.

"My son was one of those kids who fell through the cracks. When he was arrested, no one at the school notified me," said Frazier-Stephens, a volunteer at Miami Norland High. "Now he's in a correctional facility."

SOURCE: Bailey, Peter. (2006, April 29). Forum Tackles Discipline, At-Risk Youths. *Miami Herald,* page 5B.

a child experiences failure at school every day, he or she not only learns little but also becomes frustrated and unhappy. Curricula that do not promise a reasonable opportunity for every child to experience success in some area may, therefore, contribute to delinquency. Second, teaching without relating the subject matter to the needs and aspirations of the student leaves him or her with serious questions regarding the subject matter's relevancy. Third, for many lower-class children, school is a prison or a "babysitting" operation where they just pass time. They find little or no activity designed to give pleasure or indicate an interest in their abilities. Fourth, the impersonal school atmosphere, devoid of close relationships, may contribute toward the child seeking relationships in peer groups or gangs outside of the educational setting. In a similar vein, Polk (1984) contended that the number of marginal juveniles is growing and agreed that this is so not only because less successful students have unpleasant school experiences but also because their future occupational aspirations are severely limited.

In 1981, Zimmerman, Rich, Keilitz, and Broder investigated the relationship between learning disabilities and delinquency. They concluded that "proportionately more adjudicated delinquent children than public school children were learning disabled," although self-report data indicated no significant differences in the incidence of delinquent activity. They hypothesized that "the greater proportion of learning-disabled youth among adjudicated juvenile delinquents may be accounted for by differences in the way such children are treated within the juvenile justice system, rather than by differences in their delinquent behavior" (Zimmerman et al., 1981, p. 1).

In another study, Smykla and Willis (1981) found that 62% of the children under the jurisdiction of the juvenile court they studied were either learning disabled or mentally retarded. They concluded, "The findings of this study are in agreement with previous incidence studies that have demonstrated a correlation between juvenile delinquency and mental retardation. These results also forcefully demonstrate the need for special education strategies to be included in any program of delinquency prevention and control" (p. 225).

Others, including Brownfield (1990), also have concluded that poor school performance and delinquency are related. Browning and Loeber (1999) found that low IQ was related to delinquency independent of socioeconomic status, ethnicity, neighborhood, and impulsivity.

The emptiness that some students feel toward school and education demands our attention. Rebellion, retreatism, and delinquency may be a response to the false promises of education or simply a response to being "turned off" again in an environment where this has occurred too frequently. Without question, curriculum and caliber of instruction need to be relevant for all children. Social and academic skill remediation may be one means of preventing learning-disabled children from becoming involved in delinquency (Winters, 1997). Beyond these primary educational concerns, the school may currently be the only institution where humanism and concern for the individual are expressed in an otherwise bleak environment. Even this one-time sanctuary is under attack by gang members involved with drugs and guns. In some cases, the question is not whether a child can learn in school but rather whether he or she can get to school and back home alive. Armed security guards, barred windows, and metal detectors have given many schools the appearance of being the prisons that some children have always found them to be. Although student fears of being attacked at school have declined (the percentage of children who feared attack at school or on the way to and from school decreased significantly from 12% in 1995 to 6% in 2003), statistics vary among racial groups (Child Trends DataBank, 2006). As Figure 3.1 shows, larger percentages of African American and Hispanic students feared attacks than did white students. This may be a direct result of the geographic area in which these schools are located, an impersonal school atmosphere, and/or

▲ A significant number of juveniles annually report experiencing fear of attack while at school, on the way to school, or on the way home from school.

a lack of support or understanding that African American and Hispanic students feel in the school environment.

Fear of attack at school or on the way to and from school may cause some students to miss days of school and may negatively affect academic performance. Fear at school can create an unhealthy school environment, affect students' participation in class, and lead to more negative behaviors among students (Child Trends DataBank, 2006). Furthermore, students in lower grades are more likely to fear for their safety at school and on the way to and from school than are students in higher grades. In 2003, 10% of sixth-grade students had such fears, compared with 4% of twelfth-grade students (Child Trends DataBank, 2006). In addition, students in urban schools are roughly twice as likely as students in suburban or rural schools to fear being attacked at school or while traveling to and from school (Child Trends DataBank, 2006).

In another survey of American schoolchildren (Institute of Education Sciences, 2005), it was found that improvements in school safety have occurred over the past dozen or so years. The violent crime victimization rate declined from 48 per 1,000 students in 1992 to 28 per 1,000 students in 2003. Despite the decrease, violence, theft, bullying, drugs, and weapons are still widespread. In 2003, there were nearly three quarters of a million violent crimes and more than a million crimes of theft committed against students between the ages of 12 and 18 years (Institute of Education Sciences, 2005).

Events of the past few years have raised national concern about school safety. A chronology of the events leading to this concern was presented by the *Indianapolis Star* (Indystar.com, 2006):

October 2, 2006: Charles Carl Roberts, 32, took 10 girls hostage in an Amish school in Nickel Mines, Pennsylvania, killing five of them before killing himself.

September 29, 2006: Eric Hainstock, 15, took two guns into his Cazenovia, Wisconsin, school and fatally shot the principal before being captured and arrested.

September 27, 2006: Duane Morrison, 53, took six girls hostage at Platte Canyon High School in Bailey, Colorado, molesting them and holding them for hours before fatally shooting one girl and then himself.

Figure 3.1 Percentages of Students Fearing Attack at School

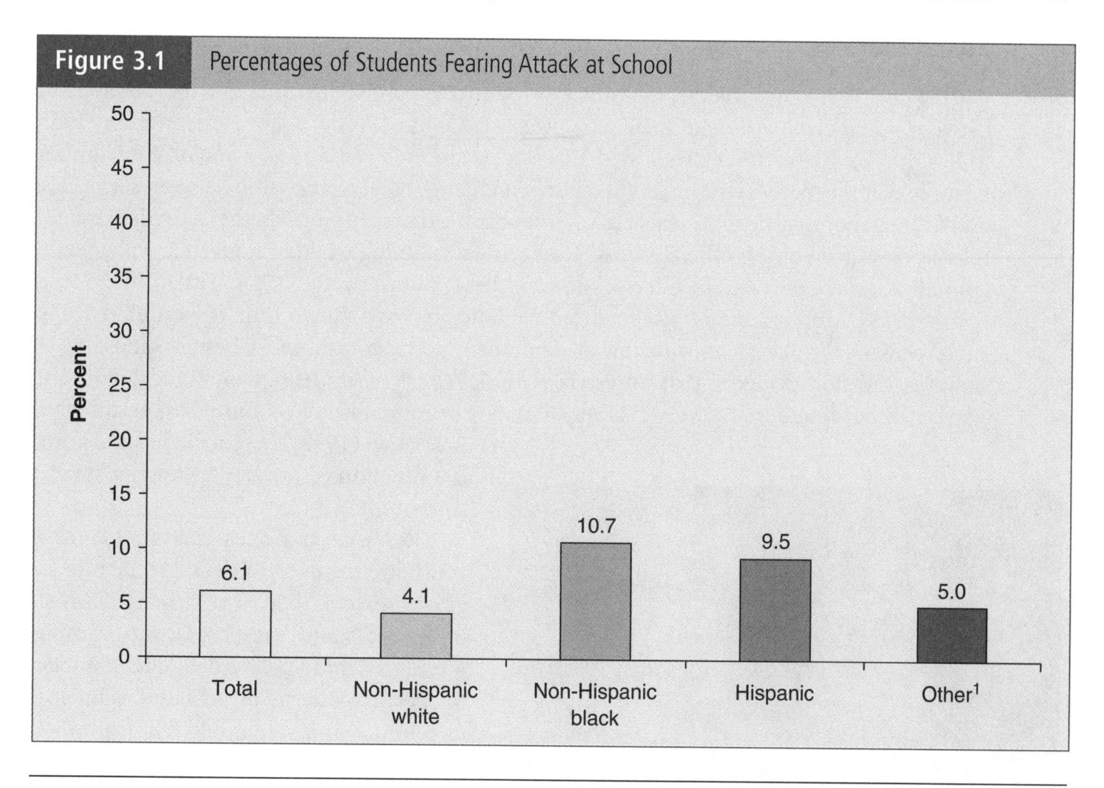

Source: DeVoe, J. F., Peter, K., Kaufman, P., Miller, A., Noonan, M., Snyder, T. D., & Baum, K. (2004). *Indicators of school crime and safety: 2004* (NCES 2005 -002/NCJ 205290, U.S. Departments of Education and Justice). Washington, DC: U.S. Government Printing Office, Table 12.1.

[1] "Other" includes those students who selected another race or more than one race. Hispanic students are excluded.

November 8, 2005: Assistant principal Ken Bruce was killed and two other administrators were seriously wounded when Kenny Bartley, a 15-year-old student, opened fire in a Jacksboro, Tennessee, high school.

August 24, 2006: Christopher Williams, 27, went to Essex Elementary School in Vermont, and when he could not find his ex-girlfriend, a teacher, he shot and killed one teacher and wounded another. Earlier, he had killed the ex-girlfriend's mother. He attempted suicide but survived and was arrested.

March 21, 2005: Jeff Weise, 16, shot to death his grandfather and his grandfather's girlfriend and then went to his high school in Red Lake, Minnesota, where he killed a security guard, a teacher, and five students, and wounded seven others, before killing himself.

These events and others emphasize the importance of events occurring at or near schools of the students involved. It is difficult to determine the impact of these events on the students actually involved and on those who become aware of the events through the national media.

Social Class

During the 1950s and 1960s, a number of studies emerged focusing on the relationship between social class and delinquency (Cloward & Ohlin, 1960; Cohen, 1955; Merton, 1955; Miller, 1958). These studies indicated that socioeconomic status was a major contributing factor in delinquency. According to further research, the actual relationship between social class and delinquency may be that social class is important in determining whether a particular juvenile becomes part of the official statistics, not in determining whether a juvenile will actually commit a delinquent act (Dentler & Monroe, 1961; Short & Nye, 1958; Tittle, Villemez, & Smith, 1978). Most studies of self-reported delinquency have shown little or no difference by social class in the actual commission of delinquent acts. Morash and Chesney-Lind (1991), however, did find evidence that lower-class youth report more delinquency, and Elliott and Ageton (1980) found that lower-class juveniles may be more likely to commit serious offenses. Ackerman (1998) also concluded that crime is a function of poverty, at least in smaller communities.

▲ Although more males than females are arrested for delinquency, the number of female delinquents has increased significantly during recent years.

Research indicates that middle-class youth are involved in delinquency to a far greater extent than was suspected previously. Scott and Vaz (1963), for example, found that middle-class delinquents adhere to specific patterns of activities, standards of conduct, and values different from their parents. Young people a generation ago had more in common with their parents, including attitudes and outlook on life. However, today's middle-class youth are securely entrenched in a **youth culture** that is often apart from, or in conflict with, the dominant adult culture. Within the youth culture, juveniles are open to the influence of their peers and generally conform to whatever behavior patterns prevail. Scott and Vaz identified partying, joyriding, drinking, gambling, and various types of sexual behavior as dominant forms of conduct within the middle-class youth culture. By participating in and conforming to the youth culture, status and social success are achieved through peer approval. Scott and Vaz argued that the bulk of middle-class delinquency occurs in the course of customary nondelinquent activities but moves to the realm of delinquency as the result of a need to "be different" or "start something new." Wooden and Blazak (2001) noted that these trends continue at the present time: "In the 1990s research began revealing what those who had survived the 1980s already knew: The safe cocoon of middle-class youth was eroding" (pp. 4–5). In *Youth Crisis: Growing Up in a High-Risk Society,* Davis (1999) pointed out that adolescence is a period of transition from childhood to adulthood. Each of the institutions of this transition (e.g., the family, education, employment) is in a state of turmoil, causing adolescents to be in a state of crisis.

Accessibility to social objects for participating in the youth culture is an important part of delinquent behavior. Social objects, such as cars, the latest styles, alcoholic beverages, and drugs, are frequently part of middle-class delinquency. Peer recognition for male middle-class youth may be a reason for senseless acts of destruction of property. Acts of vandalism in which one's bravery can be displayed for peer approval are somewhat different from the violent behavior often seen in lower-class youth, who may demonstrate their bravery by gang fights/shootings, muggings, robbery, and other crimes against people. Wooden and Blazak (2001) indicated that suburban youth are often told to act like adults but are not given the privileges of adulthood, forcing them into a subculture characterized by delinquency-producing focal concerns (p. 19). Some end up in trouble-oriented male groups, and they sometimes get involved in violent crime to conform to group norms. More typically, those in middle-class coed groups get involved in petty theft and drug use.

Although most evidence indicates that juveniles from all social classes may become delinquent (Elrod & Ryder, 2005, p. 61), the subculture theorists maintain that many delinquents grow up in lower-class slum areas. According to Cloward and Ohlin (1960), the type of delinquency exhibited depends in part on the type of slum in which juveniles grow up. The slum that produces professional criminals is characterized by the close-knit lives and activities of the people in the community. Constant exposure to delinquent and criminal processes coupled with an admiration of criminals provides the model and impetus for future delinquency and criminality. Cloward and Ohlin described this as a **criminal subculture** in which juveniles are encouraged and supported by well-established conventional and criminal institutions. Going one step further, Miller (1958), in his study of lower- and middle-class norms, values, and behavioral expectations, concluded that a delinquent subculture is inherent in lower-class standards and goals. The desirability of the achievement of status through toughness and smartness, as well as the concepts of trouble, excitement, fate, and autonomy, is interpreted differently depending on one's socioeconomic status. Miller concluded that by adhering to lower-class norms, pressure toward delinquency is inevitable and is rewarded and respected in the lower-class value system. Lawbreaking is not in and of itself a deliberate rejection of middle-class values, but it automatically violates certain moral and legal standards of the middle class. Miller believed that lower-class youth who become delinquent are primarily conforming to traditions and values held by their families, peers, and neighbors. As indicated earlier, Wooden and Blazak (2001) used this same approach to describe middle-class delinquency during the 21st century.

In summarizing the findings with respect to the relationship between social class and delinquency, Johnson (1980) concluded that some conceptualizations of social class may have been inappropriate and that a more appropriate distinction is the one between the **underclass** and the earning class. His results suggest, however, that even given this distinction, there is no reason to expect that social class will emerge as a "major correlate of delinquent behavior, no matter how it is measured" (p. 86). Current evidence presented by Wooden and Blazak (2001) seems to indicate that this may well be the case, as does the paucity of current research in this area.

Still, the concept of the underclass (the extremely poor population that has been abandoned in the inner city as a result of the exodus of the middle class) seems to attract continuing attention (Bursik & Grasmick, 1995; Jarjoura, Triplett, & Brinker, 2002). As the more affluent withdraw from inner-city communities, they also tend to withdraw political support for public spending designed to benefit those communities. They do not want to pay taxes for schools they do not use, and they are not likely to use them because they find those left behind too frightening to be around (Ehrenreich, 1990). Those left behind are largely excluded, on a permanent basis, from the primary labor market and mainstream occupations. Economically motivated delinquency is one way of coping with this disenfranchisement to maintain a short-term cash flow. Because many children growing up in these circumstances see no relationship between attaining an education and future employment, they tend to drop out of school prior to graduation. Some then become involved in theft as a way of meeting economic needs, often as members of gangs that may become institutionalized in underclass neighborhoods (Bursik & Grasmick, 1995, p. 122).

Perhaps Chambliss (1973) summed up the impact of social class on delinquency best some years ago when he concluded that the results of some delinquents' activities are seen as less serious than others as the result of class in American society:

> No representative of the upper class drew up the operations chart for the police which led them to look in the ghettoes and on street corners—which led them to see the demeanor of lower class youth as troublesome and that of upper middle class youth as tolerable. Rather, the procedures simply developed from experience—experience with irate and influential upper middle class parents insisting that their son's vandalism was simply a prank and his drunkenness only a momentary "sowing of wild oats"—experience with cooperative or indifferent, powerless lower class parents who acquiesced to the law's definition of their son's behavior. (p. 30)

Gangs

The influence of juvenile gangs is so important, and has received so much attention in the recent past, that we have devoted a separate chapter (Chapter 12) to the subject. In this section, we simply say that gangs are an important factor in the development of delinquent behavior, not only in inner-city areas but also increasingly in suburban and rural areas.

Drugs

Although drugs clearly have physical effects on those who use them, drug use is also a social act. We have more to say about drug use later in the book, but for now a brief discussion of the topic is in order.

Our society is characterized by high rates of drug use and abuse, and it should not be surprising to find such use and abuse among juveniles. The manufacture, distribution, and use of illicit drugs seem to be on the rise, and one new drug in particular, **methamphetamine** ("meth," "ice," "crystal," "glass," or "speed") has experienced a tremendous resurgence in popularity during the past few years (Scaramella, 2000). In a study reported by the National Center for Education Statistics (1997), 30% of sixth- through twelfth-grade students surveyed reported that alcohol and marijuana were available in their schools, and 20% said that other drugs were

▲ Gangs from Latin America have become increasingly common in the United States.

available. One third said that they had seen other students under the influence of alcohol at school, and 27% said that they had seen students under the influence of other drugs. Another study found that as many as 51% of high school seniors reported using illicit drugs at some time (Cohn, 1999). One should keep in mind that these figures apply to students still in school and do not include data from those who have dropped out of school. A 1985 study by Fagan and Pabon (1990) found that 54% of **dropouts** reported using illicit drugs during the past year, as compared with 30% of students. Addiction to alcohol, tobacco, prescription drugs, and illicit drugs frequently occurs during early adolescence, and Wade and Pevalin (2005), among others, have identified temporal associations between nuisance delinquency and both alcohol and marijuana use.

According to Watson (2004), research over the past 20 years has established the correlation of substance abuse to juvenile delinquency. The problem of substance use is even more pronounced among adolescents in contact with the juvenile justice system. Survey research indicates that more than half of juvenile male arrestees tested positive for at least one drug, and it appears that 60% to 87% of female offenders need substance abuse treatment. Marijuana appears to be the drug of choice among youthful offenders, growing from roughly 15% in 1991 to 62% in 1999.

There has, of course, been a good deal written about the relationship between illegal drug use and crime. This has been particularly true since the mid-1980s when **crack**, a cocaine-based stimulant drug, first appeared. As Inciardi, Horowitz, and Pottieger (1993) noted, "Cocaine is the drug of primary concern in examining drug/crime relationships among adolescents today.

It is a powerful drug widely available at a cheap price per dose, but its extreme addictiveness can rapidly increase the need for more money" (p. 48). Today, this concern has been replaced in many areas by a concern with methamphetamines:

> Methamphetamine and cocaine have similar behavioral and psychological effects on users. . . . Both psychostimulants spark a rapid accumulation in the brain of the neurotransmitter dopamine, which causes a feeling of euphoria. . . . Tests have found that . . . meth damages the neurons that produce dopamine and seratonin, another neurotransmitter. . . . Cocaine is not neurotoxic. . . . A high from smoking crack cocaine lasts about 20–30 minutes. A meth high can last more than 12 hours. . . . Heavy use can also lead to psychotic behavior such as paranoia and hallucinations. Some evidence suggests that chronic meth users tend to be more violent than heavy cocaine users. (Parsons, 1998, p. 4)

There is also considerable interest in the relationship between illegal drugs and gangs. For example, it was reported that gang members accounted for 86% of serious delinquent acts, 69% of violent delinquent acts, and 70% of drug sales in Rochester, New York (Cohn, 1999). Possession, sale, manufacture, and distribution of any of a number of illegal drugs are, in themselves, crimes. Purchase and consumption of some legal drugs, such as alcohol and tobacco, by juveniles are also illegal. Juveniles who violate statutes relating to these offenses may be labeled as delinquent or status offenders. Equally important, however, are other illegal acts often engaged in by drug users to support their drug habits. Such offenses are known to include theft, burglary, robbery, and prostitution, among others. It is also possible that use of certain drugs, such as cocaine and its derivatives and amphetamines, is related to the commission of violent crimes, although the exact nature of the relationship between drug abuse and crime is controversial. Some maintain that delinquents are more likely to use drugs than are nondelinquents—that is, drug use follows rather than precedes delinquency—whereas others argue the opposite (Dawkins, 1997; Thornton, Voight, & Doerner, 1987; Williams, Ayers, & Abbott, 1999). Whatever the nature of the relationship between drug abuse and delinquency, the two are intimately intertwined for some delinquents, whereas drug abuse is not a factor for others. Why some juveniles become drug abusers while others in similar environments avoid such involvement is the subject of a great deal of research. The single most important determinant of drug abuse appears to be the interpersonal relationships in which the juvenile is involved, particularly interpersonal relationships with peers. Drug abuse is a social phenomenon that occurs in social networks accepting, tolerating, and/or encouraging such behavior. Although the available evidence suggests that peer influence is most important, there is also evidence to indicate that juveniles whose parents are involved in drug abuse are more likely to abuse drugs than are juveniles whose parents are not involved in drug abuse. Furthermore, behavior of parents and peers appears to be more important in drug abuse than do the values and beliefs espoused (Schinke & Gilchrist, 1984; Williams et al., 1999).

There is no way of knowing how many juveniles suffering from school-, parent-, or peer-related depression and/or the general ambiguity surrounding adolescence turn to drugs as a means of escape, but the prevalence of teen suicide, combined with information obtained from self-reports of juveniles, indicates that the numbers are large. Although juvenile involvement with drugs in general apparently declined during the 1980s, it now appears that the trend has been reversed. There is little doubt that such involvement remains a major problem, particularly in light of gang-related drug operations. When gangs invade and take over a community, drugs are sold openly in junior and senior high schools, on street corners, and in shopping centers. The same is true of methamphetamines that are manufactured easily and sold inexpensively (Bartollas, 1993, p. 341; Scaramella, 2000).

Howell and Decker (1999) suggested that the relationship among gangs, drugs, and violence is complex. Pharmacological effects of drugs can lead to violence, and the high cost of drug use often causes users to support continued use with violent crimes. Finally, violence is common among gangs attempting to protect or expand drug territories.

Physical Factors

In addition to social factors, a number of physical factors are often employed to characterize juvenile delinquents. The physical factors most commonly discussed are age, gender, and race. (All of the data presented in this section are from the Federal Bureau of Investigation's [FBI] *Crime in the United States* for 2004 and 2005 [FBI, 2005, 2006].)

Age

For purposes of discussing official statistics concerning persons under the age of 18 years, we should note that little official action is taken with respect to delinquency under the age of 10 years. Rather than considering the entire age range from birth to 18 years, we are basically reviewing statistics covering an age range from 10 to 18 years. Keep in mind also our earlier observations (Chapter 2) concerning the problems inherent in the use of official statistics as we review the data provided by the FBI.

As Table 3.1 indicates, crimes committed by persons under 18 years of age (the maximum age for delinquency in a number of states) declined by roughly 3% between 2004 and 2005. However, murder and nonnegligent manslaughter arrests increased by nearly 20%, robbery arrests increased by slightly more than 11%, and forcible rape arrests decreased by approximately 11% among those under 18 years of age.

Table 3.1 also includes statistics on less serious offenses. Considering these offenses, gambling arrests increased by roughly 23% among those under 18 years of age, and weapons-related offenses increased by slightly more than 7%.

As you can see in Table 3.1, total offenses among those under 15 years of age declined by more than 7% between 2004 and 2005, while similar crimes among those 18 years of age and over increased very slightly (less than 1%).

As illustrated in Table 3.2, the total number of persons under the age of 18 years arrested for all crimes decreased 6%, the number of persons in this category arrested for murder and nonnegligent manslaughter increased 16%, and the number arrested for robbery increased roughly 14% between 2001 and 2005. The number arrested for forcible rape decreased 15%, and the number of arrests for auto theft decreased 24%. Comparable figures for those 18 years of age and over all showed some increase with the exception of forcible rape. Among offenses other than index crimes, carrying/possessing weapons (24% increase), offenses against family and children (40% decrease), gambling (37% decrease), embezzlement (40% increase), drunkenness (21% decrease), and vagrancy (109% increase) showed significant changes among those under 18 years of age.

Juveniles under the age of 18 years accounted for an estimated 25% of the 2006 U.S. population. Persons in this age group accounted for 15% of violent crime clearances and 26% of property crime clearances (cleared by arrests of suspected perpetrators). Murder (8%) and aggravated assault (13%) show the lowest percentage of juvenile involvement in violent crime, and robbery

(Text continues on page 67)

Table 3.1 Current Year Over Previous Year Arrest Trends—Totals, 2004–2005 (9,869 agencies; 2005 estimated population 194,973,254; 2004 estimated population 193,248,637)

	Number of Persons Arrested											
	Total All Ages			Under 15 Years of Age			Under 18 Years of Age			18 Years of Age and Over		
Offense Charged	2004	2005	Percentage Change	2004	2005	Percentage Change	2004	2005	Percentage Change	2004	2005	Percentage Change
Total[a]	8,975,704	8,997,831	+0.2	451,098	417,492	−7.4	1,403,555	1,360,641	−3.1	7,572,149	7,637,190	+0.9
Murder and nonnegligent manslaughter	7,698	8,259	+7.3	72	71	−1.4	593	711	+19.9	7,105	7,548	+6.2
Forcible rape	16,485	16,004	−2.9	1,050	867	−17.4	2,743	2,434	−11.3	13,742	13,570	−1.3
Robbery	63,691	65,841	+3.4	3,319	3,462	+4.3	14,099	15,713	+11.4	49,592	50,128	+1.1
Aggravated assault	274,827	278,708	+1.4	12,909	12,649	−2.0	37,298	36,995	−0.8	237,529	241,713	+1.8
Burglary	193,032	194,273	+0.6	18,667	16,715	−10.5	53,508	50,756	−5.1	139,524	143,517	+2.9
Larceny–theft	794,116	763,239	−3.9	82,359	70,698	−14.2	220,493	200,866	−8.9	573,623	562,373	−2.0
Motor vehicle theft	84,554	82,811	−2.1	5,469	4,633	−15.3	22,012	19,960	−9.3	62,542	62,851	+0.5
Arson	10,019	10,369	+3.5	3,194	3,182	−0.4	5,161	5,222	+1.2	4,858	5,147	+5.9
Violent crime[b]	362,701	368,812	+1.7	17,350	17,049	−1.7	54,733	55,853	+2.0	307,968	312,959	+1.6
Property crime[b]	1,081,721	1,050,692	−2.9	109,689	95,228	−13.2	301,174	276,804	−8.1	780,547	773,888	−0.9
Other assaults	838,946	843,739	+0.6	68,546	65,257	−4.8	160,251	158,891	−0.8	678,695	684,848	+0.9

(Continued)

Table 3.1 (Continued)

Offense Charged	Number of Persons Arrested											
	Total All Ages			Under 15 Years of Age			Under 18 Years of Age			18 Years of Age and Over		
	2004	2005	Percentage Change	2004	2005	Percentage Change	2004	2005	Percentage Change	2004	2005	Percentage Change
Forgery and counterfeiting	80,636	76,353	−5.3	503	324	−35.6	3,312	2,792	−15.7	77,324	73,561	−4.9
Fraud	217,421	209,228	−3.8	909	881	−3.1	5,089	4,991	−1.9	212,332	204,237	−3.8
Embezzlement	12,613	12,820	+1.6	39	47	+20.5	739	797	+7.8	11,874	12,023	+1.3
Stolen property (buying, receiving, possessing)	85,034	86,393	+1.6	4,327	3,827	−11.6	15,616	14,635	−6.3	69,418	71,758	+3.4
Vandalism	179,999	180,332	+0.2	30,146	28,456	−5.6	68,840	68,010	−1.2	111,159	112,322	+1.0
Weapons (carrying, possessing, etc.)	107,676	115,803	+7.5	9,010	9,253	+2.7	25,062	26,859	+7.2	82,614	88,944	+7.7
Prostitution and commercialized vice	41,761	40,686	−2.6	102	99	−2.9	859	759	−11.6	40,902	39,927	−2.4
Sex offenses (except forcible rape and prostitution)	53,672	52,464	−2.3	6,005	5,318	−11.4	11,610	10,573	−8.9	42,062	41,891	−0.4
Drug abuse violations	1,054,785	1,097,989	+4.1	20,648	19,044	−7.8	117,095	114,888	−1.9	937,690	983,101	+4.8
Gambling	3,758	3,450	−8.2	66	84	+27.3	324	398	+22.8	3,434	3,052	−11.1

(Continued)

Table 3.1 (Continued)

Offense Charged	Number of Persons Arrested											
	Total All Ages			Under 15 Years of Age			Under 18 Years of Age			18 Years of Age and Over		
	2004	2005	Percentage Change	2004	2005	Percentage Change	2004	2005	Percentage Change	2004	2005	Percentage Change
Offenses against the family and children	81,769	83,428	+2.0	1,273	1,096	–13.9	3,734	3,463	–7.3	78,035	79,965	+2.5
Driving under the influence	926,335	904,976	–2.3	280	205	–26.8	13,003	11,824	–9.1	913,332	893,152	–2.2
Liquor laws	408,373	392,438	–3.9	9,177	8,180	–10.9	88,603	86,328	–2.6	319,770	306,110	–4.3
Drunkenness	381,585	374,847	–1.8	1,422	1,264	–11.1	11,470	10,576	–7.8	370,115	364,271	–1.6
Disorderly conduct	414,172	416,240	+0.5	56,253	53,780	–4.4	132,445	131,174	–1.0	281,727	285,066	+1.2
Vagrancy	17,154	17,413	+1.5	976	1,027	+5.2	3,135	3,173	+1.2	14,019	14,240	+1.6
All other offenses (except traffic)	2,484,175	2,531,124	+1.9	68,157	63,253	–7.2	245,043	239,249	–2.4	2,239,132	2,291,875	+2.4
Suspicion	1,514	1,710	+12.9	119	76	–36.1	410	337	–17.8	1,104	1,373	+24.4
Curfew and loitering law violations	60,682	62,171	+2.5	17,730	17,645	–0.5	60,682	62,171	+2.5	—	—	—
Runaway	80,736	76,433	–5.3	28,490	26,175	–8.1	80,736	76,433	–5.3	—	—	—

SOURCE: Adapted from Federal Bureau of Investigation. (2006). *Crime in the United States, 2005.* Available: www.fbi.gov/ucr/ucr.htm#cius

a. Does not include suspicion.

b. Violent crimes are offenses of murder, forcible rape, robbery, and aggravated assault. Property crimes are offenses of burglary, larceny–theft, motor vehicle theft, and arson.

Table 3.2 Five-Year Arrest Trends, Totals, 2001–2005 (9,869 agencies; 2005 estimated population 194,973,254; 2004 estimated population 193,248,637)

	Number of Persons Arrested								
	Total All Ages			Under 18 Years of Age			18 Years of Age and Over		
Offense Charged	**2001**	**2005**	**Percentage Change**	**2001**	**2005**	**Percentage Change**	**2001**	**2005**	**Percentage Change**
Total[a]	8,288,959	8,573,824	+3.4	1,385,876	1,303,278	–6.0	6,903,083	7,270,546	+5.3
Murder and nonnegligent manslaughter	7,605	8,176	+7.5	629	728	+15.7	6,976	7,448	+6.8
Forcible rape	16,327	15,483	–5.2	2,762	2,340	–15.3	13,565	13,143	–3.1
Robbery	63,773	67,748	+6.2	14,495	16,445	+13.5	49,278	51,303	+4.1
Aggravated assault	287,870	282,224	–2.0	39,023	37,229	–4.6	248,847	244,995	–1.5
Burglary	179,626	189,547	+5.5	56,207	48,941	–12.9	123,419	140,606	+13.9
Larceny–theft	720,880	720,730	*	221,373	188,291	–14.9	499,507	532,439	+6.6
Motor vehicle theft	81,860	83,025	+1.4	26,007	19,675	–24.3	55,853	63,350	+13.4
Arson	10,842	9,949	–8.2	5,822	5,140	–11.7	5,020	4,809	–4.2
Violent crime[b]	375,575	373,631	–0.5	56,909	56,742	–0.3	318,666	316,889	–0.6
Property crime[b]	993,208	1,003,251	+1.0	309,409	262,047	–15.3	683,799	741,204	+8.4
Other assaults	790,255	808,673	+2.3	143,353	156,493	+9.2	646,902	652,180	+0.8
Forgery and counterfeiting	71,796	71,517	–0.4	3,807	2,629	–30.9	67,989	68,888	+1.3
Fraud	213,970	197,736	–7.6	5,506	4,779	–13.2	208,464	192,957	–7.4
Embezzlement	13,706	12,763	–6.9	1,317	796	–39.6	12,389	11,967	–3.4
Stolen property (buying, receiving, possessing)	75,415	83,855	+11.2	16,633	13,963	–16.1	58,782	69,892	+18.9
Vandalism	167,182	171,439	+2.5	66,826	64,660	–3.2	100,356	106,779	+6.4

(Continued)

Table 3.2 (Continued)

Offense Charged	Number of Persons Arrested								
	Total All Ages			Under 18 Years of Age			18 Years of Age and Over		
	2001	2005	Percentage Change	2001	2005	Percentage Change	2001	2005	Percentage Change
Weapons (carrying, possessing, etc.)	95,711	114,538	+19.7	21,702	26,844	+23.7	74,009	87,694	+18.5
Prostitution and commercialized vice	39,139	43,308	+10.7	687	841	+22.4	38,452	42,467	+10.4
Sex offenses (except forcible rape and prostitution)	52,520	50,378	–4.1	10,945	9,733	–11.1	41,575	40,645	–2.2
Drug abuse violations	919,547	1,062,638	+15.6	117,577	109,552	–6.8	801,970	953,086	+18.8
Gambling	4,396	3,661	–16.7	312	426	+36.5	4,084	3,235	–20.8
Offenses against the family and children	87,150	78,901	–9.5	5,652	3,401	–39.8	81,498	75,500	–7.4
Driving under the influence	847,303	829,098	–2.1	12,289	10,699	–12.9	835,014	818,399	–2.0
Liquor laws	386,431	358,417	–7.2	88,779	76,849	–13.4	297,652	281,568	–5.4
Drunkenness	412,931	373,325	–9.6	13,363	10,570	–20.9	399,568	362,755	–9.2
Disorderly conduct	359,245	367,634	+2.3	102,717	117,123	+14.0	256,528	250,511	–2.3
Vagrancy	15,720	19,886	+26.5	1,532	3,202	+109.0	14,188	16,684	+17.6
All other offenses (except traffic)	2,201,664	2,396,884	+8.9	240,466	219,638	–8.7	1,961,198	2,177,246	+11.0
Suspicion	2,240	2,386	+6.5	786	295	–62.5	1,454	2,091	+43.8
Curfew and loitering law violations	80,326	80,029	–0.4	80,326	80,029	–0.4	—	—	—
Runaway	85,769	72,262	–15.7	85,769	72,262	–15.7	—	—	—

SOURCE: Adapted from Federal Bureau of Investigation. (2006). *Crime in the United States, 2005.* Available: www.fbi.gov/ucr/ucr.htm#cius

*Less than one-tenth of 1%.

a. Does not include suspicion.

b. Violent crimes are offenses of murder, forcible rape, robbery, and aggravated assault. Property crimes are offenses of burglary, larceny–theft, motor vehicle theft, and arson.

(Text continued from page 61)

(24%) shows the highest. With respect to other index crimes, juveniles appear to be overrepresented in burglaries (26%), larceny–theft (26%), motor vehicle theft (24%), and arson (50%), especially when we consider the fact that, for all practical purposes, we are dealing only with juveniles between the ages of 10 and 18 years (approximately 17% of the nation's population).

It is sometimes interesting to compare short-term trends, such as those in Table 3.2, with trends over the longer term. Ten-year arrest trends (1996–2005) show a significant decrease in total crime rates among those under 18 years of age (25%) and also show a significant decrease in both violent crimes (25%) and property crimes (44%) (Table 3.3). Most notable here is the considerable decrease (nearly 47%) in murder/nonnegligent manslaughter. Only prostitution-related offenses showed a significant increase.

Gender

As indicated in Table 3.4, total crime in the under-18-years-of-age category declined over the 5-year period between 2001 and 2005 by roughly 7% to 8% among males and by roughly 2% among females. However, murder/nonnegligent manslaughter and robbery among both genders increased significantly during the same time period. Overall, violent crime decreased slightly among males under the age of 18 years and increased slightly among females in the same age group, and property crime decreased among both groups. Weapons offenses increased significantly among both males and females, as did gambling and vagrancy. Prostitution-related offenses also increased significantly among females under 18 years of age over the 5-year period in question.

Historically, we have observed three to four arrests of juvenile males for every arrest of a juvenile female. During the period from 2001 to 2005, this ratio changed considerably so that juvenile females now account for roughly 42% of arrests of those under 18 years of age (see Table 3.4). The total number of arrests of males under age 18 decreased 8%, and the total number of arrests of females in the same age group decreased roughly 2%. Considering the index crimes, we note that among those under age 18, arrests for violent crimes remained nearly constant for males but increased very slightly for females.

Considering all crimes, we note an increase in the number of females arrested for murder and nonnegligent manslaughter (20%), robbery (23%), other assaults (16%), weapons-related offenses (27%), prostitution-related offenses (39%), gambling (39%), and vagrancy (181%).

According to Chesney-Lind (1999), females have been largely overlooked by those interested in juvenile justice, and indeed many of their survival mechanisms (e.g., running away when confronted with abusers) have been criminalized. It appears that the juvenile justice network does not always act in the best interests of female juveniles because it often ignores their unique problems (Holsinger, 2000). Still, the number of girls engaging in problematic behavior is increasing, and it may well be that we need to develop treatment methods that address their specific problems. For example, a study conducted by Ellis, O'Hara, and Sowers (1999) found that troubled female adolescents have a profile distinctly different from that of males. The female group was characterized as abused, self-harmful, and social, whereas the male group was seen as aggressive, destructive, and asocial. The authors concluded that different treatment modalities (more supportive and more comprehensive in nature) may need to be developed to treat troubled female adolescents. Johnson (1998) maintained that the increasing number of delinquent females can be addressed only by a multiagency approach based on nationwide and systemwide cooperation. Peters and Peters's (1998) findings seem to provide support for Johnson's proposal. They

(Text continues on page 73)

Table 3.3 Ten-Year Arrest Trends, Totals, 1996–2005 (8,009 agencies; 2005 estimated population 178,017,991; 1996 estimated population 159,290,470)

Offense Charged	Number of Persons Arrested								
	Total All Ages			Under 18 Years of Age			18 Years of Age and Over		
	1996	2005	Percentage Change	1996	2005	Percentage Change	1996	2005	Percentage Change
Total[a]	8,619,699	8,244,321	−4.4	1,703,500	1,278,948	−24.9	6,916,199	6,965,373	+0.7
Murder and nonnegligent manslaughter	9,564	7,989	−16.5	1,388	739	−46.8	8,176	7,250	−11.3
Forcible rape	18,745	15,129	−19.3	3,202	2,392	−25.3	15,543	12,737	−18.1
Robbery	80,980	67,841	−16.2	25,318	16,791	−33.7	55,662	51,050	−8.3
Aggravated assault	315,405	282,003	−10.6	46,124	36,967	−19.9	269,281	245,036	−9.0
Burglary	220,798	180,973	−18.0	85,248	47,416	−44.4	135,550	133,557	−1.5
Larceny–theft	905,963	692,593	−23.6	319,161	182,813	−42.7	586,802	509,780	−13.1
Motor vehicle theft	100,318	82,160	−18.1	42,957	19,755	−54.0	57,361	62,405	+8.8
Arson	11,598	9,716	−16.2	6,506	4,915	−24.5	5,092	4,801	−5.7
Violent crime[b]	424,694	372,962	−12.2	76,032	56,889	−25.2	348,662	316,073	−9.3
Property crime[b]	1,238,677	965,442	−22.1	453,872	254,899	−43.8	784,805	710,543	−9.5
Other assaults	756,129	737,475	−2.5	137,850	142,957	+3.7	618,279	594,518	−3.8
Forgery and counterfeiting	72,103	70,738	−1.9	5,433	2,600	−52.1	66,670	68,138	+2.2
Fraud	255,162	193,539	−24.2	6,947	4,779	−31.2	248,215	188,760	−24.0
Embezzlement	10,152	12,087	+19.1	880	751	−14.7	9,272	11,336	+22.3
Stolen property (buying, receiving, possessing)	91,832	82,771	−9.9	26,647	13,902	−47.8	65,185	68,869	+5.7
Vandalism	190,069	168,366	−11.4	87,907	63,697	−27.5	102,162	104,669	+2.5

(Continued)

Table 3.3 (Continued)

Offense Charged	Number of Persons Arrested								
	Total All Ages			Under 18 Years of Age			18 Years of Age and Over		
	1996	2005	Percentage Change	1996	2005	Percentage Change	1996	2005	Percentage Change
Weapons (carrying, possessing, etc.)	123,016	112,054	−8.9	31,067	26,834	−13.6	91,949	85,220	−7.3
Prostitution and commercialized vice	48,936	41,641	−14.9	723	870	+20.3	48,213	40,771	−15.4
Sex offenses (except forcible rape and prostitution)	56,484	52,410	−7.2	10,620	10,437	−1.7	45,864	41,973	−8.5
Drug abuse violations	830,684	1,034,844	+24.6	117,400	106,150	−9.6	713,284	928,694	+30.2
Gambling	6,352	3,446	−45.7	563	395	−29.8	5,789	3,051	−47.3
Offenses against the family and children	84,459	72,623	−14.0	4,839	3,067	−36.6	79,620	69,556	−12.6
Driving under the influence	877,727	816,243	−7.0	11,000	10,550	−4.1	866,727	805,693	−7.0
Liquor laws	364,792	348,974	−4.3	95,686	76,756	−19.8	269,106	272,218	+1.2
Drunkenness	446,767	335,730	−24.9	14,821	9,094	−38.6	431,946	326,636	−24.4
Disorderly conduct	420,232	379,439	−9.7	112,697	116,422	+3.3	307,535	263,017	−14.5
Vagrancy	16,424	17,376	+5.8	1,998	1,395	−30.2	14,426	15,981	+10.8
All other offenses (except traffic)	2,062,908	2,269,707	+10.0	264,418	220,050	−16.8	1,798,490	2,049,657	+14.0
Suspicion	4,025	2,569	−36.2	1,453	360	−75.2	2,572	2,209	−14.1
Curfew and loitering law violations	119,407	87,658	−26.6	119,407	87,658	−26.6	—	—	—
Runaway	122,693	68,796	−43.9	122,693	68,796	−43.9	—	—	—

SOURCE: Adapted from FBI. (2006). *Crime in the United States, 2005.* Available: www.fbi.gov/ucr/ucr.htm#cius

a. Does not include suspicion.

b. Violent crimes are offenses of murder, forcible rape, robbery, and aggravated assault. Property crimes are offenses of burglary, larceny–theft, motor vehicle theft, and arson.

Table 3.4 Five-Year Arrest Trends by Gender (8,815 agencies; 2005 estimated population 185,294,195; 2001 estimated population 178,385,937)

Offense Charged	Male						Female					
	Total			Under 18 Years of Age			Total			Under 18 Years of Age		
	2001	2005	Percentage Change	2001	2005	Percentage Change	2001	2005	Percentage Change	2001	2005	Percentage Change
Total[a]	6,399,891	6,506,200	+1.7	986,957	913,169	–7.5	1,889,068	2,067,624	+9.5	398,919	390,109	–2.2
Murder and nonnegligent manslaughter	6,585	7,250	+10.1	563	649	+15.3	1,020	926	–9.2	66	79	+19.7
Forcible rape	16,145	15,300	–5.2	2,724	2,306	–15.3	182	183	+0.5	38	34	–10.5
Robbery	57,233	60,085	+5.0	13,223	14,875	+12.5	6,540	7,663	+17.2	1,272	1,570	+23.4
Aggravated assault	230,818	223,876	–3.0	30,034	28,426	–5.4	57,052	58,348	+2.3	8,989	8,803	–2.1
Burglary	154,572	160,826	+4.0	49,339	43,043	–12.8	25,054	28,721	+14.6	6,868	5,898	–14.1
Larceny–theft	454,670	439,286	–3.4	134,816	108,418	–19.6	266,210	281,444	+5.7	86,557	79,873	–7.7
Motor vehicle theft	68,342	68,145	–0.3	21,393	16,116	–24.7	13,518	14,880	+10.1	4,614	3,559	–22.9
Arson	9,227	8,390	–9.1	5,190	4,470	–13.9	1,615	1,559	–3.5	632	670	+6.0
Violent crime[b]	310,781	306,511	–1.4	46,544	46,256	–0.6	64,794	67,120	+3.6	10,365	10,486	+1.2
Property crime[b]	686,811	676,647	–1.5	210,738	172,047	–18.4	306,397	326,604	+6.6	98,671	90,000	–8.8
Other assaults	603,381	606,705	+0.6	98,286	104,432	+6.3	186,874	201,968	+8.1	45,067	52,061	+15.5

(Continued)

Table 3.4 (Continued)

Offense Charged	Male						Female					
	Total			Under 18 Years of Age			Total			Under 18 Years of Age		
	2001	2005	Percentage Change	2001	2005	Percentage Change	2001	2005	Percentage Change	2001	2005	Percentage Change
Forgery and counterfeiting	42,879	43,241	+0.8	2,424	1,792	−26.1	28,917	28,276	−2.2	1,383	837	−39.5
Fraud	114,367	105,797	−7.5	3,597	3,061	−14.9	99,603	91,939	−7.7	1,909	1,718	−10.0
Embezzlement	6,913	6,255	−9.5	752	442	−41.2	6,793	6,508	−4.2	565	354	−37.3
Stolen property (buying, receiving, possessing)	61,656	67,082	+8.8	13,839	11,610	−16.1	13,759	16,773	+21.9	2,794	2,353	−15.8
Vandalism	140,283	141,823	+1.1	58,080	55,657	−4.2	26,899	29,616	+10.1	8,746	9,003	+2.9
Weapons (carrying, possessing, etc.)	87,988	105,428	+19.8	19,453	23,991	+23.3	7,723	9,110	+18.0	2,249	2,853	+26.9
Prostitution and commercialized vice	14,185	13,883	−2.1	219	191	−12.8	24,954	29,425	+17.9	468	650	+38.9
Sex offenses (except forcible rape and prostitution)	49,210	47,050	−4.4	10,145	8,981	−11.5	3,310	3,328	+0.5	800	752	−6.0
Drug abuse violations	752,614	854,368	+13.5	98,127	89,890	−8.4	166,933	208,270	+24.8	19,450	19,662	+1.1
Gambling	3,797	3,081	−18.9	299	408	+36.5	599	580	−3.2	13	18	+38.5

(Continued)

Table 3.4 (Continued)

Offense Charged	Male						Female					
	Total			Under 18 Years of Age			Total			Under 18 Years of Age		
	2001	2005	Percentage Change	2001	2005	Percentage Change	2001	2005	Percentage Change	2001	2005	Percentage Change
Offenses against the family and children	67,753	60,033	−11.4	3,586	2,050	−42.8	19,397	18,868	−2.7	2,066	1,351	−34.6
Driving under the influence	704,313	670,011	−4.9	10,099	8,322	−17.6	142,990	159,087	+11.3	2,190	2,377	+8.5
Liquor laws	293,767	264,279	−10.0	59,822	49,449	−17.3	92,664	94,138	+1.6	28,957	27,400	−5.4
Drunkenness	356,632	317,209	−11.1	10,549	8,094	−23.3	56,299	56,116	−0.3	2,814	2,476	−12.0
Disorderly conduct	270,901	269,402	−0.6	71,782	78,232	+9.0	88,344	98,232	+11.2	30,935	38,891	+25.7
Vagrancy	12,495	15,612	+24.9	1,218	2,319	+90.4	3,225	4,274	+32.5	314	883	+181.2
All other offenses (except traffic)	1,728,473	1,844,968	+6.7	176,706	159,130	−9.9	473,191	551,916	+16.6	63,760	60,508	−5.1
Suspicion	1,694	2,084	+23.0	508	218	−57.1	546	302	−44.7	278	77	−72.3
Curfew and loitering law violations	55,839	56,266	+0.8	55,839	56,266	+0.8	24,487	23,763	−3.0	24,487	23,763	−3.0
Runaway	34,853	30,549	−12.3	34,853	30,549	−12.3	50,916	41,713	−18.1	50,916	41,713	−18.1

SOURCE: Adapted from FBI. (2006). *Crime in the United States, 2005.* Available: www.fbi.gov/ucr/ucr.htm#cius

a. Does not include suspicion.

b. Violent crimes are offenses of murder, forcible rape, robbery, and aggravated assault. Property crimes are offenses of burglary, larceny–theft, motor vehicle theft, and arson.

(Text continued from page 67)

concluded that violent offending by females is the result of a complex web of victimization, substance abuse, economic conditions, and dysfunctional families, and this would seem to suggest the need for a multiagency response. It is fairly common for girls fleeing from abusive parents to be labeled as runaways. Krisberg (2005) concluded, "Research on young women who enter the juvenile justice system suggests that they often have histories of physical and sexual abuse. Girls in the juvenile justice system have severe problems with substance abuse and mental health issues" (p. 123). If they are dealt with simply by being placed on probation, the underlying causes of the problems they confront are unlikely to be addressed. To deal with these causes, counseling may be needed for all parties involved, school authorities may need to be informed if truancy is involved, and further action in adult court may be necessary. If, as often happens, a girl's family moves from place to place, the process may begin all over because there is no transfer of information or records from one agency or place to another. According to Krisberg, "There are very few juvenile justice programs that are specifically designed for young women. Gender-responsive programs and policies are urgently needed" (p. 123).

Race

Official statistics on race are subject to a number of errors, as pointed out in Chapter 2. Any index of nonwhite arrests may be inflated as a result of discriminatory practices among criminal justice personnel (Benekos & Merlo, 2004, pp. 194–210). For example, the presence of a black under "suspicious circumstances" may result in an official arrest even though the police officer knows the charge(s) will be dismissed. Frazier, Bishop, and Henretta (1992) found that black juveniles receive harsher dispositions from the justice system when they live in areas with high proportions of whites (i.e., where they are true numerical minority group members). Kempf (1992) found that juvenile justice outcomes were influenced by race at every stage except adjudication. Feiler and Sheley (1999), collecting data via phone interviews in the New Orleans metropolitan area, found that both black and white citizens were more likely to express a preference for transfer of juveniles to adult court when the juvenile offenders in question were black. Sutphen, Kurtz, and Giddings (1993), using vignettes with police officers, found that blacks were charged with more offenses more often than were whites and that whites received no charges more often than did blacks. Leiber and Stairs (1999) found partial support for their hypothesis that African Americans charged with drug offenses would be treated more harshly in jurisdictions characterized by economic and racial inequality and adherence to beliefs in racial differences than in jurisdictions without such characteristics. Taylor (1994) pointed out that young black males are more likely to be labeled as slow learners or educable mentally retarded, to have learning difficulties in school, to lag behind their peers in basic educational competencies or skills, and to drop out of school at an early age. Juvenile black males are also more likely to be institutionalized or placed in foster care.

Many minority group members live in lower-class neighborhoods in large urban centers where the greatest concentration of law enforcement officers exists. Because arrest statistics are more complete for large cities, we must take into account the sizable proportion of blacks found in these cities rather than the 12% statistic derived from calculating the proportion of blacks in our society. It is these same arrest statistics that lead many to believe that any overrepresentation of black juveniles in these statistics reflects racial inequities in the juvenile and criminal justice networks.

Analysis of official arrest statistics of persons under the age of 18 years has traditionally shown a disproportionate number of blacks. Data presented in Table 3.5 show that blacks accounted for 30% of all arrests in 2005. Blacks accounted for roughly 50% of reported arrests for violent crime and 30% of the arrests for property crimes in the under-18-years-of-age category. American Indians/Alaskan Natives and Asian or Pacific Islanders accounted for very small portions of all crimes, as can be seen in Table 3.5.

With respect to specific crimes, blacks under the age of 18 years accounted for more than half (68%) of the arrests for robbery, slightly more than half (54%) of the arrests for murder and nonnegligent manslaughter, 43% of the arrests for stolen property, 40% of the arrests for "other assaults," and 56% of the arrests for prostitution-related offenses. They also accounted for some 90% of all arrests for gambling. Based on population parameters, blacks under the age of 18 years accounted for lower than expected arrest rates for driving under the influence (4%), other liquor law violations (5%), and drunkenness (8%). Other minority group arrests accounted for less than 3% of total arrests in 2005.

As indicated previously, social–environmental factors have an important impact on delinquency rates and perhaps especially on official delinquency rates (Leiber & Stairs, 1999). Race and ethnicity as causes of delinquency are complicated by social class (Bellair & McNulty, 2005). A disproportionate number of blacks are found in the lower socioeconomic class with all of the correlates conducive to high delinquency. Unless these conditions are changed, each generation caught in this environment not only inherits the same conditions that created high crime and delinquency rates for its parents but also transmits them to the next generation. It is interesting to note that, according to research, when ethnic or racial groups leave high crime and delinquency areas, they tend to take on the crime rate of the specific part of the community to which they move. It should also be noted that there are differential crime and delinquency rates among black neighborhoods, giving further credibility to the influence of the social–environmental approach to explaining high crime and delinquency rates. It is unlikely that any single factor can be used to explain the disproportionate number of black juveniles involved in some type of delinquency. The most plausible explanations currently center on environmental and socioeconomic factors characteristic of ghetto areas. Violence and a belief that planning and thrift are not realistic possibilities may be transmitted across generations. This transmission is cultural, not genetic, and may account in part for high rates of violent crime and gambling (luck as an alternative to planning).

Whatever the reasons, it is quite clear that black juveniles are overrepresented in delinquency statistics, especially with respect to violent offenses, and that inner-city black neighborhoods are among the most dangerous places in America to live. Because most black offenders commit their offenses in black neighborhoods against black victims, these neighborhoods are often characterized by violence, and children living in them grow up as observers and/or victims of violence. Such violence undoubtedly takes a toll on children's ability to do well in school, to develop a sense of trust and respect for others, and to develop and adopt nonviolent alternatives. The same concerns exist for members of other racial and ethnic groups growing up under similar conditions.

Krisberg (2005) summed up the current state of knowledge concerning the impact of the characteristics of juvenile offenders as follows:

> If you are feeling confused and getting a mild headache after considering these complexities, you are probably getting the right messages. Terms such as *race, ethnicity,* and *social class* are used imprecisely and sometimes interchangeably. This is a big problem that is embedded in the existing data and research. There is no simple solution to this conceptual quagmire except to recognize that it exists and frustrates both good research and sound public policy discussions on this topic. (pp. 83–84)

Table 3.5 Arrests by Race (10,974 agencies; 2005 estimated population 217,722,329)

Offense Charged	Arrests Under 18 Years of Age					Percentage Distribution[a]				
	Total	White	Black	American Indian or Alaskan Native	Asian or Pacific Islander	Total	White	Black	American Indian or Alaskan Native	Asian or Pacific Islander
Total	1,570,282	1,059,742	469,382	20,490	20,668	100.0	67.5	29.9	1.3	1.3
Murder and nonnegligent manslaughter	924	397	499	18	10	100.0	43.0	54.0	1.9	1.1
Forcible rape	2,851	1,834	969	31	17	100.0	64.3	34.0	1.1	0.6
Robbery	21,460	6,598	14,487	96	279	100.0	30.7	67.5	0.4	1.3
Aggravated assault	44,845	24,951	18,942	487	465	100.0	55.6	42.2	1.1	1.0
Burglary	57,054	38,287	17,663	565	539	100.0	67.1	31.0	1.0	0.9
Larceny–theft	218,383	149,754	61,407	3,176	4,046	100.0	68.6	28.1	1.5	1.9
Motor vehicle theft	27,499	14,798	11,943	349	409	100.0	53.8	43.4	1.3	1.5
Arson	5,787	4,575	1,076	63	73	100.0	79.1	18.6	1.1	1.3
Violent crime[b]	70,080	33,780	34,897	632	771	100.0	48.2	49.8	0.9	1.1
Property crime[b]	308,723	207,414	92,089	4,153	5,067	100.0	67.2	29.8	1.3	1.6
Other assaults	181,114	105,684	71,486	2,063	1,881	100.0	58.4	39.5	1.1	1.0
Forgery and counterfeiting	3,051	2,259	725	18	49	100.0	74.0	23.8	0.6	1.6
Fraud	5,796	3,723	1,981	32	60	100.0	64.2	34.2	0.6	1.0
Embezzlement	849	532	293	8	16	100.0	62.7	34.5	0.9	1.9

(Continued)

Table 3.5 (Continued)

Offense Charged	Arrests Under 18 Years of Age					Percentage Distribution[a]				
	Total	White	Black	American Indian or Alaskan Native	Asian or Pacific Islander	Total	White	Black	American Indian or Alaskan Native	Asian or Pacific Islander
Stolen property (buying, receiving, possessing)	16,305	8,941	7,019	146	199	100.0	54.8	43.0	0.9	1.2
Vandalism	76,096	59,349	14,961	955	831	100.0	78.0	19.7	1.3	1.1
Weapons (carrying, possessing, etc.)	32,949	20,154	12,161	253	381	100.0	61.2	36.9	0.8	1.2
Prostitution and commercialized vice	1,200	501	670	9	20	100.0	41.8	55.8	0.8	1.7
Sex offenses (except (except forcible rape and prostitution)	11,979	8,534	3,226	110	109	100.0	71.2	26.9	0.9	0.9
Drug abuse violations	139,776	96,207	41,076	1,301	1,192	100.0	68.8	29.4	0.9	0.9
Gambling	1,463	106	64,881	36,928	876	1,201	100.0	62.5	35.5	0.8
Offenses against the family and children	3,850	3,042	714	80	14	100.0	79.0	18.5	2.1	0.4
Driving under the influence	12,584	11,744	502	223	115	100.0	93.3	4.0	1.8	0.9
Liquor laws	91,800	84,107	4,322	2,535	836	100.0	91.6	4.7	2.8	0.9
Drunkenness	11,401	10,122	951	230	98	100.0	88.8	8.3	2.0	0.9

(Continued)

Table 3.5 (Continued)

Offense Charged	Arrests Under 18 Years of Age					Percentage Distribution[a]				
	Total	White	Black	American Indian or Alaskan Native	Asian or Pacific Islander	Total	White	Black	American Indian or Alaskan Native	Asian or Pacific Islander
Disorderly conduct	147,976	87,640	57,331	1,889	1,116	100.0	59.2	38.7	1.3	0.8
Vagrancy	3,416	2,702	676	12	26	100.0	79.1	19.8	0.4	0.8
All other offenses (except traffic)	264,643	190,241	67,221	3,429	3,752	100.0	71.9	25.4	1.3	1.4
Suspicion	400	253	144	2	1	100.0	63.3	36.0	0.5	0.3
Curfew and loitering law violations	103,886	64,881	36,928	876	1.201	100.0	62.5	35.5	0.8	1.2
Runaway	80,945	57,826	18,660	1,534	2,925	100.0	71.4	23.1	1.9	3.6

SOURCE: Adapted from Federal Bureau of Investigation. (2006). *Crime in the United States, 2005*. Available: www.fbi.gov/ucr/ucr.htm#cius

a. Percentages might not add to 100.0 because of rounding.

b. Violent crimes are offenses of murder, forcible rape, robbery, and aggravated assault. Property crimes are offenses of burglary, larceny–theft, motor vehicle theft, and arson.

Career Opportunity—Criminalist

Job Description: Includes positions of laboratory technicians who examine evidence such as fingerprints and documents. Use chemistry, biology, and forensic science techniques to examine and classify/identify blood, body fluid, DNA, fiber, and fingerprint evidence that may be of value in solving criminal cases. Often on call, work in dangerous locations and in proximity to dead bodies and chemical and biological hazards. Sometimes testify in court as to evidentiary matters.

Employment Requirements: At least a 4-year degree in chemistry, biology, physics, or forensic science. In some agencies, applicant must be a sworn police officer and must complete entry-level requirements for that position before moving to forensics. In other jurisdictions, civilians are hired as criminalists.

Beginning Salary: Between $30,000 and $40,000. Benefits vary widely depending on jurisdiction and whether or not the position requires a sworn officer.

SUMMARY

Official profiles of juvenile offenders reflect only the characteristics of those who have been apprehended and officially processed. Although they tell little or nothing about the characteristics of all juveniles who actually commit delinquent acts, they are useful in dealing with juveniles who have been officially processed. These official statistics currently lead us to some discomforting conclusions about the nature of delinquency in America as it relates to social and physical factors.

It might not be the broken home itself that leads to delinquency; instead, it may be the quality of life within the family in terms of consistency of discipline, level of tension, and ease of communication. Therefore, in some instances, it may be better to remove children from intact families that do not provide a suitable environment than to maintain the integrity of the families. In addition, it might not be necessary to automatically place juveniles from broken homes into institutions, foster homes, and so forth provided that the quality of life within the broken homes is acceptable.

We perhaps need to rethink our position on the "ideal" family consisting of two biological parents and their children. This family no longer exists for most American children. For many children, the family of reality consists of a single mother who is head of the household or a biological parent and stepparent. Although many one-parent families experience varying degrees of delinquency and abuse/neglect, children in many others are valued, protected, and raised in circumstances designed to give them a chance at success in life.

Because education is an important determinant of occupational success in our society, and because occupational success is an important determinant of life satisfaction, it is important that we attempt to minimize the number of juveniles who are "pushed out" of the educational system. Both juvenile justice practitioners and school officials need to pursue programs that minimize the number of juveniles who drop out. It may be that we are currently asking too much of educators when we require them not only to provide academic and vocational information but also to promote psychological and social well-being, moral development, and a sense of direction for juveniles (formerly provided basically by the family). At the current time, however, if educators fail to provide for these concerns, the juvenile often has nowhere else to turn except his or her peers, who may be experiencing similar problems. One result of this alienation from both the family and the educational system is the development of delinquent behavior patterns. Another may be direct attacks on school personnel or fellow students.

We have concentrated our interest and research activities on delinquency and abuse/neglect of the lower social class and have generally ignored the existence of these problems in the middle and upper classes. The importance of lower-class delinquency cannot be ignored, but we must also realize that the problem may be equally widespread, although perhaps in different forms, in the middle and upper classes. We can no longer afford the luxury of viewing delinquency as only a problem of lower-class neighborhoods in urban areas. The problem of delinquency is increasing at a rapid rate in what were commonly considered to be "quiet middle-class suburban areas" and in many rural areas as well. Because motivations and types of offenses committed by middle-class delinquents may differ from those of their lower-class counterparts, new techniques and approaches for dealing with these problems may be required.

If those working with children can develop more effective ways of promoting good relationships between juveniles and their families and of making the importance of a relevant education clear to juveniles, involvement in gang activities may be lessened. At the current time, however, understanding the importance of peer group pressure and the demands of the gang on the individual juvenile is extremely important in understanding drug abuse and related activities. If gangs could be used to promote legitimate concerns rather than illegitimate concerns, one of the major sources of support for certain types of delinquent activities (e.g., vandalism, drug abuse) could be weakened considerably. Reasonable alternatives to current gang activities need to be developed and promoted.

Finally, there is no denying that black juveniles are disproportionately involved in official delinquency. Although there are still those who argue racial connections to such delinquency, the evidence that such behavior is a result of family, school, and neighborhood conditions, and perhaps the actions of juvenile justice practitioners, rather than genetics is overwhelming. Whatever the reasons for the high rates of delinquency, and especially violent offenses, in black neighborhoods, it behooves us all to address this issue with as many resources as possible in the interests of those living in both high crime areas and the larger society.

None of the factors discussed in this chapter can be considered a direct cause of delinquency. It is important to remember that official statistics reflect only a small proportion of all delinquent activities. Profiles based on the characteristics discussed in this chapter are valuable to the extent that they alert us to a number of problem areas that must be addressed if we are to make progress in the battle against delinquency.

Attempts to improve the quality of family life and the relevancy of education, and attempts to change discriminatory practices in terms of social class, race, and gender, are needed badly. Improvements in these areas will go a long way toward reducing the frequency of certain types of delinquent activity.

Note: Please see the Companion Study Site for Internet exercises and Web resources. Go to www.sagepub.com/juvenilejustice6study.

Critical Thinking Questions

1. What is the relationship between profiles of delinquents based on official statistics and the actual extent of delinquency?
2. Discuss the relationships among the family, the educational system, drugs, and delinquency.
3. Discuss some of the reasons for the overrepresentation of black juveniles in official delinquency statistics. What could be done to decrease the proportion of young blacks involved in delinquency? How do area of the city, race, and social class combine to affect delinquency? Is delinquency basically a lower-class phenomenon? If so, why should those in the middle and upper classes be concerned about it?
4. Discuss the methamphetamine "crisis." How does it differ from other drug-related crises we have faced in the past? What do you think can be done to deal with this crisis?

Suggested Readings

Bellair, P. E., & McNulty, T. L. (2005). Beyond the bell curve: Community disadvantage and the explanation of black–white differences in adolescent violence. *Criminology, 43,* 1135–1169.

Carter, P. L. (2003). "Black" cultural capital, status positioning, and schooling conflicts for low-income African American youth. *Social Problems, 50,* 136–155.

Curry, G. D., Decker, S. H., & Egley, A., Jr. (2002). Gang involvement and delinquency in middle school. *Justice Quarterly, 19,* 275–293.

De Coster, S., Heimer, K., & Wittrock, S. M. (2006). Neighborhood disadvantage, social capital, street context, and youth violence. *Sociological Quarterly, 17,* 723–753.

Demuth, S., & Brown, S. L. (2004). Family structure, family processes, and adolescent delinquency: The significance of parental absence versus parental gender. *Journal of Research in Crime and Delinquency, 41,* 58–81.

Ginwright, S. A. (2002). Classed out: The challenges of social class in black community change. *Social Problems, 49,* 544–562.

Hango, D. W. (2006). The long-term effect of childhood residential mobility on educational attainment. *Sociological Quarterly, 17,* 631–664.

Holzman, H. R. (1996). Criminological research on public housing: Toward a better understanding of people, places, and spaces. *Crime & Delinquency, 42,* 361–378.

McNulty, T. L., & Bellair, P. E. (2003). Explaining racial and ethnic differences in adolescent violence: Structural disadvantage, family well-being, and social capital. *Justice Quarterly, 20,* 1–31.

Osgood, D. (2004). Unstructured socializing and rates of delinquency. *Criminology, 42,* 519–550.

Stewart, E. A. (2003). School social bonds, school climate, and school misbehavior: A multilevel analysis. *Justice Quarterly, 20,* 575–604.

Sutphen, R., Kurtz, D., & Giddings, M. (1993). The influence of juveniles' race on police decision-making: An exploratory study. *Juvenile and Family Court Journal, 44*(2), 69–76.

Thornberry, T. P., Huizinga, D., & Loeber, R. (2004). The Causes and Correlates studies: Findings and policy implications. *Juvenile Justice, 9*(1). Available: www.ncjrs.gov/html/ojjdp/203555/jj2.html

Wade, T. J., & Pevalin, D. J. (2005). Adolescent delinquency and health. *Canadian Journal of Criminology and Criminal Justice, 4,* 619–655.

Watson, D. W. (2004). Juvenile offender comprehensive reentry substance abuse treatment. *Journal of Correctional Education, 55,* 211–224.

THEORIES OF CAUSATION

CHAPTER LEARNING OBJECTIVES

On completion of this chapter, students should be able to

- *Recognize the requirements for a good theory*
- *Understand and discuss the strengths and limitations of various theories*
- *Recognize and discuss the importance of the relationship between theory and practice*
- *Evaluate research relating to theories of causation*

KEY TERMS

Conceptual schemes
Scientific theory
Demonology
Trephining
Classical theory
Free will approach
Neoclassical approach
Rational choice theory
Postclassical theory
Deterrence theory
Routine activities theory
Positivist school of criminology
Biological theories
Atavists
Anomalies
Phrenology
Somatotypes
XYY chromosome
Biosocial criminology
Psychological theories
Personality inventories
Psychoanalytic approach
Id, ego, and superego
Psychopath
Learning theory
Behaviorists
Conditioning
Sociological theories
Anomie theory

Strain theory

Delinquency and drift

Techniques of neutralization

Illegitimate opportunity structure

Ecological/social disorganization approach

Concentric zone theory

Theory of differential association

Theory of differential anticipation

Labeling theory

Conflict/radical/critical/Marxist theories

Feminism

Control theories

Integrated theories

Let us now examine some of the theories that have been developed in an attempt to explain offenses by and against juveniles. It is important to note from the outset that numerous studies over the past 50 years have suggested links between delinquency and child abuse/neglect. For example, Scudder, Blount, Heide, and Silverman (1993) noted that the results of their research "suggest that children who break the law, especially through acts of violence, often have a history of maltreatment as children" (p. 321). The results of their research indicate further that "a child abused at a young age is at higher risk for subsequent delinquent behaviors than a nonabused child" (p. 321). Other researchers have arrived at similar conclusions (Siegal & Williams, 2003).

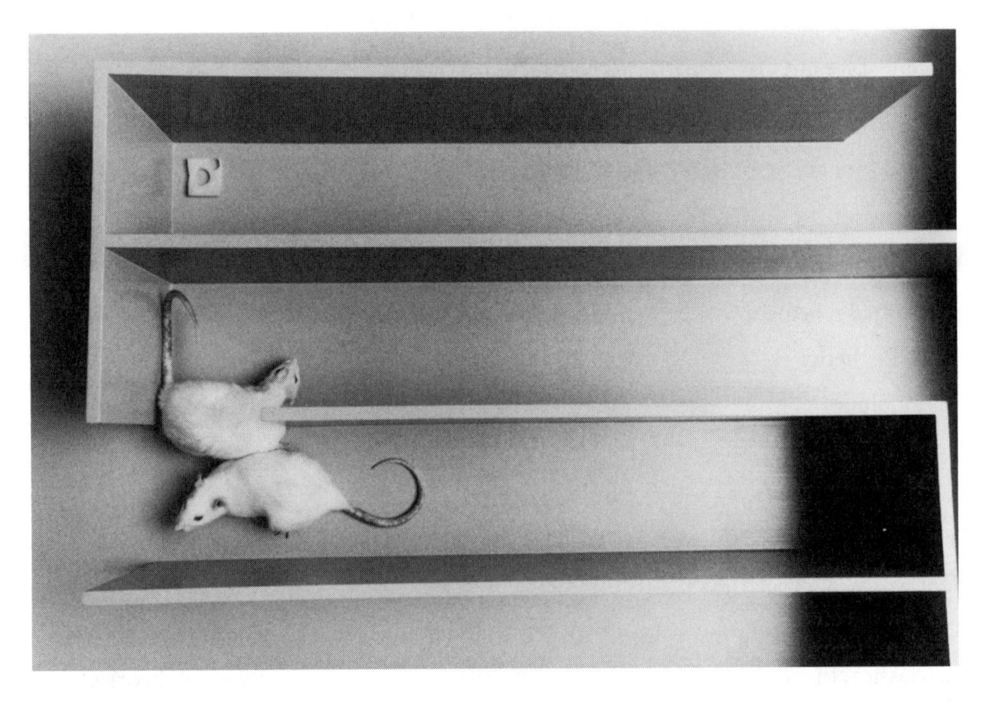

▲ Behaviorists believe that many of the principles learned in the study of animal behavior can be applied to humans.

Scientific Theory

Although dozens of **conceptual schemes** have been proposed in attempts to specify the causes of crime and delinquency, only a few of the more prominent attempts are discussed here.

A scientific theory may be defined as a set of two or more related, empirically testable assertions (statements of alleged facts or relationships among facts about a particular phenomenon [Fitzgerald & Cox, 2002, p. 47]). Although this definition may sound complex, it is really quite simple if we look at it one part at a time. A testable assertion or proposition is simply a statement of a relationship between two or more variables. In an acceptable theory, these assertions or propositions are related in a logical manner so that some other assertions or propositions can be derived (deduced) from others. Here is an example:

Proposition 1: All delinquents are from broken homes.

Proposition 2: Harry is a delinquent.

Proposition 3: Harry is from a broken home.

In this case, Proposition 3 is derived from Propositions 1 and 2; that is, Proposition 3 is said to be explained by Propositions 1 and 2 and is logically correct. Our definition of a theory, however, requires that at least some of the propositions be empirically testable. To be acceptable, then, a theory must be logically correct and must accurately describe events in the real world. Suppose that Harry is not, in fact, from a broken home. Clearly, our explanation of delinquency is erroneous and our theory must be revised or rejected.

Although conceptual schemes that suggest relationships between variables but do not meet our requirements for theory may be useful stepping-stones in describing delinquency or abuse, only a logically correct and empirically accurate theory will enable us to explain these phenomena. Explanation is central to preventing or controlling delinquency and abuse. All policy and practice in juvenile justice is shaped, intentionally or not, by theory. For example, "get tough" policies are an offshoot of classical theory and hedonism, policies relying on individual or group therapy are based in psychological or social psychological theory, and policies stressing neighborhood improvement, better education, and job opportunities are based on sociological theories.

As we discuss some theories and conceptual schemes, you may find it useful to assess the extent to which they meet the requirements of our definition and the extent to which they are useful in helping us to understand offenses by and against juveniles. As Klofas and Stojkovic (1995) indicated:

> Our ideas about crime—what it means and why it happens—have varied considerably over the past several hundred years. We have changed from (1) viewing crime as the work of the devil to (2) describing it as the rational choice of free-willed economic calculators to (3) explaining it as the involuntary causal effects of biological, mental, and environmental conditions, and then back to (2). (p. 37)

Assessing the state of criminological theory toward the end of the 20th century, Bernard (1990) concluded that not much progress had been made during the prior 20 years in weeding out theories that cannot be supported or in verifying other theories. We hope that we have made progress during the past 15 years, but efforts to develop and empirically test theories remain sporadic.

Our intent in this chapter is simply to familiarize you with some of the numerous conceptual schemes and to note some of the strengths and weaknesses of these schemes. For those who desire more detailed information about specific theories, the Suggested Readings list at the end of the chapter should prove to be useful.

Some Early Theories

Demonology

Early attempts to explain various forms of deviant behavior (e.g., crime, delinquency, mental illness) focused on demon or spirit possession (**demonology**). Individuals who violated societal norms were thought to be possessed by some evil spirit that forced them to commit evil deeds through the exercise of mysterious supernatural power (Moyer, 2001, p. 13). Deviant behavior, then, was viewed not as a product of free will but rather as determined by forces beyond the control of the individual; thus, the demonological theory of deviance is referred to as a deterministic approach. To cure or control deviant behavior, a variety of techniques were employed to drive the evil spirits from the mind and/or body of the perceived deviant.

One process that was employed was **trephining,** which consisted of drilling holes in the skulls of those perceived as deviants to allow the evil spirits to escape. Various rites of exorcism, including beating and burning, were practiced to make the body of the perceived deviant such an uncomfortable place to reside that the evil spirits would leave or to make the deviant confess his or her association with evil spirits. As might be expected, such torture of the body often resulted in death or permanent disability to the individual who was allegedly possessed. In addition, either confession or failure to confess could be taken as evidence of possession. Tortured sufficiently, many individuals undoubtedly confessed simply to prevent further torture. Those who persisted in claiming innocence were often thought to be so completely under the control of evil spirits that they could not tell the truth. Needless to say, the consequences for both categories of accused were frequently very unpleasant.

Many observers believe that belief in spirit possession as a cause of deviance is rare today, but our analysis of news articles over the past few years has turned up numerous articles on ritual abuse of children by persons or groups claiming to have been instructed by deities, typically God or Satan, to commit the acts in question (Charton, 2001; "In This Church," 2001; "NYC Mom Arrested," 2006, p. A8; Stearns & Garcia, 2001). As Klofas and Stojkovic (1995) noted, supernatural bases for crime have not been totally rejected, although they have been largely supplanted by more scientific explanations (p. 39).

Perhaps demonology as an explanation of deviance persists because, in some respects, attempts to deal with deviance thought to be caused by spirits are logical if the basic premise is accepted as true; that is, if one believes that spirit possession causes deviance, it makes sense to drive the spirits away if possible. As is the case with all theories of deviance, this one implies a method of cure or control. Although such an explanation of deviance seems simplistic to criminologists today, it cannot be scientifically disproved and is still clearly accepted as valid by significant numbers of people in a substantial number of countries. Precisely because it cannot be scientifically tested, however, this attempt to explain deviant behavior is of little value from a theoretical perspective.

Classical Theory

During the last half of the 18th century, the classical school of criminology (**classical theory,** often referred to as a **free will approach**) emerged in Italy and England in the works of Cesare Beccaria and Jeremy Bentham, respectively. This approach to explaining and controlling crime was based on the belief that humans exercise free will and that human behavior results from rationally calculating rewards and costs in terms of pleasure and pain. In other words, before an individual commits a specific act, he or she determines whether the consequences of the act will be pleasurable or painful. Presumably, acts that have painful consequences will be avoided. To control crime, then, society simply needed to make the punishment for violators outweigh the benefits of their illegal actions. Thus, penalties became increasingly more severe as offenses became increasingly more serious. Under classical theory, threat of punishment is considered to be a deterrent to criminals who rationally calculate the consequences of their illegal actions.

By the early 1800s, Beccaria's approach had been modified in recognition of the fact that not all individuals were capable of rationally calculating rewards and costs. The modified approach, generally referred to as the **neoclassical approach,** called for the mitigation of punishment for the insane and juveniles (Conklin, 1998, p. 41; Moyer, 2001, p. 27). By definition, the insane were not capable of rational calculation, and juveniles (up to a certain age at least) were thought to be less responsible than adults.

It is important to understand the classical approach because its propositions (punishment deters crime, the punishment should fit the crime, and juveniles and the insane should be treated differently from sane adults) are basic to our current criminal and juvenile justice system.

Rational Choice Theory

The **rational choice theory** or **postclassical theory** of the 20th century also involves the notion that before people commit crimes, they rationally consider the risks and rewards. A burglar noting no lights on and no police presence at an expensive mansion over several nights might rationally conclude that the risk is relatively low and the potential rewards are worth pursuing and, therefore, may commit the crime. According to the rational choice model, focusing on the development of rational thought and the application of scientific laws, as well as using empirical research, might help the state to develop policies that better control crime and deviance and thereby improve quality of life (Bohm, 2001, p. 15; Lanier & Henry, 1998, p. 72; Reid, 2006, pp. 77–78).

This view, that delinquents exercise free will and rationally calculate the consequences of their behavior, fits well with the conservative ideology and the "get tough" approach to delinquency. If delinquency is a product of free will and not predetermined by social conditions, the delinquent may best be deterred by the threat of punishment rather than by the promise of treatment. Gang members who go into the drug business with the clear intent of making a profit by outwitting both their competitors and law enforcement officials may be described as using rational choice theory.

Deterrence Theory

Deterrence theory is another extension of the classical approach. It focuses on the relationship between punishment and misbehavior at both the individual and group levels. Specific deterrence refers to preventing a given individual from committing further crimes, whereas general deterrence refers to the effect that punishing one wrongdoer has on preventing others from committing offenses. When we attempt to measure the extent of deterrence, we are actually measuring perceived deterrence—what individuals believe will happen to them (will they be caught? will they be punished? will the punishment be severe?) if they commit offenses. Most authorities appear to agree that the deterrent effects of punishment are greater if the punishment is swift and certain. It appears that the deterrent effect of severe punishment is moderated by celerity (swiftness) and certainty (Reid, 2006, pp. 74–78). With respect to delinquents, we might ask whether the increasingly severe punishments suggested by "get tough" policies are likely to have significant impact on juveniles who do not believe they will be apprehended for their delinquent acts or who do not believe they will be punished if apprehended. Another concern with respect to deterrence theory has to do with the fact that in some instances, those about to commit delinquent acts do not consider the possibility of being apprehended or punished (e.g., those under the influence of intoxicants, those who strike out in a passionate or angry state of mind).

Routine Activities Theory

Routine activities theory is yet another extension of the belief that rational thought and sanctions largely determine criminal behavior. According to this approach, crime is simply a function of people's everyday behavior. One's presence in certain types of places, frequented by motivated offenders, makes him or her a suitable target and, in the absence of capable guardians, is likely to lead to crime (Conklin, 1998, p. 319; Cote, 2002, p. 286; Lanier & Henry, 1998, p. 82). Plass and Carmody (2005) studied the effect of engaging in risky activities on the violent victimization experiences of delinquent and nondelinquent juveniles. Their results showed that there are some modest differences in the effects of routinely engaging in risky behaviors and the likelihood of violent victimization. There is also research that supports the existence of "hot spots" or areas in which crimes occur repeatedly over time (Buerger, Cohn, & Petrosino, 2000; Sherman & Weisburd, 1995). In other cases, however, victims' absence may be critical to the crime in question (e.g., burglary is easier if no one is home).

Unfortunately, the classical approach to controlling crime has never been very successful. Although there seems to be some logic to the approach, the premise that the threat of punishment deters crime, at least as currently employed, is inaccurate. There are a variety of possible sources of error in this premise. First, it may be that humans do not always rationally calculate rewards and costs. An individual committing what we commonly refer to as a "crime of passion" (as in the case of the murder of a spouse caught in an adulterous act or excessive corporal punishment of a child in a moment of anger) might not stop to think about the consequences. If this individual does not stop to make such calculations, the threat of punishment (no matter how severe) will not affect that person's behavior. Second, an individual may calculate rewards and costs in a way that appears rational to him or her (but perhaps not to society) and may decide that certain illegal acts are worth whatever punishment he or she will receive if apprehended (as in the case of a starving person stealing food). Finally, the individual may rationally calculate rewards and costs but have no fear of punishment because he or she believes that the chances of apprehension are slight (as in the case of many juveniles involved with alcohol and minor vandalism). If the individual believes that he or she will not be apprehended for his or her illegal acts, the threat of punishment has little meaning. In addition, the individual may believe that even if he or she is caught, punishment will not be administered (as in the cases of juveniles who are aware that most juvenile cases never go to court and of parents who abuse their children in the name of discipline).

For whatever reasons, the classical approach to explaining and controlling crime has not proved to be successful. It would appear that whatever possibility of success this approach has rests with delivering punishment relatively immediately and with a great deal of certainty. Because our society largely continues to rely on the classical approach, and because neither immediacy of punishment nor certainty of apprehension exists, it is not surprising that we are unsuccessful in our attempts to control crime and delinquency.

In spite of the fact that severe punishment does not appear to lead to desirable behavior, many child abusers obviously believe that such punishment will lead to improved behavior on behalf of their children. Thus, when a child fails to meet the expectations of abusive parents, whether in the area of toilet training, eating habits, schoolwork, or showing proper deference to the parents, emotional and/or physical abuse results. This often leads to lowered self-esteem on behalf of the child, whose performance then suffers even more, leading to more severe punishment on the part of the parents and so forth. This "cycle of violence," once begun, is difficult to break, and there is at least some evidence that the abused child may later abuse his or her own children in the same ways (Knudsen, 1992, pp. 61–63).

The Positivist School

The **positivist school of criminology** emerged during the second half of the 19th century. Cesare Lombroso is recognized as the founder of the positivist school and also as the "father" of modern criminology. Lombroso, with other positivists such as Raffaele Garofalo and Enrico Ferri, believed that criminals should be studied scientifically and emphasized determinism as opposed to free will (classical school) as the basis of criminal behavior. Although a number of positivists believed that heredity is the determining factor in criminality, others believed that the environment determined, in large measure, whether or not an individual became a criminal.

The positivists emphasized the need for empirical research in criminology, and some stressed the importance of environment as a causal factor in crime. Although their methodology was unsophisticated by modern standards, their contributions to the development of modern criminology are undeniable. Lombroso may also be considered, earlier in his career at least, as one of the founders of the biological school of criminology.

▲ Cesare Lombroso is recognized as one of the founding fathers of criminology.

Biological Theories

Biological theories of delinquency were initially based on the assumption that delinquency (criminality) is inherited. Over the past century, the approach has tended to emphasize more the belief that offenders differ from nonoffenders in some physiological way (Conklin, 1998, p. 146). This approach has offered a number of different explanations of delinquency, ranging from glandular malfunctions to learning disabilities, to racial heritage, to nutrition. Rafter (2004) noted that today biological explanations are again gaining credibility and are joining forces with sociological explanations in ways that may make them partners in explaining crime and delinquency. She advised (and we agree) students of crime and delinquency to become familiar with the biological tradition that includes physiognomists, phrenologists, Lombroso, Goddard, Hooton, the Gluecks, and Sheldon, among others. By studying where these forerunners of contemporary biological theories came from, we can determine how they developed, what they contributed, and where they went astray. As we examine some of these explanations, keep in mind our definition of an acceptable theory.

Cesare Lombroso's "Born Criminal" Theory

Lombroso (1835–1909) became known for the theory of the "born criminal." As a result of his research, he became convinced at one point in his career that criminals were **atavists** or throwbacks to more primitive beings. According to Lombroso, these born criminals could be recognized by a series of external features such as receding foreheads, enormous development of their jaws, and large or handle-shaped ears. These external traits were thought to be related to personality types characterized by laziness, moral insensitivity, and absence of guilt feelings.

Individuals with a number of these criminal features or **anomalies** were thought to be incapable of resisting the impulse to commit crimes except under very favorable circumstances. Many of Lombroso's assumptions can be traced to the influence of Darwinism (which provided a means of ranking animals as more or less primitive) at the end of the 19th century and to the influence of **phrenology** (the study of the shape of the skull) and physiognomy (the study of facial features) as they related to deviance (Conklin, 1998, pp. 146–147; Reid, 2006, pp. 62–63).

Later in his career, Lombroso modified his approach by recognizing the importance of social factors, but his emphasis on biological causes encouraged many other researchers to seek such causes. Lombroso remains important today largely because of his attempts to explain crime scientifically rather than as a result of his particular theories.

Other Biological Theories

Following Lombroso, there have been a number of attempts over the years to find biological or genetic causes for crime and delinquency. Identical twin studies were conducted based on the belief that if genetics determines criminality, when one twin is criminal, the other will also be criminal. In general, these studies provide evidence that genetic structure is not the sole cause of crime given that none of them indicates that 100% of the twins studied were identical with respect to criminal behavior. Research on the relationship between genetics and crime in twins continues nonetheless. The results of twin studies conducted over the past 75 years do seem to indicate that there may be a genetic factor in delinquency/crime, but the exact nature of the relationship remains undetermined (Fishbein, 1990).

The next logical step in studying the relationship between heredity and crime involved studies of children adopted at an early age who had little or no contact with biological parents. Would the offense rates and types of the children more closely resemble those of the adoptive parents or the biological parents? Evidence suggests a hereditary link, but it is very difficult to separate the effects of heredity and environment (Bohm, 2001, pp. 36–41). Jones and Jones (2000) concluded that the similar behavior of the twins they studied might have more to do with the contagious nature of antisocial behavior than with heredity. They noted that the more antisocial behavior present in a family or community in which boys grew up, the greater the risk that boys will be affected. Unnever, Cullen, and Pratt (2003) studied the relationships among attention deficit hyperactivity disorder (ADHD), parenting, and delinquency and concluded that the effects of ADHD on delinquency are affected by low self-control. Wright and Beaver (2005) noted that genetic research has demonstrated that ADHD and other deficits in the frontostriatal system of the brain are related to heredity. Their research tested whether the role of parents in creating low self-control was important once genetic influences are taken into account. Based on a sample of twins, they found that parenting activities demonstrate a weak and inconsistent effect. These authors concluded that researchers have often failed to address genetic influences in parenting studies.

Richard Dugdale made the Jukes family a famous test case for inherited criminality during the late 1800s when he demonstrated that over generations this family had been characterized by criminality. Dugdale believed that crime and heredity were related, but his own admission that over the years the family had established a reputation for deviant behavior points to the possibility that other factors (e.g., learning and labeling) might be of equal or greater importance in explaining his observations (Dugdale, 1888).

Other researchers, including Kretschmer (1925), Sheldon (1949), and Glueck and Glueck (1950), turned to studies of the relationship between **somatotypes** (body types) and delinquency/criminality. Causes of delinquency and body type were thought by Sheldon to be biologically determined, for example, and selective breeding was suggested as a solution to delinquency. The Gluecks continued the body type tradition of explaining delinquency but included in their analysis a variety of other factors as well. The basic conclusion of the Gluecks' work with respect to body type and crime is that a majority of delinquents are muscular as opposed to thin or obese. One possible explanation for this conclusion, which does not require any assumptions about biological determination, is that juveniles who are not particularly physically fit recognize this fact and, therefore, consciously tend to avoid at least those delinquent activities that might require strength and fitness. In addition, measurements of body type are rather subjective, and the data presented by the body typists do not account for different individuals with the same body type being delinquent, on the one hand, and nondelinquent, on the other.

Over time, emphasis in the biological school has shifted. Studies examining the relationships among learning disabilities, chromosomes, chemical imbalances, and delinquency have emerged. We have already discussed some of the literature on the relationship between learning disabilities and delinquency in Chapter 3. Here we simply state that many learning disabilities, as typically conceived, are psychosocial (as opposed to biological) in nature. Others are more clearly organic in nature, and there is some evidence that brain dysfunctions and neurological defects are more common among violent individuals than among the general population. Such individuals seem to have defects in the frontal and temporal lobes of the brain, and these may lead to loss of self-control. Other dysfunctions include dyslexia (the failure to attain language skills appropriate to intellectual level), aphasia (problems with verbal communication and understanding), and attention deficit disorder (manifested in hyperactivity and inattentiveness).

Satterfield (1987) found that children who are hyperactive are several times more likely to be arrested during adolescence than are children without the disorder. None of these disorders, at this point in time, has been shown to be directly causally related to delinquency. In fact, Satterfield found that arrest rates for hyperactive children were affected by social class, with those from the lower social class being more likely to be arrested. In addition, many learning-disabled children adapt and find ways to overcome the handicap. Perlmutter (1987) suggested that there is little middle ground and indicated that those who are not able to overcome the disability appear to be at risk for developing emotional and behavioral difficulties as adolescents. Fishbein (1990) summarized the relationship between learning disabilities and delinquency by stating that low IQ and/or learning disabilities are not inherently determinants of delinquency. However, without proper intervention, juveniles may become frustrated in attempting to pursue mainstream goals without the skills to achieve them and eventually succumb to delinquent behavior.

During the 1960s, a number of researchers explored the relationship between the presence of an extra Y chromosome in some males and subsequent criminal behavior. Mednick and Christiansen (1977) found that roughly 42% of the **XYY chromosome** cases identified in Denmark had criminal histories, compared with only 9% of the XY population. Research is still being conducted on the possible relationship between chromosomes and criminality, although little if any work has been done specifically on the relationship between delinquency and chromosomes.

Currently, it is safe to say that a direct relationship between chromosome structure and criminality has not been scientifically established and that many of the studies conducted to date are characterized by serious methodological problems.

Jeffery (1978, 1996), Booth and Osgood (1993), and Denno (1994) viewed behavior as the product of interaction between a physical environment and a physical organism and believed that contemporary criminology should represent a merger of biology, psychology, and sociology. The basis for this argument, **biosocial criminology,** is that most contemporary criminologists believe that criminal/delinquent behavior is learned but neglect the fact that learning involves physical (biochemical) changes in the brain. These researchers contend that although criminality is not inherited, the biochemical preparedness for such behavior is present in the brain and will, given a particular type of environment, produce criminal behavior (Fishbein, 1990; Turkheimer, 1998; Walsh, 2000).

There have been numerous other attempts to explain both delinquency and crime in terms of biology, genetics, and biochemistry. As early as 1939, Ernest Hooton wrote of the consequences of biological causes of crime for rehabilitation and control of offenders. According to Hooton (1939), if criminality is inherited, the solutions to crime lie in isolation and/or sterilization of offenders to prevent them from remaining active in the genetic pool of a society. A third alternative is extermination (which Hooton opposed), and a fourth is the practice of eugenics (Rafter, 2004). At various times, European societies have isolated (e.g., Devil's Island, the Colonies), sterilized, and exterminated offenders. Experiments with eugenics are certainly possible but raise serious ethical and moral issues. The extent to which genetic engineering becomes acceptable as a means of dealing with a wide variety of social problems will likely determine its use in controlling criminality if genetic deficiencies or abnormalities are shown to be causes of crime and delinquency (see In Practice 4.1). Recent developments that have made it possible to create human genetic blueprints, hailed as one of the greatest scientific contributions of the 21st century, make it likely that if there is a genetic link to crime, it will be discovered (Friend, 2000).

In Practice 4.1

Chapter of History Painful to Recall: Forced Sterilization Is Today Unthinkable—but Could It Tempt Future Generations?

America saw itself as a promised land, a place where the hard-working and faithful could come and build a nation of prosperity, free from the imperfections they had left behind. But that ideal also sowed a bad seed that germinated over generations into a shameful chapter in American history.

From the early 20th century through the 1970s, seemingly well-meaning scientists, theologians, and lawmakers worked to create a racially pure society by using hereditary traits as an excuse to prevent those deemed mentally and morally deficient from having children. It was a new science called "eugenics," or the "self-direction of human evolution," according to a 1931 leaflet by the United States Eugenic Records Office. (One cannot escape the irony of the root of the word "eugenics," the Greek "eugenes," meaning "well-born" or "good genes.")

More than 30 U.S. states passed laws that resulted in the forced sterilizations of at least 65,000 Americans, and there were untold others in states without laws. The movement became a model for similar laws in Canada, Denmark, Finland, France, Sweden, and, perhaps most troublingly, Germany. Hitler enacted a comprehensive sterilization law in 1933, sterilizing more than 150,000 Germans within two years.

Perhaps the heaviest burden of the story of eugenics and forced sterilization is this American connection to the master race theories that culminated in the Holocaust, writes Harry Bruinius, author of *Better for All the World.* His new book chronicles America's history in the eugenics movement, using trial transcripts, personal letters and diaries, and other original documents.

Bruinius, who studied theology at Yale and journalism at Columbia, takes us into the minds of the thought leaders of the time, showing us how otherwise well-respected people could countenance the now-unthinkable act of forced sterilization. Eugenics was supported by a Who's Who of society, among them Theodore Roosevelt, Margaret Sanger, Winston Churchill, and George Bernard Shaw.

Oliver Wendell Holmes Jr., associate justice of the U.S. Supreme Court, wrote the majority opinion in the landmark case *Buck v. Bell* that made eugenic sterilization a constitutionally sanctioned method to fight poverty and crime.

In the opinion, which still stands today, Holmes wrote, "It is better for all the world, if instead of waiting to execute degenerate offspring for crime, or to let them starve for their imbecility, society can prevent those who are manifestly unfit from continuing their kind."

Buck v. Bell sits at the heart of the American eugenics movement and Bruinius's book. It was a test case meant to challenge the constitutionality of Virginia's forced sterilization act before the Supreme Court. It focused on the first person sterilized under that act, Carrie Buck, who was suing the physician who sterilized her, John H. Bell.

Carrie was a 21-year-old unwed mother who, with her own mother Emma, was a resident of the Virginia Colony for Epileptics and Feebleminded. Carrie's illegitimate infant daughter, Vivian, was already claimed to be showing signs of feeblemindedness.

(Continued)

(Continued)

The Colony's board earlier had approved the request for the operation by Dr. Albert Priddy, the first superintendent of the Colony. But those involved in the decision were aiming for far broader approval to continue eugenic practices. They orchestrated a lawsuit on behalf of Carrie against themselves, first at the state level against Dr. Priddy and then before the Supreme Court with Priddy's successor, Dr. Bell.

Not much of a defense was put up on Carrie's behalf. Even her own lawyer supported eugenics. Holmes's decision failed to take into account school records showing that Carrie's daughter Vivian, the "third-generation imbecile" referenced in his opinion, had made the honor roll.

Little has been done to make amends for the harm caused to Americans whose only crime may have been their poverty or lack of education. Some states have since issued apologies to the victims, but those who tried to sue for damages have had little success.

Bruinius, who is a contributor to *The Christian Science Monitor* and a journalism professor at Hunter College, presents a compelling and readable narrative of the people and motivations behind the eugenics movement. And while the practice was not as secret as the book's subtitle would suggest, many Americans today are unaware of it, which is a key part of the book's value.

Bruinius uses extensive endnotes citing sources for readers who want to learn more. The illustrations, from eugenics pamphlets to evidence used to show defective family trees, are chilling and enliven the text.

At the end of the book, Bruinius draws parallels to today's advances in genetics and poses questions about whether the notion of "better breeding" could resurface in modern bioengineering technologies. The basic premise of eugenics, he writes, is still valid. This section of the book leaves the reader hungry for more specifics.

But Bruinius aptly points out that genetic engineering is something to be watched. Disturbingly, he writes, "It is not an irresponsible prophecy to say that ideas of better breeding could again lead to the horrors witnessed in the twentieth century.".

SOURCE: Valigra, Lori. (2006, March 7). A chapter of history painful to recall: Forced sterilization is today unthinkable. But could it tempt future generations? *The Christian Science Monitor*, 16. Article reprinted by permission of the author. This article first appeared in *The Christian Science Monitor*.

Psychological Theories

The human mind has long been considered a source of abnormal behavior and, therefore, crime (Lanier & Henry, 1998, p. 113). Early varieties of **psychological theories** of delinquency and crime focused on lack of intelligence and/or personality disturbances as major causal factors. Several of the early pioneers in the psychological school were convinced that biological factors played a major role in determining intelligence; therefore, they could be considered proponents of both schools of thought. Goddard's (1914) studies of the Kallikak family and the intellectual abilities of reformatory inmates, for instance, led him to conclude that feeblemindedness, which he believed to be inherited, was an important contributing factor in criminality. He suggested that "eliminating" a large proportion of mental defectives would reduce the number of criminals and other deviants in society. Similarly, Goring (1913) focused on defective intelligence and psychological characteristics as basic causes of crime in his attempt to refute Lombroso and the other positivists. As we indicated previously, research concerning the relationship between defective intelligence, IQ, or learning disabilities and delinquency continues. Problems concerning the reliability and validity of IQ tests and **personality inventories,** as well as other methodological shortcomings, continue to plague such research, and the psychological school as a whole has taken other directions. Still, many believe that those who commit heinous crimes must be emotionally disturbed—different from the rest of us in some identifiable way.

Sigmund Freud's Psychoanalytic Approach

Sigmund Freud, born in 1856, spent most of his life in Vienna, Austria. He is regarded as the founder of psychoanalysis, a **psychoanalytic approach** to explaining behavior that relies heavily on the techniques of introspection (looking inside one's self) and retrospection (reviewing past events). Freud's theories were introduced in the United States during the early 1900s. Freud divided personality into three separate components: the **id, ego, and superego.** The function of the id, according to Freud, is to provide for the discharge of energy that permits the individual to seek pleasure and reduce tension. The id is also said to be the seat of instincts in humans and not thought to be governed by reason. The ego is said to be the part of the personality that controls and governs the id and the superego by making rational adjustments to real-life situations. For example, the ego might prevent the id from causing the individual to seek immediate gratification of his or her desires by deferring gratification to a later time. The development of the ego is said to be a product of interaction between the individual's personality and the environment and is thought to be affected by heredity as well. The superego is viewed as the moral branch of the personality and may be equated roughly with the concept of conscience. Both the ego and the superego are thought to develop out of the individual's interactions with his or her environment, whereas the id is said to be a product of evolution.

In general, deviance is viewed as the product of an uncontrollable id, a faulty ego, or an underdeveloped superego or some combination of the three. Therefore, those who commit a criminal or delinquent act do so as the result of a personality disturbance. To correct or control this behavior, the causes of the personality disturbance are located primarily through introspection and retrospection, with a particular emphasis on childhood experiences, and then are eliminated through therapy.

Freud is one of the most important figures (if not the most important figure) in the history of psychology. There are, no doubt, many cases where psychoanalytic techniques prove to be effective in therapeutic treatment. As a system for explaining the causes of deviance, however, Freudian psychology has several shortcomings. First, the existence of the id, ego, and superego cannot be demonstrated empirically. Second, instincts, which Freud viewed as the driving forces in the id, are thought by many behavioral scientists to be extremely rare or nonexistent in humans. Third, there seems to be faulty logic among practitioners using Freud's system. They accept the premise that those who commit deviant acts must be experiencing personality disturbances; that is, they employ circular reasoning rather than logical deduction (Akers, 1994, p. 85; Lanier & Henry, 1998, p. 117). In response to the question, "How do you know X has a disturbed personality?" they might answer, "Because he committed a deviant act, he must have been experiencing a personality disturbance." Such a response is more a statement of faith than a matter of fact. Currently, it is safe to say that the psychoanalytic approach is of very little value in explaining crime and delinquency (or any other form of deviance, for that matter).

Nonetheless, the Freudian approach has remained popular in much of the Western world, and Freud has had many disciples who have applied his techniques directly to delinquency.

Among those who emphasized the psychoanalytic perspective were Healy and Bronner (1936), who believed that the delinquent was a product of a personality disturbance resulting from thwarted desires and deprivations that led to frustration and a weak superego. Healy and Bronner interviewed numerous juvenile offenders and came to the conclusion that 90% of them were emotionally disturbed. Adler (1931), Halleck (1971), and Fox and Levin (1994) concluded that those who are frustrated, believe the world is against them, and feel inferior may turn to crime as a compensatory means of expressing their autonomy.

Others, using a variety of personality inventories (e.g., the Minnesota Multiphasic Personality Inventory [MMPI], the California Personality Inventory [CPI]), have concluded that such inventories do appear to discriminate between delinquents and nondelinquents, but the reasons for such discrimination are not at all clear-cut, and neither are the numerous definitions of "abnormal" personality employed (Bohm, 2001, pp. 56–57). Akers and Sellers (2004) concluded, "The research using personality inventories and other methods of measuring personality characteristics has not been able to produce findings to support personality variables as major causes of criminal and delinquent behavior" (p. 47).

Psychopathology

One of the terms most commonly employed to describe certain types of criminals and delinquents is **psychopath.** Typically, the term is used to describe aggressive criminals who act impulsively with no apparent reason. Sutherland and Cressey (1978) indicated that some 55 descriptive terms are consistently linked with the concept of psychopathy (sociopathy or antisocial personality). Bohm (2001) listed 16 characteristics ranging from "unreliability" to "fantastic and uninviting behavior" to "failure to follow any life plan" (p. 54). Attempts have been made to clarify the concept of psychopathology, but such attempts have helped little in understanding the relationship between psychopathology and criminality because criminality is typically included in the symptomatic basis for psychopathology. In other words, the two conditions are often perceived as being one and the same.

Although the concept of psychopathology is generally considered to be too vague and ambiguous to distinguish psychopaths from nonpsychopaths, there have been attempts to operationalize the concept in more meaningful fashion. Gough (1948, 1960) conceptualized psychopathy as the inability to take the role of the other (the inability to identify with others). The scales he developed to measure role-taking ability generally result in lower scores for offenders than for nonoffenders. Whether or not such differences could have been detected before the offenders committed offenses is another matter.

Research in this area continues. Martens (1999) reported a case in which psychopathy appeared to have been cured as a result of therapeutic psychosocial influences and life events. In this case, the individual began a career in delinquency at 15 years of age and went on to commit offenses, including fraud, theft, rape, and assaults, until 26 years of age. Following life-changing events and therapy, the individual had remained crime free for more than 20 years and appeared to be leading a "normal" life.

Poythrees, Edens, and Lilienfeld (1998) administered the Psychopathic Personality Inventory (PPI), a self-report measure of psychopathic personality features, and the Psychopathy Checklist–Revised (PCL-R) to youthful offender prison inmates. They found that the PPI could be used to accurately predict PCL-R classifications of psychopath and nonpsychopath, raising the possibility that the PPI could be used for clinical purposes to detect psychopathic personalities.

Lynam (1998) hypothesized that there is a developmental relationship between adult psychopathy and children with symptoms of hyperactivity, impulsivity, attention problems, and conduct problems (HIA–CP). Using a large sample of adolescent boys, Lynam found that boys who were hyperactive and impulsive, with attention disorders and conduct problems, scored high on a measure of psychopathic personality. These boys were the most antisocial, were the most disinhibited, and tended to be the most neuropsychologically impaired of the groups studied. Further support for the relationship between adolescent behavior patterns of this type and adult psychopathy comes from Gresham, MacMillan, and Bocian (1998), who found marked differences between third- and fourth-grade students with HIA–CP and other students on peer measures of rejection and friendship and teachers' ratings of social skills. The notion of the "fledgling psychopath" appears to emerge from these recent studies. Still, as Akers (1994) concluded, based on the research currently available, the term *psychopath* appears to be so broad that it could be applied to anyone who violates the law (p. 87).

After reviewing attempts to relate psychopathy to child abuse, Knudsen (1992) concluded that there is little evidence of such a relationship. Wolfe (1985) also found no relationship between underlying personality attributes and child abuse beyond general descriptions of stress-related complaints and displeasure in the parenting role. The exact nature of the relationship between psychopathology and child abuse remains unclear (Walsh et al., 2002).

Further research on the relationship between psychopathology and delinquency and abuse is clearly needed. On the one hand, it may turn out that behavior patterns involving hyperactivity, impulsivity, and inattention, combined with conduct problems, are forerunners of psychopathology. On the other hand, most children exhibit one or more of these behaviors periodically but do not turn out to be psychopaths.

Behaviorism and Learning Theory

During the latter 19th century, a number of psychologists became increasingly concerned about weaknesses in the theory and techniques developed by Freud and his followers and those of the biological school emphasizing heredity. Tarde, by contrast, thought that crime was learned by normal people in the process of interacting in specific environments (Bohm, 2001, p. 82). He and others called for a change in focus from genetics and the internal workings of the mind to observable behavior. Although the major work on this **learning theory** model as it relates to delinquency has been done by sociologists and is discussed under that topic, the psychological underpinnings are discussed here.

As indicated previously, **behaviorists** called for a change of techniques from the subjective speculative approach based on introspection and retrospection to a more empirical objective approach based on observing and measuring behavior. Perhaps the most important individual in the behaviorist tradition was B. F. Skinner, who directed his attention toward the relationship between a particular stimulus and a given response and to the learning processes involved in connecting the two. Skinner (1953) viewed human social behavior as a set of learned responses to specific stimuli. Criminal and delinquent behaviors are viewed as varieties of human social behavior, learned in the same way as other social behaviors. Through the process of **conditioning** (rewarding for appropriate behavior and/or punishing for inappropriate behavior), any type of social behavior can be taught (see In Practice 4.2). Therefore,

In Practice 4.2

Attempt the "Impossible": Be Role Models, Expect More, Work as a Community to Help Children

From Charlotte native Lauren Izard, a rising high school senior at the Lawrenceville School in Lawrenceville, N.J., and a summer intern for the advocacy group, Council for Children's Rights:

Before I began working as an intern at the Council for Children's Rights, my first inclination when I learned about kids committing crime was to think they were "bad kids." Since working here, my perspective has changed. When I go to court with staff advocates and attorneys, I've discovered that most of these kids are victims of extreme poverty, often living without role models, surrounded by adults and peers who are drug users, sex offenders, gang members, etc. These kids are born 10 steps behind. Now when I hear about a child's criminal record, the first thing I think about is how badly role models have failed [the child].

Children live to the expectations that people have of them. They are influenced by the examples they observe in their environment, and they listen to what is said both to and about them. Children are not born with the ability to distinguish right from wrong; they must be taught. When my little sisters used to steal things from my room, I would get angry and my parents would say, "They are too young to understand." My sisters were taught through "time-outs" that stealing is wrong.

Many children aren't taught these lessons. Parents are our primary advisors and counselors; even if they are drug users or criminals, we still look to them for guidance. Not all kids who come from broken homes are criminals, and not all criminals come from broken homes. Regardless, those who escape unscathed had someone who taught them right from wrong. We should not blame the children but [rather] look to their environment for evidence of their behavior.

One environmental example is the correlation between domestic violence and its effect on children. In *Observer* reporter Cleve Wooten's article, he reports that "violent crime arrests jumped 25 percent in the first six months of 2006 compared with the same period last year." However, let's also consider that demands for domestic violence services have jumped 27 percent over the past five years. That percentage only covers the recorded incidents, while only 1/7 of all domestic assaults come to the attention of the police. Children who are raised in violent environments accept it as routine and are five times as likely to commit or suffer violence as adults.

When these kids have been told they are inferior from birth and have been given no positive guidance, what options are they left with but to assume the role that society has given them as "violent offenders"? In Mr. Wooten's article, he writes, "Decreasing violent juvenile crime can be a nearly impossible task, experts say." However, it is possible to give kids options other than drug abuse and theft. It is possible for them to be raised in loving and positive homes. It is possible to teach them how to be productive citizens in our community.

The problem is that so few attempt what has been deemed "the impossible." To end juvenile delinquency and violence, we must focus our attention on what is possible: working as a community to improve children's support systems (families, role models, education, health care, etc.). We must fight this "impossible" battle because, as author Pearl S. Buck once said, "If our American way of life fails the child, it fails us all."

SOURCE: Izard, Lauren. (2006, August 14). Attempt the "Impossible": Be Role Models, Expect More, Work as a Community to Help Children. *The Charlotte Observer,* 12A. Reprinted with permission from *The Charlotte Observer.*

when an individual behaves in a delinquent manner (exhibits an inappropriate response in a given situation), his or her behavior can be modified using conditioning. To control and rehabilitate delinquents, then, the therapist employs behavior modification techniques to extinguish inappropriate behavior and replace it with appropriate behavior.

▲ B. F. Skinner believed that social behaviors can be taught through the process of conditioning (rewarding desired behavior and/or punishing undesirable behavior).

Although behaviorists do not seek to explain the ultimate causes of social behavior except in the sense that they are learned, their approach holds considerably more promise for understanding and controlling delinquent behavior than does the psychoanalytic approach. The behaviorist approach forces us to focus on the specific problem behavior and to recognize that it is learned, so it can—hypothetically at least—be unlearned (Reid, 2006, p. 106). With this focus, we are dealing with observable behavior that can be measured, counted, and perhaps modified. Success in modifying behavior in the laboratory has been noted (Echeburua, Fernandez-Montalvo, & Baez, 2000; Krasner & Ullman, 1965; Martin & Peas, 1978; Paul, Marx, & Orsillo, 1999). The extent to which this success can be transferred to the world outside the laboratory remains an empirical question (Florsheim, Shotorbani, & Guest-Warnick, 2000; Ross & McKay, 1978; Shelton, Barkley, & Crosswait, 2000). Think about the difficulties of transferring desirable behavior from the laboratory to the street in the following hypothetical case.

Joe Foul Up, a juvenile, is repeatedly apprehended for fighting. Finally, he is turned over to a therapist who, over a period of several weeks, eliminates the undesirable behavior by punishing Joe (e.g., with electric shock) when he begins to exhibit the undesirable behavior and by rewarding him when he exhibits appropriate alternative behavior. After therapy ends, Joe's behavior has been modified, and he returns home to his old neighborhood and his old street gang. When Joe refuses to fight, the gang thinks that it is appropriate to punish him by calling him a coward and excluding him from gang activities. When he does fight, they reward him by treating him like a hero. What are the chances that the behavior modification that occurred in the laboratory will continue to exist?

We have more to say about the learning theory or behaviorist approach in the section on learning theory.

Sociological Theories

There have been a number of different **sociological theories** of delinquency causation, some dealing with social class and/or family differences (Cloward & Ohlin, 1960; Cohen, 1955; Miller, 1958; Quinney, 1975), some dealing with blocked educational and occupational goals (Merton, 1938), some dealing with neighborhood and peers (Miller, 1958; Shaw & McKay, 1942; Thrasher, 1927), and some dealing with the effects of official labeling (Becker, 1963). Most of these theories share the notion that delinquent behavior is the product of social interaction rather than the result of heredity or personality disturbance. For sociologists, delinquency must be understood in social context. Thus, we must consider time, place, audience, and nature of the behavior involved when studying delinquency.

Anomie and Strain Theory

Beginning in the 1930s in the United States, a number of theorists focused on a systems model to explain crime and delinquency. Adapting Durkheim's **anomie theory** (a breakdown of social norms or the dissociation of the individual from a general sense of morality of the times), Merton (1938) focused on the discrepancy between societal goals and the legitimate means of attaining those goals. He argued that strain is placed on those who wish to pursue societal goals but lack the legitimate means of doing so (**strain theory**). According to Merton, people adapt to this strain in different ways; some attempt to play the game, some retreat (and may become addicts and outcasts), some develop innovative responses (including the illegitimate responses of crime and delinquency), and some rebel (another potential source of crime).

During the 1950s, Cohen (1955) adapted Merton's theory in an attempt to explain juvenile gangs. He argued that lower-class juveniles experience the strain of being unsuccessful in middle-class terms, especially in the school setting. Because many lower-class youth find success in school difficult to achieve, they reject middle-class values and seek to gain status by engaging in behaviors contrary to middle-class standards. Thus they establish their own anti-middle-class value system and, through mutual recruitment, form delinquent gangs. Miller (1958) disagreed with Cohen's theory that lower-class youth act in terms of inverted middle-class values; instead, Miller focused on what he called the "focal concerns" (toughness, trouble, smartness, fate, autonomy, and excitement) of the lower social class as the sources of delinquent behavior.

Sykes and Matza (1957) argued in their theory of **delinquency and drift** that firm commitment to subcultural values was not necessarily a precursor of delinquent behavior (unlike the view of Cohen, Miller, and others). Sykes and Matza viewed delinquency as being based on an extension of defenses to crimes in the form of justifications for deviant behavior that are accepted by delinquents but not necessarily by the legal system or larger society. These defense were called **techniques of neutralization** and included (a) a denial of responsibility (for the consequences of delinquent actions), (b) a denial of injury (to the victim or larger society), (c) the denial of a victim (the victim "had it coming"), (d) condemnation of the condemners (as hypocrites or spiteful people), and (e) an appeal to higher loyalties (e.g., to the gang). Using these techniques, juveniles drift in and out of the delinquent subculture over time.

Cloward and Ohlin (1960) extended anomie or strain theory by focusing on the differential opportunities that exist among juveniles. If an **illegitimate opportunity structure** is readily available, they argued, juveniles who are experiencing strain or anomie are attracted to that structure and are likely to become involved in delinquent activities.

In 1985, Agnew again revised strain theory. He discussed three types of strain that may produce deviant behavior. The first is the individual's failure to achieve goals, the second involves loss of a source of stability (e.g., death of a loved one), and the third occurs when the individual is confronted by negative stimuli (e.g., lack of success in school). Furthermore, Agnew (1985) suggested that, rather than pursuing specific goals, many people are simply interested in being treated justly based on their own efforts and resources. People who do not perceive themselves to be treated fairly experience strain, according to Agnew. Reactions to this perception of unfair treatment may lead to crime and delinquency. Later, Agnew (2001) argued that criminal victimization might be among the most consequential strains experienced by adolescents and, therefore, might be an important cause of delinquency. Subsequently, Hay and Evans (2006) examined predictions from general strain theory about the effects of victimization on later involvement in delinquency. They concluded that violent victimization significantly predicted later involvement in delinquency, even when controlling for the individual's earlier involvement in delinquency, and that the effects of victimization were slightly greater for juveniles with weak emotional attachment to their parents and significantly greater for those low in self-control.

Using a sample of homeless street youth, Baron (2004) examined how specific forms of strain, including emotional abuse, physical abuse, sexual abuse, homelessness, being a victim of robbery, being a victim of violence, being a victim of theft, relative deprivation, monetary dissatisfaction, and unemployment, are related to crime and drug use. He also explored how strain is conditioned by deviant peers, deviant attitudes, external attributions, self-esteem, and self-efficacy. He concluded that all 10 types of strain examined can lead to criminal behavior either as main effects or when interacting with conditioning variables.

There are numerous criticisms of strain/anomie theory. It tends to focus almost exclusively on lower-class delinquency. It also ignores the effects of labeling and fails to explain why many juveniles who undoubtedly experience strain do not turn to delinquency as a means of attaining their goals.

The Ecological/Social Disorganization Approach

The **ecological/social disorganization approach** to explaining crime and delinquency was developed during the 1930s and 1940s and is one of the oldest interest areas of American criminologists. This approach focuses on the geographic distribution of delinquency. Shaw and McKay (1942), and later others, found that crime and delinquency rates were not distributed equally within cities. They mapped the areas marked by high crime and delinquency rates along with the socioeconomic problems of those areas. Using Burgess's (1952) **concentric zone theory** of city growth, the ecological/social disorganization studies generally found that zones of transition between residential and industrial neighborhoods consistently had the highest rates of crime and delinquency. These zones are characterized by physical deterioration and are located adjacent to the business district of the central city. The neighborhoods in this zone are marked by deteriorating buildings and substandard housing with accompanying overcrowdedness, lack of sanitation, and generally poor health and safety conditions. In addition, the area is marked by a transient population, high unemployment rates, poverty, broken homes, and a high adult crime rate. In short, the area is characterized by a general lack of social stability and cohesion or social disorganization.

▲ Sociologists often look at environmental factors and their relationship to delinquent behavior.

Wilks (1967) best summarized early ecological/social disorganization studies and their findings on the distribution of delinquency. Her conclusions were as follows.

1. Rates of delinquency and crime vary widely in different neighborhoods and within a city or town.
2. The highest crime and delinquency rates generally occur in the low-rent areas located near the center of the city, and the rates decrease with increasing distance from the city center.
3. High delinquency rate areas tend to maintain their rates over time, although the population composition of the area may change radically within the same time period.
4. Areas that have high rates of truancy also have high rates of juvenile court cases and high rates of male delinquency and usually have high rates of female delinquency. The differences in area rates reflect differences in community background. High-rate areas are characterized by things such as physical deterioration and declining population.
5. The delinquency rates for particular nationality and ethnic groups show the same general tendency as the entire population; namely, they are high in the central area of the city and low as the groups move toward the outskirts of the city.
6. Delinquents living in areas with high delinquency rates are the most likely to become recidivists and are likely to appear in court several times more often than are those living in areas with low delinquency rates.
7. In summary, delinquency and crime follow the pattern of social and physical structures of the city, with concentration occurring in disorganized, deteriorated areas.

According to Wilks (1967), to predict delinquency using the ecological/social disorganization approach, it is necessary to be aware of the existing social structure, social processes, and population composition, as well as the area's position within the large urban societal complex, because these variables all affect the distribution of delinquency. In general, this approach found that family and neighborhood stability were lacking and that the street environment was the prevailing determinant of behavior. If delinquent behavior is learned behavior, this learning would be maximized in environments such as those in transitional zones. In transitional zones, those agencies or institutions that traditionally produce stability, cohesion, and organization have often been replaced by the street environment of adult criminals and delinquent gangs.

The ecological/social disorganization approach to explaining delinquency has been challenged on the grounds that using only one variable to explain delinquency is not likely to lead to success. In Lander's (1970) study of Baltimore, for example, he found anomie, or normlessness, to be a more appropriate explanation of delinquency rates than socioeconomic area. Nonetheless, follow-up studies by Shaw and McKay (1969) in other American cities (Boston, Philadelphia, and Cleveland) support their contention that official delinquency rates decrease from the central city out to the suburbs. Similarly, Lyerly and Skipper (1981) found that significantly less delinquent activity was reported by rural youth than by urban youth in their study of juveniles in detention. Stark (1987) concluded that certain geographic areas (those characterized by high population density, poverty, transience, dilapidation, etc.) attract deviant people who drive out those who are not so deviant, and these places then become "deviant places" with high crime rates and weak social control. Whatever the cause, the fact remains that high official delinquency rates are found in certain areas or types of areas where serious and repetitive misconduct not only is common but also appears to have become traditional and more or less acceptable (Lowencamp, Cullen, & Pratt, 2003). There is a real danger here, however, of drawing false conclusions based on what has been called the "ecological fallacy." This term refers to false conclusions drawn from analyzing data at one level (e.g., the group level) and applying those conclusions at another level (e.g., the individual level). In short, group crime rates tell us nothing about whether a particular individual is likely to become involved in crime (Bohm, 2001, p. 71). In spite of these criticisms, Moyer (2001) found that "one can find the early development of the interactionist perspective, control theory, and conflict theory in their works" (p. 118).

Edwin Sutherland's Differential Association Theory

Sutherland (1939) developed what is known as the **theory of differential association.** Sutherland's approach combines some of the principles of behaviorism (or learning theory) with the notion that learning takes place in interaction within social groups. For Sutherland, the primary group (family or gang) is the focal point of learning social behavior, including deviant behavior. In this context, individuals learn how to define different situations as appropriate for law-abiding or law-violating behavior. Therefore, seeing an unattended newsstand might be defined as a situation appropriate to the theft of a newspaper by some passersby but not by others. The way a given individual defines a particular situation depends on that individual's prior life experiences. An individual who has a balance of definitions favorable to law-violating behavior in a given situation is likely to commit a law-violating act. The impact of learned definitions on the individual depends on how early in life the definitions were learned (priority), how frequently the definitions are reinforced (frequency), the period of time over which such definitions are reinforced (duration), and the importance of the definition to the individual (intensity) (Sutherland, Cressey, & Luckenbill, 1992, pp. 88–90).

Sutherland's approach has the advantage of discussing both deviant and normal social behavior as learned phenomena. The approach also indicates that the primary group is crucial in the learning process. In addition, Sutherland suggested some important variables to be considered in determining whether behavior will be criminal or noncriminal in given situations. Finally, Sutherland suggested that it is not differential association with criminal and noncriminal types that determines the individual's behavior; rather, it is differential association with, or exposure to, definitions favorable or unfavorable to law-violating behavior.

The learning theory and differential association approaches have been used to try to explain child abuse and neglect as well as delinquency. According to these approaches, abusive parents learned abusive behavior when they were abused as children. Thus child abuse is said to be an intergenerational phenomenon. Kaufman and Zigler (1987), after reviewing self-report data, concluded that the rate of abuse by individuals with a history of abuse is six times higher than that in the general population (p. 190). This finding supports the belief that abusive behavior is learned in primary groups that define it as acceptable behavior (as Sutherland suggested is the case with other forms of deviance). However, other researchers have criticized Kaufman and Zigler and have failed to find a relationship between being abused and abusing. Knudsen (1992) also concluded that the cycle of violence appears to be a minor factor in explaining child abuse (p. 63). Still, as we indicated earlier, there is evidence to the contrary. Scudder and colleagues (1993) concluded that children who break the law often have a history of maltreatment as children (p. 321). These researchers and Siegel and Williams (2003) indicated that a child abused at a young age is at higher risk for subsequent delinquent behaviors than is a nonabused child.

There are a number of criticisms of Sutherland's approach. It is clearly difficult to operationalize the terms *favorable to* and *unfavorable to.* There are serious problems with trying to measure the variable intensity. How many exposures to definitions favorable to law violation are required before definitions unfavorable to law violation are outweighed and the individual commits the illegal act? These and other weaknesses have been pointed out over the years by critics of differential association. Nonetheless, there is a certain logic to Sutherland's approach. Some of the propositions are empirically testable, and the description of the learning process seems to be relatively accurate. Sutherland's approach has sensitized us to an approach to understanding crime and delinquency that has been built on by other theorists and researchers (Akers, 1998; Burgess & Akers, 1968; Curran & Renzetti, 1994; Glaser, 1960).

One attempt to improve on Sutherland's theory was made by Glaser (1978). Glaser referred to his theory as the **theory of differential anticipation,** which, in his view, combines differential association and control theory and is compatible with biological and personality theories. Differential anticipation theory assumes that a person will try to commit a crime wherever and whenever the expectations of gratification from it—as a result of social bonds, differential learning, and perceptions of opportunity—exceed the unfavorable anticipations from these sources (pp. 126–127). In short, expectations determine conduct, and expectations are determined by social bonds, differential learning, and perceived opportunities. Burgess and Akers (1968) also expanded on the learning theory approach developing differential association–differential reinforcement theory. Akers (1985, 1992) later referred to his theoretical approach as social learning theory. This theory holds that social sanctions of engaging in (deviant) behavior may be perceived differently by different individuals. However, so long as these sanctions are perceived as more rewarding than alternative behavior, the deviant behavior will be repeated under similar circumstances. Progression into sustained deviant behavior is promoted to the extent that reinforcement, exposure to deviant models, and definitions are not offset by negative sanctions and definitions. These theories are eclectic in the sense that they extend Sutherland while being compatible with most of the approaches we have discussed and with labeling theory, to which we now turn our attention.

Labeling Theory

A number of social scientists have contributed to what might be called the **labeling theory** school of crime/delinquency causation. Becker (1963) discussed the process of labeling deviants as outsiders. Erikson (1962) pointed out the importance of what he called the labeling "ceremony" for deviants. These authors and others shifted the focus of attention from the individual deviant (e.g., delinquent, criminal, mentally ill) to the reaction of the audience observing and labeling the behavior as deviant. As we have indicated repeatedly, it is clear that many individuals commit deviant acts, but only some are dealt with officially. The time at which the act occurs, the place where it occurs, and the people who observe the act all are important in determining whether or not official action will be taken. Thus, the juvenile using heroin in the privacy of his gang's hangout in front of other gang members is not subject to official action. If, however, he used heroin in a public place in the presence of a police officer who was observing his behavior, official action would be likely.

From the labeling theorist's point of view, then, society's reaction to deviant behavior is crucially important in understanding who becomes labeled as deviant. Erikson (1962) discussed the ceremony that deviants typically go through once the decision to take official action has been made. First, the alleged deviant is apprehended (arrested or taken into custody). Second, the individual is confronted, generally at a trial or hearing. Third, the individual is judged (a verdict, disposition, or decision is rendered). Finally, the individual is placed (imprisoned, committed to an institution, or put back into society on probation). The result is that the individual is officially labeled as deviant.

One of the consequences of labeling in our society is that, once labeled, the individual may never be able to redeem himself or herself in the eyes of society. Therefore, John Q. Convict does not become John Q. Citizen on release from prison. Instead, he becomes John Q. Ex-Convict. Having been labeled may make it extremely difficult for the rehabilitated deviant to find employment and establish successful family ties. The more difficult it becomes for the rehabilitated deviant to succeed in the larger society, the greater the chances that he or she will return to old associates and old ways. Of course, these are often the very associates and ways that led the individual to become officially labeled in the first place. Thus, the individual may be more or less forced to continue his or her career in deviance, partially as a result of the labeling itself.

Research by Blankenship and Singh (1976) indicated that a juvenile's prior career of delinquent behavior (the extent to which he or she has been officially labeled previously) is indeed an important determinant of official action. These authors, as well as Covington (1984), pointed out that labeling comes in different forms (e.g., legalistic vs. peer group) and has different consequences for different types of offenders (e.g., whites vs. blacks). If we could assume that society never makes a mistake in attaching the label of deviant and that rehabilitation programs never succeed, we might regard the consequences of labeling as somewhat less alarming. As we have already seen in Chapters 2 and 3, the assumption that society never makes a mistake is unwarranted. We see later that there is at least some hope that rehabilitation programs do succeed. If the result of official labeling forces the labeled individual back into a deviant career, in the case of juveniles at least, we are accomplishing exactly the opposite of what we intended when we created a separate juvenile justice system designed to protect, educate, and treat juveniles rather than to punish them. One of the consequences of negative societal reaction to the label of delinquent may be the changing of the delinquent's self-concept, so the individual, like society, begins to think about himself or herself in negative terms. Possibilities for rehabilitation may be lessened as a result.

An interesting contribution to labeling theory was made by Braithwaite (1989). He discussed what he referred to as "disintegrative shaming" (negative stigmatization) and noted that it is destructive of social identities because it morally condemns and isolates people but involves no attempt to reintegrate the shamed people at some later time. He contrasts this harmful approach to stigmatization with "reintegrative shaming" in which there is an attempt to reconnect the stigmatized person to the larger society.

The labeling approach accurately describes how individuals become labeled, why some maintain deviant careers, and some of the possible consequences of labeling (Krisberg, 2005, p. 184). It does not deal with the issue of why some individuals initially commit acts that lead them to be labeled; rather, it deals only with what is referred to as secondary deviance. In addition, those who support the approach often lose sight of the fact that the individual is in some way responsible for the actions that are viewed as unacceptable; that is, social audiences do not appear to attach negative labels haphazardly. They are responding to some stimulus presented by the individual committing a crime for which he or she must accept some responsibility (unless we return to a completely deterministic concept of deviance).

Despite some weaknesses, the labeling approach contributes significantly to our understanding of deviance. Through this approach, deviance is viewed as a product of social interaction in which the actions of both the deviant and his or her audience must be considered.

Conflict/Radical/Critical/Marxist Theories

Chambliss (1984) described conflict theories of crime as focusing on whole political and economic systems and on class relations in those systems. Conflict theorists argue that conflict is inherent in all societies, not just capitalist societies, and focus on conflict resulting from gender, race, ethnicity, power, and other relationships. Conflict results from competition for power among many groups. Those who are successful in this competition define criminality at any given time. Thus, criminal behavior is viewed not as universal or inherent but rather as situational and definitional. This view does not account for individual acts of criminality occurring outside of the group context but serves basically to alert us to the social factors that may be related to criminality. Why, for example, do we pass laws with severe sanctions for use of marijuana but deal with tobacco use among teens much less harshly? Is it because the tobacco lobby is powerful and able to convince legislators that tobacco use among juveniles should, at most, be regulated but not outlawed?

The Marxist approach to criminology and delinquency finds the causes of such phenomena in the repression of the lower social classes by the "ruling class." In short, laws are passed and enforced by those who monopolize power against those who are powerless (e.g., the poor and minorities). The causal roots of crime are assumed, by many proponents of this approach, to be inherent in the social structure of capitalistic societies. Crime control policies are developed and implemented by those who have power (e.g., own the means of production, have wealth), and these policies serve to criminalize those who threaten the status quo (Beirne & Quinney, 1982; Chambliss & Mandoff, 1976; Platt, 1977; Quinney, 1970, 1974; Turk, 1969; Vold, 1958). Labeling the discontented as criminals and delinquents allows the ruling class to call on law enforcement officials to deal with such individuals without needing to grant legitimacy to their discontent. Although there are a number of variations on the theme as discussed here, these are the essential components of most radical or critical explanations of delinquency and crime.

Radical criminology became relatively popular in the United States during the 1970s and 1980s, but its popularity has declined over the years and some of its most important spokespersons have abandoned this approach, at least in part, as an explanation of crime and delinquency. Little empirical research that supports the radical/critical approach has been done (Moyer, 2001, p. 238).

As we indicated earlier, delinquency appears to be rather uniformly distributed across social classes, contrary to the teachings of the Marxist approach. In addition, as we indicated earlier, this approach fails to recognize that the legal order serves the purpose of maintaining the system in all known types of societies, including those that claim to be Marxist/Communist/Socialist (Cox, 1975). As Klockars (1979) noted, "The leading figures of American Marxist criminology have not raised the details of Gulag or Cuban solutions to the problems of crime in America, nor have they seriously examined such solutions in states which legitimate them" (p. 477). Bohm (2001) added, "Today, it probably makes little sense to speak of capitalist and socialist societies anyway, because no pure societies of either type exist. (They probably never did.)" (p. 119).

Feminism

One reaction to conflict theory, which concentrated largely on crimes committed by males, is **feminism.** Feminism as an approach to studying crime and delinquency focuses on women's experiences, typically in the areas of victimization, gender differences in crime, and differential treatment of women by the justice network. Some feminists focus on equal rights and equal participation for women, some focus on the ills of capitalist society, and others focus on the issue of patriarchal oppression (in the form of male control over sex, money, and power) that has resulted in second-class citizenship for women in our society. Traditional criminology has certainly largely ignored female crime, raising the issue of whether any of the theories of crime apply directly to women. Furthermore, there are clearly differences in the extent and nature of crime by gender, and there is a question as to whether or not current theories can explain these differences (Daly & Chesney-Lind, 1988; Naffine, 1996). Still, the focus on gender as a major determinant of delinquent and criminal behavior has been seriously questioned because there appears to be limited empirical support for the approach (Akers, 1994, p. 177; Bohm, 2001, p. 122).

Control Theories

Control theories assume that all of us must be held in check or "controlled" if we are to resist the temptation to commit criminal or delinquent acts. The types of systems used to control or check delinquent behavior fall into two categories: personal (internal) and social (external). The containment theory of Reckless (1961, 1967), for instance, emphasizes the importance of both inner controls and external pressures on self-concept. A poor self-concept is thought to increase the chances that a juvenile will turn to delinquency; a positive self-concept is seen as insulating the juvenile from delinquent activities. Negative self-concepts and low self-esteem have also been frequently noted as characteristics of those who abuse or neglect children (Marshall et al., 1999; Shorkey & Armendariz, 1985).

Hirschi's (1969) control theory places more emphasis on social factors (bonds and attachments) than on inner controls. For example, the term *attachment* is used to refer to the feelings one has toward other persons or groups. The stronger one's attachment to nondelinquent others, the less likely one is to engage in delinquency. The same type of argument is applied to commitment (profits associated with conformity vs. losses associated with nonconformity), involvement (in conforming vs. nonconforming activities), and beliefs (in the conventional value system vs. some less conventional value system). Although these four components of control theory may vary independently, Hirschi maintained that in general they vary together. Strong positive ties in each of these four areas minimize the possibility of delinquency, whereas strong negative ties maximize the likelihood of delinquency. Hirschi's formulation has encouraged considerable research, and although there is some empirical evidence to support portions of the control theory approach, this approach leaves unanswered a number of important questions. What is the exact nature of the relationship between self-concept and labeling? How is it that some juveniles who appear to be well insulated from negative attachments and bonds commit delinquent acts? Do such bonds and attachments themselves actually inhibit delinquent behavior, or are the bonds and attachments perceived by law enforcement and criminal justice personnel simply used to determine whether or not to take official action? Are there longitudinal data that support the approach? Attempts to answer some of these questions are ongoing. In a reanalysis of Hirschi's original data, Costello and Vowell (1999) found support for Hirschi's theory. May (1999) found that social control theory had a significant association with juvenile firearms possession in school. But Greenberg's (1999) reanalysis of Hirschi's data found that social control theory has only limited explanatory power.

In 1990, Hirschi collaborated with Michael Gottfredson to develop what they referred to as a "general theory of crime" in which they sought to examine criminal conduct in the more general context of deviant behavior that they regarded as simply one form of behavior, not a distinct category (Gottfredson & Hirschi, 1990). From this perspective, crime and delinquency are viewed as routine behaviors that are poorly planned, not very lucrative, and largely localized geographically. In general, these authors viewed crime as a result of low self-control that results in a desire for immediate gratification. Furthermore, they indicated that the degree of self-control one possesses is determined largely by child-rearing practices.

The general theory developed by Gottfredson and Hirschi (1990) has been criticized on several grounds but has provoked a good deal of empirical research. For example, Piquero, Gomez-Smith, and Langton (2004) used Gottfredson and Hirschi's notion of self-control to examine whether an individual perceives sanctions as fair or unfair and how perceptions of sanctions and low self-control influence the perceived anger that may result from being singled out for sanctioning. Piquero and colleagues also examined the relationship among self-control, perceptions of fairness, and anger. Their results suggest that individuals with low self-control are more likely to perceive sanctions as being unfair and that this combination leads to anger for being singled out for punishment.

Integrated Theories

Numerous attempts have been made to combine two or more preexisting theories in an attempt to provide more comprehensive explanations (**integrated theories**) of criminal and delinquent behavior. The resulting theories or conceptual schemes are far too numerous to discuss here, but we mention a few of the more prominent attempts. Developmental and life course theory (DLC) attempts to explain how antisocial behavior develops, how different risk factors exist at different stages of life, and the differential effects of life events on antisocial behavior (Farrington, 2003; Reid, 2006, pp. 195–198).

Interactional theory represents an attempt to combine social learning, social bonding, and social–structural theories (Thornberry, 1987). This theory holds that, like all other human social behavior, delinquency is the result of interactions among individuals and is the result of the learning and exchanges that occur in such interaction. Thus, understanding interaction among juveniles and their parents, siblings, peers, gang members, school personnel, and others is critical. Interaction with gang members, for example, may increase the level of delinquent behavior among new members, but those who leave the gang and interact with others who may be less criminally inclined become less likely to engage in behaviors encouraged by the gang.

Other integrated theories that you may wish to examine further include network analysis, control balance theory, and strain/control theories. As noted, integrated theories represent attempts to improve on our understanding of delinquent behavior and have inspired considerable research. Ultimately, those proposing such theories are searching for commonalities among existing theories that will form the background for a more comprehensive theory. This is important because, as we mentioned at the beginning of this chapter, all criminological theories have implications for criminal justice policy and practice (Akers & Sellers, 2004).

Career Opportunity— Criminal Justice Professor

Job Description: Teach courses in the field of criminal justice, criminology, and juvenile justice with respect to causes, consequences, extent, and nature of crime/delinquency, justice system responses, and problems. Conduct research related to these issues.

Employment Requirements: At least a master's degree, and preferably a Ph.D., in criminology, criminal justice, sociology, psychology, law, or a related field.

Beginning Salary: Between $35,000 and $50,000. Benefits include tenure (job security) for those who are successful as teachers, as researchers, and in publishing the results of their research. Other benefits include health and life insurance for faculty members and their families as well as solid retirement plans.

SUMMARY

We have provided a brief overview of some of the attempts to explain delinquency. It should be clear at this point that, using our definition of theory, few if any of these attempts have resulted in explanations that are scientifically sound. Many have been more or less discarded over time, and others continue to provide leads that need to be pursued. Bridging the gap between theory and practice is crucial to controlling delinquency and to improving the juvenile justice network. The input of practitioners is extremely useful in testing our theoretical statements. The benefits to be reaped, if and when a sound theoretical base is established, are considerable. We can no longer afford to ignore the importance of theory, nor can we continue to rely on commonsense notions of causation that are, as we have seen, very often inaccurate.

Unlike demonology, which has been largely discounted as an explanation of delinquency today, the classical school of criminology remains important as a basis of our current criminal and juvenile justice networks. Public opinion continues to indicate a belief that severe punishment will deter crime/delinquency, and legislatures around the country continue to pass "get tough" measures in the hope of meeting public expectations. As a result, there is pressure for more arrests, more convictions, and more severe punishment, none of which seem to have accomplished the desired goal, perhaps because of the lack of certainty and swiftness of punishment. Even capital punishment, which certainly deters the subject, has been shown to have little effect on others, and the procedures currently employed are fraught with difficulties that have led to moratoria in some states.

Biological theories of causation raise some important issues. Although biological factors do not appear to be a direct cause of delinquency, we must remain constantly alert to the possibility that physiological malfunctions or abnormalities may be important in assessing juveniles' behaviors. For example, a juvenile who has become increasingly aggressive, irrational, and uncooperative with others could conceivably be suffering from brain damage (e.g., tumor, lesion) that causes these symptoms. In cases where physical ailments or the use of intoxicants might be related to delinquency, it is obviously best to provide for appropriate medical intervention.

There is always, of course, the possibility that some emotional or psychological difficulty may be present in a specific delinquent. The evidence in support of personality disturbances as causes of delinquency is ambiguous at best, due in part to measurement and definitional difficulties. Nevertheless, the psychological approach to explaining delinquency remains important because psychotherapy of some type—individual or group therapy or counseling—is often prescribed as treatment within correctional facilities. Whether or not such treatment is likely to help remains an empirical question, but some successes are reported.

The sociological school views delinquency as a result of social interaction, learned in much the same way as nondelinquent behavior. According to this approach, much of the juvenile justice network makes sense, but some does not. For example, if labeling is an important factor in delinquency, attempts to keep juvenile proceedings confidential makes sense. However, it does not make sense, within this theoretical context, to house minor or first-time delinquents in large institutions with more serious delinquents from whom they are likely to learn additional delinquent behaviors. This may account for our failure to rehabilitate many delinquents in such settings. In addition, the sociological approach looks for causes of delinquency in society as well as in the individual. It may be that the only way to significantly reduce delinquency rates is to change some social policies such as those leading to educational and racial discrimination and unemployment. Finally, the sociological approach suggests methods of control and rehabilitation that do not require the death penalty, the practice of eugenics, or complete restructuring of the individual's personality. This approach suggests that positive reinforcement, administered in surroundings where the juvenile lives and by those with whom the juvenile regularly interacts, may provide more positive results than do many techniques currently employed. Although the sociological approach is not a panacea, it does provide a number of leads for future research and treatment that may prove to be beneficial provided that public and agency cooperation can be obtained.

Finally, the search for new and better theories of delinquency continues in attempts to combine the tenets of different theories into more comprehensive theories that do a better job of explaining delinquent behavior. Ultimately, the success or failure of policies and practices in the field of juvenile delinquency is determined by the accuracy of explanations for its existence and persistence.

Note: Please see the Companion Study Site for Internet exercises and Web resources. Go to www.sagepub.com/juvenilejustice6study.

Critical Thinking Questions

1. What is a scientific theory, and why is the development of such theories crucial to our understanding and control of delinquency?
2. What are the strengths and weaknesses of our current juvenile justice network in terms of the learning theory and labeling theory approaches? Discuss some of the reasons why the classical approach to the control of delinquency has been, and continues to be, ineffective. Why do you think the approach has remained popular in spite of its ineffectiveness? What contemporary theories are extensions of the classical approach?
3. What are the major strengths and weaknesses of the psychological approach to understanding and controlling delinquency? What has been Freud's impact on the treatment of delinquency?
4. Is there evidence in support of the biological school of delinquency causation? Discuss some of the attempts to demonstrate a relationship between biology and delinquent behavior.
5. What is your overall assessment of the sociological approach to understanding and controlling delinquency? Which of the various attempts in this school do you think do the best job of explaining delinquency? The worst job?

Suggested Readings

Baron, S. W. (2004). General strain, street youth, and crime: A test of Agnew's revised theory. *Criminology, 42,* 457–484.

Cornish, D., & Clarke, R. (Eds.). (1986). *The reasoning criminal: Rational choice perspectives on offending.* New York: Springer-Verlag.

Cote, S. (Ed.). (2002). *Criminological theories: Bridging the past to the future.* Thousand Oaks, CA: Sage.

Farrington, D. P. (2003). Developmental and life-course criminology: Key theoretical and empirical issues—The 2002 Sutherland Address. *Criminology, 41,* 221–255.

Greenberg, D. F. (1999). The weak strength of social control theory. *Crime & Delinquency, 45,* 66–81.

Lynam, D. R. (1998). Early identification of the fledgling psychopath: Locating the psychopathic child in the current nomenclature. *Journal of Abnormal Psychology, 107,* 566–575.

Naffine, N. (1996). *Feminism and criminology.* Cambridge, MA: Polity Press

Paul, R. H., Marx, B. P., & Orsillo, S. M. (1999). Acceptance-based psychotherapy in the treatment of an adjudicated exhibitionist: A case example. *Behavior Therapy, 30,* 149–162.

Plass, P. S., & Carmody, D. C. (2005). Routine activities of delinquent and non-delinquent victims of violent crime. *American Journal of Criminal Justice, 29,* 235–246.

Rafter, N. (2004). Earnest Hooton and the biological tradition in American criminology. *Criminology, 42,* 735–772.

Walsh, A. (2000). Behavior genetics and anomie/strain theory. *Criminology, 38,* 1075–1107.

Wright, J. P., & Beaver, K. M. (2005). Do parents matter in creating self-control in their children? A genetically informed test of Gottfredson and Hirschi's theory of low self-control. *Criminology, 43,* 1169–1202.

PURPOSE AND SCOPE OF JUVENILE COURT ACTS

CHAPTER LEARNING OBJECTIVES

On completion of this chapter, students should be able to

- *Discuss the purpose and scope of juvenile court acts*
- *Compare and contrast adult and juvenile justice systems*
- *Understand the concepts of stigmatization, jurisdiction, waiver, and double jeopardy*
- *Discuss the constitutional rights of juveniles in court proceedings*
- *List and discuss the various categories of juveniles covered by juvenile court acts*

KEY TERMS

Purpose statement of a juvenile court act

Scope of a juvenile court act

Delinquency, neglect, abuse, and dependency cases

Adjudicatory hearing

Dispositional hearing

Petition

Uniform Juvenile Court Act

Due process

Beyond a reasonable doubt

In need of supervision

Unruly children

Status offenses

Standard of preponderance of evidence

Exclusive jurisdiction

Concurrent jurisdiction

Automatic waiver

Discretionary waiver

Double jeopardy

Since the inception of the juvenile court in 1899, some critics have argued that the court ought to be abandoned. Some believe that the court is now far removed from the original concepts on which it was based or too limited in scope to be viable today. Others believe that it is currently incapable of meeting the purposes for which it was created. In this chapter, we review the purpose and scope of a variety of juvenile court acts in terms of both constitutional requirements and legislative differences among several states.

Every juvenile court act contains sections that discuss purpose and scope. The **purpose statement of a juvenile court act** spells out the intent or basic philosophy of the act. The **scope of a juvenile court act** is indicated by sections dealing with definitions, age, jurisdiction, and waiver. In this chapter, we discuss and refer to the Uniform Juvenile Court Act (National Conference of Commissioners on Uniform State Laws, 1968), which was developed in an attempt to encourage uniformity of purpose, scope, and procedures in the juvenile justice network. (A copy of the act is included in Appendix A.) For purposes of comparison, sections of various state juvenile codes are presented and analyzed. Revisions of most states' juvenile court acts are now in accord with the recommendations of the Uniform Juvenile Court Act.

As stated previously, many of the examples we provide are taken from the Illinois Juvenile Court Act. We do this because Illinois has been regarded as a progressive juvenile justice state and because, as mentioned previously, it is difficult for us to know all changes in every state. States customize their juvenile court acts to their own needs, and revisions occur on a continuous basis. We have chosen to illustrate our points using the law with which we are most familiar. Those using the text in other states are encouraged to provide comparable examples from their juvenile court acts. You can use the Internet to find state-specific juvenile court acts.

Purpose

As indicated previously, the first juvenile court act in the United States was passed in Illinois in 1899. By 1945, all U.S. states had juvenile court acts within their statutory enactments or constitutions (Tappan, 1949). Juvenile court acts typically authorize the creation of a juvenile court with the legal power to hear designated kinds of cases such as **delinquency, neglect, abuse, and dependency cases** as well as other special cases numerated in the acts.

Typically, a juvenile court act establishes both procedural and substantive law relative to juveniles within the court's jurisdiction. Historically, the law was administered in a general atmosphere of rehabilitation and parental concern rather than with punitive overtones. However, recent trends at the federal and state levels appear to have somewhat deemphasized traditional rehabilitative philosophy in favor of a more punitive approach. This change is largely in response to more serious, and often violent, youth crime and a desire to hold those committing such acts accountable for their actions.

Juvenile court proceedings were originally conceptualized as civil, not criminal, proceedings (Davis, 2001, sec. 1.3). As a result of reformers' interests in divorcing the juvenile court from the criminal court in 1899, a separate nomenclature was developed based on the philosophy underlying juvenile courts as opposed to criminal courts. This nomenclature is still followed today in spite of the "get tough" approach that has been suggested for serious delinquents.

Comparison of Adult Criminal Justice and Juvenile Justice Systems

As Table 5.1 indicates, we find a petition alleging that the respondent may have committed a delinquent act instead of a complaint charging a defendant with a crime. We find an **adjudicatory hearing** and a **dispositional hearing** instead of a criminal trial and a sentencing hearing, respectively. The entire proceeding is initiated by a **petition** in the interests of the juvenile rather than by an indictment against him or her. The juvenile, therefore, may not be found guilty in juvenile court but may be adjudicated as delinquent. Juvenile court acts are predicated of the basic assumption that all personnel involved in the juvenile justice system act in the best interests of the juvenile. There are, however, differences of opinion concerning how to best ensure the interests of the juvenile. In August 1968, in the hope of bringing some uniformity to legal definitions of delinquency and delinquency proceedings, the National Conference of Commissioners on Uniform State Laws drafted and recommended for enactment in all states a **Uniform Juvenile Court Act.** This act was approved by the American Bar Association at its annual meeting during the same year. Since that time, the Uniform Juvenile Court Act has served as a model for states to follow in developing their own acts.

In essence, Section 1 of the Uniform Juvenile Court Act reaffirms the basic philosophy of all juvenile court acts by stating specifically that the major purpose of the act is "to provide for the care, protection, and wholesome moral, mental, and physical development of children coming within its provisions" (National Conference of Commissioners on Uniform State Laws, 1968, sec. 1). This basic philosophy was first stated in the Cook County Juvenile Court Act of 1899 and has been stated in each state's juvenile court act adopted or revised since then. The philosophy has been controversial because of the questionable ability of the juvenile

Table 5.1 Comparison of Adult Criminal Justice and Juvenile Justice Systems

Adult	Juvenile
Arrest	Taking into custody
Preliminary hearing	Preliminary conference/detention hearing (both optional)
Grand jury/information/indictment	Petition
Arraignment	—
Criminal trial	Adjudicatory hearing
Sentencing hearing	Dispositional hearing
Sentence	Disposition (probation, incarceration, etc.)
Appeal	Appeal

justice network to provide the specified benefits to juveniles. Considerable documentation exists on the deficiencies of the state's ability to provide for the welfare of juveniles. A week seldom passes where a column in a newspaper or an article in a journal or magazine does not relate an instance of neglect by the state in its parental role (see In Practice 5.1). Although the philosophy of providing "care, guidance, and protection" is entrenched in the juvenile justice network, it would appear that a reevaluation of the state's effectiveness in adhering to that philosophy is in order.

Other basic themes expressed in the Uniform Juvenile Court Act include protecting juveniles who commit delinquent acts from the taint of criminality and punishment and substituting treatment, training, and rehabilitation; keeping juveniles within their families whenever possible and separating juveniles from their parents only when necessary for their welfare or in the interests of public safety; and providing a simple judicial procedure for executing and enforcing the act yet that ensures a fair hearing with constitutional and other legal rights recognized and enforced. We now discuss each of these philosophical themes.

Protecting the Juvenile From Stigmatization

For a long time, some states allowed a wide variety of activities to be labeled as delinquent. However, a majority of states have revised their juvenile codes and changed their legal definitions of acts considered delinquent. At issue is the difference between unthinking mischievous misbehavior of a nonserious nature and vicious intentional conduct that endangers life and property. It is difficult to ascertain exactly when mischievous behavior ends and vicious conduct begins because it is often left to individual perception of each. As a result, we sometimes encounter cases where hardcore delinquents have benefited from the treatment/rehabilitation philosophy of the juvenile court to the point where any concept of justice or accountability has been eliminated. The same is true for those who abuse or neglect their children to an extent that raises concern but where it is difficult to determine whether the legal standards required for abuse or neglect have been satisfied. Similarly, we sometimes note that mischievous juveniles are treated as hardcore delinquents. Clearly, rehabilitation and treatment might be helpful to both mischievous offenders and hardcore delinquents as well as to children who are abused or neglected. For mischievous offenders, a variety of rehabilitative/treatment programs have been developed as alternatives to punishment and are more or less effective in community-based agencies.

For delinquents who commit serious offenses, rehabilitative/treatment programs have typically been located in institutions. Some serious juvenile offenders learn how to "play the game" and are able to shift all responsibility for their actions to others or to society and, therefore, escape accountability under the rehabilitative/treatment philosophy. Such offenders and/or their attorneys are able to persuade juvenile court judges, prosecutors, and the police to "give them a break." Others, who are less skilled at "playing the game" or unable to retain private counsel, may be unable to escape more serious consequences for acts that may be less serious. The dilemma facing reformers of the juvenile court revolves around the obvious: Avoid labeling juveniles who do not deserve the label of delinquent and, at the same time, prevent the juvenile court from becoming so informal that those who are a threat to the community remain at large. This is a much bigger challenge than it might initially appear to be.

In Practice 5.1

Florida Justice System Harsher to Underage Girls, Study Finds

Florida's juvenile justice system locks up a higher percentage of underage girls than 46 other states, hands out stiffer punishment to girls than boys, and doesn't provide the kind of treatment girls need, according to a study released Tuesday.

On any given day, there are roughly 1,530 underage juvenile delinquent girls locked away in Florida, according to researchers with the National Council on Crime and Delinquency.

Two-thirds of them are in long-term residential programs, where they stay for months to get treatment while serving sentences for a variety of crimes. The rest are in detention centers.

The council study—the biggest ever of an all-girl juvenile offender population—found many of Florida's treatment programs inadequate.

"Girls' programs are often boys' programs painted pink," said Barry Krisberg, council president. "What we have found is that these programs, by and large, don't work."

That's because girl criminals are far different from boy criminals, he said.

The study found that Florida's girl offenders often have more emotional problems and, therefore, different treatment needs, he said.

Researchers interviewed 319 Florida girls in juvenile programs. They found:

- 49 percent were self-mutilators.
- 34 percent had attempted suicide.
- 35 percent were pregnant or had been [pregnant].
- 46 percent had an alcohol or substance abuse problem.

Those problems are at the root of many of the girls' crimes. In contrast, boys more frequently broke the law because of peer pressure or gang activity, Krisberg said.

A spokeswoman at the Florida Department of Juvenile Justice said officials had not seen the report and couldn't comment on the findings.

"At DJJ, we're committed to serving the unique needs of girls in our care," spokeswoman Tara Collins said.

The report concluded that Florida locks up too many girls—172 out of every 100,000 girls between the ages of 10 and 18 during 2003—when it should, instead, place some in home- or community-based programs, Krisberg said.

According to the council's report, the number of Florida girls being arrested and locked up is on the rise, just the opposite of the national trend for underage boys.

The study also found that Florida punishes girl offenders more harshly than boys, according to researcher Angela Wolf.

She said a girl is locked up longer and in a more restrictive facility than a boy even if the girl commits the same offense, the crime involves the same level of violence, and [the girl] has the same number of prior arrests.

(Continued)

(Continued)

One reason may be that there are fewer treatment alternatives for girls and judges are simply picking from a short but restrictive list of options, Wolf said.

The council, based in Oakland, Calif., and a second group, Children's Campaign Inc., a Tallahassee child advocacy group, jointly announced plans Tuesday to push the Florida Legislature to allocate more money for treatment of girl juvenile offenders.

"The standards for care of girls around the country are very poor," Krisberg said.

Using grants from the Jessie Ball duPont Fund and the Florida Bar Foundation, the two groups said they intend to come up with a plan to overhaul the way Florida rehabilitates its underage girl offenders.

That is not something Florida officials have been willing to do on their own, said Roy Miller, president of Children's Campaign.

SOURCE: Stutzman, Renee. (2006). Florida justice system harsher to underage girls, study finds. *The Orlando Sentinel,* Item: 2W62W62732998870.

Maintaining the Family Unit

The concept that a child should remain in the family unit whenever possible is another basic element of the Uniform Juvenile Court Act. The child and family are not to be separated unless there is a serious threat to the welfare of the child or society. However, once there is an established necessity for removing the child, the juvenile court must have the power to move swiftly in that direction. Determining exactly when it is necessary to remove the child is not, of course, an easy task. Careful investigation of the total family environment and its effect on the juvenile is typically required in cases of suspected abuse, neglect, and delinquency. Removal may be permanent or may include an option to return the child if circumstances improve. Careful consideration is given to the family's attitudes toward the child and the past record of relationships among other family members.

Although most of us would agree that it is generally desirable to maintain the family unit, there are certainly circumstances when removal is in the best interests of both the minor and society. The welfare of the child is clearly jeopardized by keeping him or her in a family where gross neglect, abuse, or acts of criminality occur. The emphasis placed on maintaining the integrity of the family unit at times seems to be taken so seriously by juvenile court judges and other juvenile justice practitioners that they maintain family ties even when removal is clearly the better alternative (see In Practice 5.2).

Preserving Constitutional Rights in Juvenile Court Proceedings

The Uniform Juvenile Court Act provides judicial procedures so that all parties are assured of fairness and recognition of legal rights. The early philosophy of informal hearings void of legal procedures and evidentiary standards has a limited place in the modern juvenile justice

In Practice 5.2

New Law Targets Chronic Neglect: Measure May Save Children's Lives

Washington's child welfare system was well acquainted with the troubles plaguing the Hanley family.

Over the course of 4-1/2 years, the state's Child Protective Services received a dozen referrals expressing concern about the welfare of the three children of Mathew T. and Barbara Hanley, who lived in Deer Park.

The family's file contained numerous allegations of drug abuse as well as concerns about physical abuse and gun safety in the home. In two instances, the agency conclusively determined the Hanleys had neglected their children.

Yet state workers felt legally powerless.

In September 2005, state attorneys said they lacked the evidence to remove the children. A team of experts recommended the children stay in the home.

Three months later, authorities say, the Hanleys' 11-year-old son accidentally shot his younger brother in the head while they were playing with their father's guns. Bradley M. Hanley, 6, died the following day.

For decades, social workers, state attorneys, and child welfare experts have wrestled with how to handle children living in chronic neglect where more obvious signs of abuse—such as broken bones or bruises—may not be present.

But next year, a new state law will specifically allow social workers to intervene in cases that demonstrate "a pattern of neglect." It will also allow the state to take legal actions against parents such as the Hanleys who don't follow through with treatment plans.

"It has been a huge problem," said Rep. Mary Lou Dickerson (D-Seattle), who introduced the original bill on the topic. "Caseworkers up until now haven't had the legal authority to protect children who are suffering serious harm because of chronic neglect. This legislation gives them the tools to save children's lives."

In the past 10 years, reports of neglect to the Children's Administration have increased dramatically, topping more than 45,000 in 2004—the last year for which the numbers are available. State caseworkers review more cases of neglect than physical abuse, sexual abuse, and abandonment combined.

But even in cases with dozens of complaints about neglect, social workers have been cautious about using the state's most definitive action: removing children from the home.

Child welfare experts say state and national systems cling to the outdated notion that chronic neglect may not be as injurious to children as physical abuse.

"The system hasn't really changed for 30 or 40 years," said Roy Harrington, a former state administrator now with the Child and Family Research Institute at Washington State University in Spokane. "There hasn't been any overall focus on child neglect from any state. [The law] puts Washington as a leader in looking at this. That's a really sad statement."

(Continued)

(Continued)

A History of Neglect

While physical abuse is relatively easy to distinguish, neglect can fall under a slew of categories such as failure by a parent to provide enough food, clothing, shelter, or medical care.

"You've seen cases where there are 15, maybe 20 referrals before there has been a filing," said Steve Hassett, an assistant attorney general in Olympia who oversees dependency litigation. "You don't have a clear-cut bruise where a medical expert may establish it or a law enforcement officer can bring in a photo. But the pattern of behavior is putting this child at extreme risk."

Research has shown that gross neglect not only causes children to suffer emotionally and behaviorally but can actually restrict a young brain's growth and development. Its prevalence—neglect accounts for more than two-thirds of all referrals in Washington and Idaho—has also raised the question of whether child welfare systems should act more quickly and more often.

Washington's new threshold—which could substantially increase the number of cases accepted by the agency—will come at a cost. The Legislature has budgeted $10 million per year to handle the influx of new cases.

State social workers are struggling to manage the 9,600 children who need out-of-home placement on any given day. In the past three years, the number of licensed foster homes has gradually declined, reaching the lowest numbers since 1998.

State officials must now balance the protection of thousands of children living in neglectful homes with the risk of overwhelming an already fragile child welfare system.

"It just seems that in this state and others, there is an awful lot of tolerance for these lengthy histories—particularly if they involve neglect rather than abuse," said Dee Wilson, executive director of the University of Washington's Northwest Institute for Children and Families and a former state administrator. "However, what if the agency was to take some firm stance? You would have the potential of wiping out the child welfare system."

It's not clear how many new cases could be added to worker caseloads as a result of the law, in part because the Legislature failed to explicitly state how many referrals constitute chronic neglect.

In a fiscal note attached to the original House bill, staffers suggested that chronic neglect be defined as five referrals in three years, four referrals in two years, or three referrals in one year.

But absent that definition, which did not make it into the new law, social workers and attorneys for the state must make the decision to file a dependency petition locally, without the benefit of a statewide standard, and likely after evaluating the disposition of local courts and judges.

State leaders have not yet crafted a policy on chronic neglect but have started a massive training program for social workers on the new law, said Connie Lambert-Eckel, deputy regional administrator for the Department of Children and Family Services in Spokane.

"It's going to teach us about this population that has really exhausted a lot of our resources," Lambert-Eckel said. "The focus has shifted, I think very appropriately, to the development of a model of approach. How do we support neglected families?"

The failure to act in cases of chronic neglect has resulted in several high-profile tragedies.

In November 2004, police found two baby boys dead and their mother passed out in a Kent, Wash., apartment littered with hundreds of beer cans. Marie Robinson had been reported to state social workers 10 times, including six complaints in the two years before the deaths of Justice and Raiden Robinson.

The new legislation notes that evidence of a parent's substance abuse should be given "a great weight."

(Continued)

(Continued)

The Hanley Case

At the time of his death, Bradley Hanley lived with his father "due to his mother's methamphetamine use," according to an internal child fatality review by the Children's Administration.

The report lists 12 referrals about the Hanleys that raised a host of concerns. Time and again, the parents simply avoided social workers and refused recommended treatment programs until the investigations were closed, according to the report.

On at least two occasions, law enforcement placed the children in protective custody, only to have them returned to the home.

When police officers responded to the family's home in March 2004, they found Barbara Hanley unable to account for the whereabouts of her children. Inside the "dirty disorderly" home, officers noted a puddle of urine on the floor, according to the documents.

Six months later, one of the boys told a school counselor that his mother "sniffs broken light bulbs, has pot in her purse, and has cocaine at her boyfriend's house."

For the next three months, neither parent responded to agency social workers. With no way to contact the parents, the agency said, it marked the referral as "inconclusive" and closed the investigation.

Had the new law been in place, state workers would have had the authority to remove the children based on the Hanleys' failure to follow through on treatment.

In September 2005, a Child Protective Team—typically composed of experts both within and outside the agency—recommended the children be kept in the Hanley home. The Attorney General's office warned there "was not sufficient information" to pursue legal action to remove the children from the home, according to notes from the meeting.

The shooting occurred three months later.

Spokane County detectives determined that Mathew Hanley left his sons alone in the house while he was at work and left a set of keys to the gun safe.

Neither Mathew Hanley nor his wife could be reached by the *Spokesman Review.*

Detectives believe the boys were playing with the guns when a .22-caliber rifle went off. Bradley Hanley was shot in the head. His older brother called 911, attempted CPR on Bradley, and put the firearms back in the gun safe before deputies arrived. Bradley died on Dec. 7.

The Spokane County Prosecutor's Office decided not to press charges.

CPS removed the surviving Hanley children from the home after the shooting. Today, they remain in state care.

Source: Shors, Benjamin. (2006). New Law Targets Chronic Neglect: Measure May Save Children's Lives. *The Spokesman-Review,* Item: 2W62W62988061877.

network. The application of due process standards has not deterred the court from its rehabilitative pursuits. If the issue is delinquency and the act for which the child has been accused is theft, the procedural rules of evidence should support the allegation, and the result would be an adjudication of delinquency. If the evidence does not support the allegation, no adjudication of delinquency should occur. In an informal hearing where there is an absence of established guilt and where an adjudication of delinquency is based on the attitude of the child, the types of peers with whom that child associates, or his or her family's condition, the rights of the juvenile and perhaps other parties have been violated. The philosophy of a fair hearing, where constitutional rights are recognized and enforced and where a high standard of proof for establishing delinquency is strictly enforced, has been generally established in juvenile court acts since 1967, when the U.S. Supreme Court decided in the *Gault* case (*In re Gault*, 387 U.S. 1, 78–81 [1967]) that **due process** (observing constitutional guarantees and rules of exclusion) was generally required in juvenile court adjudicatory proceedings. Informality is generally accepted in postadjudicatory hearings on disposition of the juvenile and is often permitted in prehearing stages. The adjudicatory hearing for delinquency must, however, be based on establishing **beyond a reasonable doubt** (with as little doubt as possible) that the allegations are supported by the admissible evidence.

The general purpose of juvenile court acts, then, is to ensure the welfare of juveniles while protecting their constitutional rights in such a way that removal from the family unit is accomplished only for a reasonable cause and in the best interests of the juvenile and society. A review of your state's juvenile court act should reflect these basic goals.

Scope

In addition to the basic themes discussed previously, all juvenile court acts define the ages and subject matter (conduct) within the scope of the court.

Age

Section 2 of the Uniform Juvenile Court Act defines a child as a person who is under the age of 18 years; who is under the age of 21 years but who committed an act of delinquency before reaching the age of 18 years; or who is under the age of 21 years and committed an act of delinquency after becoming 18 years of age but who is transferred to the juvenile court by another court having jurisdiction over him or her (National Conference of Commissioners on Uniform State Laws, 1968, sec. 2).

As stated in Chapter 2, both upper and lower age limits vary among the states (see your state's code). The Uniform Juvenile Court Act establishes the age of 18 as the legal age at which actions of an illegal nature will be considered criminal and the wrongdoer will be considered accountable and responsible as an adult. Prior to the 18th birthday, illegal activities will be considered acts of delinquency, with the wrongdoer processed by the juvenile court in a way that removes the taint of criminality and punishment and substitutes treatment, training, and rehabilitation in its place. The Uniform Juvenile Court Act allows two exceptions regarding the legal jurisdictional age of 18 years. Section 2(1)(iii) states that a person under the age of 21 years who commits an act of delinquency after becoming 18 years of age can be transferred to the juvenile court by another court having jurisdiction and, therefore, would be accorded all of the protection and procedural guidelines of the juvenile court. Section 34 allows for a transfer to other courts of a child under 18 years of age if serious acts of delinquency are alleged and the child was 16 years of age or older at the time of the alleged conduct (National Conference of Commissioners on Uniform State Laws, 1968, sec. 34). There are stringent guidelines to follow before a waiver to adult court jurisdiction may be permitted. Waivers of juvenile jurisdiction are occurring more and more frequently and are discussed later in this chapter.

In establishing the age of 18 years as the legal break point between childhood and adulthood, two thirds of the states are consistent with the Uniform Juvenile Court Act, as noted in Table 5.2.

Table 5.2 Oldest Age for Original Juvenile Court Jurisdiction in Delinquency Matters

Age (years)	States
15	Connecticut, New York, North Carolina
16	Georgia, Illinois, Louisiana, Massachusetts, Michigan, Missouri, New Hampshire, South Carolina, Texas, Wisconsin
17	Alabama, Alaska, Arizona, Arkansas, California, Colorado, Delaware, District of Columbia, Florida, Hawaii, Idaho, Indiana, Iowa, Kansas, Kentucky, Maine, Maryland, Minnesota, Mississippi, Montana, Nebraska, Nevada, New Jersey, New Mexico, North Dakota, Ohio, Oklahoma, Oregon, Pennsylvania, Rhode Island, South Dakota, Tennessee, Utah, Vermont, Virginia, Washington, West Virginia, Wyoming

Source: Office of Juvenile Justice and Delinquency Prevention. (2003). *State statutes define who is under juvenile court jurisdiction* (Juveniles in Court, OJJDP National Report Series Bulletin). Washington, DC: U.S. Department of Justice.

States may also establish higher age limits in cases of status offenders and abuse, neglect, and dependency—typically through the age of 20 years. In addition, courts may retain jurisdiction after the age of adulthood if the child is serving a disposition in juvenile court. A total of 35 states allow juvenile court to maintain jurisdiction until the child's 21st birthday in cases where the child is under juvenile court supervision for delinquency at the time of the 18th birthday (Office of Juvenile Justice and Delinquency Prevention [OJJDP], 2003).

As we have indicated elsewhere, there is no clearly established minimum age set by juvenile courts with respect to their jurisdiction, although 16 states have attempted to identify a limit. In Table 5.3, we see that children as young as 6 years of age are allowed into the juvenile justice system in North Carolina.

Table 5.3 Youngest Age for Original Juvenile Court Jurisdiction in Delinquency Matters

Age (years)	States
6	North Carolina
7	Maryland, Massachusetts, New York
8	Arizona
10	Arkansas, Colorado, Kansas, Louisiana, Minnesota, Mississippi, Pennsylvania, South Dakota, Texas, Vermont, Wisconsin

Source: Office of Juvenile Justice and Delinquency Prevention. (2003). *State statutes define who is under juvenile court jurisdiction* (Juveniles in Court, OJJDP National Report Series Bulletin). Washington, DC: U.S. Department of Justice.

Other states rely on case law or common law in determining the lower age limit. They presume that children under a certain age cannot form *mens rea* and are exempt from prosecution and sentencing (OJJDP, 2003). Just as there have been few clear guidelines for processing youth in matters of delinquency, there have been vague guidelines for determining the age youth may be found to be abused or neglected.

Delinquent Acts

The Uniform Juvenile Court Act clearly limits the definition of delinquency by stating in essence that a delinquent act is an act designated as a crime by local ordinance, state law, or federal law. Excluded from acts constituting delinquency are vague activities, such as incorrigibility, ungovernability, habitually disobedient, and other status offenses, which are legal offenses applicable only to children and not to adults. At the time when the Uniform Juvenile Court Act was drafted in 1968, many states legally defined delinquency as encompassing a broad spectrum of behavior. The proposal by the drafters of the Uniform Juvenile Court Act excluded the broader definition of activities labeled as delinquent and focused only on violations of laws that are applicable to both adults and minors. This narrow interpretation was consistent with the legalistic trend occurring during the latter 1960s. By narrowing the legal definition of delinquency, the Uniform Juvenile Court Act did not ignore other types of activities that fall within the court's jurisdiction but placed these activities outside the realm of delinquent acts. A minor who is "beyond the control of his parents," "habitually truant from school," or "habitually disobedient, uncontrolled, wayward, incorrigible, indecent, or deports himself or herself as to injure or endanger the morals or health of themself [*sic*] or others" was at one time considered to be delinquent in some states (*Indiana Code Annotated,* 31-37-1-1 to 31-37-2-6, 1997). The number of states with such a broad definition of delinquency is decreasing. A major difficulty with including these vague activities within the delinquent behavior category concerns the issue of who defines what is incorrigible, indecent, or habitual misconduct and the nature of the standard used to determine this behavior. These statutory expressions and a number of others like them have invited challenge on the grounds that they are unconstitutionally vague. There are no standardized definitions for habitual, wayward, incorrigible, and so on. As a result, such charges in conjunction with delinquency will inevitably be challenged in the courts.

It is interesting to note that prior to the development of the Uniform Juvenile Court Act in 1968, several states had already started restricting the definition of delinquency to include only those activities that would be punishable as crimes if committed by adults. For example, in New York under the pre-1962 Children's Court Act, the term *juvenile delinquency* included ungovernability and incorrigibility. However, in 1962, the Joint Legislative Committee on Court Reorganization, which drafted the Family Court Act (*New York Sessions Laws,* vol. 2, 3428, 3434, McKinney, 1962), developed the concept of a person **in need of supervision** to cover noncriminal status offenses and the term *juvenile delinquent* was narrowed to include only persons over 7 and under 16 years of age who commit any act that, if committed by an adult, would constitute a crime. With a more specific definition of delinquency, it was inevitable that due process procedures, rules of evidence, and constitutional rights would emerge as important issues in Supreme Court decisions involving the rights of juveniles in delinquency proceedings. As the states moved toward a more specific definition of delinquency, additional appellate decisions were rendered regarding "due process and fair treatment." The effect of this narrow interpretation of delinquency has been the advent of an adjudicatory process that is more formalized and that ensures and protects the juvenile's procedural and constitutional rights. This trend is clearly consistent with the spirit behind the creation of the Uniform Juvenile Court Act. Some states even list all forms of conduct subject to juvenile court jurisdiction in one general category (*Michigan Compiled Laws Annotated,* 712A.2 Supp., 1999; *Nebraska Revised Statutes,* 43-247, 1998; *Utah Code Annotated,* 78-3a-104 (1) amended, 2006).

Section 2(3) of the Uniform Juvenile Court Act indicates that an adjudicated delinquent is in need of "treatment or rehabilitation." The development of narrower definitions of delinquency and more formalized "due process models" is not intended to cause the juvenile court to abandon rehabilitation and treatment. This philosophy was stated as early as 1909, when it was pointed out that "the goal of the juvenile court is not so much to crush but to develop, not to make the juvenile a criminal but a worthy citizen" (*Consolidated Laws of New York Annotated,* bk. 29A, art. 7, McKinney, 1975). This initial concept of rehabilitation and treatment has been affirmed in many decisions and is summarized briefly by the Supreme Court case *in re Gault,* where the Court reaffirms the original juvenile court philosophy that "the child is to be 'treated' and 'rehabilitated' and the procedures, from apprehension through institutionalization, are to be 'clinical' rather than 'punitive'" (Faust & Brantingham, 1974, pp. 369–370). It is important to remember that although the juvenile court operates under the "treatment and rehabilitation" concept, the court is also charged with protecting the community against unlawful and violent conduct. To fulfill this obligation, the court may resort to incarceration or imprisonment. This clash between the rehabilitative ideal and the clear, present necessity to protect the community in certain situations has been described as the "schizophrenic nature" of the juvenile court process (*Consolidated Laws of New York Annotated,* bk. 29A, art. 7, McKinney, 1975).

It is clear that a majority of the states have moved toward a narrower definition of delinquency. Inherent in this trend is the movement toward formalizing the legal procedures and processes accorded to the accused delinquent. The importance of this trend is twofold. First, legal definitions of delinquency have become more standardized and by law require a violation or attempted violation of the criminal code. Second, the process of proving the allegation of delinquency may include only the same types of evidentiary materials that would be admitted if the same charges were leveled against an adult. This is a considerable change from past practices in many juvenile courts, where much of the evidentiary material that was introduced to prove an act of delinquency was basically irrelevant material concerning the juvenile's family, peers, school behavior, and other information about his or her environment. The establishment of reasonable proof that the juvenile did violate the law was lost in the process. The case was often weighed and decided on factors other than establishing, beyond a reasonable doubt, that the juvenile committed the act of which he or she had been accused. The juvenile court is a court of law. The juvenile adjudicatory process and the juvenile court must be totally dedicated to working within a legal framework that is conducive to reaching the truth and serving the ends of justice. To do otherwise would result in what is best described in an often-quoted passage of the *Kent* decision where the U.S. Supreme Court stated, "There is evidence . . . that the child receives the worst of both worlds; that he gets neither the protections accorded to adults nor the solicitous care and regenerative treatment postulated for children" (Justice Fortas in *Kent v. United States,* 383 U.S. 541, 546, 1966).

Without a doubt, there is a place in the juvenile justice network for consideration of the adjudicated delinquent's family and his or her environment. However, such consideration should be given only after an adjudication of delinquency rather than used as the basis for adjudication. For instance, suppose that as an adult you have been accused of "breaking and entering" and that throughout the pretrial process and during the course of the trial nearly all of the evidence and information introduced centers on your family, your associations, your attitude, and your overall environment. Furthermore, only a minimum amount of court time and effort is devoted to establishing beyond a reasonable doubt that you did in fact violate the law by breaking and entering, and even then most of this evidence is hearsay, not subjected to cross-examination, and based on belief rather than proof. Yet you are convicted. Such cases were fairly common in the juvenile justice system until the *Gault* decision in 1967. The focus on due process to protect the accused juvenile's constitutional rights is as important as determining whether the act was committed by the accused. The legal issue of delinquency must be determined not on the basis of a social investigation describing the minor's environment but rather on the basis of whether the evidence supports or denies the allegation of delinquent acts.

Unruly Children

Section 2(4) of the Uniform Juvenile Court Act defines an unruly child as a child who

(i) while subject to compulsory school attendance is habitually and without justification truant from school;
(ii) is habitually disobedient of the reasonable and lawful commands of his parent, guardian, or other custodian and is ungovernable; or
(iii) has committed an offense applicable only to a child; and
(iv) in any of the foregoing is in need of treatment or rehabilitation.

At one time, a majority of states included these activities in the delinquent behavior category, and that often resulted in the official label of delinquent and led to the possibility of being incarcerated in a juvenile correctional institution for treatment and rehabilitation. The Uniform Juvenile Court Act recognizes that such activities may require the aid and services provided by the juvenile court but also recognizes that these minors should not be included in the delinquent category. According to Section 32 of the Uniform Juvenile Court Act, **unruly children** cannot be placed in a correctional institution unless the court finds, after a further hearing, that they are not amenable to treatment or rehabilitation under a previous noncorrectional disposition.

The unruly child is generally characterized by activities that are noncriminal or minor violations of law. Types of offenses, such as curfew violations and running away from home, are referred to as **status offenses** (acts that are offenses only because of the age of the offender). If the same acts were committed by an adult, they would not be violations of law. A substantial number of states have separated the types of activities described as unruly by the Uniform Juvenile Court Act from delinquency and have placed them in the nondelinquent category of in need of supervision (*District of Columbia Code,* 16–2301 (8), 1997; *New York Family Court Act,* 712(a) McKinney, 1999; *South Dakota Codified Laws Annotated,* 26-8B-02, 1999).

Regardless of the title, the importance of the development of this category lies in separating the delinquent from the nonserious violator and in realizing that the behavioral activities included in the unruly child and child in need of supervision/assistance categories are often symptomatic of problems in the juvenile's home life and environment and might not indicate criminal tendencies. The unruly child category allows the juvenile court to be involved with the minor who needs supervision and allows the court flexibility and options short of the label of delinquent. Still, the labels of unruly child and in need of supervision may become terms of disrepute and produce a stigmatizing effect on the juvenile similar to the label of delinquent. As a result, one of the major benefits of the distinction is lost.

To further distinguish the differences between the delinquent and the unruly child, most states have developed different procedural requirements. These requirements allow the civil **standard of preponderance of evidence** in the adjudicatory hearing for the latter, where the bulk of the evidence, but not necessarily all of it, must support the charges. They also provide for different dispositional options and for different upper ages for the unruly and in need of supervision categories. Again, reviewing your state's juvenile code will provide information on how these issues are addressed in your jurisdiction.

In distinguishing between juveniles whose misconduct is criminal and those whose misconduct is not criminal, it is assumed that the unruly child's behavior may be of a pre-delinquent nature and that early remedial treatment might prevent the incipient delinquency. However, it may be that the unruly child has more intense emotional and behavioral problems than do some delinquents who commit a single criminal act or a series of minor criminal acts.

The unruly child and in need of supervision categories are generally written without specificity because it is difficult to define and describe all of the noncriminal (delinquent) conduct that could ultimately fall within these categories. The term *habitually* is frequently used to distinguish between isolated incidents and a recurring pattern of incorrigibility, ungovernability, or disobedience. The flagrant repetitive nature of these behaviors often serves as the basis for filing a petition.

It was noted earlier that in some instances the behavior engaged in by the juvenile and alleged in a petition (often filed by the parents) may actually reflect neglect rather than an unruly child. A lack of parental supervision, whether due to unwillingness or inability of the parents, may have created a situation within the family that resulted in the juvenile's behavior. This behavior, although alleged to be unruly in the petition, may have been precipitated by a family crisis resulting in the minor rebelling against the family.

Deprived, Neglected, or Dependent Children

In Section 2(5) of the Uniform Juvenile Court Act, a "deprived" child is defined as a child under the age of 18 years who

(i) is without proper parental care or control, subsistence, education as required by law, or other care or control necessary for his physical, mental, or emotional health, or morals, and the deprivation is not due primarily to the lack of financial means of his parents, guardian, or other custodian;
(ii) has been placed for care or adoption in violation of law; or
(iii) has been abandoned by his parents, guardian, or other custodian; or
(iv) is without a parent, guardian, or legal custodian.

A number of jurisdictions use a single classification to describe a child who is without a parent, who has been abandoned or abused, or who is without adequate parental care or supervision. Such a child is variously referred to as a "dependent child" (*Illinois Compiled Statutes* [*ILCS*], ch. 705, art. 2, sec. 405/2–4, 1999), a "deprived child" (*Georgia Code Annotated,* 15-11-2 (8), 1999), or a "neglected child" (*District of Columbia Code,* 16-2301 (9), 1997; *Wyoming Statutes Annotated,* 14-3-402(a)(xii) Michie, 1999).

Some states separate deprived children into several categories with specific labels. For example, the neglected minor category includes any minor who is neglected as to proper or necessary support, education as required by law, medical or other remedial care recognized under state law, or other care necessary for his or her well-being, including food, clothing, and shelter; who is abandoned by his or her parents, guardian, or custodian; or whose environment is injurious to his or her welfare and also includes a newborn infant whose blood or urine contains any amount of a controlled substance not as a result of medical treatment (*ILCS,* ch. 705, art. 2, sec. 405/2-3-1, 1999).

Within the neglect language of some codes is a special section on abused children who are minors under a given age whose parent or immediate family member, custodian, or any person living in the same family or household, or a paramour of the minor's parent, (a) inflicts, causes to be inflicted, or allows to be inflicted physical injury that causes death, disfigurement, impairment of physical or emotional health, or loss or impairment of any bodily function; (b) creates a substantial risk of physical injury; (c) commits or allows to be committed any sex offense; (d) commits or allows to be committed acts of torture; or (e) inflicts excessive corporal punishment (*ILCS,* ch. 705, art. 2, sec. 405/2-(i)-(v), 1999).

Frequently, juvenile court acts have a special "dependent child" provision for children under a specified age who have no living parent, have been abandoned, or lack adequate parental care or supervision (*Alabama Code,* 12-15-1 (10), 1995; *District of Columbia Code,* 16–2301, 1997; *Georgia Code Annotated,* 15-11-2 (8) Supp., 1999; *North Dakota Century Code,* 27-20-02 (8) Supp., 1999; *Wyoming Statutes Annotated,* 14-3-402(a)(xii) Michie, 1999). Illinois, for example, has a separate dependent minors category that generally includes any juvenile below a given age who (a) is "without a parent, guardian, or legal custodian"; (b) is "without proper care because of the physical or mental disability of his or her parent, guardian, or custodian"; (c) is without proper medical or remedial care or other care necessary for his or her well-being through no fault, neglect, or lack of concern by his or her parents, guardian, or custodian; or (d) has "a parent, guardian, or legal custodian who with good cause wishes to be relieved of all residual parental rights and responsibilities, guardianship, or custody and who desires the appointment of a guardian of the person with power to consent to the adoption of the minor" (*ILCS,* ch. 705, art. 2, sec. 405/2–4, 1999).

Even though the Uniform Juvenile Court Act specifically disallows "a lack of financial means" as a basis for alleging that a minor is a "deprived child," some states, under circumstances where the deprivation is so extreme that it seriously endangers the well-being of the child, provide for handling these cases under the "neglected child" portion of the juvenile court act. Deprivation may be considered "gross neglect" if the amount of parental income is sufficient but is misappropriated and jeopardizes the well-being of the children within the family. Appropriate juvenile court remedies are generally available for this type of deprivation. According to Fox (1984), "Where a statutory distinction is made between a neglected child and one who is dependent, the difference generally is a matter of the presence of some parental fault in the former case and its absence in the latter" (p. 58). The Illinois distinction between the neglected child and the dependent child clearly illustrates the difference between the "parental fault concept" for the neglected child and the "no fault concept" for the dependent child. Regardless of the statutory definitions of deprived, neglected, abused, and dependent child, it is quite clear that the situations described in these statutes exist basically through no fault of the child.

Jurisdiction

The jurisdiction of a court concerns persons, behavior, and relationships over which the court may exercise authority. The word *jurisdiction* also may be used to describe geographical areas or to describe the process through which the juvenile court acquires authority to make orders concerning particular individuals. As Regoli and Hewitt (1994) pointed out, the question of jurisdiction is of basic importance to the juvenile court judge; without jurisdiction over the subject matter and the subject, that judge's court has no power to act. The term *jurisdiction* means "the legal power, right, or authority to hear and determine a cause or causes" (p. 390). Jurisdiction is created and defined in juvenile court acts.

There is a distinction between the juvenile court's inherent jurisdictional powers and its discretion to exercise jurisdiction over a case. For example, the statutory law creating the juvenile court in a state may give that court exclusive jurisdiction in any proceeding involving cases of delinquency, unruly children, dependency, or neglect provided that the respondent is within the age range and geographical area specified by the court. However, unless a petition is duly filed and the respondent receives a copy or summary of the petition as well as adequate notification of when and where the allegations against him or her will be presented and heard, the court has not exercised proper jurisdiction over the case.

In some states, the juvenile court acts have been repealed and broader family court acts have been created, allowing for broader jurisdictional powers over virtually all problems directly involving families (*New York Family Court Act,* 301.2 (1) McKinney, 1999; *Texas Family Code Annotated,* 51.02 West, 1996). Adoptions, divorces, proceedings concerning mentally retarded or mentally ill children, custody and support of children, paternity suits, and certain criminal offenses committed by one family member against another all are within the jurisdiction of some family court acts. It is important to note, however, that for the most part, those adults who abuse or neglect their children are subject to prosecution not in juvenile courts but rather in criminal courts. The children who are abused or neglected may nonetheless be removed from their homes and placed in shelter care or other living arrangements by the juvenile court judge and/or the state department of children and family services.

Age is obviously an important factor in determining jurisdiction in all states. As stated previously, age limits for delinquency vary among the states. The majority of juvenile court acts are silent on the lower age limits; however, in some states the common law age of 7 years has been established by statute as the lower age limit for delinquency. Statutes in 16 states define the minimum age for delinquency. In the remaining states, it is technically possible that a child could be adjudicated delinquent from birth. Such adjudication is unlikely given that the juvenile court requires a reasonable degree of capacity such as the ability to understand the act and to know or appreciate its consequences (*In re William A.,* 1988; *In re Register,* 1987).

The unruly child or child in need of supervision has been generally subjected to the same upper age limit for jurisdictional purposes as the delinquent. Because common law does not deal directly with this category, the common law age of 7 years has not traditionally been recognized as the minimum age for the unruly child.

Determining the upper and lower age limits in delinquency raises difficult questions about the factor of responsibility and accountability in law. For example, a 6-year-old who is fully aware of the wrongfulness of a criminal act and its consequences and still commits the act will be immune from prosecution if the jurisdictional age of 7 years is part of the state's juvenile court act. Another child who is less mature at 7 years may commit the same act while being unaware of its consequences and may need to face juvenile court. The question becomes whether either child is fully and completely responsible for his or her actions.

States differ about whether a juvenile who commits a delinquent act while within the age jurisdiction of the juvenile court, but who is not apprehended until he or she has passed the maximum age of jurisdiction, can be handled as a juvenile. Some states have determined through court decisions or previous statutory enactments that it is the age at the time of the offense, rather than the age at the time of apprehension, that determines jurisdiction. In the Uniform Juvenile Court Act, Section 2(1)(iii) allows a person under 21 years of age who commits an act of delinquency before reaching the age of 18 years to be considered a child and within the juvenile court's jurisdiction for delinquency proceedings.

States differentiate between the upper ages for delinquency and other categories; they believe that a minor might still need the care and protection of the family even though he or she is beyond the age for an adjudication of delinquency. Similarly, the deprived, neglected, abused, or dependent child is generally not subject to a lower age limit because a younger child may have a greater need for the protection of the juvenile court than does an older counterpart. Currently, some states have set one age for all categories included in the juvenile court act, whereas others have different ages for each category. For example, Texas defines a child as a person 10 years of age or over but under 17 years of age and over 17 but under 18 years of age for wardship petition and delinquency petition, respectively (*Texas Family Code Annotated,* 51.02 (2b) Title 3 Juvenile Justice Code, 2005). Illinois continues to follow different ages for delinquency (up to the 17th birthday) and for a "minor requiring authoritative intervention," "dependency," "abuse," and "neglect" (up to the 18th birthday) (*ILCS,* ch. 705, sec. 405/2-3; 405/2-4; 405/3-3, and 405/5-105 (3), 1999). Section 2 of the Uniform Juvenile Court Act recommends the establishment of an upper age of 18 for all categories.

Concurrent or Exclusive Jurisdiction

The issue of concurrent or exclusive jurisdiction of the juvenile court is generally determined by the legislature and specifically stated in the juvenile court act. Section 3 of the Uniform Juvenile Court Act provides the juvenile court with exclusive jurisdiction of certain proceedings listed in that section. In effect, **exclusive jurisdiction** means that the juvenile court will be the only tribunal legally empowered to proceed and that all other courts are deprived of jurisdiction. In some juvenile court acts, **concurrent jurisdiction** may be present when certain specified situations exist. For example, certain criminal acts may be concurrently under the jurisdiction of the juvenile court and the criminal court (*Arkansas Code Annotated,* 9-27-318 Michie Supp., 1999; *Colorado Revised Statutes Annotated,*19-2-517 (1), 1999; *Florida Statutes Annotated,* 985.225 West Supp., 1999; *Michigan Compiled Laws Annotated,* 712 A2d West Supp., 1999). The court that acts first may exercise jurisdiction over a case not because the court has exclusive jurisdiction but simply because it exercises its jurisdiction before the other court acts. In some states, juvenile court acts may allow exclusive jurisdiction over adults who play a role in encouraging a minor to violate a law. In other states, this jurisdiction may be concurrent with the criminal court. In still other states, the juvenile court may have no jurisdiction over such adults, so exclusive jurisdiction rests with the criminal courts. To determine whether the juvenile court has exclusive or concurrent jurisdiction over the subject matter and the subject, it is necessary to refer to the juvenile court act of the state in question. Concurrent jurisdiction is at times awkward (*Iowa Code Annotated,* 232.8 (1)(c) West Supp., 1999; *Maryland Courts and Judicial Procedures Code Annotated,* 3-804(e) (1), (4), 1998; *Mississippi Code Annotated,* 43-21-105(j), 43-21-159 (4) Supp., 1999). Every state has a statutory scheme for waiving jurisdiction in the best interests of the minor and/or in the best interests of the community.

Some juvenile courts are also using blended sentencing by sharing jurisdiction with adult courts. Blended sentencing allows juvenile and/or adult courts to impose adult sanctions and juvenile correctional sanctions on certain types of juveniles. In this case, both courts would share jurisdiction of the child. The adult sanction may be suspended so long as the child remains in compliance with the juvenile court. If the child violates the juvenile sanction, the adult sentence would be imposed. A blended sentence allows for more choices in treatment and sentencing for the judge while giving the child an additional chance to avoid adult criminal court (Sickmund, 2003).

In Section 4 of the Uniform Juvenile Court Act, provision is made for the juvenile court to have concurrent jurisdiction with another court where the proceedings are to treat or commit a mentally retarded or mentally ill child.

▲ Death row inmate Percy Lee poses sitting in prison. At 17 years of age, he killed two women. The U.S. Supreme Court has now decided that the death penalty cannot be applied to teen killers.

Waiver

As stated previously, statutory provisions in juvenile court acts have given juvenile courts original and exclusive jurisdiction over certain cases if the subject is within the defined jurisdiction. However, juvenile court acts contain provisions for the waiver of the juvenile court's jurisdiction over certain offenses committed by minors of certain ages. Policies regarding waiver of juveniles to the criminal justice system differ from state to state. In some states, the juvenile judge is the decision maker. In others, the prosecutor has been given the discretion to file certain types of cases in criminal court.

The waiver should not be confused with concurrent jurisdiction, where two courts have simultaneous jurisdiction over the subject matter and the subject. Waiver, in this case, refers to the process by which a juvenile over whom the juvenile court has original jurisdiction is transferred to adult criminal court. Most authorities agree that the waiver represents a critical stage of the juvenile justice process. At this point, the juvenile may lose the *parens patriae* protection of the juvenile court, including its emphasis on treatment and rehabilitation as opposed to punishment. Once transferred (waived) to the criminal justice network, the juvenile is subjected to contact with adult offenders, may obtain a criminal record, and finds himself or herself in a generally vulnerable position. In some states, an **automatic waiver** of the exclusive jurisdiction of the juvenile court occurs when specific offenses are allegedly committed by a juvenile. For example, in Illinois, "any minor alleged to have committed a traffic, boating, or fish and game violation, or an offense punishable by fine only, may be prosecuted therefore and if found guilty punished under any statute or ordinance relating thereto without reference to the procedures set out in the Juvenile Court Act of this state" (*ILCS,* ch. 705, art. V, sec. 405/5-125, 1999).

Among jurisdictions that permit waivers, the provisions setting forth the circumstances under which such waivers may be granted are quite varied. Most jurisdictions–29 as of 1999 (Puzzanchera, 2003)—require that the child be over a certain age and that he or she be charged with a particularly serious offense before jurisdiction may be waived (*Colorado Revised Statutes,* 19-2-518 (1), 1999; *Connecticut General Statutes Annotated,* 46b-127 West Supp., 1999; *Louisiana Children's Code Annotated,* art. 857 West Supp., 1999; *Michigan Compiled Laws Annotated,* 712A 4 (1) West Supp., 1999; *New Jersey Statutes Annotated,* 2A:4A-27 West Supp., 1999). Other states allow the prosecutor to file directly with adult criminal court, whereas others (all but four states: Massachusetts, Nebraska, New Mexico, and New York) provide for juvenile court judge authorization before waiving a case to adult court (Puzzanchera, 2003, p. 1).

For the most part, automatic waivers are restricted to the more serious offenses and to lesser offenses such as traffic violations. Even in the most serious offenses, an automatic waiver may occur only if the juvenile involved is over a certain age. For example, in Indiana the juvenile must be over the age of 10 years before a waiver is possible for murder (*Indiana Code Annotated,* 31-30-3-2-0-6 Michie, 1997). In Illinois, a waiver may occur if the minor is 13 years of age or over and the alleged act constitutes a crime under the laws of the state (*ILCS,* ch. 705, sec. 405/5-805 (1)-(3), 1999). Other states authorize waivers similarly (Table 5.4) if the jurisdictional age is established and met and the specific offense is within the statutory allowance for such a waiver. In some states with statutory exclusion provisions, certain types of juvenile cases originate in criminal rather than juvenile court.

Another type of waiver is the **discretionary waiver.** A number of states permit waivers of jurisdiction over children over a certain age without regard to the nature of the offense involved. Where the juvenile court finds that the minor is not a fit and proper subject to be dealt with under the juvenile court act and the seriousness of the offense demands that the best interests of society be considered, the juvenile court judge may order criminal proceedings to be instituted against the minor (Bilchik, 1999a, p. 16).

Discretionary determination of waivers may be left to juvenile court judges to decide after a petition for a waiver has been filed and a hearing has been conducted on the advisability of

Table 5.4 State Juvenile Justice Profiles

State	Most States Had Multiple Ways to Impose Adult Sanctions on Juveniles by the End of 1999							
	Judicial Waiver							
	Discretionary	Presumptive	Mandatory	Concurrent Jurisdiction	Statutory Exclusion	Reverse Waiver	Once an Adult, Always an Adult	Blended Sentencing
Total states	46	16	15	15	29	24	34	22
Alabama	✓				✓		✓	
Alaska	✓	✓			✓			✓
Arizona	✓	✓		✓	✓	✓	✓	
Arkansas	✓			✓		✓		✓
California	✓	✓			✓		✓	✓
Colorado	✓	✓		✓		✓		✓
Connecticut			✓			✓		✓
Delaware	✓		✓		✓	✓	✓	
District of Columbia	✓	✓		✓			✓	
Florida	✓			✓	✓		✓	✓
Georgia	✓		✓	✓	✓	✓		
Hawaii	✓						✓	
Idaho	✓				✓		✓	✓
Illinois	✓	✓	✓		✓		✓	✓
Indiana	✓		✓		✓		✓	
Iowa	✓				✓	✓	✓	✓
Kansas	✓	✓					✓	✓
Kentucky	✓		✓			✓		
Louisiana	✓		✓	✓	✓			
Maine	✓	✓					✓	
Maryland	✓				✓	✓	✓	

(Continued)

Table 5.4 (Continued)

State	Most States Had Multiple Ways to Impose Adult Sanctions on Juveniles by the End of 1999							
	Judicial Waiver							
	Discretionary	Presumptive	Mandatory	Concurrent Jurisdiction	Statutory Exclusion	Reverse Waiver	Once an Adult, Always an Adult	Blended Sentencing
Massachusetts				✓	✓			✓
Michigan	✓			✓			✓	✓
Minnesota	✓	✓			✓		✓	✓
Mississippi	✓				✓	✓	✓	
Missouri	✓						✓	✓
Montana	✓			✓	✓	✓		
Nebraska				✓		✓		
Nevada	✓	✓			✓	✓	✓	
New Hampshire	✓	✓					✓	
New Jersey	✓	✓	✓					
New Mexico					✓			✓
New York					✓	✓		
North Carolina	✓		✓				✓	
North Dakota	✓	✓	✓				✓	
Ohio	✓		✓				✓	
Oklahoma	✓			✓	✓	✓	✓	✓
Oregon	✓				✓	✓	✓	
Pennsylvania	✓	✓			✓	✓	✓	
Rhode Island	✓	✓	✓				✓	✓
South Carolina	✓		✓		✓	✓		✓
South Dakota	✓				✓	✓	✓	
Tennessee	✓					✓	✓	

(Continued)

Table 5.4 (Continued)

State	Most States Had Multiple Ways to Impose Adult Sanctions on Juveniles by the End of 1999							
	Judicial Waiver							
	Discretionary	Presumptive	Mandatory	Concurrent Jurisdiction	Statutory Exclusion	Reverse Waiver	Once an Adult, Always an Adult	Blended Sentencing
Texas	✓						✓	✓
Utah	✓	✓			✓		✓	
Vermont	✓			✓	✓	✓		✓
Virginia	✓		✓	✓		✓	✓	✓
Washington	✓				✓		✓	
West Virginia	✓		✓					✓
Wisconsin	✓				✓	✓	✓	
Wyoming	✓			✓		✓		

Source: Adapted from Griffin, P. (2003). *Trying and sentencing juveniles as adults: An analysis of state transfer and blended sentencing laws* (Technical Assistance to the Juvenile Court: Special Projects Bulletin). Washington, DC: U.S. Department of Justice, Office of Juvenile Justice and Delinquency Prevention; and Sickmund, M. (2003). *Juvenile offenders and victims* (National Report Series). Washington, DC: U.S. Department of Justice, Office of Juvenile Justice and Delinquency Prevention.

Note: In states with a combination of provisions for transferring juveniles to criminal court, the exclusion, mandatory waiver, or concurrent jurisdiction provisions generally target the oldest juveniles and/or those charged with the most serious offenses, whereas those charged with relatively less serious offenses and/or younger juveniles may be eligible for discretionary waiver.

granting the waiver. In general, the criteria used by juvenile court judges to determine the granting or denial of waivers of juveniles to criminal courts are rather vague and, for the most part, quite subjective. As stated previously, if the minor is not a fit and proper subject to be dealt with under the juvenile court, an order instituting criminal proceedings may be rendered by the juvenile court. Factors typically cited by the courts as weighing heavily in the decision to waive jurisdiction include the seriousness of the offense, the age of the juvenile, and the past history of the juvenile. However, some jurisdictions confer on the prosecutor the authority to decide which court (juvenile or criminal) should hear the case. Wyoming empowers the prosecutor to make this decision when a juvenile is charged with a misdemeanor or if the juvenile is 17 years of age or over and charged with a felony, is 14 years of age or over and charged with a violent felony, or is charged with a felony and has been adjudicated delinquent twice previously for felonies (*Wyoming Statutes Annotated,* 14-6-203 (c)–(f), 14-6-211, 1999). Likewise, in Arizona the prosecutor has discretion to file charges in criminal court against any child 14 years of age or over who is alleged to have committed a Class 1 or 2 felony or a certain offense of the Class 3, 4, 5, and 6 felonies (*Arizona Revised Statutes Annotated,* 13-501 (B), 1999). In Louisiana, the prosecutor has discretion in cases involving any child 15 years of age or over who is charged with attempted first- or second-degree murder, manslaughter, armed robbery, aggravated burglary, or one of numerous other offenses, including some drug-related offenses (*Louisiana Children's Code Annotated,* Art. 305 (B), 1999). Similar prosecutorial discretion can be seen in numerous other states and the District of Columbia.

With respect to waivers, in the *Kent* case, the U.S. Supreme Court ruled that to protect the constitutional rights of the juvenile, the juvenile is entitled to the following:

1. A full hearing on the issue of a waiver
2. The assistance of legal counsel at the hearing
3. Full access to the social records used to determine whether such transfer should be made
4. Statement of the reasons why the juvenile judge decided to waive the juvenile to (adult) criminal court (*Kent v. United States,* 383 U.S. 541, 1966)

In *Kent,* the Court held that a waiver of jurisdiction is a critically important stage in the juvenile process that must be considered in terms of due process and fair treatment as required by the Fourteenth Amendment. Although the *Kent* decision applied only to the District of Columbia, most states that allow waivers have incorporated the waiver procedures of *Kent* into their juvenile court acts. A clear majority of states statutorily guarantee a waiver hearing.

Some states have attempted to establish at least some criteria that would aid the juvenile court judge in making a determination on a motion to waive the juvenile court's jurisdiction. For example, in Illinois the court must consider the following:

1. The seriousness of the alleged offense
2. Whether there is evidence that the alleged offense was committed in an aggressive and premeditated manner
3. The age of the minor
4. The previous delinquency history of the minor
5. The culpability of the minor
6. Whether there are facilities particularly available to the juvenile court for the treatment and rehabilitation of the minor
7. Whether the best interests of the minor and the security of the public may require that the minor continue in custody or under supervision for a period extending beyond his minority
8. Whether the minor possessed a deadly weapon when committing the alleged offense (*ILCS,* ch. 705, art. V, sec. 405/5-805 (3)(b), 2006)

The juvenile court judge, as well as the prosecuting officials, must weigh the consequences of a waiver for the future of the juvenile. The question concerning a waiver of a juvenile to the adult criminal court for prosecution of an offense that might result in a felony record is extremely important due to the lasting effects that a felony record might have. To justify a waiver for criminal prosecution, the juvenile court must agree to accept the more punitive, retributive, and punishment-oriented approach of the adult court. In such cases, the juvenile court judge must act not only in the best interests of the minor but also in the best interests of the community by protecting the community against further unlawful, and perhaps violent, conduct by the juvenile offender. Juvenile court judges, realizing the full effect of a felony record (e.g., in terms of future employment) generally permit a waiver for criminal prosecution only when the offense is so serious that relegating the offense to the realm of delinquency would be unconscionable and would result in a mockery of justice and when the offense is not an isolated act but rather a series of acts showing a trend toward becoming more serious.

Double Jeopardy

The Fifth Amendment states that no person shall be subject to being tried twice for the same offense. Courts in the United States at one time held that the **double jeopardy** clause did not prohibit a juvenile adjudicated delinquent from subsequently being tried for the same offense in criminal court. In *Breed v. Jones* (421 U.S. 519, 1975), the U.S. Supreme Court unanimously ruled that the Fifth Amendment's prohibition against double jeopardy precludes criminal prosecution of a juvenile subsequent to proceedings in juvenile court involving the same act.

After dealing with scope and purpose, most juvenile court acts go on to describe in detail the procedures to be employed by various components of the juvenile justice system in handling the juvenile. We discuss these procedural requirements in the following chapter.

Career Opportunity—Court Administrator

Job Description: Assist the chief judge in scheduling the court calendar. Ensure correct assignment of cases. See that appropriate transfer of cases occurs. Coordinate judges' schedules. May help to plan court security.

Employment Requirements: Must have a 4-year degree. Prior administrative experience may be required.

Beginning Salary: Between $20,000 and $40,000 depending on jurisdiction. Benefits vary widely but are typically included.

SUMMARY

A thorough understanding of both the purpose and scope of juvenile court acts is crucial because the intent of the juvenile court acts cannot be carried out without this understanding.

The primary purpose of juvenile court acts is to ensure the welfare of the juvenile within a legal framework while maintaining the family unit and protecting the public. Most of us would agree that this is an admirable goal. At the same time, however, we should be aware of the inherent difficulties involved in achieving this goal. Consider, for example, the police officer who has apprehended a particular juvenile a number of times for increasingly serious offenses. Repeated attempts at enlisting the aid of the juvenile's family in correcting the undesirable behavior have failed. If the officer decides that protection of the public is now of primary importance, the officer may feel compelled to arrest the juvenile even though this action may result in the juvenile being sent to a detention facility. As a result, the family unit is broken up and the welfare of the juvenile has been, to some extent, sacrificed by placing him or her in detention.

Also consider the dilemma of the juvenile court judge who must make the final decision concerning what is in the best interests of both the juvenile and the public. If the judge adheres to the philosophy of the juvenile court, the judge may be tempted to leave the juvenile with his or her family even though the public may suffer. In addition, the judge and prosecutor are faced with the difficult task of making distinctions between unruly and delinquent juveniles. These distinctions are crucial given that different types of treatment, correctional, and rehabilitation programs are available depending on the label attached.

A thorough understanding of the scope of juvenile court acts is equally important. The police officer on the street must be aware of both the age limits and the different categories into which juveniles are separated if the requirements of the juvenile court act are to be met. Prosecutors and judges must be certain that jurisdictional requirements have been met and must understand the consequences of requesting or granting waivers. In short, the purposes of juvenile court acts cannot be achieved without thorough knowledge of the subjects and behaviors dealt with in the scope of such acts.

The purposes of juvenile court acts are, in general, to create courts with the authority to hear designated kinds of cases, to discuss the procedural rules to be used in such cases, and to provide for the best interests of juveniles while at the same time protecting the interests of the family and society. Unfortunately, it is not always possible to achieve all of these purposes in any one case. For example, it might be in the best interests of society to send a particular juvenile to a correctional facility, but this action is not likely to be in the best interests of the juvenile.

Sections in juvenile court acts dealing with scope generally include information on age requirements, geographical requirements, types of behaviors covered by the acts, and waivers.

The Uniform Juvenile Court Act requires legal accountability, narrows the definition of delinquency (excludes status offenses), and attempts to ensure the best interests of juveniles while maintaining the family unit and protecting the public.

In 1967, the President's Commission on Law Enforcement and Administration of Justice recommended that serious thought be given to completely eliminating from juvenile court jurisdiction children who commit noncriminal acts or status offenses. Consistent with this recommendation, two national commissions (the American Bar Association's [1977] Standards Project and the Twentieth Century Task Force on Sentencing Policy Toward Youthful Offenders [1987]) proposed the elimination of juvenile court jurisdiction over status offenders, and most states have followed this recommendation.

Note: Please see the Companion Study Site for Internet exercises and Web resources. Go to www.sagepub.com/juvenilejustice6study.

Critical Thinking Questions

1. In addition to protecting the community from youthful offenders, what are the three major purposes or goals of juvenile court acts?
2. How and why did the Uniform Juvenile Court Act (see Appendix A) come into existence? Has this act had much impact on the various state juvenile court acts? Give some examples to support your answer.
3. What are some of the considerations of jurisdiction that fall within the scope of juvenile court acts? Why are these considerations important?
4. Suppose that a 15-year-old is taken before juvenile court in the county in which he resides for allegedly repeatedly refusing to obey his parents' orders to be home before 10 o'clock at night. Would such behavior fall within the scope of most juvenile court acts? Would the juvenile be dealt with as a delinquent under Uniform Juvenile Court Act recommendations? If not, why?
5. What are the legal requirements surrounding waivers to adult court?

Suggested Readings

Anderson, D. C. (1998, May). When should kids go to jail? *American Prospect, 9*, 72–78.

Davis, S. M. (2006). *Rights of juveniles: The juvenile justice system.* (2nd ed.). Eagan, MN: Thomson/West.

Drizin, S. A. (1999, July/August). The juvenile court at 100. *Judicature, 83*, 8–15.

Foster, L. (2000). School shootings and the over-reliance upon age in choosing criminal or juvenile court. *Vermont Law Review, 24*, 537–540.

Gondles, J. A., Jr. (1997, June). Kids are kids, not adults. *Corrections Today, 59*, 6.

National Council of Juvenile and Family Court Judges. (2005). *Juvenile delinquency guidelines: Improving court practice in juvenile delinquency cases.* Washington, DC: U.S. Department of Justice, Office of Juvenile Justice and Delinquency Prevention.

Palmer, E. A. (2000). Weary of juvenile justice logjam, members move provisions separately (Aimee's Law). *CQ Weekly, 58*(29), 1727–1728.

Puzzanchera, C. M. (2000). *Delinquency cases waived to criminal court, 1988–1997* (OJJDP Fact Sheet). Washington, DC: U.S. Department of Justice.

Redding, R. E. (1999). Examining legal issues: Juvenile offenders in criminal court and adult prison. *Corrections Today, 61*(2), 92–95.

Schwartz, I. M., Weiner, N. A., & Enosh, G. (1999). Myopic justice? The juvenile court and child welfare systems. *Annals of the American Academy of Political & Social Science, 564,* 126–141.

Snyder, H., & Sickmund, M. (2000). *Juvenile transfers to criminal court in the 1990's: Lessons learned from four studies.* Washington, DC: U.S. Department of Justice, Office of Juvenile Justice and Delinquency Prevention.

Urban, L., Cyr, J., & Decker, S. (2003). Goal conflict in the juvenile court: The evolution of sentencing practices in the United States. *Journal of Contemporary Criminal Justice, 19,* 454-479.

Winner, L., Lanza-Kaduce, L., Bishop, D., & Frazier, C. (1997). The transfer of juveniles to criminal court: Re-examining recidivism over the long term. *Crime & Delinquency, 43,* 548–563.

JUVENILE JUSTICE PROCEDURES

CHAPTER LEARNING OBJECTIVES

On completion of this chapter, students should be able to

- *Understand and discuss juvenile court procedures*
- *Discuss the rights of juveniles at various stages, from taking into custody through appeals*
- *Understand requirements for bail, notification, and filing of petitions*
- *Discuss procedures involved in detaining juveniles*

KEY TERMS

Stationhouse adjustment

Preliminary conference

Fourth Amendment

Totality of circumstances

Guardian ad litem

Bail

Taking into custody

Interrogation

Detention hearing

Detention

Shelter care

Notification

Clear and convincing evidence

Sixth Amendment

Continuance under supervision

Social background investigations

Appeals

Juvenile court acts discuss not only the purposes and scope of the juvenile justice system but also the procedure the juvenile courts are to follow. Proceedings concerning juveniles officially begin with the filing of a petition alleging that a juvenile is delinquent, dependent, neglected, abused, in need of supervision, or in need of authoritative intervention. Most juvenile court acts, however, also discuss the unofficial or diversionary activities available as remedies prior to the filing of a petition such as a **stationhouse adjustment** and a **preliminary conference.** A stationhouse adjustment occurs when a police officer negotiates a settlement with a juvenile, often with his or her parents, without taking further official action. A preliminary conference is a voluntary meeting arranged by a juvenile probation officer with the victim, the juvenile, and typically the juvenile's parents or guardian in an attempt to negotiate a settlement without taking further official action. Juvenile court acts clearly indicate those persons who are eligible to file a petition. For example, in Illinois any adult person (21 years of age or over), agency, or association by its representative may file a petition, or the court on its own motion may direct the filing through the state's attorney of a petition in respect to a minor under the act (*Illinois Compiled Statutes* [*ILCS*], ch. 705, art. 1, sec. 405/1-3, 1999).

▲ Are stationhouse adjustments more common for juveniles of some ages, races, and gender than for others?

Although it is true that a petition may be filed by any eligible person by going directly to the prosecutor (state's attorney or district attorney), a large proportion of petitions are filed following police action or by social service agencies dealing with minors. To understand the step-by-step procedures involved in processing juveniles, we discuss the typical sequence of events occurring after the police take a juvenile into custody. We rely heavily on the procedures given in the Uniform Juvenile Court Act and the Illinois Juvenile Court Act, which closely resemble similar acts in many states. Although a general discussion of juvenile justice procedures is given, some states differ with respect to specific requirements. You should consult the juvenile court act or code relevant to your state for exact procedural requirements.

Rights of Juveniles

Regardless of the particular jurisdiction, juveniles in the United States have been (since the 1967 *Gault* decision) guaranteed a number of basic rights at the adjudicatory stage. Thus a juvenile who is alleged to be delinquent has the following rights (*In re Gault,* 1967):

1. The right to notice of the charges and time to prepare for the case
2. The right to counsel
3. The right to confront and cross-examine witnesses
4. The right to remain silent in court

As a direct result of the *Gault* decision, the constitutional guarantees of the Fifth and Sixth Amendments are applicable to states through the Fourteenth Amendment and not only apply to delinquency matters but also have been extended to some cases involving the need for supervision or intervention. The question remaining after the *Gault* decision concerned the extent to which its mandate logically extended to other stages of the juvenile justice process, particularly the police investigatory process. *Gault* and the *Kent* decision (*Kent v. United States,* 1966) have been interpreted to require the application of the **Fourth Amendment** and the exclusionary rule to the juvenile-justice process. The most difficult issue has revolved around the juvenile's competency to waive his or her rights under Miranda. In general, the courts have relied on a **totality of circumstances** approach in determining the validity of the waiver. Circumstances considered include the age, competency, and educational level of the juvenile; his or her ability to understand the nature of the charges; and the methods used in, and length of, the interrogation (Davis, 2001, sec. 3.13, pp. 3-86–3-90).

The Uniform Juvenile Court Act (National Conference of Commissioners on Uniform State Laws, 1968, sec. 26) provides that all parties to juvenile court proceedings are entitled to representation by counsel. Many jurisdictions currently provide for representation by counsel in neglect, abuse, and dependency proceedings, extending the *Gault* decision to such cases (*Georgia Code Annotated,* 15-11-31(b), 1999; *Montana Code Annotated,* 41-5-331 (1), 1999; *New Mexico Statutes Annotated,* 32-A-Z-14 (c) Michie, 1995). In a neglect and/or abuse case, legal counsel for the minor may be the state's attorney, who represents the state that has a duty to protect the child. The court may also appoint a **guardian ad litem** for a juvenile if the juvenile has no parent or guardian appearing on his or her behalf or if the parent's or guardian's interests conflict with those of the juvenile, as is often the case in abuse and neglect cases.

The protection afforded by the Fourth Amendment against illegal search and seizure extends to juveniles. All courts that have specifically considered the issue of the applicability of the Fourth Amendment to the juvenile justice process have found it to be applicable, or more correctly, no court has found it to be inapplicable (Davis, 2001, 3–17). The Uniform Juvenile Court Act (National Conference of Commissioners on Uniform State Laws, 1968, sec. 27(b)) states that evidence seized illegally will not be admitted over objection. Similarly, a valid confession made by a juvenile out of court is, in the words of the Uniform Juvenile Court Act, "insufficient to support an adjudication of delinquency unless it is corroborated in whole or in part by other evidence." This extends some protection to juveniles not normally accorded to adults. In addition, the Uniform Juvenile Court Act (sec. 27(a)) recommends that a party be entitled to introduce evidence and otherwise be heard in his or her own behalf and to cross-examine adverse witnesses. Furthermore, a juvenile accused of a delinquent act need not be a witness against, or otherwise incriminate, himself or herself. A majority of juvenile court acts do not spell out a detailed code of evidence. However, most do specify whether the rules permit only competent, material, and relevant evidence and whether the rules of evidence that apply in criminal or civil cases are applicable in juvenile cases. A number of states provide that the rules of evidence applicable in criminal cases apply in delinquency proceedings and that the rules of evidence applicable in civil cases apply in other proceedings (i.e., neglect, dependency, and "in need of supervision" cases).

The Children's Bureau of the U.S. Department of Health and Human Services recommended many years ago that, unless a child is advised by counsel, the statements of the child made while in the custody of the police or probation officers, including statements made during a preliminary inquiry, predisposition study, or consent decree, should not be used against the child prior to the determination of the petition's allegations in a delinquency or "in need of supervision/intervention" case or in a criminal proceeding prior to conviction (Children's Bureau, 1969, sec. 26). In abuse and neglect cases, however, the courts have eased restrictions on the admission of spontaneous utterances, statements made in the totality of circumstances, and so on (see Chapter 11).

It should be noted that some rights guaranteed to adults are not guaranteed to juveniles in most jurisdictions. As a result of the *McKeiver* decision (*McKeiver v. Pennsylvania,* 1971), juveniles are not generally guaranteed the right to a trial by jury or a public trial. The U.S. Supreme Court, in deciding *McKeiver,* indicated that a jury was not necessary for fact-finding purposes and left the issue of trial by jury up to the individual states. Although the majority of jurisdictions provide for hearings without juries, some provide for jury trials by statute or judicial decision (*Colorado Revised Statutes Annotated,* 19-2-107, 1999; *Massachusetts General Laws Annotated,* ch. 119, sec. 55A, West Supp., 1999; *Michigan Compiled Laws Annotated,* 712A.17 (2) West Supp., 1999; *Montana Code Annotated,* 41-5-1502 (1), 1999; *West Virginia Code Annotated,* 49-5-6, 1999). In addition, the *McKeiver* decision left open the question of whether juvenile court proceedings are necessarily adversarial in nature and left on the states the burden of establishing that a separate justice system for juveniles represents a useful alternative to criminal processing. There is a clear-cut trend toward treating all juvenile court procedures as adversarial.

Bail

The issue of **bail** (release from custody pending trial after payment of a court-ordered sum) for juveniles is also controversial. Some jurisdictions permit bail, whereas others do not on the grounds that the juvenile has not been charged with a crime and, therefore, is not entitled to bail. Because of special release provisions for juveniles (to the custody of parents or a guardian), bail has not been a question of paramount concern in terms of litigation. A number of states forbid the use of bail with respect to juveniles (*Connecticut General Statutes Annotated,* 466-133b, West, 1991; *Hawaii Revised Statutes,* 571-32 (h), 1993; *Oregon Revised Statutes,* 419C. 179, 1995), several states authorize release on bail at the discretion of a judge (*Minnesota Statutes Annotated,* 260.171 (1) West, 1998; *Nebraska Revised Statutes,* 43-253, 1998; *Tennessee Code Annotated,* 37-1-117(e), 1996; *Vermont Statutes Annotated,* tit. 33. sec. 5513(c), 1991), and some states allow the same right to bail enjoyed by adults (*Colorado*

Revised Statutes, 19-2-509, 1999; *Georgia Code Annotated,* 15-11-19 (d), 1999; *Oklahoma Statutes Annotated,* tit. 10, sec. 7301-4.3 (c) West Supp., 1998).

Finally, most jurisdictions require that official records kept on juveniles be maintained in separate and confidential files. These may be opened only by court order or following stringent guidelines established by state statutes.

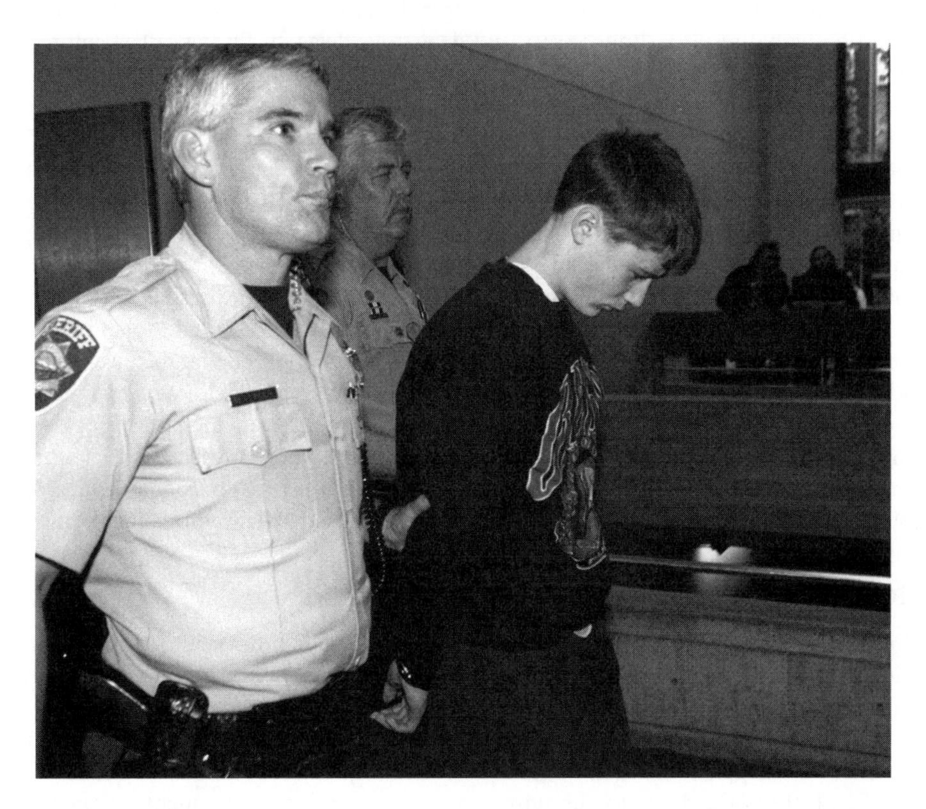

▲ A series of court decisions and some constitutional amendments help to protect the rights of juveniles.

Taking Into Custody

The Uniform Juvenile Court Act (National Conference of Commissioners on Uniform State Laws, 1968) states,

> A child may be taken into custody pursuant to an order of the court under that Act, or pursuant to the laws of arrest; or by a law enforcement officer if there are reasonable grounds to believe that the child is suffering from illness or injury or is in immediate danger from his surroundings and that his removal is necessary; or by a law enforcement officer if there are reasonable grounds to believe that the child has run away from his parents or guardian. (sec. 13)

The broad jurisdictional scope of the juvenile courts generally provides that any juvenile can be taken into custody (detained) without a warrant if the law enforcement officer reasonably believes the juvenile to be delinquent, in need of supervision, dependent, abused, or neglected as defined within that state's juvenile court act. However, some states have recognized that removing a juvenile from home before there has been any trial is a power to be used on a limited basis. For truancy, disobedience, and even neglect, the legal process should begin with a summons unless there is "imminent danger" involved and unless waiting for the court's permission would result in unnecessary and dangerous delay. In Illinois, a law enforcement officer may, without a warrant, take into temporary custody a minor whom the officer, with reasonable cause, believes to be delinquent and requiring authoritative intervention, dependent, abused, or neglected as defined within that state's juvenile court act (*ILCS*, ch. 705, sec. 405/2-5, 3-4, 4-4, 5-401, 1999). In addition, the officer may take into custody any juvenile who has been adjudged a ward of the court and has escaped from any commitment ordered by the court. The officer may also take into custody any juvenile who is found on any street or in any public place suffering from any sickness or injury requiring care, medical treatment, or hospitalization. The taking into temporary custody under the Uniform Juvenile Court Act does not constitute an official arrest. Although statutes in various states provide that **taking into custody** is not deemed an arrest, this is somewhat a legal fiction given that the juvenile is often held in involuntary custody. In light of recent court decisions, when delinquency is the alleged reason for taking into custody, law enforcement officers must adhere to appropriate constitutional guidelines. For categories other than delinquency, the *parens patriae* concern for protecting minors from dangerous surroundings will suffice constitutionally as reasonable grounds for taking minors into custody when it is not abused by law enforcement officers.

Interrogation

While in custody, the juvenile has rights similar to those of an adult with respect to **interrogation.** To determine whether a confession or statement was given freely and voluntarily, the totality of circumstances surrounding the giving of the statement is to be considered. Even prior to *Gault,* the U.S. Supreme Court in *Haley v. Ohio* (1948) and *Gallegos v. Colorado* (1969) used the voluntariness test to determine the admissibility of statements made by juveniles to the police. If the police desire to question a juvenile concerning a delinquent act, the juvenile should be given the Miranda warning and should be clearly told that a decision to remain silent will not be taken as an indication of guilt. This can be problematic, as seen in In Practice 6.1, when police or others are not aware of the policies, procedures and requirements with regard to juveniles. Many police administrators, prosecutors, and juvenile court judges believe that it is best not to question the juvenile unless his or her parents or a counselor are present. In Colorado, for example, "no statement or admission of a child made as a result of interrogation by law enforcement officials . . . shall be admissible . . . unless a parent, guardian, or custodian was present . . . and the child was advised of his right to counsel and to remain silent" (*Colorado Revised Statutes Annotated,* 19–2–511 (1), 1999). Any confession obtained without these safeguards might be considered invalid on grounds that the juvenile did not understand his or her rights or was frightened. Other state statutes and the Uniform Juvenile Court Act (sec. 27(b)) contain similar provisions (*California Welfare and Institutional Code,* 625, 627.5 West, 1998; *Connecticut General Statutes Annotated,* 46 (b)-137 (a) West, 1999; *North Carolina General Statutes,* 7B-2101, 1999).

In Practice 6.1

Teen Interrogation Not a Textbook Case

According to police documents, it was a textbook arrest.

Three suspects were interrogated separately after being advised of their rights to remain silent, and all confessed to the alleged crime.

But in this case, the suspects were 13- and 15-year-old students at Fox Chapel Middle School. The offense was drawing their initials and profanities in wet cement at a subdivision construction site.

Officials say the interrogations and arrests took place on campus without the knowledge of administrators, staff members, or their own parents—a direct violation of the school system's own policy.

The parents have filed a complaint with the Florida chapter of the American Civil Liberties Union, saying their sons' rights to remain silent weren't adequately protected.

"We would have at least been able to be there for them, and we weren't," said Joyce Thompson, mother of one of the boys. "They're supposed to be protected at school."

No one disputes the fact that the boys committed the offense of marking their initials, profanities, and bicycle tire marks in the wet cement of a residential driveway under construction at 5172 Mentmore Ave. on Jan. 28, actions that caused more than $1,000 worth of damages.

Prosecutors have since reduced the charges to a misdemeanor so the boys can attend teen court and potentially keep their records clean.

The controversy lies in how the boys' confessions were obtained Jan. 31.

Arrest reports show Hernando deputies arrived at the school shortly after noon and talked with the school resource officer, Deputy Wendy Tolbert.

The report said the boys—who were identified by parents as 13-year-olds Mitchell Thompson and Frank Macchione and 15-year-old Luis Rosario—were interrogated individually and gave statements "post-Miranda" admitting their involvement.

Parents say they weren't notified until the boys were handcuffed and on the way to jail, each charged with felony criminal mischief for damages in excess of $1,000.

And principal David Schoelles said he and other administrators didn't learn of the investigation taking place in the building until it was all over.

"This is the first time this has happened, where students have been interviewed [by police] without our being notified," said Schoelles. "This happened without our even knowing the deputies were on campus."

In 2000, the Hernando County School Board revised its policy on police contact with students twice—first allowing unsupervised interviews of students by police, on the advice of then-board attorney Karen Gaffney, and then reversing itself after an uproar of dissent from parents and community members.

Under the current district policy, a school principal must attempt to contact parents before interrogations take place.

If [parents] can't be reached, the principal or a staff member must be present except if a high school student requests that they be excluded.

But Schoelles, who has since taken a new district assignment as a curriculum specialist, said those policies weren't followed in January since deputies didn't notify him of their presence.

(Continued)

(Continued)

He described that as an oversight by deputies who didn't understand the policy and said he was comfortable that officers now understand the rules and are willing to abide by them.

"They're not legally obligated to contact parents," Schoelles added. "But when they're here [in school], they have to play by our rules."

Donna Black, a spokeswoman for the Sheriff's Office, had no immediate comment on the Fox Chapel arrests, saying officials needed to contact the arresting officer for more information on what happened.

Under the law, police aren't required to round up administrators and parents in order to interview juvenile suspects. But they must immediately notify parents when a child is taken into custody.

Other districts in the Tampa Bay area have crafted detailed policies to protect the rights of minors, considered to be less capable of understanding their rights to remain silent and get advice from an attorney before speaking to police.

And several police departments have recognized the need to cooperate with those policies.

Lt. Brian Moyer, school district safety officer for the Pasco County Sheriff's Department, said his officers must make every effort to contact a parent or guardian before interrogating a minor. And deputies must request the presence of a school staff member and notify administrators if the minor is to be removed from the school.

"That certainly would not be the procedure or process we would follow in Pasco County," he said, referring to the Fox Chapel incident. "Even if the administrator weren't present in the interview, he would know there were special investigators around."

In Citrus County, the district and sheriff's department have crafted a detailed memorandum outlining their joint policies. Among them, efforts must be made to notify parents of plans to interrogate their child unless the potential crime involves abuse by a family member.

And except in such cases, a school representative must sit in on all interviews in which the student is considered a potential suspect or witness to a crime.

Under the Hernando County Sheriff's Office policy, deputies must make an attempt to notify parents prior to interviewing juveniles. No more than two deputies may take part in those interviews and must provide juveniles with "reasonable breaks and/or rest periods."

Rebecca Steele, director of the ACLU's Tampa office, said prosecutors recently lost a similar Florida case on appeal. In that situation, too, a middle school student was interrogated without the knowledge and participation of parents or the school staff.

Steele said the court recognized that a minor's ability to understand the court-mandated Miranda warning varies, with some children being more able than others to understand their rights to silence and legal counsel.

"Also, the police officers' violation of school policy on notifying parents, much less the school principal, is quite troubling," she added. "Parents have rights to protect their children from losing their rights without adequate protection."

Neither Superintendent Wendy Tellone nor board Chairman Jim Malcolm said they'd heard anything about interrogations taking place in potential violation of their policy.

"Personally, as a parent, I want to be there when you're interrogating my kid," Malcolm said. "Particularly if they're younger than 18 years old. You can't just have people coming on campus and grabbing kids."

SOURCE: Marshall, Tom. (2006, June 21). Teen interrogation not a textbook case. *St. Petersburg Times,* 1. Record Number: 0606219465000030566283.

Juveniles taken into custody may be either detained or released to the custody of their parents or guardian. Most model juvenile court acts and the Uniform Juvenile Court Act (sec. 15) dictate that the police make an "immediate" and "reasonable" attempt to notify the juvenile's parents or guardian of his or her custody. The maximum length of time considered to be immediate is usually established by statute. The definition of *reasonable* usually includes attempts to phone and/or visit the residence of the juvenile's parents, place of employment, and any other known "haunts."

The Detention Hearing

If a juvenile is not released to his or her parents soon after being taken into custody, most states require that a **detention hearing** (a court hearing to determine whether **detention** is required) be held within a specified time period. Sufficient notification must, of course, be given to all parties concerned before the proceeding. Section 17 of the Uniform Juvenile Court Act indicates that if a juvenile is brought before the court or delivered to a detention or **shelter care** facility, the intake or other authorized officer of the court will immediately begin an investigation and release the juvenile unless it appears that further detention or shelter care is warranted or required. If the juvenile is not released within 72 hours after being placed in detention, an informal detention hearing is held to determine whether further detention is warranted or required.

Reasonable notice of the hearing must be given to the juvenile and to the parents or guardian. In addition, notification of the right to counsel and of the juvenile's right to remain silent regarding any allegations of delinquency or unruly conduct must also be given by the court to the respondents. States vary with respect to the criteria used to determine the need for further detention, but they usually focus on the need to ensure the protection of society and the juvenile and on the possibility of the juvenile fleeing the jurisdiction (see In Practice 6.2). For example, in Illinois, after a minor has been delivered to the place designated by the court,

> the in-take personnel shall immediately investigate the circumstances of the minor and the facts surrounding his being taken into custody. The minor shall be immediately released to the custody of his parents unless the in-take officer finds that further detention is a matter of immediate and urgent necessity for the protection of the minor, or of the person or property of another, or that he is likely to flee the jurisdiction of the court. (*ILCS*, ch. 705, sec. 5-501 (2), 1999)

Detention can be authorized by the intake officer (generally a designated juvenile police officer or the juvenile probation officer) for up to 36 hours, at which time the minor is either released to his or her parents or brought before the court for a detention hearing. Failure to file a petition or to bring the juvenile before the court within 40 hours will result in a release from detention (*ILCS*, ch. 705, sec. 405/5-415, 1999). In Illinois, the detention hearing focuses first on whether there is probable cause to believe that the minor is within the category of delinquency, in need of intervention, abused/neglected, or dependent. The court then decides, using the same criteria as the intake officer, whether further detention is a matter of immediate and urgent necessity (*ILCS*, ch. 705, sec. 405/5-501 (2), 1999). For a sample temporary custody order, see Figure 6.1.

(Text continues on page 152)

In Practice 6.2

Boy Charged With Arson Locked Up: A Judge Orders the Detention of a 10-Year-Old Accused With Three Other Juveniles in Several Fires

A 10-year-old boy accused of setting a four-alarm fire at the Chicago Lumber Co. was ordered into detention Thursday by a Douglas County juvenile court judge. The judge said the boy is a danger to himself, the community, and property.

The 10-year-old is one of four boys, ages 10 and 11, who are facing juvenile charges of second-degree arson in connection with six fires, including the Chicago Lumber Co. fire, which caused an estimated $2 million in damage.

The 10-year-old and an 11-year-old are suspected of setting the July 13 blaze that destroyed three Chicago Lumber Co. warehouses, north of Pierce Street near 14th Street, and threatened nearby homes.

Fire investigators said Thursday afternoon that they suspect the boys set six of 17 to 20 fires that were reported earlier this year in the boys' neighborhood, an area bounded by 13th, Vinton, 24th, and Pierce streets. Officials still are trying to determine who was responsible for the other fires.

Three of the boys were arrested last week. The fourth was arrested Wednesday.

Fire officials said the boys all knew each other and referred to themselves as a gang.

Deputy Douglas County Attorney Melissa Stanosheck said in court Thursday that the boys called their group "KOB"—Killing Off Bitches.

Stanosheck said when the 10-year-old accused in the Chicago Lumber fire was interviewed, he told investigators he had smoked marijuana before setting the fire.

The boys, who are in the fourth, fifth, and sixth grades, sometimes would tell their parents or guardians that they were going to stay the night with their friends, said Omaha Fire Capt. Joe Gibilisco, the lead fire investigator in the case. Other times, he said, they would meet up after sneaking out of their homes at night.

"Most of these fires were done under the cover of darkness in the early-morning hours," said Battalion Chief Dave Adolf, another fire investigator.

Fire officials wouldn't give details about how the boys started the fires. They said only that accelerants sometimes were used.

At Thursday's juvenile court hearing for the 10-year-old accused in the Chicago Lumber fire, Judge Douglas Johnson expressed concern about whether the boy understood the seriousness of the allegations.

Before the hearing, the boy, dressed in a dark navy jumpsuit, played with a swivel chair, spinning it around.

"The court is concerned about public safety," Johnson said. "It's fortunate that no one was hurt. It does strike the court that he is very young."

Stanosheck said the boy was charged with three felony arson counts. The three charges are in connection with the Chicago Lumber fire and two other fires, on June 10 and Aug. 3, that caused $10,000 in damage at a vacant single-family home at 2015 Dupont St. An 11-year-old also is accused in the Dupont Street fires.

Stanosheck said the 10-year-old said in a signed statement that he knew who lived at the Dupont Street address and that he hated the person. Stanosheck added that the boy said one of the fires was set at the Dupont Street address because the group was bored.

(Continued)

(Continued)

Public Defender Katie Conrad said the boy would not fight the detention order. However, Jeffrey Wagner, the court-appointed guardian for the boy, said his client was at serious risk at the facility because of his size.

A woman identified as the boy's grandmother said the child needed to be in a more structured environment.

At the end of the hearing, the boy kissed his grandmother before being escorted from the courtroom.

The lumber company blaze the boy was accused of setting was by far the worst of the six fires in which the group is accused. Crews responding to the inferno along the west side of 14th Street between Pierce and Pacific streets contended with extreme heat, exploding propane tanks, and downed electrical wires. It took firefighters three hours to contain the fire, which was reported at 1:18 a.m.

Chicago Lumber plans to rebuild at the site of the fire.

Fire officials said the boys also are suspects in the following arsons, which caused a total of $56,000 in damage:

- A July 30 fire reported at 8:07 a.m. in a portable classroom at Castelar Elementary School, 2316 S. 18th St., that caused an estimated $6,000 in damage. A 10-year-old and an 11-year-old are accused of setting it.
- A July 1 fire reported at 2:39 a.m. in a detached garage at 2111 Martha St. that caused an estimated $10,000 in damage. An 11-year-old is accused of starting the fire.
- A May 25 fire reported at 12:41 a.m. in a home being remodeled at 2713 S. 17th St. Damage was estimated at $40,000. An 11-year-old is accused of setting the fire.

Each of the boys has been charged with second-degree arson for each fire he has been accused of starting.

Fire investigators said a man, whose name they wouldn't release, called last week and tipped them off to the group. The man told fire officials that some children had threatened to set fire to his house because he stopped allowing his child to play with them.

"He was the key that led us in a little direction of where to go with this," Gibilisco said.

Fire officials said the boys were arrested in their homes, at school, or at Central Police Headquarters. They were taken to the Douglas County Youth Detention Center and then released to their parents or guardians.

At first, Gibilisco said, the juveniles showed no remorse. He said it seemed like it never occurred to them that they would be caught.

"Only when they found out that they had to go to the detention center did they show any remorse," he said. "They cried when they realized they were going to get taken out of their homes and away from their families."

Gibilisco said investigators had not indicated whether the group had a leader. "They kind of like to point fingers at each other, but all of them stated their involvement," he said.

Adolf said he wasn't surprised that children are suspected because many children have a fascination with fire.

"When you recognize more than 50 percent of arson fires are committed by juveniles," he said, "it wasn't that surprising."

But Adolf said the case is unusual because of the number of fires and amount of damage.

SOURCE: Alexander, Deborah, & Deering, Tara. (2000, October 27). Boy charged with arson locked up: A judge orders the detention of a 10-year-old accused with three other juveniles in several fires. *Omaha World-Herald*, p.1.

Figure 6.1 Temporary Custody Hearing Order

SS:
IN THE COURT OF THE 9th JUDICIAL CIRCUIT McDONOUGH COUNTY, ILLINOIS
NO.

IN THE INTEREST OF: Ex Parte ☐
______________________ Without Prejudice ☐
Minor(s)

TEMPORARY CUSTODY HEARING ORDER
(705) ILCS 405/2-10)

THIS CAUSE coming to be heard upon the motion of ____________________ for ____________________.
The Court having jurisdiction over the matter and parties, and being fully advised in the premises.

THE COURT FINDS:

1. The minor's

Mother received	☐ notice	☐ no notice and was	☐ present	☐ not present
Father received	☐ notice	☐ no notice and was	☐ present	☐ not present
Guardian/ custodian/ relative received	☐ notice	☐ no notice and was	☐ present	☐ not present

2. Probable cause

☐ A. does not exist that the minor is (abused/neglected/dependent); or

☐ B. does exist that the minor is (abused/neglected/dependent)

The basis of the finding is:

____

____

____

3. Immediate and urgent necessity

☐ A. does not exist to support removal of the minor from the home; or

☐ B. does exist to support the removal of the minor from the home and remaining in the home is contrary to the child's welfare, safety, or best interest.

The basis of the finding is:

____

____

____

4. Reasonable efforts

☐ A. have been made but have not eliminated the immediate and urgent necessity to remove the child (children) from the home; or

☐ B. cannot reasonably be made at this time, for good cause, prevent or eliminate the necessity of removal of the minor(s) from the home; or

☐ C. have been made and have eliminated the immediate and urgent necessity to remove the child (children) from the home.

☐ D. have not been made.

The basis of the finding is:

____

____

____

(Continued)

Figure 6.1 (Continued)

5. Consistent with the health, safety and best interest of the minor
   - ☐ A. the minor shall be released to the parent; or
   - ☐ B. the minor shall be placed in shelter care.

Case #____________________

IT IS ORDERED:

- ☐ A. The petition is dismissed.
- ☐ B. Consistent with the health, safety and best interests of the minor, the minor shall (be returned to/remain) in the custody of the (mother/father/parents/guardian/custodian/responsible relative).
- ☐ C. Consistent with the health, safety, and best interests of the minor, the minor shall be removed from the home; and
  1. Temporary custody of the minor is granted to
     - ☐ a. private custodian/guardian ____________________ whose relationship to the minor is ____________________.
     - ☐ b. DCFS Guardianship Administrator with the right to place the minor.
  2. The temporary custodian is authorized to consent to:
     - ☐ a. ordinary and routine medical care AND major medical care* on behalf of the minor (temporary custody with the right to consent to major medical care).
     - ☐ b. only ordinary and routine medical care on behalf of the minor (temporary custody without the right to consent to major medical care.

       *Major medical care is defined as those medical procedures which are not administered or performed on a routine basis and which involve hospitalization, surgery, or use of anesthesia (e.g., appendectomies, blood transfusion, psychiatric hospitalization).
     - ☐ c.

       ____

       ____
     - ☐ d. A 405/2-25 or 405/2-20 Order of Protection entered this date is incorporated herein against

       ____

       ____
     - ☐ e. DCFS shall investigate the need for services in the following areas:

       ____

       ____

       ____
     - ☐ f. If there is a finding of no reasonable efforts under paragraph 4 above, DCFS shall make all reasonable efforts to ameliorate the causes contributing to the finding of probable cause or the immediate and urgent necessity which led to the removal of this child from the home. These efforts shall include

       ____

       ____

       ____
     - ☐ g. CFS shall prepare and file with the court on or before __________, 20 __, a 45 day Case Plan pursuant to 705 ILCS 405/2-10.1.
     - ☐ h. The Order on Visiting entered this date or on subsequent dates is incorporated herein by reference.

(Continued)

Figure 6.1 (Continued)

☐ i. A Social Investigation shall be filed by _______________, 20 __.

☐ j. The next hearing is set on ____________________, 20__, at ________a.m./p.m. for

☐ presentation of an affidavit of diligent efforts to notify ☐ progress report

☐ status ☐ adjudication hearing ☐ court family conference

☐ before the judge ☐ before the hearing officer

☐ The first permanency hearing date is _________________, 20__.

DATED: ____________________

ENTERED: ___________________________ ___________________________

JUDGE JUDGE'S NO.

(Text continued from page 147)

A substantial number of juvenile cases are "unofficially adjusted" by law enforcement personnel at the initial encounters as well as at the stationhouse. Among those juveniles who are turned over to the court's intake personnel, a substantial number are disposed of at the intake stage and at the detention hearings. In many instances, the intake personnel, the minor and his or her family, and the injured party are able to informally adjust the differences or problems that caused the minor to be taken into custody. Only the most serious cases of delinquency, cases of unruly behavior, and cases involving serious abuse or neglect result in processing through the entire juvenile justice system. There are both legal and ethical questions about unofficial dispositions at the intake stage and the assumption of guilt that often leads to some prescribed treatment program. Although most practitioners make it clear that participation in informal dispositions is voluntary and that following advice or referrals is not mandatory, there may still be some official pressure perceived by the juvenile or the juvenile's parents that violates the presumption of innocence.

Detention/Shelter Care

Under the Uniform Juvenile Court Act (National Conference of Commissioners on Uniform State Laws, 1968),

> A child taken into custody shall not be detained or placed in shelter care prior to the hearing on the petition unless such detention is required to protect the person or property of others or of the child or because the child may abscond [flee] or be removed from the jurisdiction of the court or because he has no parent or guardian who is able to provide supervision and to return him to the court when required or an order for detention or shelter care has been made by the court pursuant to this Act. (sec. 14)

The absence of any of these conditions must result in the child's release to his or her parents or guardian with their promise to bring the child before the court as requested (sec. 15(1)). Failure to bring the child before the court will result in the issuance of a warrant directing that the child be taken into custody and brought before the court (sec. 15(b)).

The Uniform Juvenile Court Act (National Conference of Commissioners on Uniform State Laws, 1968) requires that the "person taking a child into custody, with all reasonable speed and without first taking the child elsewhere, shall release the child to his parents or guardian . . . . unless detention or shelter care is warranted or required" (sec. 15(a)(1)). This section of the Uniform Juvenile Court Act is designed to reduce the number of children in detention by specifying criteria that would "require and warrant" further detention.

If reasonable cause for detention cannot be established, the juvenile should be released to his or her parents. In practice, and according to most juvenile court acts, the juvenile is taken to a police or juvenile facility, at which time the parents or guardian are contacted. However, the Uniform Juvenile Court Act implies that the juvenile should be taken immediately to his or her parents or guardian unless detention appears to be warranted. This policy spares the juvenile the experience of being held in the most depressing and intimidating of all custodial facilities—the jail or police lockup.

In some states, if the juvenile is not released to his or her parents or guardian, the juvenile must be taken without unnecessary delay to the court or to a place designated by the court to receive juveniles (*ILCS,* ch. 705, sec. 405/5-405, 1999). The Uniform Juvenile Court Act does allow detention in a local jail if, and only if, a detention home or center for delinquent children is unavailable (sec. 16(a)(4)). If the juvenile is confined in a jail, detention must be in a room separate and removed from the rooms for adults. This required separation from confined adults is commonly found in statutes and extends to cell, room, and yard and sometimes even to any sight or sound. In all categories other than delinquency, the child is normally taken to a designated shelter care facility, meaning a "physically unrestricted facility," according to the Uniform Juvenile Court Act (sec. 2(6)). The procedures for contacting the parents or guardian and the criteria used to maintain custody in such a facility are the same as for the delinquent child. Shelter care facilities are generally licensed by the state and designated by the juvenile court to receive children who do not require the physically restrictive surroundings of a jail or juvenile detention center.

Maximum time limits for detention are set forth in the various juvenile court acts so that a juvenile will not be detained for lengthy periods without a review by the courts. In some cases, the issue of bail may arise (see discussion earlier in this chapter).

Once the juvenile has been taken into custody and either released to his or her parents or guardian or, with just cause, placed in a detention facility, an officer of the court may attempt to settle the case without a court hearing by arranging for a preliminary conference.

The Preliminary Conference

The Uniform Juvenile Court Act (sec. 10) includes a provision that allows a probation officer or other officer designated by the court to hold a preliminary conference so as to give counsel or advice with a view toward an informal adjustment without filing a petition. This preliminary conference is in order only if the admitted facts bring the case within the jurisdiction of the court and if such an informal adjustment, without an adjudication, is in the best interests of the public and the child. The conference is to be held only with the consent of the juvenile's parents or guardian. However, such a conference is not obligatory (sec. 10(a)). A similar provision is found in the Illinois Juvenile Court Act, which states that "the court may authorize the probation officer to confer in a preliminary conference with any person seeking to file a petition . . . concerning the advisability of filing the petition, with a view to adjusting suitable cases without the filing of a petition" (*ILCS*, ch. 705, sec. 405/5-305, 1999). If agreement between the parties can be reached at the preliminary conference, no further official action may be necessary. If judicial action seems necessary, the probation officer may recommend the filing of a petition. However, if the injured party demands that a petition be filed, that demand must be satisfied. Although the preliminary conference or informal adjustment may be of value in diverting cases that could be better settled outside of juvenile court, it has been subject to criticism as a method of engaging in legal coercion without trial (Tappan, 1949, pp. 310–311). In general, information or evidence presented at the preliminary conference is not admissible at any later stage in the juvenile court proceedings.

The Petition

As indicated earlier, juvenile court proceedings begin with the filing of a petition naming the juvenile in question and alleging that this juvenile is delinquent, dependent, abused, neglected, or a minor in need of intervention/supervision. A copy of a sample petition is shown in Figure 6.2. Although states vary regarding who is eligible to file a petition, similarities do exist concerning the content of petitions and the initiation of follow-through activities as a result of the petition. In some states, a preliminary inquiry may be conducted by juvenile court personnel to determine whether the best interests of the child or the public will require that a petition be filed. In other states, this inquiry is accomplished after the petition has been filed and may result in the petition being dismissed by the court if the alleged facts are not supported. Regardless of whether the inquiry is conducted before or after the filing of a petition, a stipulation that is commonly found is one in which a court authorizes a person to endorse the petition as being in the best interests of the public and the child. The Uniform Juvenile Court Act (National Conference of Commissioners on Uniform State Laws, 1968) specifies that "a petition may be made by any person who has knowledge of the facts alleged or is informed and believes that they are true" (sec. 20). The act also states that "the petition shall not be filed unless the court or designated person has determined and endorsed upon the petition that the filing is in the best interest of the child and the public" (sec. 19). It should be noted that the signing of a petition and the authority to file the petition may be separate and distinct acts. This has led to some confusion. Some states require designated court personnel to sign the petition to establish some sufficiency of the allegations at the outset.

The contents of the petition are governed by statutory requirements in each juvenile court act. The petition may be filed on "information and belief" rather than on verified facts necessary for an adjudicatory hearing. The petition is generally prefaced with the words "in the interests of." The petition continues by giving the name and age of the child and frequently giving the names and addresses of the parents. It typically indicates whether the minor is currently in detention. Also included in the petition is the statement of facts that bring the child within the jurisdiction of the juvenile court. This particular requirement has been a troublesome area because questions are often raised about whether sufficient facts have been stated and about the specificity of the charges. According to the Uniform Juvenile Court Act (sec. 21(1)), the petition must also contain allegations that relate to the child's need of treatment or rehabilitation if delinquency or unruly conduct is alleged. Once the petition has been filled out, it is filed with the prosecutor, who then decides whether or not to prosecute. If the prosecutor decides to go ahead with the case, proper notice must be given to all concerned parties.

Figure 6.2 Petition for Adjudication of Wardship

In the Circuit Court for the Ninth Judicial Circuit, McDonough County, Illinois

IN THE INTEREST OF:)

)

PETITION FOR ADJUDICATION OF WARDSHIP

I, ________________, State's Attorney, on oath state on information and belief:

1. That ________________ is a male/female minor, born on ________________, who resides or may be found at ________________________, McDonough County, Illinois.
2. The names and residence addresses of the minor's parents are:

   ____

   The minor and the persons named in this paragraph are designated respondents.
3. That the minor is delinquent by reason of the following:

   ____
4. The minor is/is not in detention custody;
5. It is in the best interests of the minor and the public that the minor be adjudged a ward of the Court. I ask that the minor be adjudged a ward of the Court and for other relief under the Juvenile Court Act.

I have read the aforesaid Petition for Adjudication of Wardship and do hereby swear that the facts contained herein are true and correct to the best of my knowledge and belief.

____

Assistant State's Attorney

Subscribed and sworn to before me this _____ day of ________, 2007.

____

Notary Public

Assistant State's Attorney

McDonough County

McDonough County Courthouse

Macomb, Illinois 61455

Notification

In establishing a **notification** requirement (all interested parties be given official notice of time, places, and changes), the U.S. Supreme Court in *Gault* set forth two conditions that must be met: timeliness and adequacy. Although petitions might not need to meet all of the legal requirements of an indictment, they do need to describe the alleged misconduct with some particularity so that all parties involved are clear as to the nature of the charges involved.

Delinquency petitions, for example, must contain sufficient factual details to inform the juvenile of the nature of the offense leading to allegation of delinquency and must be sufficient to enable the accused to prepare a defense to the charges.

Once a petition has been filed, the court will issue a summons to all concerned adult parties informing them of the time, date, and place of the adjudicatory hearing and of the right of all parties to counsel. In addition, many states direct a separate summons to the child who is over a certain age and is within a designated category such as delinquent or unruly child. A copy of the petition will accompany the summons unless the summons is served "by publication" (printed in a newspaper of reasonable circulation). States vary regarding the length of time required between the serving of the summons and the actual proceedings. However, in accordance with the *Gault* decision, a reasonable amount of time should be allowed to provide the parties with sufficient time to prepare. However, unnecessary and long delays should be avoided, particularly in those cases where a child is held in detention or shelter care. For example, Illinois allows at least 3 days before appearance when the summons is personally served to the parties, 5 days when notification is by certified mail, and 10 days when notification is by publication. If it becomes necessary to change dates, notice of the new dates must be given, by certified mail or other reasonable means, to each respondent served with a summons (*ILCS*, ch. 705, sec. 405/5-525, 1999). Illinois law and the Uniform Juvenile Court Act (sec. 23(a,b)) provisions on service of summons are similar. The Uniform Juvenile Court Act allows at least 24 hours before the hearing when the summons is personally served and 5 days when certified mail or publication is used.

Service of the summons may be made by any person authorized by the court, usually a county sheriff, coroner, or juvenile probation officer. If the information received by the court indicates that the juvenile needs to be placed in detention or shelter care, the court may endorse on the summons an order that the child should be taken into immediate custody and taken to the place of detention or shelter care designated by the court.

Following the filing of the petition and proper notification, the adjudicatory hearing is held. In delinquency cases, this is the juvenile court's equivalent of an adult criminal trial.

The Adjudicatory Hearing

The adjudicatory hearing is a fact-finding hearing to determine whether the allegations in the petition are valid. In delinquency cases, it is the rough equivalent of a criminal trial. In cases of dependency, neglect, or authoritative intervention, the adjudicatory hearing more closely resembles a civil trial. Although the U.S. Supreme Court has extended the legalistic principle of due process to the juvenile justice system, not all rights accorded under the Constitution and its amendments have been incorporated into the juvenile system. For example, in 1971 the Court held that juveniles had no constitutional right to a jury trial because the juvenile proceeding had not yet been held to be a criminal prosecution within the meaning and reach of the Sixth Amendment (*McKeiver v. Pennsylvania,* 1971). The Court reiterated that the due process standard of "fundamental fairness" should be applied to juvenile court proceedings. However, the Court further stated that it was unwilling to "remake the juvenile proceeding into a full adversary process." As indicated previously, some states do currently allow trial by jury. However, most cases are tried by a juvenile judge. The Uniform Juvenile Court Act (sec. 24(a)) recommends that hearings be conducted by the court without a jury. The Supreme Court was clear in its holding that when the state undertakes to prove a child delinquent for committing a criminal act, it must do so beyond a reasonable doubt (*In re Winship*, 1970). The Uniform Juvenile Court Act not only advocates this standard of proof for the delinquency issue but also extends this standard to the unruly category (sec. 29(b)). Some states have adopted this recommended standard (*Georgia Code Annotated*, 15-11-33 (c), 1999; *New York Family Court Act*, 342.2 (2), 744 (b) McKinney, 1999; *North Dakota Century Code*, 27-20-29 (2), 1991; *Texas Family Code Annotated*, 54.03 (f), West, 1996). The standard applicable to categories such as deprived, abused/neglected, and dependent is usually the civil standard of preponderance of evidence or **clear and convincing evidence.** For example, the Uniform Juvenile Court Act (sec. 29(c)) requires "beyond a reasonable doubt" to determine delinquency but allows the civil standard of "clear and convincing evidence" to determine whether the adjudicated delinquent is in need of treatment or rehabilitation. In general, of course, it is more difficult to establish guilt beyond a reasonable doubt (no reasonable doubt in the mind of the judge) than to determine guilt based on a preponderance of evidence.

The adjudicatory hearing is generally closed to the public. If the juvenile court judge agrees, certain persons, agencies, or associations that have a direct interest in the case may be admitted. Although the **Sixth Amendment** declares that "in all criminal prosecutions, the accused shall enjoy the right to a speedy and public trial," juvenile court acts prohibit these public hearings on the grounds that opening such hearings would be detrimental to the child. Although the application of the "public trial" concept of the Sixth Amendment has not been adopted in most juvenile court acts, other due process provisions of the amendment have been incorporated into juvenile court acts as a result of the *Gault* decision. The Uniform Juvenile Court Act (sec. 24(d)) states that the general public shall be excluded except for parties, counsel, witnesses, and other persons requested by a party and approved by the court as having an interest in the case or in the work of the court. Those persons having an interest in the work of the court include members of the bar and press who may be admitted on the condition that they will refrain from divulging any information that could identify the child or family involved.

As discussed previously, the due process concept of "speedy trial" contained in the Sixth Amendment has been incorporated into juvenile court acts. Specific time frames are contained in most acts designating the length of time between custody, detention, adjudicatory, and disposition hearings. Requests for delay are entertained by the juvenile court whenever reasonable and justifiable motions are submitted. Unfortunately, it has been common in some jurisdictions for juvenile court judges to ignore the time limits established by the statute, so a speedy trial might not result. Some judges appear to ignore the statutory requirement of an adjudicatory hearing within 30 days of the time the petition is filed (without detention) even when there is no motion for a continuance by defense counsel (Butts, 1997; Schwartz, Weiner, & Enosh 1999). Although this practice has been overturned in the New York Court of Appeals (*In re George T.*, 2002), it is still a fairly common practice that the juvenile might not be brought before the court for an adjudicatory hearing for as long as 6 months—a clear violation of the statutory requirement. It is possible, of course, for defense counsel to move for dismissal or to appeal, but very seldom are such actions taken. When motions to dismiss based on procedural irregularities are made, they are almost routinely overruled. Once again, the gap between theory and practice comes to light.

According to the Uniform Juvenile Court Act (sec. 29), after hearing the evidence on the petition, the court will make and file its findings about whether the child is deprived, delinquent, abused, neglected, or unruly as alleged in the petition. If the evidence does not support the allegation, the petition will be dismissed, and the child will be discharged from any detention or other restrictions. If the court finds that the allegation is supported by evidence using the appropriate standard of proof for that hearing, the court may proceed immediately or hold an additional hearing to hear evidence and decide whether the child is in need of treatment or rehabilitation. In the absence of evidence to the contrary, the finding of delinquency where felonious acts were committed is sufficient to sustain a finding that the child is in need of treatment or rehabilitation. However, even though the court may find that the child is within the alleged criteria of the petition, it might not find that the child is in need of treatment or rehabilitation. The court may then dismiss the proceeding and discharge the child from any detention or other restrictions (sec. 29(a,b)).

It should also be noted that juvenile court judges in many states may decide prior to or in the early stages of the adjudicatory hearing to "continue the case under supervision." An example of an order for **continuance under supervision** is shown in Figure 6.3. This usually means that the judge postpones adjudication and specifies a time period during which the judge (through court officers) will observe the juvenile. If the juvenile has no further difficulties during the specified time period, the petition will be dismissed. If the juvenile does get into trouble again, the judge will proceed with the original adjudicatory hearing.

Continuance under supervision may benefit the juvenile by allowing him or her to escape adjudication as delinquent. It is generally used by juvenile court judges for precisely this purpose. However, if the juvenile did not commit the alleged delinquent act, he or she may be unjustly subjected to court surveillance. If the juvenile's parents or counselor object to the procedure and request the judge to proceed with the adjudicatory hearing, the judge must, in most jurisdictions, comply with those wishes.

In the adjudicatory hearing, the Uniform Juvenile Court Act and the juvenile court acts of many states separate the issues of establishing whether the child is within the defined category and whether the state should exercise wardship or further custody. The determination of further custody or wardship is usually made on the basis of what type of treatment or rehabilitation the court believes is necessary.

The term *ward of the court* means simply that the court, as an agency of the state, has found it necessary to exercise its role of *in loco parentis.* The decisions that are normally made by the parents are now made by a representative of the court, usually the juvenile probation officer in consultation with the juvenile court judge. As indicated in the Uniform Juvenile Court Act (sec. 29(c,d)), the determination for continued custody for treatment or rehabilitation purposes may be made as part of the adjudicatory hearing or in a separate hearing. The court, in determining wardship, will receive both oral and written evidence and will use this evidence to the extent of its probative value even though such evidence might not have been admissible in the adjudicatory hearing. The standard of clear and convincing evidence is recommended by the Uniform Juvenile Court Act (sec. 29(c)) in determining wardship. The Uniform Juvenile Court Act (sec. 29(e)) also permits a continuance of hearings for a reasonable period to receive reports and other evidence bearing on the disposition or the need of treatment or rehabilitation. The child may be continued in detention or released from detention and placed under the supervision of the court during the period of continuance. Priority in wardship or dispositional hearings will always be given to those children who are in detention or have been removed from their homes pending a final dispositional order.

To avoid giving a child a record, it has become a common practice in some jurisdictions for juvenile courts to place a child under probation supervision without reaching any formal finding. This practice may be engaged in without filing any formal petition. Placing children

Figure 6.3 Continuance Under Supervision Form

IN THE CIRCUIT COURT FOR THE NINTH JUDICIAL CIRCUIT, McDONOUGH COUNTY, ILLINOIS

IN THE INTEREST OF:)
)
a MINOR.)

CONTINUANCE UNDER SUPERVISION

(Before Adjudication)

This cause coming before the Court on the Motion of the Petitioner for an Order of Continuance Under Supervision (Before Adjudication) pursuant to Chapter 705, Act 405, Section 5-19 of the Illinois Compiled Statutes.

And the Court having been fully advised in the premises and there being no objection made in Open Court by the minor, his counsel, parents, guardian, or responsible relative, the Court finds that the Petition has been proved by stipulation of the parties in the manner and form as alleged in the Petition for Adjudication of Wardship signed and sworn on______________________, 2007.

NOW THEREFORE IT IS ORDERED that this matter is continued until ____________________, at _______ p.m. The minor shall be subject to the following conditions during the period of said continuance:

1. That the minor shall not violate any criminal statute or city ordinance of any jurisdiction;
2. That the minor shall not possess a firearm or any other dangerous weapon;
3. That the minor shall not leave the State of Illinois without written permission of the State's Attorney's Office and the Probation Officer;
4. That the minor shall attend school while it is in session without any absences unless excused by the school; shall abide by all school rules; and shall cooperate with school officials;
5. That the minor shall report to the Juvenile Probation Officer as directed by the officer and shall permit the officer to visit him at any time or place, with or without prior notice, and he shall at all times abide by the directives of the Probation Officer;
6. That the minor shall notify the Probation Officer within twenty-four (24) hours of a change of address or of any arrest or traffic ticket;
7. That the minor shall follow his parents' rules of supervision;
8. That the minor shall write a letter of apology to ________________________ apologizing for his actions;
9. That the minor shall pay probation fees in the amount of $ _________;
10. That the minor shall consent to having his photograph taken by the Probation Officer to be placed in the Probation file;
11. That the minor shall participate in and successfully complete the LIFT Program if accepted into said program;
12. That the minor shall obtain a mental health evaluation at the Fulton/McDonough County Community Mental Health Center and shall successfully complete all recommendations of said evaluation;

DATED: _________________ ____________________________

JUDGE

under probation supervision should not be confused with continuances granted by the court to complete investigations for wardship or disposition proceedings. Although "unofficial probation or supervision" may help to divert less serious cases from adjudication and thus avoid stigmatizing the child involved, it has been subject to much criticism as the result of disregarding due process requirements.

The Social Background Investigation

After a determination in the adjudicatory hearing that the allegations in the petition have been established and that wardship is necessary, a dispositional hearing is set to determine final disposition of the case. There are differences among the states as to whether the dispositional hearing must be separated from the adjudicatory hearing (*California Welfare and Institutional Code,* 701, 702 West, 1998; *Georgia Code Annotated,* 15-11-33 (b), (c), 1999; *ILCS,* 705, sec. 405/2-22 (1), 1999). In some states, the two hearings are separate because different procedures and rights are involved. For example, in some states in an adjudicatory hearing on delinquency, the standard of proof and the rules of evidence in the nature of criminal proceedings are applicable; however, the civil rules of evidence and standard of proof are applicable to adjudicatory hearings on neglect, dependent, abuse, and minor requiring authoritative intervention (in need of supervision) cases (*California Welfare and Institutional Code,* 405/2-18 (1) West, 1998; *ILCS,* ch. 705, sec. 405/2-18(1), 1999; *Iowa Code Annotated,* 232.47 (5) West, 1999). Yet in the Illinois dispositional hearing for all categories, all evidence helpful in determining the disposition, including oral and written reports, may be admitted and relied on to the extent of its probative value even though it might not be competent for the purposes of the adjudicatory hearing (*ILCS,* sec. 405/5-22, 1999). Similar wording and evidentiary concepts are contained in the Uniform Juvenile Court Act's (sec. 29(d)) references to determination of whether the adjudicated child requires treatment and rehabilitation and to the dispositional stage of the case.

Between the adjudicatory hearing and the dispositional hearing, the court's staff members (usually probation officers) are engaged in obtaining information useful in aiding the court to determine final disposition of a case. This information is obtained through **social background investigations** and is premised on the belief that individualized justice is a major function of the juvenile court. Social background investigations typically include information about the child, the child's parents, school, work, and general peer relations as well as other environmental factors. This information is gathered through interviews with relevant persons in the community and is compiled in report form to aid the judge in making a dispositional decision. The probative value of some information collected is questionable and can certainly be challenged in the dispositional hearing. Some juvenile judges delegate the court's staff to make recommendations and to justify the elimination of some options or alternatives from consideration. Unfortunately, social background investigations have been used by some courts prior to the adjudicatory hearings, and this can result in an adjudication of delinquency without proving that the accused juvenile did commit the acts of delinquency alleged in the petition. As a result of the *Kent* decision (*Kent v. United States,* 1966), counsel for the juvenile has been extended the right to review the contents of staff social background investigations used in waiver hearings because there is no irrefutable presumption of accuracy attached to staff reports. This principle has been extended by most juvenile court acts to legal counsel representing the child in dispositional hearings.

The Dispositional Hearing

Whereas the adjudicatory hearing determines whether the allegations are supported by the evidence, the dispositional hearing is concerned only with what alternatives are available to meet the needs of the juvenile. In fact, some states specify by statute that the rules of evidence do not apply during dispositional proceedings (*District of Columbia Code,* 16-2316 (b), 1997; *Georgia Code Annotated,* 15-11-33 (d), 1999; *ILCS,* 705 sec. 405/2-22 (1), 1999). Dispositional alternatives are clearly stated in each state's juvenile court act. The state may differ in the dispositional alternatives available to juveniles in the separate categories. An option available for the deprived child might not be available for the delinquent child. According to the Uniform Juvenile Court Act (sec. 30), the deprived child may remain with his or her parents, subject to conditions imposed by the court, including supervision by the court. Also according to Section 30, the deprived child may be temporarily transferred legally to any of the following:

(i) any individual . . . found by the court to be qualified to receive and care for the child;
(ii) an agency or other private organization licensed or otherwise authorized by the law to receive and provide care for the child;
(iii) the Child Welfare Department of the [county] [state] [or other public agency authorized by law to receive and provide care for the child]; or
(iv) an individual in another state with or without supervision.

For the delinquent child, the Uniform Juvenile Court Act (sec. 31) states that the court may make any disposition best suited to the juvenile's treatment, rehabilitation, and welfare, including the following:

(1) any order authorized by Section 30 for the disposition of a "deprived child";
(2) probation under the supervision of the probation officer . . . under conditions and limitations the court prescribes;
(3) placing the child in an institution, camp, or other facility for delinquent children operated under the direction of the court [or other local public authority]; or
(4) committing the child to [designate the state department to which commitments of delinquent children are made or, if there is no department, the appropriate state institution for delinquent children].

According to the Uniform Juvenile Court Act (sec. 32), the unruly child may be disposed of by the court in any authorized disposition allowable for the delinquent except commitment to the state correctional agency. However, if the unruly child is found to be not amenable to treatment under the disposition, the court, after another hearing, may make any disposition otherwise authorized for the delinquent.

A general trend occurring in juvenile court acts is to refrain from committing all categories, other than delinquents, to juvenile correctional institutions unless the unruly or "in need of supervision" child warrants such action after other alternatives have failed. Commitment to an institution is generally regarded as a last resort.

Most juvenile court acts also provide for transferring a juvenile demonstrating mental retardation or mental illness to the appropriate authority within the state. A similar section is included in the Uniform Juvenile Court Act (sec. 35). With the advent of a multiplicity of community treatment programs and child guidance centers, many of the current dispositions contain conditions for attendance at these centers. Dispositions of probation or suspended sentence often require compulsory attendance at a community-based treatment or rehabilitation program. Violation of these conditions may result in revocation of probation or a suspended sentence. This is accomplished through a revocation hearing. Most states now specify the maximum amount of time for confinement of a juvenile. Extensions of the original disposition generally require another hearing with all rights accorded in the original dispositional hearing. The court may, under some circumstances, terminate its dispositional order prior to the expiration date if it appears that the purpose of the order has been accomplished. Juvenile court acts generally terminate all orders affecting the juvenile on reaching the age of majority in those states. This termination results in discharging the juvenile from further obligation or control. If the disposition is probation, both the conditions of probation and its duration are spelled out by the court. For copies of dispositional and sentencing court orders, see Figures 6.4 and 6.5.

Figure 6.4 Dispositional Order

IN THE CIRCUIT FOR THE NINTH JUDICIAL CIRCUIT, McDONOUGH COUNTY, ILLINOIS

IN THE INTEREST OF:)
)
both MINORS.)

DISPOSITIONAL ORDER

This cause coming to be heard for the purposes of a Dispositional

Hearing, ________________________, Assistant State's Attorney for McDonough County present, with ________________________ of the Illinois Department of Children and Family Services; Guardian ad Litem present for the minor(s), ______________; Attorney present with respondent father, ______________; and Attorney present with respondent mother, ______________.

The Court, having received the evidence and heard the arguments of counsel, having jurisdiction and being fully advised in the premises FINDS:

1. That it is in the best interests of the minors that Guardianship shall be granted to the Guardianship Administrator of the Illinois Department of Children and Family Services.

IT IS HEREBY ORDERED:

1. That Guardianship of the children shall be granted to the Guardianship Administrator of the Illinois Department of Children and Family Services, with the Department having the right to consent to medical and dental care and the right to place;
2. That ________________________ shall successfully complete the ________________________ program, individual counseling, and all other services as outlined in the client service plan;

(Continued)

Figure 6.4 (Continued)

3. That ________________________ shall successfully complete counseling, participate in Victim Services programs, secure and maintain safe and appropriate housing, and complete all other services as outlined in the client service plan;
4. That ________________________ shall adhere to his/her safety plan regarding contact between ______________ and the minor(s), ______________;
5. That________________________shall cooperate with the Illinois Department of Children and Family Services, shall comply with the client service plan, and correct the conditions which led to the Department's involvement or shall risk loss of custody and possible termination of parental rights;
6. That the Permanency Goal shall be ______________________________;
7. That ___________________________ shall comply with all early childhood education services for ______________;
8. That a Status Hearing shall be held on _____________, at ____________ a.m./p.m.

DATED: ___________________ ____________________________

JUDGE

Figure 6.5 Sentencing Order

STATE OF ILLINOIS IN THE CIRCUIT COURT FOR THE NINTH JUDICIAL CIRCUIT COUNTY OF McDONOUGH

CASE NO. ___ JD ___

IN THE INTEREST OF:

a MINOR.

Date of hearing:

Parties present for hearing:

Assistant State's Attorney:

Minor: Attorney for Minor:

Mother: Attorney for Mother:

Father: Attorney for Father:

SENTENCING ORDER

THIS MATTER comes before the Court for hearing on the date noted above with the parties indicated being present. The parties have been advised of the nature of the proceedings as well as their rights and the dispositional alternatives available to the Court. The minor admits the allegations of Count I _________ of the Petition filed _________. The Court makes the following FINDINGS:

___ 1. The Court has jurisdiction of the subject matter.

___ 2. The Court has jurisdiction of the parties.

___ 3. The admission by the minor is knowingly and voluntarily made.

___ 4. The minor has signed an Admission form.

(Continued)

Figure 6.5 (Continued)

___ 5. There is a factual basis for the admission by the minor.
___ 6. The parties have agreed to a sentencing recommendation with regard to this matter.

THEREFORE, it is the ORDER OF THIS Court that the request of the parties for an immediate sentencing hearing is GRANTED:

THIS MATTER then proceeds to sentencing hearing. Both parties waive the preparation of a social investigation. The agreement of the parties is heard. The Court makes the following FINDINGS:

1. The agreement of the parties is in the best interests of the minor.
2. The agreement of the parties should be affirmed and incorporated in the Sentencing Order of this Court.

THEREFORE, it is the ORDER of this Court that:

1. The minor is adjudicated to be a delinquent minor.
2. The minor is made a Ward of this Court.
3. The minor is placed on probation pursuant to 705 ILCS 405/5-23 for a period of _________.
4. This probation is conditioned upon the following terms and conditions:

___ The respondent minor shall obey his/her parents rules of supervision; and
___ The respondent minor shall attend school regularly and put forth his or her best efforts; and
___ The respondent minor shall maintain a 9:00 p.m. to 7:00 a.m. curfew unless he or she is accompanied by a parent or responsible adult. Discretion shall be left to the Juvenile Probation Officer to adjust the respondent minor's curfew; and
___ The respondent minor shall not possess any firearm or other dangerous weapon; and
___ The respondent minor and parents shall cooperate with the Juvenile Probation Officer in any and all programs deemed to be in the minor's best interest; and
___ The respondent minor shall meet with the Juvenile Probation Officer as directed; and
___ The respondent minor shall notify the Juvenile Probation Officer within twenty-four (24) hours of a change in address; and
___ The respondent minor shall reside in McDonough County unless authorized to reside elsewhere by the Probation Officer, the State's Attorney's Office and the Court; and
___ The respondent minor shall not leave the State of Illinois without the approval of the Probation Officer, the State's Attorney's Office, and the Court; and
___ The respondent minor shall obey all federal, state, and local laws; and
___ The respondent minor shall obtain a Drug and Alcohol evaluation and follow any and all recommendations of the evaluator; and
___ The respondent minor shall obtain a Mental Health evaluation and follow any and all recommendations of the evaluator; and
___ The respondent minor shall refrain from having in his or her body the presence of any illicit drug prohibited by the Cannabis Control Act or the Illinois Controlled Substance Act, unless prescribed by a physician. Furthermore, if the Juvenile Probation Officer receives any report of illicit drug or alcohol use by the minor, the minor shall submit to random drug screens to determine the presence of any illicit drug. The minor shall be responsible for all related costs to the drug screens; and

(Continued)

Figure 6.5 (Continued)

___ The respondent minor shall pay $ ________ in restitution to the McDonough County Circuit Clerk's Office for the benefit of the victim in this case; and

___ The respondent minor shall draft a letter of apology to __________. The letter shall be approved and sent by the Probation Officer. The minor shall continue to revise the letter until the Probation Officer is satisfied with the content of the letter; and

___ Other terms and conditions shall be:

THIS ORDER MAY BE MODIFIED BY THE COURT.

Entered this ___ day of __________, 2007.

____

JUDGE

Career Opportunity—Magistrate

Job Description: Determine whether probable cause exists when the police make arrests. Determine whether, and ensure that, defendants have been properly advised of their rights. Decide whether or not to detain defendants. Supervise preliminary hearings, hold trials, and sentence offenders.

Employment Requirements: Must have a law degree and be admitted to the bar.

Beginning Salary: Varies widely depending on jurisdiction. Benefits also vary widely.

SUMMARY

It is essential that those involved in the juvenile justice network be completely familiar with appropriate procedures for dealing with juveniles and with the rules governing other members of the juvenile justice system. This awareness helps to ensure that the interests of juveniles will be protected within the guidelines established by society. Otherwise, juveniles' rights may be violated, practitioners may be put in a position where they cannot take appropriate actions, and society might not be protected as a result of ignorance of proper procedures.

For example, a police officer may take a juvenile into custody for a serious delinquent act (e.g., robbery). The officer may, on interrogation, obtain a confession from the juvenile. It may be impossible for the prosecutor to prosecute if the police officer failed to warn the juvenile of his or her rights according to Miranda, if a reasonable attempt to contact the juvenile's parents was not made, if the juvenile was frightened into confessing when his or her parents or legal representative were not present, or if the evidence in the case was obtained illegally. Of course, there will be no adjudication by the judge, and rehabilitation/corrections personnel will have no chance to rehabilitate, correct, or protect through detention. In the long run, then, neither the best interests of society nor those of the juvenile will be served.

Every state has a juvenile court act spelling out appropriate procedures for dealing with juveniles from the initial apprehension through final disposition. In looking at several juvenile court acts, we have seen that there are many uniformities in these acts as well as many points of disagreement. Uniformities are often the result of U.S. Supreme Court decisions, whereas differences often result from legislative efforts in the individual states. It is crucial, therefore, for all juvenile justice practitioners to become familiar with the juvenile court act under which they operate so that the best interests of juveniles, other practitioners, and society may be served to the maximum extent possible.

Note: Please see the Companion Study Site for Internet exercises and Web resources. Go to www.sagepub.com/juvenilejustice6study.

Critical Thinking Questions

1. What are the constitutional rights guaranteed to adults in our society that are not always guaranteed to juveniles in juvenile court proceedings? What is the rationale for depriving juveniles of these rights?
2. What are the benefits of the current trend toward a more legalistic stance in juvenile court proceedings? Are there any disadvantages for juveniles in this trend? If so, what are they?
3. What are the strengths and weaknesses of informal adjustments, unofficial probation, and continuance under supervision?
4. Discuss the pros and cons of allowing jury trials in juvenile court.
5. Assume that a 15-year-old male has been caught shoplifting a small transistor radio at a local discount store. The security guard at the store calls the police. The police officer arriving on the scene has settled similar disputes between this particular juvenile and the management of the chain store on several previous occasions. In addition, the police officer knows that, besides shoplifting frequently, the juvenile frequently runs away from home and is gone for days at a time. The security guard and store manager are determined to prosecute the juvenile. Based on the juvenile court act in your particular state, answer the following questions:
   a. What steps must the police officer take after he or she has taken the juvenile into custody?
   b. What steps may the probation officer take in an attempt to settle the dispute?
   c. If further custody is a consideration, what steps must be taken to continue such custody?
   d. Assuming that a petition has been filed, what are the juvenile court's obligations with respect to all parties to the proceedings?

e. What are the two major findings to be determined at the adjudicatory hearing? What degree of proof is required? What are the juvenile's rights during the hearing?
f. At the dispositional hearing, what alternatives are available to the juvenile court judge, and what information does the judge have at hand to help to arrive at the proper disposition?

Suggested Readings

Cohn, A. W. (2000). Juvenile focus. *Federal Probation, 64*(1), 73–75.

Davis, S. M. (2006). *Rights of juveniles: The juvenile justice system.* (2nd ed.). Eagan, MN: Thomson/West.

Fagan, J. (2005, September). Adolescents, maturity, and the law. *American Prospect, 16,* A5–A7.

Gahr, E. (2001). Judging juveniles. *American Enterprise, 12*(4), 26–28.

In the matter of George T. (2002). *New York Law School Law Review, 47*(2/3). Available: www.nyls.edu

Mohr, W., Gelles, R. J., & Schwartz, I. M. (1999). Shackled in the land of liberty: No rights for children. *Annals of the American Academy of Political & Social Science, 564,* 37–55.

Myers, W. (2006). *Roper v. Simmons:* The collision of national consensus and proportionality review. *Journal of Criminal Law and Criminology, 96,* 947–994.

National Council of Juvenile and Family Court Judges. (2005). *Juvenile delinquency guidelines: Improving court practice in juvenile delinquency cases.* Washington, DC: U.S. Department of Justice, Office of Juvenile Justice and Delinquency Prevention.

Schiraldi, V., & Drizin, S. (1999). 100 years of the children's court: Giving kids a chance to make better choices. *Corrections Today, 61*(7), 24.

JUVENILES AND THE POLICE

CHAPTER LEARNING OBJECTIVES

On completion of this chapter, students should be able to

- *Discuss the importance of police discretion in juvenile justice*
- *Compare and contrast unofficial and official police procedures in dealing with juveniles*
- *Discuss the importance of training police officers to deal with juveniles*
- *Describe police–school liaison programs*
- *Discuss the impact of community policing on the relationships among the police, school authorities, and juveniles*

KEY TERMS

Police discretion

Demeanor

Overpolicing

Street corner adjustment

Mandated reporters

Official procedures

Police–school liaison officers

DARE

GREAT

Youth-Focused Community Policing

SHOCAP

One of the first specializations in police departments following World War II was the juvenile bureau. Juvenile bureaus grew in number during the 1940s, 1950s, and 1960s until virtually all police departments of all sizes had them by the 1970s (Mays & Winfree, 2000, pp. 61–62). Historically, the police are the first representatives of the juvenile justice network to encounter delinquent, dependent, and abused or neglected children (Kratcoski & Kratcoski, 1995). The importance of the police in the juvenile justice system is considerable for this very reason. If the police decide not to take into custody or arrest a particular juvenile, none of the rest of the official legal machinery can go into operation. In fact, although the police often decide not to take official action when dealing with juveniles, roughly 85% of all cases referred to juvenile court are referred by the police (Drowns & Hess, 1990, p. 138).

▲ Are police officers' decisions concerning whether or not to take official action with respect to juveniles affected by the demeanor of the juvenile involved?

Police Discretion in Encounters With Juveniles

It is well established that a considerable amount of **police discretion** (individual judgment concerning the type of action to take) is exercised in handling juveniles. Although the exercise of discretion is a necessary and normal part of police work, the potential for abuse exists because there is no way to routinely review this practice. Police officers are often inconsistent in the decision-making process because, as Meehan (1992) stated,

> What is most revealing about the bulk of police problems with juveniles is the obvious ambiguity with respect to whether any formal rule of law applies in the instances where officers are summoned by the citizen's call for service. Yet the officer is nonetheless held organizationally accountable for the reported problem. (p. 475)

There are a number of cues to which most police officers respond in making decisions about whether to take official action against a particular juvenile. These cues include the following:

1. The wishes of the complainant
2. The nature of the violation
3. The race, attitude, and gender of the offender
4. Knowledge about prior police contacts with the juvenile in question
5. The perceived ability and willingness of the juvenile's parents to cooperate in solving the problem
6. The setting or location (private or public) in which the encounter occurs (Black & Reiss, 1970; Mays & Winfree, 2000; Piliavin & Briar, 1964; Regoli & Hewitt, 1994; Werthman & Piliavin, 1967)

In general, the wishes of the complainant and the nature of the offense weigh heavily on police officers' decision to arrest. If the offense is serious (e.g., a violent robbery), the police are generally expected by their department and by the public to arrest, and under most circumstances they do so. There is some evidence, however, that the police might not arrest, even for a serious offense, if the complainant does not wish to pursue the matter (Davis, 1975). If the offense is minor and the complainant does not desire to pursue the matter, the police will often handle the case unofficially (Allen, 2005). Again, in the case of a minor offense, the police will often intervene on behalf of the juvenile to persuade the complainant not to take official action. It should be noted, however, that in most jurisdictions the police cannot prevent a complainant from filing a petition if he or she insists. Furthermore, research by the Office of Juvenile Justice and Delinquency Prevention (1992) found that the police are more likely to refer juveniles to the juvenile court than they were in the past.

Historically, research has shown that juveniles who show proper respect for the police, have few if any known prior police contacts, and are perceived as having cooperative parents are more likely to be dealt with unofficially than are those who show little respect, have a long history of encounters with the police, and are perceived as having uncooperative parents (Allen, 2005; Black & Reiss, 1970). Most authorities agree that those juveniles who are most likely to have a "police record of arrest are those who conform to police preconceptions about delinquent types, who are perceived as a threat to others, and who are most visible to the police" (Morash, 1984, p. 110). Morash (1984) indicated that she found a "convincing demonstration of regular tendencies of the police to investigate and arrest males who have delinquent peers regardless of these youths' involvement in delinquency" (p. 110). Moyer (1981), while indicating that gender and race are not critical factors in the police decision-making process with respect to adults, indicated that the nature of the offense and the demeanor of the offender when confronting the police are important in determining the type of action taken by the police (Allen, 2005; Walker, Spohn, & DeLone, 2004). Biases on behalf of the police may lead to more informal adjustments for certain types of juveniles. This is largely a matter of speculation given that records of such dispositions have not been routinely kept, although there is currently a trend to formalize such dispositions. It is clear, however, that based on their perceptions of a number of cues, police officers make decisions as to whether official action is in order or whether a particular juvenile can be dealt with unofficially (Suthpen, Kurtz, & Giddings, 1993; Walker et al., 2004).

Research on the relationship between the police and juveniles was sparse during the 1970s and 1980s. However, recent research by Engel, Sobol, and Worden (2000) on 24 police departments in three metropolitan areas indicated that in most situations "officers do not treat hostile adults and juveniles, males and females, and blacks and whites differently" (p. 256), as was speculated by Klinger (1996, p. 76). The authors found that the police are likely to take official action in cases where there are disrespectful suspects who are intoxicated by use of alcohol or other drugs and in circumstances where disrespect is demonstrated in front of other officers. The effects of **demeanor,** then, were not contingent on suspects' personal characteristics, at least in this study.

In contrast, Sealock and Simpson (1998), in analyzing data collected from a 1958 Philadelphia birth cohort, found that race, gender, and socioeconomic status significantly affect the arrest decision. They also noted that within gender categories, officers consider the seriousness of the offense and the number of prior police contacts in making arrest decisions. Similarly, Walker and colleagues (2004) contended that age, race, and gender have been found to be significant variables with arrest decisions. However, the authors provided an extensive analysis of public racial perceptions of police officers, showing that Latino Americans have a view less favorable than that of Caucasian Americans but that this is not as negative a view as tends to exist among the African American population. These researchers pointed out that these negative attitudes—and behavior that belies these negative attitudes—can be the actual reason that racial variables appear to be significant. Walker and colleagues noted that "individuals who are less respectful or more hostile are more likely to be arrested" and that "African Americans were more likely to be arrested" (p. 332). They concluded that the hostility that is transmitted during these encounters results in higher arrest rates.

In minority communities where police–community relations have typically been impaired, it is perhaps to be expected that citizens will have a negative and distrustful view of the police. This is particularly true among juvenile offenders who have been exposed to the effects of such negative encounters (Shusta, Levine, Wong, & Harris, 2005). With this in mind, it becomes clear that police must be sensitive to this issue—when and where feasible—to prevent encounters from escalating (Shusta et al., 2005). This is a particularly relevant point if one is concerned with citizen perceptions of the police agency and if one hopes to build a genuine rapport and/or connection with juveniles from diverse backgrounds (Shusta et al., 2005).

Bazemore and Senjo (1997), looking at the relationship of police and juveniles from the community policing viewpoint, analyzed data collected from field research and ethnographic interviews over a 10-month period. Among the community police officers they studied, they found a distinctive style of interaction with young people, different attitudes toward juveniles, and unique views of the appropriate role of officers in response to youth crime. The authors concluded that the officers' efforts to enhance prevention, creative diversion, and advocacy provide at least partial support for the belief that community policing (as discussed previously) can lead to positive outcomes.

Although the exact nature of the relationships among personal characteristics, demeanor, and police decisions remains unclear, it is likely that all of these factors and others continue to play a role in police–juvenile encounters. On numerous occasions, for example, we have seen police officers respond differently to male and female juveniles. As we pointed out in Chapter 3, it appears that male officers seldom search ("pat down") juvenile females even in circumstances where the juveniles are likely to be carrying drugs and/or weapons for their male companions (Hurst, McDermott, & Thomas, 2005). Wooden and Blazak (2001) discussed the emergence of the "mall rat" as a type of delinquent, noting that petty theft is the most frequent crime committed by these juveniles and that females are as likely to be caught as are males, although the latter are more likely to be arrested (pp. 32–33).

Although there has been increased attention during recent years to juvenile perceptions of the police, few studies have focused specifically on the attitudes of young girls (Hurst et al., 2005). In response to the lack of research that examines the perceptions of police held by juvenile girls, Hurst and colleagues (2005) conducted a study in Ohio and examined one key determinant that tends to be relevant to many young females—the fear of victimization. Their research found generally low support for police, particularly when respondents expressed more concern with potential victimization (Hurst et al., 2005). It is interesting to point out that although girls in this survey did find police to be helpful in service roles (e.g., providing aid when a car breaks down, helping persons who are sick or disabled), they did not view the police as effective in general law enforcement functions (e.g., curbing drug activity, preventing violence). Overall, therefore, it was determined that girls tended to have little trust or support for the police (Hurst et al., 2005). Girls may actually benefit more from other agencies such as transitional living programs and counseling programs (see In Practice 7.1).

Furthermore, using multivariate analyses, Hurst and colleagues (2005) did find significantly different attitudes toward police when examined by race. As in previously discussed research, they found that African American girls expressed much more negative perceptions of the police than did Caucasian girls. In fact, fewer than 30% of African American girls agreed with any of the Likert scale items related to attitude measures for liking the police, trusting the police, or being satisfied with the police (Hurst et al., 2005). Throughout their analysis, it was found that race was a significant predictor of attitudes toward the police (Hurst et al., 2005). These researchers pointed toward differences in racial socialization, the analysis of social situations in terms of relationship power between African Americans and Caucasians, and other psychological and sociological factors. From this, it is thought that many of the attitudes toward police are part and parcel of racial socialization, particularly in communities that already have poor police–community relations (Hurst et al., 2005). This again points to the need for police training in diversity and cultural differences with juveniles, just as would be expected with the adult population (Hurst et al., 2005; Shusta et al., 2005).

In inner-city neighborhoods, police beat officers often arrive at a kind of "working peace" with groups of young black males hanging out on street corners (Anderson, 1990). They may allow juveniles to get away with certain minor violations for which they could take official action so as to "keep the peace." The police face a dilemma in such neighborhoods. On the one hand, the police are accused of **overpolicing** in black neighborhoods; on the other hand, they are accused of failing to provide sufficient protection (Shusta et al., 2005; Walker et al., 2004).

In Practice 7.1

Girls Driven From Homes Find Safe Haven: When Family Sends Teens Into the Street, The Harbour Provides Shelter and Nurturing

Jennifer Mendoza, by her own account, was trouble as a teenager. There were fights with her mother, drugs and drinking, skipping school, and being "wild and crazy."

At 17, she was living on the streets or with friends—anywhere but home.

"I was just a rebel," Mendoza said of her adolescent years in Park Ridge.

Mendoza had run-ins with authorities, but it took a helpful police officer to get her on the right path.

He brought her to The Harbour Inc., a Park Ridge–based social service agency that provides emergency shelter and transitional living programs for adolescent girls in the north and northwest suburbs.

Today, Mendoza, 24, is a success story. She has a job at a software company in Chicago, [has] her own apartment with her 4-year-old son, and has nearly finished work on a criminal justice degree at Harper College in Palatine. Her goal: to be a police officer.

"I have my own place, my own car," Mendoza said. "I take care of my son. I'm a working mom back in school. The Harbour drilled into me [the need] to finish up school."

The Harbour is a beneficiary of Chicago Tribune Holiday Giving, a campaign of Chicago Tribune Charities, a McCormick Tribune Foundation fund.

The Harbour traces its beginnings to 1968, when a teenage girl was locked out of her home in Wilmette by her father, who had sexually abused her.

At the time, the only option for police was to take the girl to the Audy Home, now known as the Cook County Temporary Juvenile Detention Center.

Social workers and youth advocates were outraged, said Randi Gurian, The Harbour's executive director. "This girl was twice victimized," she said.

The incident spurred community members to form the social service agency, which began providing emergency housing to adolescent girls in 1975.

The organization now provides three programs to assist homeless, runaway, abused, and neglected adolescent girls and women ages 12 to 21.

With a $2 million annual budget and a staff of 38, The Harbour has about 40 clients at a time. In 2005, the group's shelter program placed 38 girls in safe environments, 83 percent with family members, officials said. A follow-up check found 88 percent of these girls remained in the same stable placement three months later.

The goal is to ensure that homeless girls are safe, receive necessary services, and can continue schooling and be discharged to a safe environment, preferably with family.

In addition to the shelter, the group also provides transitional living programs for girls 16 to 21.

Younger girls are placed in a supervised living situation and work 10 to 20 hours a week while attending school. Older clients are helped to make the move into their own apartments.

The issues that force adolescent girls out of their homes tend to focus on family stress, Gurian said. It can start with trouble in school, an unacceptable boyfriend, sexual abuse by a family member, or drug or alcohol use.

(Continued)

(Continued)

Gurian blames "family isolation" as a key social ill. Oftentimes, single parents don't have the resources—family, friends, even neighbors—to help in a crisis.

"No one is aware there's a problem. There's a lot of shame," Gurian said.

On any day in the U.S., Gurian said, there are 25,000 homeless youths on the streets.

Of the ranks of homeless girls, 70 to 80 percent of them will end up prostituting themselves, she said.

Homeless teens tend to fall into four categories, Gurian said: runaways, throwaways, lockouts and push-outs.

"Parents set [the youths] up to run away," Gurian said. "Sometimes the rules are so onerous that a child can't and isn't going to adapt to them."

The organization attempts to work collaboratively, hooking family members up with a variety of agencies to resolve conflicts and stresses.

Mendoza was never able to reconcile her family situation. Her mother moved out of state, so Mendoza remained in the transitional living program, years she called some of the best of her life.

"I'm a beacon of success," Mendoza said. "I've referred a couple of girls to The Harbour. I've even donated money when I can. Everything I have is because of The Harbour."

SOURCE: Wronski, R. (2006, December 6). Girls driven from homes find safe haven: When family sends teens into the street, The Harbour provides shelter and nurturing. *The Chicago Tribune,* p. 2 (Metro, Zone C). Available: www.lexis.com/research/retrieve/frames?_m=836ef33efb12e8433ae5c13dc35d7487&csvc=bl&cform=bool&_fmtstr=XCITE&docnum=1&_startdoc=1&wchp=dGLbVzb-zSkAz&_md5=60756f5e01851ee7fd11563d05a98003.

Complaints concerning the former are typically voiced by young black males who are often stopped, frisked, and questioned on the streets; complaints concerning the latter often arise when the police fail to act against street corner juveniles or in domestic violence situations where the police fail to make arrests (Walker et al., 2004).

According to Lardiero (1997), race makes a difference at all stages of the juvenile justice process but may be most important at the initial point of contact with the police. He maintained that minority representation throughout the juvenile justice network would drop if the police used arrest as a last rather than as a first resort. This is also the contention of other experts who study minority citizens' perceptions of the police and/or cross-cultural training needs for police officers (Shusta et al., 2005; Walker et al., 2004).

Unofficial Procedures

As Piliavin and Briar (1964) pointed out, police officers who encounter juveniles involved in delinquent activities have a number of alternatives available for handling such juveniles. Basically, police officers may simply release the juvenile in question, release the juvenile and submit a "juvenile card" briefly describing the encounter, reprimand the juvenile and release him or her, take the juvenile into custody to make a stationhouse adjustment, or arrest the juvenile and request that the state attorney file a petition in juvenile court. Only the last two alternatives involve official action. Each of the other alternatives may occur either on the street or in a police

▲ Informal contacts between the police and juveniles occur frequently and help to shape the perceptions each group has of the other.

facility. These informal adjustments are commonly referred to as a **street corner adjustment** or a stationhouse adjustment. A typical street corner adjustment might occur when the police have been notified by a homeowner that a group of juveniles have congregated on his property and have refused to leave when asked to do so. Because the offense is not serious, and because the homeowner is likely to be satisfied once the juveniles have left, the officer may simply tell the juveniles to leave and not return. If, for some reason, the police officer is not satisfied that the orders to move on and not return will be obeyed, the officer may take the juveniles to the police station and request that the juveniles' parents meet with them there. If an agreement can be reached among the juveniles and their parents that the event leading to the complaint will not recur, the officer may release the juveniles in what is commonly referred to as a stationhouse adjustment (*Illinois Compiled Statutes* (*ILCS*], 2006; Meehan, 1992). In either case, there is no further official action taken by others in the juvenile justice network, and often no official record of the encounter is kept. Although information cards and/or computer notes kept on juveniles do not constitute official records, they are sometimes used by juvenile officers to determine the number of prior contacts between a particular juvenile and the police and, therefore, may be used to determine whether official action will be taken. In some states, formal records of stationhouse adjustments are now required for certain offenses (*ILCS*, ch. 705, 2006).

Informal adjustments such as these usually cause little controversy so long as all parties (complainant, police, parents, and juveniles) are reasonably satisfied. In fact, some states have attempted to formalize the stationhouse adjustment process by spelling out exactly what the police officer's alternatives are in such adjustments (*ILCS*, ch. 705, 2006). For example, in Illinois a police officer may, with the consent of the minor and his or her guardian, require the minor to perform public or community service or make restitution for damages. Although police officers often see solutions of this type as being better for the juvenile than official processing, some serious objections have been raised by parents, the courts, and sometimes the juveniles involved.

Suppose that a juvenile was allegedly involved in vandalism where he or she spray-painted some derogatory comments on the front of a school building. Also suppose that, as a condition of not taking official action, a police officer instructs the juvenile to spend every night after school cleaning the paint off the school building with paint remover and brushes that are provided at the expense of the juvenile or his or her parents. Finally, suppose that the juvenile persists in maintaining his or her innocence. The implications of this type of "treatment without trial" should be relatively clear. First, it has not been demonstrated that the juvenile did commit the delinquent act in question; that is, the juvenile has not been adjudicated delinquent in a court of law. Second, because it has not been demonstrated that the juvenile committed the vandalism, there is no legal basis for punishment. Third, even if the juvenile did in fact commit the offense, the police generally have no legal authority to impose punishment on alleged offenders unless, of course, such offenders voluntarily agree to the punishment. But how voluntary is such agreement?

Although many police officers who employ informal adjustments realize that their actions might not be strictly legal, they justify the use of informal adjustments on the basis that the juvenile and/or parent (guardian) entered into it voluntarily. These officers reason that because the treatment or punishment is not mandatory and is in the juvenile's best interests, there does not need to be prior adjudication of delinquency or finding of guilt. Many of these officers fail to recognize that the extent to which their "suggested" treatment or punishment programs are voluntary is highly questionable. The threat of taking official action, if unofficial suggestions are not acceptable to the offenders involved, largely removes any element of voluntarism and is coercive. In cases of this type (which are not atypical), the juvenile may be upset about being punished for an act that he or she did not commit, the parents may be upset because their child did not receive a fair trial, and the juvenile court judge may be upset because the functions of the court have been usurped (taken over) by the police. Of course, not all stationhouse adjustments are negative. Some can be very successful in resolving minor instances of delinquency through proper referral to competent counselors by officers skilled in accurately assessing the needs of the juveniles.

With respect to abused or neglected children, police options are technically more restricted and require more training and expertise (Figure 7.1). As **mandated reporters** (those required to report suspected cases of abuse to the state), police are often required to report suspected incidents of child abuse or neglect to the state department of children and family services even though they might not have enough evidence to arrest the suspected abusers. Investigators from the children and family services unit are typically required to contact the parties involved within 72 hours of the time of notification. If the investigators are convinced that neglect or abuse is occurring, or if the original investigating officer is convinced it is occurring, the child may be taken into protective custody until further hearings can be held. It is the responsibility of the local law enforcement department to develop the procedures to handle abuse and neglect situations, to ensure that law enforcement officials are properly trained in identifying cases of abuse or neglect, to objectively investigate abuse or neglect cases, and to interview victims and perpetrators of abuse or neglect.

Figure 7.1 Considerations for Child Abuse Investigations

When you receive the referral:

- Identify personal or professional biases with child abuse cases. Develop the ability to desensitize yourself to those issues and maintain an objective stance.
- Know the department guidelines and state statutes.
- Know what resources are available in the community (e.g., therapy, victim compensation) and provide this information to the child's family.
- Introduce yourself, your role, and the focus and objective of the investigation.
- Ensure that the best treatment will be provided for the protection of the child.
- Interview the child alone, focusing on corroborative evidence.
- Do not rule out the possibility of child abuse with a domestic dispute complaint. Talk with the children at the scene.

Getting information for the preliminary report:

- Inquire about the history of the abusive situation. Dates are important to set the time line for when the abuse may have occurred.
- Cover the elements of the crime necessary for the report. Inquire about the instrument of abuse or other items on the scene.
- Do not discount children's statements about who is abusing them, where and how the abuse is occurring, or what types of acts occurred.
- Save opinions for the end of the report and provide supportive facts. Highlight the atmosphere of disclosure and the mood and demeanor of participants in the complaint.

Preserving the crime scene:

- Treat the scene as a crime scene (even if abuse has occurred in the past) and not as the site of a social problem.
- Secure the instrument of abuse or other corroborative evidence that the child identifies at the scene.
- Photograph the scene and, when appropriate, include any injuries to the child. Re-photograph injuries needed to capture any changes in appearance.

Follow-up investigation:

- Be supportive and optimistic to the child and the family.
- Arrange for a medical examination and transportation to the hospital. Collect items for a change of clothes if needed.
- Make use of appropriate investigative techniques.
- Be sure that the child and family have been linked to support services or therapy.
- Be sure that the family knows how to reach a detective to disclose further information.

During the court phase:

- Visit the court with the child to familiarize him or her with the courtroom setting and atmosphere before the first hearing. This role may be assumed by the prosecutor or, in some jurisdictions, by victim/witness services.
- Prepare courtroom exhibits (e.g., pictures, displays, sketches) to support the child's testimony.
- File all evidence in accordance with state and court policy.
- Unless they are suspects, update the family about the status and progress of the investigation and stay in touch with them throughout the court process. Depending on the case, officers should be cautious about the type and amount of information provided to the family because they may share the information with others.
- Provide court results and case closure information to the child and the family.
- Follow up with the probation department for preparation of the presentence report and victim impact statement(s).

Source: Office of Justice Programs. (1997, May). *Law enforcement response to child abuse.* Washington, DC: Office of Juvenile Justice and Delinquency Prevention.

The major concerns of police officers when dealing with abused or neglected children are, of course, the safety and well-being of the minors involved. Still, there are officers who, for a variety of reasons, prefer not to take formal action in cases that they conclude do not involve serious abuse or neglect. "Rarely are abusive and neglectful parents arrested. Exceptions exist when the injury to the child is extremely severe or obviously sadistically inflicted, when a crime has been committed, when the parents present a danger to others, or when arrest is the only way to preserve the peace" (Tower, 1993, p. 275). Official action is more likely to occur today because of mandated reporting laws. Still, even though they are mandated reporters, officers sometimes hesitate to take official action. This is sometimes the case because police officers are concerned about the possibility of false allegations or of being used by one party involved in a hostile divorce or separation to cause trouble for the other party through implanting false allegations in the mind of the child or by falsely reporting abuse or neglect (Goldstein & Tyler, 1998).

In spite of the difficulties just mentioned, it is estimated that as many as 85% of all police–juvenile contacts are resolved informally (Black & Reiss, 1970; Mays & Winfree, 2000). The proportion of child abuse or neglect cases handled unofficially is unknown but is probably considerable. Police officers who use informal dispositions often see such dispositions as more desirable than official processing, which is certain to leave the offender with a record and may lead to detention for some period of time. Most police officers agree that neither juvenile records nor attempts to rehabilitate juveniles who are detained are beneficial to juveniles. The latter holds true for child abusers as well, although when the abuse is severe, officers are typically more than willing to take official action (Willis & Welles, 1988). When police officers act informally, they often sincerely believe they are doing so in the best interests of the parties involved. This may be the case if we assume that all of the persons apprehended did commit a delinquent or criminal act and if we assume that treatment and rehabilitation are of little or no value. However, if we recognize that sometimes the police do make mistakes, that some juveniles and some parents do need and might benefit from treatment of some type, that the police have no mandate to impose punishment or treatment, and that the juvenile court judge often has no way of knowing how many times a particular juvenile or abusive parent has been dealt with informally, the problems inherent in informal adjustments become very apparent (Portwood, Grady, & Dutton, 2000).

Official Procedures

The **official procedures** to be followed when processing juveniles are clearly spelled out in juvenile court acts. It is important to note that police procedures for juvenile offenders differ from adult procedures in most jurisdictions. As a rule, these procedures are tailored specifically toward implementing the juvenile court philosophy of treatment, protection, and rehabilitation rather than punishment. As a result, to carry out proper procedures, specialized training is necessary. It has been our observation that many officers in most jurisdictions believe that being assigned as a juvenile officer is not particularly desirable (Mays & Winfree, 2000, p. 62). We have heard juvenile officers referred to as "kiddie cops" and seen distinctions made between "real" police officers and "juvenile" officers. These traditional police attitudes have slowed the development of a professional corps of juvenile officers. Nonetheless, being an effective juvenile police officer requires more skill than being a good patrol officer. In addition to learning the basics of policing, the juvenile officer is required to learn a great deal about the special requirements of juvenile law, about the nature of adolescence, about the nature of parent–child relationships, and about the social service agencies (public and private) to which juveniles may be referred for assistance (Tower, 1993, p. 275). These skills are not easy to acquire, and those who have mastered them should take pride in their accomplishments. In addition, police organizations should reward those who possess and actively employ these skills in terms of both salary and promotional opportunities.

Although the development of effective juvenile officers and juvenile bureaus is highly desirable, most initial contacts between juveniles and the police involve patrol officers. It would appear logical to provide at least minimal training in the area of juvenile law for all patrol personnel to safeguard the rights of juveniles and to ensure proper legal processing by the police. It does little good, either for the juvenile or for the prosecutor's case, to have a competent juvenile officer if the initial encounter between the juvenile or abusive parent and the police has been mishandled.

Police officers who are involved in the official processing of juveniles need to be aware that all of the guarantees in terms of self-incrimination and searches and seizures characteristic of adult proceedings also hold for juveniles. In addition, juveniles are, in most jurisdictions, extended even further protection by law. Thus, the police are required to notify a juvenile's parents about their child's whereabouts and are required to release the juvenile to his or her parents unless good cause exists for detention. Detention in a lockup routinely used for adult offenders is often illegal, and the police must, in these cases, make special arrangements to transport and detain juveniles if further detention is necessary. In Illinois, for example, juveniles may not be detained in an adult jail or lockup for more than 6 hours, and while they are detained they must not be permitted sight or sound contact with adults in custody (*ILCS,* ch. 705, 2006).

Similarly, police records concerning juveniles must, in most jurisdictions, be kept separate from adult records and are more or less confidential (see Chapter 6). Although fingerprints and photographs of juvenile offenders may be taken, there are often restrictions placed on their use; that is, they may not be transmitted to other law enforcement agencies without a court order in many jurisdictions. There is, however, a trend toward making juvenile records more readily available to interested parties in the interests of more effective law enforcement efforts (*ILCS,* ch. 705, 2006).

Court decisions indicate that a juvenile charged with a delinquent act has a right to counsel prior to placement in a police lineup. There is some concern that a juvenile's waiver of his or her right to remain silent during interrogation without a parent or lawyer present is of questionable value (Dorne & Gewerth, 1998, p. 34). As a result, many police departments delay interrogation until either a parent and/or an attorney is present.

In many jurisdictions, police officers who have been designated juvenile officers have the task of ensuring that juveniles are properly handled. These juvenile officers are, presumably, specially trained in juvenile law and procedures.

Training and Competence of Juvenile Officers

For roughly the past 75 years, there have been repeated calls for professionalization of the police through increased education and training (Shusta et al., 2005). The number of 2- and 4-year college programs in criminal justice and law enforcement has increased dramatically during the past two decades, as has the number of special institutes, seminars, and workshops dealing with special police problems. Because juvenile cases present special problems for the police, one might expect considerable emphasis on training for juvenile officers. Indeed, the number of police officers qualified by training to serve in juvenile bureaus has increased dramatically during recent years, especially in large metropolitan departments. In these departments, promotion within the juvenile bureau is possible, and both male and female officers deal with juvenile offenders and victims. The possibilities of promotion and recognition for a job well done provide incentive and rewards for those choosing to pursue a career in juvenile law enforcement.

The situation of juvenile officers in smaller cities has also improved during recent years. More jurisdictions require compliance with laws mandating special training for juvenile officers, although personnel shortages and reduced financial resources sometimes make both training and specific assignment to purely juvenile matters difficult. There are still many smaller police departments with no female officers, so male officers must deal with juveniles of both genders (Shusta et al., 2005). Some rural departments have no officers specifically trained to deal with juveniles, and others, to conform to statutory requirements, simply select and designate an officer, often one who has no prior training in juvenile matters, as juvenile officer. Considering the fact that juvenile officers are frequently expected to speak to civic action groups about juvenile problems, run junior police programs, visit schools and preschools, form working relationships with personnel of other agencies, and investigate cases of abused and missing children, this lack of training is a very serious matter. Police departments with 10 or fewer sworn officers still face difficulties in providing adequately trained officers for 24-hour-a-day service. When these departments do train and appoint officers to handle juvenile offenders, they can seldom afford to relieve these officers of other duties. This, in effect, makes it impossible for the appointed officers to become specialized in juvenile matters. This also eliminates the possibility of developing a stable juvenile bureau and of advancing one's career as a juvenile officer. One result of these difficulties is that officers have little incentive to volunteer for service in juvenile bureaus. Consequently, juvenile officers are frequently appointed on the basis of a perceived affinity for "getting along" with juveniles. Unfortunately, this affinity is not a substitute for proper training, although it may appear to be to police administrators who regard handling juvenile offenses as something less than real police work.

It is essential that police departments train officers to handle juvenile cases. In Illinois, for example, a juvenile police officer is defined by statute as

> a sworn police officer who has completed a Basic Recruit Training Course, has been assigned to the position of juvenile officer by his or her chief law enforcement officer, and has completed the necessary juvenile officers training as described by the Illinois Law Enforcement Training Standards Board or, in the case of a State police officer, juvenile officer training approved by the Director of the Department of State Police. (*ILCS,* ch. 705, sec. 405/1–3, 1999)

Police–School Consultant and Liaison Programs

Over the past four decades, police departments and schools have worked together to develop programs to help prevent delinquency and improve relationships between juveniles and the police (Brown, 2006; Ervin & Swilaski, 2004). These programs involve more than simply providing security through police presence in the schools. Rather, the programs attempt to foster a more personal relationship between juveniles and the police by using police officers in counseling settings, by improving communications between the police and school officials, and by increasing student knowledge of the law and the consequences of violations (Brown, 2006; Ervin & Swilaski, 2004).

One early police–school consultant or liaison program was developed in Flint, Michigan, in 1958. **Police–school liaison officers** are located in schools and serve as sources of information and counselors for students. They are often funded, at least in part, by school districts even though they work for police agencies. A 1972 evaluation of this program concluded that the police officers assigned had difficulties in being both authority figures and counselors/confidants. Since then and more recently, similar programs have shown similar results in Tucson, Arizona; Montgomery, Alabama; Woodburn, Oregon; and Tampa, Florida, to mention just a few.

More recent programs have focused on the latter role for assigned officers, who act as additional resource persons in the school setting and who generally have been evaluated positively by school officials, although not always by students (Brown, 2006). These programs have proliferated based on these evaluations and the belief that the closer the relationship between police and juveniles in nonthreatening situations (those other than investigatory or crime intervention), the better in terms of improving the image of the police, uncovering information concerning abuse and neglect, and decreasing delinquency (Brown, 2006; Gandhi, Murphy-Graham, Anthony, Chrismer, & Weiss, 2007). **DARE** (Drug Abuse Resistance Education) programs, in which police officers teach children how to avoid use of illicit drugs, are widespread in the United States and abroad (Gandhi et al., 2007). A number of schools and police agencies throughout the country are now involved with such programs, and they appear to be having more positive than negative effects on officers, juveniles, and school authorities, particularly when the officers involved have received special training to prepare them for their assignments (Brown, 2006; Martin, Schulze, & Valdez, 1988). Even though a good deal of research shows that DARE is ineffective at preventing drug use among those who have gone through the program (Aniskievicz & Wysong, 1990; Ennett, Tobler, Ringwalt, & Flewelling, 1994; Gandhi et al., 2007; National Institute of Justice, 1994), the program may still improve understanding and relationships between juveniles and the police officers involved. This alone is thought to be a beneficial outcome given that such positive experiences can reduce the likelihood that juveniles will engage in delinquent behavior (Goldberg, 2003). This is much more likely to be true if high school educators ensure that the material is developmentally appropriate so as to be positively received by peer groups at various age ranges (Goldberg, 2003). In addition, it has been found that juveniles are more receptive to an emphasis on short-term negative social consequences as opposed to physiological consequences (Goldberg, 2003). Lastly, as mentioned earlier in this chapter, it is important to consider race, gender, and cross-cultural effects when educating on the topic of drugs, thereby necessitating that programs/curricula be culturally sensitive (Goldberg, 2003).

Another program, **GREAT** (Gang Resistance Education and Awareness Training), is based on similar assumptions and has been subject to similar criticisms (Palumbo & Ferguson, 1995). Although these programs may have been subject to some degree of criticism, there is no doubt that GREAT programs are very popular throughout the United States (Valdez, 2005). Similar to DARE programs, these gang resistance programs train police officers to conduct comprehensive antigang education programs for children who are not yet in high school (Valdez, 2005). As with DARE programs, there is a need for such programs to be appropriate for the age range of the peer group. Valdez (2005) noted that one way to overcome this challenge is to provide student training and allow the students to share in the formal leadership roles that the educator and/or police presenter might have. This empowers the students and likewise provides for more internalization of the anti-gang (and anti-drug) values that are being transmitted (Valdez, 2005).

Another police–school-based approach to prevention is known as the SANE Gang (Substance Abuse and Narcotics Education Gang) program (Valdez, 2005). This program originated in southern California and is taught by both the classroom educator and a law enforcement officer. The lessons in this curriculum include issues such as the development of positive self-identity, specific techniques for counteracting violent encounters, conflict resolution, and prosocial values that likewise deemphasize materialism (Valdez, 2005). The key strength of this program is that it uses interactive forms of teaching and training rather than relying on films and overstructured lecture formats (Valdez, 2005). From these examples just discussed, it is clear that there are various means of addressing drugs, gangs, and any other issues that might face juveniles while in and out of public schools. For information on how some Michigan schools are responding to youth violence, see In Practice 7.2.

In Practice 7.2

CMU Evaluation of Police School Program Shows Positive Impact on Student Attitudes

A collaborative training program by Central Michigan University and the Michigan State Police is changing high school students' attitudes about a variety of issues relating to the presence of police and safety in schools.

A 2002 study by CMU education professor Joseph Rivard sampled 2,377 high school students in assorted grades in almost a dozen schools between Lansing and Traverse City.

The study shows a positive impact from the Teaching, Educating, and Mentoring program in which Rivard and specialists from the Michigan State Police train liaison officers at specially designed seminars for the law enforcement community. These trained officers are assigned to schools to establish trust and understanding between police, youth, and the community.

"This type of collaboration between the Michigan State Police and CMU is the first of its kind," said Rivard. "The TEAM model is under consideration for adoption by several other states. One of the strengths of the TEAM program has been the combined emphasis of presenting contemporary and relevant content with the desire to measure the overall effectiveness of the program."

The TEAM program includes lessons for kindergarten through twelfth grade that provide an understanding of laws affecting youth, the consequences of actions, and how to build a community spirit.

Rivard administered pre- and post-program tests to high school students to determine if TEAM helped them understand the concepts presented and if the program was successful in changing attitudes. Their comprehension of issues showed improvement for each unit studied, said Rivard.

"In addition, several attitudes showed significant changes—having police officers in the schools, feeling OK about reporting a student lawbreaker, the dangers of meeting strangers from Internet communications, victim mentality, how police work differs from the media portrayals, and what officers do on a daily basis," he said.

(Continued)

(Continued)

Conducted at the Michigan State Police Academy, the law enforcement training seminars are attended by a large variety of community police agencies. State troopers and county sheriff officers combine with city and village police personnel to equip schools with liaison officers who constructively try to alter attitudes about violence, crime, gangs, drug abuse, and the role of the police officer. Liaison officers are visible in corridors, cafeterias, and other areas as part of their routine schedules. Students get to know officers by name, and officers have a personal impact on the daily lives of students.

"The partnership has resulted in an incredibly positive and powerful program for Michigan schools," said Rivard. "TEAM is all about prevention. While we will always be made vividly aware of the kinds of things that can go wrong in a school environment, it is also somewhat comforting to realize there are programs out there that are helpful in preventing bad things from happening within the world of kindergarten through twelfth-grade students."

Source: Housley, P., & Rivard, J. (2004, February 14). CMU evaluation of police school program shows positive impact on student attitudes. *CMU News.* Retrieved from: www.news.cmich.edu/archived/index.asp?id=30. Reprinted with permission of Central Michigan University.

▲ Oklahoma City Police Officer Terry Yeakey talks with young students about DARE (Drug Abuse Resistance Education) in school.

Community-Oriented Policing and Juveniles

Community-oriented policing is a trend that deserves mention here. Community-oriented policing refers to a strategy that relies on identification of problems by police and members of the community they serve and shared ownership of law enforcement/order maintenance duties (Glensor, Correia, & Peak 2000; Shusta et al., 2005; Walters, 1993; Webber, 1991). There are a number of community-oriented policing programs currently in existence, with many more on the drawing table. Although community-oriented policing is a general police strategy, it certainly has applications in police work with juveniles given that it requires joint community–police identification of, and efforts to solve, problems (Shusta et al., 2005). Thus, police officers and school, probation, civic action, neighborhood, and political groups work together to find solutions to problems rather than asking the police to handle incidents as and after they occur (Brown, 2006). One example of a program of this type, sponsored by the Department of Justice, is **Youth-Focused Community Policing** (YFCP). This program provides information-sharing activities that promote proactive partnerships among the police, juveniles, and community agencies cooperating to identify and address juvenile problems in a manner consistent with community policing philosophy. A number of communities have established such programs, and initial evaluations are mixed (Mays & Winfree, 2000, p. 85).

Still other programs have been introduced to improve the relationship between schools and the police when concerning juvenile offenders. One example is **SHOCAP** (Serious Habitual Offender Comprehensive Action Program), initiated in Illinois (*ILCS*, ch. 705, 2006). SHOCAP is a multidisciplinary interagency case-management and information-sharing system intended to help the agencies involved to make informed decisions about juveniles who repeatedly engage in delinquent acts. Each county in Illinois is encouraged to develop a SHOCAP committee (Cook County may develop several committees) consisting of representatives of law enforcement, schools, the state attorney's office, and probation. Committees are then to develop written interagency agreements that will allow for sharing of information about serious habitual offenders with other committee agencies in such a way that the information is kept confidential by the agencies involved. It is hoped that the sharing of such information on a need-to-know basis will result in better coordination of efforts to intervene and deal appropriately with repeat offenders.

Police and Juvenile Court

The police are the primary source of referral to juvenile court (Mays & Winfree, 2000, p. 86), and juvenile court judges rely heavily on the police for background information concerning juveniles who come before them. Because the police and the court may have different goals with respect to juveniles (e.g., control vs. treatment), this might not always be in the best interests of juveniles. On the one hand, the juvenile court may become overly concerned with control; on the other hand, the police officer who believes that the court is unfair to the police or too lenient with offenders may fail to report cases to the court because, in his or her opinion, nothing will be gained by official referral (Bartollas 1993, p. 394). In some cases, the police may attempt to resolve the case at hand by themselves, and this, as we pointed out earlier, may or may not be in the best interests of the juvenile involved. In short, whether or not a particular juvenile is referred to juvenile court depends in part on the police officer's attitude toward the court.

Career Opportunity—Municipal Police Officer

Job Description: Enforce laws and maintain order. Patrol the community, control traffic, make arrests, prevent crime, investigate criminal activity, and work with the public and representatives of other agencies to improve the quality of life in the community.

Employment Requirements: Must be a U.S. citizen, have a valid driver's license, and be at least 20 years old at the time of application. Must not have any felony convictions. Must be able to pass physical, written, medical, polygraph, psychological, and background investigations. Must possess good communication skills. Must be a high school graduate, but preference is frequently given to those with 2- or 4-year degrees. Appointment contingent on completion of basic training.

Beginning Salary: Typically range between $25,000 and $55,000 depending on jurisdiction; average is probably roughly $35,000. Additional benefits package (insurance, vacation, sick leave, and pension).

Note: To become a juvenile police officer, an applicant typically needs to serve as a patrol officer first and then may be required to participate in special training to be certified as a juvenile officer.

SUMMARY

To implement proper juvenile procedures and benefit from theoretical notions concerning prevention, causes, and correction of delinquent behavior and child abuse and neglect, juvenile officers must first know proper procedures and understand theories of causation. Because both types of knowledge are specialized, it is imperative that juvenile officers receive special training in these areas. This specialized training is advantageous for the police department, juveniles, the justice network, the social service network, and the community. The police department benefits in terms of creating a more professional image and in terms of efficiency because mistakes in processing should be reduced. Juveniles benefit in that trained personnel can better carry out the intent of juvenile court acts that were developed to protect the best interests of juveniles. The justice system benefits from the proper initial processing of juveniles and abusive adults who are to be processed further (e.g., prosecuted) in that system. Finally, the community and social service network benefit from decisions made by police officers who are properly trained. In return for these benefits, it is essential to reward juvenile officers who perform well through recognition and promotional opportunities.

The majority of police juvenile contacts result in unofficial dispositions in the form of street corner or stationhouse adjustments. It is important that decisions concerning proper disposition of juvenile cases by police officers be based on a thorough knowledge of procedural requirements and the problems of juveniles and abusive or neglectful adults. When trained, competent officers make such decisions, the imposition of punishment by the officers handling cases unofficially is reduced, and the rights of all parties are better protected. In cases that require official disposition, further processing is facilitated by proper initial processing. To ensure that police officers handle juvenile cases properly, specialized training programs need to be developed and used, and incentives for good performance by juvenile officers need to be provided.

Note: Please see the Companion Study Site for Internet exercises and Web resources. Go to www.sagepub.com/juvenilejustice6study.

Critical Thinking Questions

1. List and discuss some of the cues frequently used by police officers in deciding whether to handle a case officially or unofficially. What are some of the dangers in relying on these cues from the point of view of the juvenile offender? From the point of view of the victim of abuse or neglect?
2. Joe Foul Up, a 13-year-old white male, has just been apprehended by a police officer for stealing a bicycle. Joe admits taking the bicycle but says that he only intended to go for a joyride and was going to return the bicycle later in the day. Joe has no prior police contacts of which the officer is aware. The bicycle has been missing for only an hour and is unharmed. The owner of the bicycle is undecided about whether or not to proceed officially. Discuss the various options available to the police officer in handling this case. What options do you consider to be most appropriate and why?
3. Why do you think that juvenile officers handle the majority of contacts with juveniles unofficially even when they could clearly proceed officially? Why are police officers often hesitant to take official action in cases involving abuse or neglect even though they are mandated reporters? What are some of the advantages and disadvantages of unofficial dispositions to both juveniles and society?
4. Locate the Web site for the Department of Justice, and see what information you can obtain on the Youth-Focused Community Policing program. Is there a recent evaluation of the program? If so, what conclusions can you draw about the program based on the evaluation?

Suggested Readings

Allen, T. T. (2005). Taking a juvenile into custody: Situational factors that influence police officers' decisions. *Journal of Sociology and Social Welfare, 32,* 121–129.

Black, D. J., & Reiss, A. J., Jr. (1970). Police control of juveniles. *American Sociological Review, 35,* 63–77.

Breen, M. D. (2001). A renewed commitment to juvenile justice. *Police Chief, 68*(3), 47–52.

Engel, R. S., Sobol, J. J., & Worden, R. E. (2000). Further exploration of the demeanor hypothesis: The interaction effects of suspects' characteristics and demeanor on police behavior. *Justice Quarterly, 17,* 235–258.

Ervin, J. D., & Swilaski, M. (2004). Community outreach through children's programs. *Police Chief, 71*(9), 42–45.

Gandhi, A. G., Murphy-Graham, E., Anthony, P., Chrismer, S. S., & Weiss, C. H. (2007). The devil is in the details: Examining the evidence for "proven" school-based drug abuse prevention programs. *Evaluation Review, 31,* 43–74.

Hurst, Y. G., Frank, J., & Browning, S. L. (2000). The attitudes of juveniles toward the police: A comparison of black and white youth. *Policing: An International Journal of Police Strategies and Management, 23,* 37–53.

Hurst, Y. G., McDermott, M. J., & Thomas, D. L. (2005). The attitudes of girls toward the police: Differences by race. *Policing: An International Journal of Police Strategies and Management, 28,* 578–594.

Lundman, R. L., Sykes, E. G., & Clark, J. P. (1978). Police control of juveniles: A replication. *Journal of Research in Crime and Delinquency, 15,* 74–91.

Moriarty, A., & Fitzgerald, P. (1992, May). A rationale for police–school collaboration. *Law and Order, 40,* 47–51.

Piliavin, I., & Briar, S. (1964). Police encounters with juveniles. *American Journal of Sociology, 70,* 206–214.

Sealock, M. D., & Simpson, S. S. (1998). Unraveling bias in arrest decisions: The role of juvenile offender type-scripts. *Justice Quarterly, 15,* 427–457.

Walker, S., Spohn, C., & DeLone, M. (2004). *The color of justice: Race, ethnicity, and crime in America* (3rd ed.). Belmont, CA: Wadsworth/Thomson Learning.

KEY FIGURES IN JUVENILE COURT PROCEEDINGS

CHAPTER LEARNING OBJECTIVES

On completion of this chapter, students should be able to

- *Explain the roles of the prosecutor, defense counsel, judge, and probation officer in juvenile court*
- *Discuss differences between private and state-appointed defense counsel*
- *Discuss conflicting views of the relationship between the prosecutor and defense counsel*
- *Explain plea bargaining*
- *Discuss the roles of child and family services and court-appointed advocates in juvenile court proceedings*

KEY TERMS

Prosecutor/State's attorney

Unofficial probation

Defense counsel

Private counsel

Court-appointed counsel

Juvenile court judge

"Parent figure" judge

"Lawgiver" judge

Juvenile probation officer

Role identity confusion

Electronic monitoring

Children and family services

Court-appointed special advocates

Preservice/inservice training

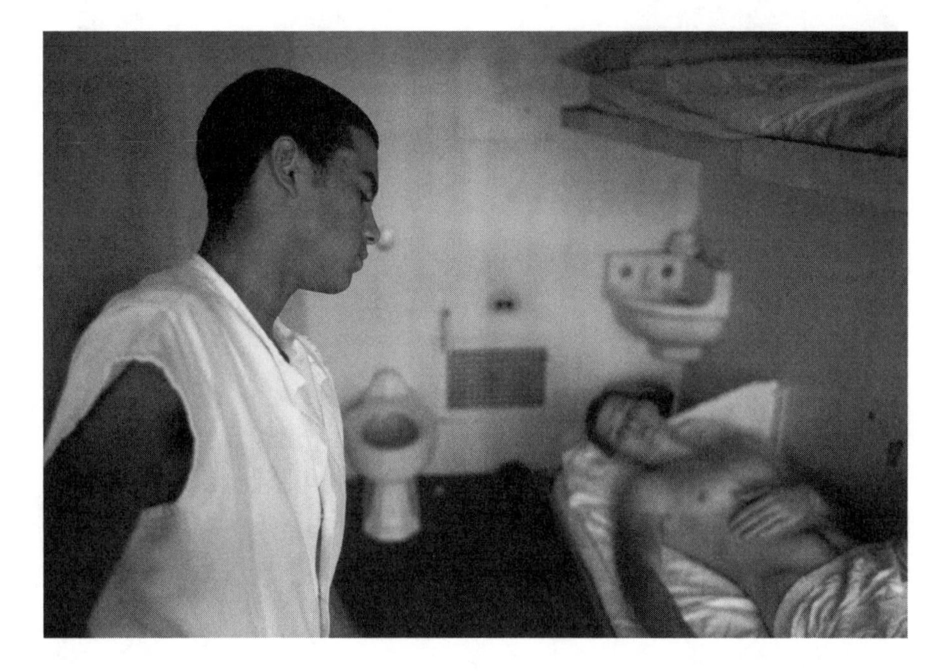

▲ Chris Bay, a teenager sentenced to 40 years for murder, shares a small cell with his roommate at the Clemens Unit Prison in Brazoria County, Texas. The state of Texas, like many other states, sometimes sentences juvenile offenders convicted of serious crimes to adult prisons run by the Department of Corrections.

One of the alternatives available to the police in dealing with juvenile offenders or adults who commit offenses against children involves taking official action that can result in further processing through the juvenile justice network or, in the case of adult perpetrators, the adult justice network. Once the decision to take official action has been made, juvenile court personnel become involved in the case. We use the term *juvenile court personnel* in a broad sense to include the prosecutor, defense counsel, the judge, the juvenile probation officer, and (in abuse and neglect cases) representatives from the department of children and family services (also known as children's protective services).

The Prosecutor

The final decision about whether a juvenile will be dealt with in juvenile court rests with the prosecutor. Regardless of the source of the referral (e.g., police officer, teacher, parent), the prosecutor may decide not to take the case to court and, for all practical purposes, no further official action may be taken on the case in question. The prosecutor, then, exercises an enormous amount of discretion in the juvenile (and adult) justice system (Stuckey, Roberson, & Wallace, 2004). Although the police officer may "open the gate" to the juvenile justice system, the prosecutor may close that gate. The prosecutor may do this without accounting for his or her reasons to anyone else in the system (except, of course, to the voters who elect the prosecutor to office, with the next election often occurring long after the case in question has been dismissed).

Clearly, there are some circumstances under which the prosecutor would be foolish to proceed with court action. For example, lack of evidence, lack of probable cause, or lack of due process may make it virtually impossible to prosecute a case successfully. There are, however, a number of somewhat less legitimate reasons for failure to prosecute. There have been instances where prosecutors have failed to take cases to court for political or personal reasons (e.g., when the juvenile in question is the son or daughter of a powerful and influential citizen) or because the caseload of the prosecutor includes an important or serious case in which successful prosecution will result in favorable publicity. As a result, the prosecutor may screen out or dismiss a number of "less serious" cases such as burglary and assault (Atkins & Pogrebin, 1978; Neubauer, 2002). In short, the prosecutor is the key figure in the justice system and is recognized as such by both defendants and defense counsel (Ellis & Sowers, 2001, p. 40; Laub & MacMurray, 1987; Mays & Winfree, 2000).

During recent years, however, the prosecutor has lost some discretion historically afforded to him or her because of discretionary controls enacted within state legislation. These controls have been designed to decrease the amount of discretion a prosecutor has in determining whether or not a case remains in the jurisdiction of the juvenile court or is waived to adult court. In Illinois, for example, it is mandated that the prosecutor request to transfer a juvenile to adult court if the child is 15 years of age or over, commits an act that is a forcible felony, and has previously been adjudicated delinquent or committed the act in conjunction with gang-based activity (*Illinois Compiled Statutes* [*ILCS*], ch. 705, sec. 405/5-805, 1999). There are also presumptive transfers that deal with violence involving firearms and other clearly stated legislative policies on when prosecutors may use their discretion to transfer juveniles to adult criminal court. The discretionary controls have not been designed to take away from the prosecutor's role in court or to undermine the duties placed on the prosecutor but rather are in place to ensure that the prosecutor is not abusing the position and power given to him or her by the court system. The discretionary controls are also a political response to the public's recent outcries against juvenile violence. Despite the discretionary controls, prosecutors are still key figures in the juvenile court system (Neubauer, 2002; Siegel, Welsh, & Senna, 2003; Viljoen, Klaver, & Roesch, 2005).

The prosecutor's key role in the American juvenile justice system has emerged slowly over time. Initially, the **prosecutor/state's attorney** was seen as both unnecessary and harmful in juvenile court proceedings that were supposedly nonadversarial proceedings "on behalf of the juvenile" (U.S. Department of Justice, 1973). The *Gault* decision, along with the decisions in *Kent* and *Winship,* brought about a number of changes in juvenile court proceedings. Among these changes was a growing recognition of the need for legally trained individuals to represent both the state and the juvenile (and, in some instances, the juvenile's parents) at all stages of juvenile justice proceedings. The need resulted from increased emphasis on procedural requirements and the adversarial nature of the proceedings.

Today the prosecutor is a key figure in juvenile justice because he or she determines whether or not a case will go to court, most waiver decisions, the nature of the petition, and (to a large extent) the disposition of the case after adjudication (the judge seldom imposes more severe punishment than is recommended by the prosecutor). Siegel and colleagues (2003) noted that it is likely that the prosecutor will continue to play a primary role in the juvenile justice system due to the constitutional safeguards provided to youthful offenders and to the publicity associated with juvenile crime.

In addition, there is a tendency on the part of some prosecutors to impose **unofficial probation.** The prosecutor indicates that he or she has a prosecutable case but also indicates that prosecution will be withheld if the suspect in question agrees to behave according to certain guidelines. These are often the same guidelines handed down by probation officers subsequent to an adjudication of delinquent, abused, or neglected. This amounts to a form of continuance under supervision without proving the charges in court and may result from an admission of the facts by the minor or a lack of objection to this procedure by the minor, his or her parents, and legal counsel. In essence, this procedure provides an alternative to official adjudication as a delinquent and is regarded as beneficial in that sense. However, although the use of unofficial probation is clearly beneficial to the prosecutor because it eliminates the need to prepare a case for court and may be beneficial for the juvenile court by reducing the number of official cases, unofficial probation has the same potential disadvantages as do informal adjustments by the police. In short, unofficial probation imposed by the prosecutor amounts to punishment without trial, and the voluntary nature of this probation is highly questionable. Informal agreements may also work to the disadvantage of juveniles who are suspected of being abused or neglected and who are allowed to remain in their homes as a result of such agreements.

In addition to enforcing the law and representing the state, the prosecutor has many of the following duties in juvenile justice proceedings:

- Investigates possible violations of the law
- Cooperates with the police, intake officers, and probation officers regarding the facts alleged in petitions
- Authorizes, reviews, and prepares petitions for the court
- Plays a role in the initial detention or temporary placement process
- Represents the state in all pretrial motions, probable-cause hearings, and consent decrees
- Represents the state at transfer and waiver hearings
- May recommend physical or mental examinations
- Seeks amendments or dismissals of filed petitions if appropriate
- Represents the state at adjudication of cases
- Represents the state at disposition of cases
- Enters into plea bargaining discussions with defense attorneys
- Represents the state on appeals or in *habeas corpus* proceedings
- Is involved in hearings dealing with violations of probation (Siegel & Senna, 1994, p. 551)

The attorney for the state (prosecutor), then, participates in every proceeding of every stage of every case subject to the jurisdiction of the family court in which the state has an interest.

Defense Counsel

The American Bar Association (ABA, 1977) described the responsibility of the legal profession to the juvenile court in Standard 2.3 of *Standards Relating to Counsel for Private Parties.* The ABA stated that legal representation should be provided in all proceedings arising from, or related to, a delinquency or in-need-of-supervision action—including mental competency, transfer, postdisposition, probation revocation and classification, institutional transfer, and

▶ The courtroom work group often negotiates the outcome of cases.

disciplinary or other administrative proceedings related to the treatment process—that may substantially affect the juvenile's custody, status, or course of treatment.

Juvenile court proceedings involving delinquency and abuse are adversarial in nature in spite of the intent of the early developers of juvenile court philosophy. It is for this reason that the role of **defense counsel** (the attorney representing the defendant) has become increasingly important. Today in most jurisdictions, all juveniles named in petitions are represented by counsel. In Illinois, for example, no proceeding under the state's juvenile court act may be initiated unless the juvenile is represented by counsel (*ILCS,* ch. 705, sec. 405/1–5, 1999). In many cases, the juvenile's parents also have legal representation. In some cases, a guardian ad litem may be appointed by the court. The guardian ad litem is a person appointed by the court as a third party to protect the interests of the child both in court and while placed in social services (Davidson, 1981). In general, the guardian ad litem is used in abuse, neglect, and dependency cases where the minor is in need of representation because of immaturity (Sedlak, Doueck, Lyons, & Wells, 2005; Siegel et al., 2003).

There are two basic categories of defense counsel: **private counsel** and **court-appointed counsel.** Private counselors are sometimes retained or appointed to represent the interests of juveniles in court. Frequently, however, juveniles are represented by court-appointed counsel (attorneys or public defenders). The former are typically drawn from a roster of practicing attorneys in the jurisdiction, whereas the latter are full-time salaried employees. Both are paid by the county or state (or by both) to represent defendants who do not have the money to retain private counsel. For many young lawyers interested in criminal law, the position of public defender represents a stepping-stone (Neubauer, 2002). In most areas, the public defender is paid a relatively low salary, but the position guarantees a minimal income that can be supplemented by private practice (Stuckey et al., 2004). For example, the most recent information available on defense systems for the indigent found that the average cost per case to state and local government for indigent defense was $5.37 per capita, ranging from a low of $0.11 per case in West Virginia to a high of $11.23 per case in Alaska (Barlow, 2000, p. 374).

In addition to the low personal pay, many public defender programs are inadequately funded (Wice, 2005). This makes the job of public defenders even more difficult because, in addition to being underpaid personally, they must work with fewer agency resources at their disposal (Wice, 2005). This low pay and inadequate agency funding have led to a reputation of providing low-quality representation (Neubauer, 2002; Wice, 2005) that is further compounded by the fact that, understandably, many public defenders have short job tenures (Botch, 2006; Wice, 2005). These factors have contributed to a public image of ineptness that has become a virtual stigma for persons working in the role of public defender (Botch, 2006; Wice, 2005).

As a rule, public defender caseloads are heavy, investigative resources are limited, and many clients are, by their own admission, guilty or delinquent (Barlow, 2000; Stuckey et al., 2004). The public defender, therefore, spends a great deal of time negotiating pleas and often very little time talking with clients. In fact, sometimes a public defender in juvenile court will indicate to the judge that he or she is ready to proceed and then ask someone in the courtroom which of the several juveniles present is the client. As a result, public defenders often enjoy a less than favorable image among their clients (Barlow, 2000, pp. 377–379).

Some public defenders seem to have little interest in using every possible strategy to defend their clients (Botch, 2006; "Too Poor," 1998; Wice, 2005). On occasion, prosecutors and juvenile court judges make legal errors to which public defenders raise no objections. Appeals initiated by public defenders in cases tried in juvenile court are relatively rare even when the chances of successful appeals seem to be good. There are also public defenders who pursue their clients' interests with all possible vigor, but on the whole it appears that juveniles who have private counsel often fare better in juvenile court than do those who are represented by public defenders. There is little doubt that the office of public defender is frequently underfunded and that such underfunding is a major factor in most of the criticisms leveled at the office.

Whether defense counsel is private or public, his or her duties remain essentially the same. These duties are to see that the client is properly represented at all stages of the system, that the client's rights are not violated, and that the client's case is presented in the most favorable light possible regardless of the client's involvement in delinquent or criminal activity (Pollock, 1994, pp. 145–152). To accomplish these goals, the defense counsel is expected to battle the prosecutor, at least in theory, in adversarial proceedings. However, the quality of representation afforded is not guaranteed. The public defender's office is frequently understaffed, and private counsel is often too expensive to be considered an option. As Siegel and Senna (1994) noted, "Representation should be upgraded in all areas of the juvenile court system" (p. 557).

Relationship Between the Prosecutor and Defense Counsel: Adversarial or Cooperative?

In theory, adversarial proceedings result when the "champion" of the defendant (defense counsel) and the "champion" of the state (prosecutor) do "battle" in open court, where the "truth" is determined and "justice" is the result. In practice, the situation is quite often different due to considerations of time and money on behalf of both the state and the defendant (Stuckey et al., 2004).

The ideal of adversarial proceedings is perhaps most closely realized when a well-known private defense attorney does battle with the prosecutor. The *O. J. Simpson* case of the 1990s is an excellent example. Prominent defense attorneys often have competent investigative staffs and considerable resources in terms of time and money to devote to a case. Thus the balance of power between the state and the defendant may be nearly even. This is generally not the case when defense counsel is a public defender who is often paid less than the prosecutor, often has less experience than the prosecutor, and generally has more limited access to an investigative staff than the prosecutor. For a variety of reasons, then, both defense counsel and the prosecutor may find it easier to negotiate a particular case rather than to fight it out in court because court cases are costly in terms of both time and money. The vast majority of adult criminal cases in the United States are settled by plea bargaining. A substantial proportion of delinquency and abuse/neglect cases are disposed of in this way as well. In fact, it has been suggested that justice in the United States is not the result of the adversarial system but rather the result of a cooperative network of routine interactions among defense counsel, the prosecutor, the defendant, and (in many instances) the judge (Barlow, 2000, p. 349; Blumberg, 1967; Sudnow, 1965).

In plea bargaining, both the prosecutor and defense counsel hope to gain through compromise (Neubauer, 2002; Viljoen et al., 2005). The prosecutor wants the defendant to plead guilty—if not to the original charge, then to some less serious offense. Defense counsel seeks to get the best deal possible for his or her client, and this may range from an outright dismissal to a plea of guilty to some offense less serious than the original charge (Neubauer, 2002). The nature of the compromise depends on conditions such as the strength of the prosecutor's case and the seriousness of the offense. Most often, the two counselors arrive at what both consider a "just" compromise, which is then presented to the defendant to accept or reject (Siegel et al., 2003). As a rule, the punishment to be recommended by the prosecutor is also negotiated. Thus, the nature of the charges, the plea, and the punishment are negotiated and agreed on before the defendant actually enters the courtroom. The adversarial system, in its ideal form at least, has been circumvented (Edwards, 2005; Stuckey et al., 2004). Perhaps a hypothetical example will help to clarify the nature and consequences of plea bargaining.

Suppose that our friend, Joe Foul Up, is once again in trouble. This time, Joe is seen breaking into a house. The break-in is reported to the police, who apprehend Joe in the house with a watch and some expensive jewelry belonging to the homeowner. This time, the police decide to take official action. Because Joe is over 13 years of age and the offense is fairly serious, the prosecutor threatens to prosecute Joe as an adult in adult court. She also indicates that she intends to seek a prison sentence for Joe. Joe's attorney, realizing that the prosecutor has a strong case, knows that he cannot get Joe's case dismissed. He argues with the prosecutor that this is Joe's first appearance before the juvenile court and that Joe is, after all, a juvenile. After some discussion, the prosecutor agrees to prosecute Joe in juvenile court provided that the allegation of delinquency is not contested. Joe's attorney agrees provided that the prosecutor recommends only a short stay in a private detention facility in the community. Joe's attorney then presents the deal to Joe and perhaps to Joe's parents, indicating that it is the best he can do and recommending that Joe accept because he could be found guilty and sentenced to prison if he is tried in adult court. Joe accepts and the bargain is concluded. The case has been settled in the attorney's offices. All that remains is to make it official during the formal court appearance. Most judges will concur with the negotiated plea.

The benefits of plea bargaining to the prosecutor, defense counsel, and the juvenile court are clear. The prosecutor is successful in prosecuting a case (she obtains an adjudication of delinquency), defense counsel has reduced the charges and penalty against his client, and all parties have saved time and money by not contesting the case in court. The juvenile may benefit as well given that he might have been convicted of burglary in adult court (if the judge had accepted the prosecutor's motion to change jurisdiction) and ended up in prison with a felony record. The dangers of plea bargaining, however, should not be overlooked. First, there is always the possibility that the motion to change jurisdiction might have been denied. Second, Joe might have been found not guilty even if he had been tried in adult court or might have been found not delinquent if his case had been heard in juvenile court. Third, because negotiations most often occur in secret, there is a danger that the constitutional rights of the defendant might not be stringently upheld. For example, Joe did not have the chance to confront and cross-examine his accusers. Finally, the juvenile court judge is little more than a figurehead, left only to sanction the bargain, in cases settled by plea bargaining. The juvenile court judge has the responsibility to see that the hearings are conducted in the best interests of both the juvenile and society and has the responsibility to ensure due process. Neither of these can be guaranteed in cases involving plea bargaining. A final concern in all plea bargaining processes, whether adult or juvenile, is that the victim seldom feels good about the bargain.

The Juvenile Court Judge

Theoretically, the **juvenile court judge** is the most powerful and central figure in the juvenile justice system, although he or she does not always exercise this power (Edwards, 2005). Noting that this is *theoretically* the case in the courthouse underlies the fact that there are many actors who are involved within the courtroom work group that processes a juvenile case (Edwards, 2005; Neubauer, 2002). This courtroom work group tends to develop a sense of shared informal norms and understandings, with a strong organizational emphasis being placed on effective case processing (Neubauer, 2002; Viljoen et al., 2005). Indeed, while consisting of the typical members of the adult courtroom work group, the juvenile court will also typically rely heavily on professional judgments of nonlawyers in assessing both the background of the juvenile and other circumstances such as the quality of family supervision (Hanser, 2007a; Viljoen et al., 2005). In many cases, the input of various mental health workers may weigh heavily in the judge's decision (Hanser, 2007a; Viljoen et al., 2005).

In the end, however, it is the juvenile court judge who decides whether a juvenile will be adjudicated delinquent, abused, in need of intervention, dependent, or neglected. Because there is no jury in most instances, the decision of the judge is final unless an appeal overturns the judge's decision. In addition, the judge makes the final determination about the disposition of the juvenile. Therefore, the juvenile court judge decides matters of law, matters of fact, and the immediate futures of those who come before the bench. Juvenile judges likewise tend to have a wide degree of discretion when fulfilling their role (Leiber & Fox, 2005; Neubauer, 2002). Despite this flexibility, assignment to the juvenile court is often not considered to be a highly desired position among many judges, and many may seek rotation as a means of advancing their judicial careers (Stuckey et al., 2004).

In many states, hearing officers known as referees or commissioners are appointed to assist juvenile court judges (see In Practices 8.1 and 8.2). These hearing officers typically submit recommendations that must be certified by a judge before they have the effect of law (Roberts, 1989, p. 114). Within the confines of legislative mandates, juvenile judges rule on pretrial motions involving issues such as arrest, search and seizure, interrogation, and lineup identification. They make decisions about the continued detention of children prior to hearings, and they make decisions about plea bargaining agreements and informal adjustments (Siegel et al., 2003). They hold bench hearings, rule on appropriateness of conduct, and settle questions concerning evidence and procedure. They guide the questioning of witnesses. They decide on treatment for juveniles. They preside over waiver hearings, and they handle appeals where allowed by statute (Siegel & Senna, 1994; Siegel et al., 2003).

Although judges in some jurisdictions are assigned to juvenile court on a full-time basis, there are also many juvenile court judges who serve on a part-time basis. The latter are circuit judges who perform judicial functions in civil, criminal, probate, and other divisions of the court and are occasionally assigned to juvenile court. It is difficult for such judges to become specialists in juvenile court proceedings, and some are not as well versed in juvenile law as they could be, although many perform well.

Juvenile court judges may be placed along a continuum ranging from those who see themselves largely as parent figures to those who are concerned mainly about the juvenile court as a legal institution. The **"parent figure" judge** is often genuinely concerned about the total well-being of juveniles who appear before the court. He or she is likely to overlook some of the formalities of due process in an attempt to serve as a parent figure who both supports and disciplines juveniles. This judge's primary concern is serving what he or she perceives as the best interests of the juveniles who appear in court, based on the assumption that they must have problems even though they might not have committed the specific acts that led to the filing of the petitions or been victims of abuse or neglect in the specific instances in question (Ford, Chapman, Mack, & Pearson, 2006). Often these judges talk to the juveniles and/or parents involved in an attempt to obtain expressions of remorse or regret (Edwards, 2005; Ford et al., 2006; Viljoen et al., 2005). Once these expressions are

In Practice 8.1

Referees in Juvenile Court

It is possible, however, for much of the work [of the juvenile court] to be done under the supervision of the judge by individuals who have not had legal training. Many cases are now settled by intake officers who are not lawyers. In more than twenty-five states, the law gives juvenile court judges the authority to appoint referees, who make tentative disposals of the cases petitioned for hearing, subject to the judge's approval. This power to appoint referees makes it possible to extend the court to rural districts, far from the sites where court sessions are regularly held. If no such arrangement is made, juvenile offenses may be passed over, handled by a justice of the peace, or dealt with by some other unsatisfactory method because of the inconvenience of attending juvenile court sessions.

SOURCE: Sutherland, E. H., Cressey, D. R., & Luckenbill, D. F. (1992). *Criminology* (11th ed.). Dix Hills, NJ: General Hall, p. 4,438.

In Practice 8.2

The Role of Referees in District Court

Hearing and resolving thousands of cases a year, referees play a key role in the administration of justice in Hennepin County District Court. Lawyers who regularly work in the specialty courts are very familiar with the referees and how they function, but for many lawyers and the public at large, the role of referees is new territory. This article explains how referees function in our court system. In addition, it highlights an important change in the law regarding the review of referee decisions in Family Court. One sidebar accompanying this article profiles the three referees most recently appointed to the Hennepin County District Court, while another reacquaints us with the 13 other referees who have already been serving in the court.

The Referee Position

Currently, 16 district court referees serve in the Fourth Judicial District, including six in Family Court, three in Juvenile Court, two in Housing Court, four in Probate/Mental Health Court, and one Court Trials referee. The referee position is a creature of statute, with the general authority arising from Minn. Stat. § 484.70 (2004). The statute authorizes the chief judge of the judicial district to appoint one or more suitable persons to act as referees. Referees hold office at the pleasure of the judges of the district court and must be "learned in the law." The statute enumerates the duties and powers of a referee. A referee is to hear and report all matters assigned by the chief judge and recommend findings of fact, conclusions of law, temporary and interim orders, and final orders for judgment. Thus, a referee has broad authority to make both procedural and substantive decisions in a case to which the referee is assigned. However, a referee may not hear a contested trial, hearing, motion, or petition if a party or attorney for a party objects in writing to the assignment of a referee to hear the matter. A party who objects to a referee hearing a contested matter must serve and file the objection within 10 days of notice of the assignment of the referee but not later than the commencement of any hearing before a referee.

Referee Decisions

All recommended orders and findings of a referee are subject to confirmation by a judge. Upon the conclusion of a hearing, the statute requires that the referee transmit the court file to a judge together with written recommended findings and orders. Once confirmed, the referee's recommended findings and orders become the findings and orders of the court. In general, a party may seek judicial review of any recommended order or finding of a referee by serving and filing notice within 10 days of effective notice of the recommended order or finding. The notice of review must specify the grounds for review and the specific provisions of the recommended findings or orders in dispute. The court, upon receipt of a notice of review, sets a time and place for a review hearing.

Source: Chawla, M. J. (2006, December 20). The role of referees in district court. *The Hennepin Lawyer*. Hennepin County, MN: TimberLake Publishing. Retrieved from: www.hennepin.timberlakepublishing.com/article.asp?article=1079&paper=1&cat=148. Reprinted with permission of the author and *The Hennepin Lawyer*, membership publication of the Hennepin County Bar Association, Vol. 76, #1.

given, the acts involved can often be "forgiven" and attention centers on how to best help the juveniles to avoid future trouble or victimization. If these expressions of remorse or regret are not given, the judge frequently resorts to a role as disciplinarian, sometimes overlooking the facts in the case.

There is a tendency among parent figure judges to continue juvenile cases under supervision for various lengths of time. These judges apparently assume that an adjudication of delinquency, abuse, neglect, or minor requiring authoritative intervention is less desirable than using the threat of adjudication in an attempt to induce acceptable behavior. Although most juvenile court acts provide for judicial continuance, this action can be carried to the extreme in situations where the case against the juvenile, parent, or guardian is weak and the continuance period is long. These continuances amount to punishment without trial much as informal adjustments and unofficial probation do. It is also not unlikely that during this period the child will be caught for another offense and may be brought to court again. This creates a revolving door effect.

At the other end of the continuum is the **"lawgiver" judge** who is concerned primarily that all procedural requirements are fulfilled. This type of judge has less interest in the total personality of the juvenile than in the evidence of the case at hand. The lawgiver judge dismisses cases that the prosecutor cannot prove beyond a reasonable doubt (or, in abuse and neglect cases, cannot demonstrate a preponderance of evidence for) and does not believe that it is his or her duty to prescribe treatment for juveniles who have not committed the offenses of which they have been accused or who cannot be shown to have been victims of abuse or neglect. The dispositions of the lawgiver judge are based on statutory requirements more than on the personal characteristics of the parties involved.

Most juvenile court judges fall somewhere between the two extremes, reflecting the lack of consensus about the proper role of the juvenile court discussed in Chapter 1. Most judges make a sincere effort to maximize legal safeguards for juveniles while attempting to act in the best interests of both the juveniles and society. They ensure that legal counsel is available, they try to arrive at objective decisions during adjudicatory hearings, and they try to ensure that the disposition of each case takes into account the needs of the juvenile involved. Tower (1993) described the efforts of the juvenile court judge in abuse and neglect cases as follows: "Deprived of the support of a jury (in most cases), the judge must base the final decision on the report of the investigator, on what has been heard in the courtroom, on the judge's own experience, and often on the assumption of what will be best for all concerned" (p. 293).

Tower (1993) concluded, "Since people's motivations are never predictable, the juvenile court judge realizes there is no assurance that a child will be safe when returned home or happy in placement. Using only best judgment and the hope that it is correct, the judge renders the decision" (p. 293). In a study of serious child maltreatment cases brought before the Boston Juvenile Court in 1994, Bishop, Murphy, and Hicks (2000) concluded that despite some improvements during the past decade, "the system still fails to promptly find permanent placements for seriously maltreated children" (p. 610). In attempting to arrive at an acceptable disposition, the juvenile court judge frequently relies heavily on the recommendations of the juvenile probation officer (for one type of recommendation, see In Practice 8.3) (MacDonald & Baroody-Hard, 1999) and, in abuse and neglect cases, on the recommendations of the representatives from the department of children and family services.

In Practice 8.3

Teens Deal Out Justice Their Way in Maryvale

Crystal Dorosky was the first teenager to appear in the Maryvale Teen Court when it opened its doors in west Phoenix last week.

At this court, defendant Dorosky's attorney was younger than she is. The prosecutor was not old enough to vote. The four-member jury had not graduated from high school. And the court clerk plays guitar in a budding rock band called Third Right Turn.

The only adult was Maryvale Justice of the Peace, Maryvale Precinct, Judge Hercules Dellas, who oversaw the proceedings.

The jury ordered Dorosky, 17, to do 10 hours of community service as her punishment for possessing alcohol at a New Year's Eve party in December. A traditional juvenile court judge might have issued a series of trips to a probation officer, a monetary fine, or community service and a potential record for the same offense.

Dorosky believes the sentence was steep. Her court-appointed attorney, Diane Villafana, a Maryvale High School student, had asked for eight hours of community service. Then again, Dorosky can't imagine being judged by an adult in a traditional juvenile court.

"I think a teen jury knows how it is to be a kid in this day and age. We like to party, and we like to be with friends," Dorosky said. "Maybe if the jury were adults, they would have given me a harsher consequence."

Maryvale Teen Court is the most recent addition to the Maricopa County Teen Court Youth Diversion Program. The others are in Phoenix, Tempe, Fountain Hills, Glendale, and Gilbert. To make the court a reality in Maryvale, Judge Dellas asked Phoenix Union High School District's Maryvale High for help, and a high school business law class served as prosecutors, legal counsel, and jurors.

Students take responsibility for their mistakes and understand the consequences. On the flip side, they gain experience "serving as a juror or as a courtroom participant," Dellas said.

Maryvale Teen Court is expected to hear two to eight cases per session once a month during the school year.

Teen Court is designed after a traditional adult courtroom. It benefits students who are younger than 17 whose offenses range from alcohol possession, [to] theft valued at less than $250, [to] disorderly conduct.

With a probation officer's approval and a parent's permission, a teen's case could end up in Teen Court. A child also accepts responsibility for the offenses before making an appearance at Teen Court, where a jury delivers the punishment with a deadline. Students must complete the orders within two months after the hearing. If a teen fails to comply, the case is returned to the juvenile probation officer for action. If a defendant successfully meets all of the requirements, his or her case is closed without a criminal record.

The success rate of Teen Court impressed Maricopa County Supervisor Mary Rose Wilcox, who asked Dellas to add the program to his schedule. Dellas agreed.

Studies show that a high percentage of teens sentenced at Teen Court complete the jury's orders, Wilcox said. At least 92 percent of juveniles who complete Teen Court are not referred to Juvenile Court within a year, she said.

(Continued)

(Continued)

Maryvale Teen Court, at 4622 W. Indian School Road, also is closer to home for teens than a trip to Juvenile Justice Courts at Durango.

"When peers judge peers, you get a whole different outcome," Wilcox said. "The hardest critics are your peers. It also exposes them to court's legal system. They don't want to be on one side of the table but be a lawyer and prosecutor."

Tom Camp, who teaches business law at Maryvale High, watched his 13 students become prosecutors, jurors, legal counsel, court bailiff, and clerk.

"They were scared," he said. "I guess what made them nervous were the people. When kids are held accountable, it makes them very nervous. That is kids in general. They do not like to be held responsible."

Derek Penne, 16, said he was so nervous as a Teen Court clerk that he mispronounced the judge's name. He introduced a Judge George to the audience and corrected himself.

The Maryvale teen isn't aspiring to be a lawyer, at least for now. He is exerting his energy studying music and playing guitar for the Third Left Turn band. "It was nerve-racking because there were too many people watching," Penne said.

SOURCE: Reid, B. (2006, April 7). Teens deal out justice their way in Maryvale. *The Arizona Republic.* Available: www.azcentral.com/arizonarepublic/centralphoenix/articles/0407ext-teencourt0407Z4.html. Used with permission. Permission does not imply endorsement.

The Juvenile Probation Officer

Probation is the oldest and most widely used disposition, with more than 18,000 juvenile probation officers in the United States (Torbet, 1996). Probation is a disposition by the juvenile court in which the minor is placed and maintained in the community under the supervision of a duly authorized officer of the court, the **juvenile probation officer.** "Probation may be used at the 'front-end' of the juvenile justice system for the first-time, low-risk offenders or at the 'back-end' as an alternative to institutional confinement for more serious offenders" (p. 1) Either way, it allows the minor to remain with the family or a foster family under conditions prescribed by the court to ensure acceptable behavior in a community setting.

The juvenile probation officer is a key figure at all levels of the juvenile justice system (Siegel et al., 2003). He or she may arrange a preliminary conference between interested parties that may result in an out-of-court settlement between an alleged delinquent and the injured party or between parties in cases of abuse or neglect. After an adjudicatory hearing, the juvenile probation officer is often charged with conducting a social background investigation (Neubauer, 2002; Siegel et al., 2003). This investigation will be used to help the judge make a dispositional decision. Probation officers are also charged with supervising those juveniles who are placed on probation and released into the community and with supervising parents deemed to have committed neglect or abuse (Ford et al., 2006; Goodkind, Ng, & Sarri, 2006; Siegel et al., 2003). Probation officers have the power to request a revocation of probation if violations of the conditions of probation occur.

The duties of chief probation officers generally include assignment of cases and supervision of subordinates (Stuckey et al., 2004). Chief probation officers may or may not handle cases themselves, depending on available staff. In addition, they normally serve as a liaison between judges and other department heads. The better the rapport they are able to establish with the juvenile court judge, and the more effective they are in transmitting information to subordinates, other juvenile justice practitioners, and the judge, the better the opportunity to serve the interests of juveniles and the community.

The role of juvenile probation officers is an ambiguous one. They are officers of the court who occasionally must act as authority figures and disciplinarians. At the same time, they are charged with helping juveniles in trouble by attempting to keep the juveniles out of court, by recommending the most beneficial dispositions, by protecting juveniles from abusive parents while counseling those parents, and by being available to help probationers solve problems encountered during their probationary periods. If they are to be effective in their role as helping professionals, they must encourage open interaction and trust among the juveniles and parents or guardians they encounter (Parker-Jimenez, 1997; Gardner, Rodriguez, & Zatz, 2004). If they seem too authoritarian, they may receive little cooperation. If they become too friendly, they may find it difficult to take disciplinary steps when necessary.

Juvenile probation officers may find that they are integral in coordinating a variety of services for juveniles. A range of skills, services, and resources are often brokered by juvenile probation officers to aid juveniles in reintegrating into the community and improving their ability to meet the conditions of their probation (Champion, 2002; Hanser, 2007a; Siegel et al., 2003). Juvenile probation officers may coordinate a number of services such as mental health counseling, drug and alcohol counseling, academic achievement, vocational and employment training, alternative education programs, Big Brother/Big Sister programs, and foster parent/grandparent programs (Champion, 2002; Hanser, 2007a)

As a result of the ambiguous role requirements, several different types of juvenile probation officers exist. Some think of themselves largely as law enforcement officers whose basic function is to detect violations of probation. Others see themselves as juvenile advocates whose basic function is to ensure that the rights of juveniles are not violated by the police or potential petitioners. Still others view themselves basically as social workers whose function is to facilitate treatment and rehabilitation. Champion (2002) pointed out that none of these approaches is ideal. Rather, each has its time and place, depending on the circumstances. Therefore, the most effective juvenile probation officers exercise all of these options at different times under differing circumstances. However, it should be pointed out that the balancing of these different orientations in supervision can lead to a sense of **role identity confusion** (Hanser, 2007a). This is a primary source of burnout among community supervision officers, including those assigned to juvenile offenders (Hanser, 2007a). Role identity confusion occurs when officers are unclear about the expectations placed on them when they attempt to balance the competing interests of their "policing" role and their "reform"-oriented role (Champion, 2002; Hanser, 2007a).

Perhaps the most difficult task for most juvenile probation officers is the supervision of probationers. Many have excessive caseloads and have little actual contact with their clients other than short weekly or monthly meetings (Champion, 2002). High caseloads have been defined as 50 or more juvenile offenders, with caseloads actually going as high as 300 juveniles or more in some jurisdictions (Champion, 2002). Obviously, not a great deal of counseling or supervision can occur under these circumstances. When field contacts are made with probationers, probation officers are often considerably concerned about further stigmatizing their clients. Parents who have problems with their children sometimes try to use juvenile probation officers' official position to frighten the children into compliance with their demands. As a result of these difficulties, most juvenile probation officers, in discussing probation conditions with their clients, make it clear that they are available to discuss whatever problems probationers believe are significant. Some juvenile probation officers using this technique allow clients to choose the time and place for conferences to minimize stigmatization.

Juvenile probation officers must also work daily to overcome several issues, including job safety, rising caseloads, a lack of resources, and feelings of failure (Champion, 2002). In a study by the Office of Juvenile Justice and Delinquency Prevention (OJJDP) in 1996, it was reported that more than one-third of juvenile probation officers had been assaulted on the job and that 42% stated they were usually or always concerned about their personal safety while working (Torbet, 1996). In response to these concerns, some jurisdictions have implemented intensive supervision and school-based programs into local schools. Along with safety concerns, rising caseloads have also become a problem for juvenile probation officers. Respondents to the 1996 OJJDP survey stated that their caseloads ranged between 2 and more than 200, with the typical caseload at roughly 41 probationers. It was also reported that probation caseloads are involving more violent juveniles than in previous years. However, the number of resources available to juvenile probation officers has not increased with the number of probationers. Juvenile probation officers are still limited in the types of placements available to probationers and in the amount of funding they can receive from the jurisdiction for treatment of juveniles on their caseloads. This often means that juvenile probation officers must be creative in their approach to their probationers' treatments and rehabilitative efforts. In the same 1996 OJJDP survey, it was reported that "although [juvenile probation officers] chose this line of work 'to help kids,' their greatest sources of frustration are an inability to impact the lives of youth, the attitudes of probationers and their families, and difficulties in identifying successes" (p. 1).

Technological innovations such as **electronic monitoring** are of some help to probation officers in supervising their clients. These supervision tools often work to augment the supervision of juveniles who are processed through juvenile intensive supervision programs (JISPs), which typically accommodate violent and/or repeat youthful offenders (Champion, 2002; Hanser, 2007a). Recently, the Florida Department of Corrections began a pilot project using the global positioning satellite (GPS) system to track the movements and locations of probationers, warn prior victims if necessary, and determine whether probationers are in "off-limits" locations (Mercer, Brooks, & Bryant, 2000).

Children and Family Services Personnel

Although personnel from departments of **children and family services** (child protective services) do not actually work for the juvenile courts, they play major roles in investigation, presentation of evidence, and dispositional recommendations in abuse and neglect cases (Sedlak et al., 2005). Typically when law enforcement officers believe they have discovered a case of abuse or neglect, they are required to report the case to children and family services (Sedlak et al., 2005). Departments of children and family services usually maintain a central register of abuse and neglect cases. On receiving a report of suspected or confirmed abuse or neglect, personnel from the child protective agency begin an investigation of the allegation (Sedlak et al., 2005). In emergency cases, such investigations are to be conducted immediately, in theory at least. In other cases, investigators are normally required to conduct an investigation within a specified time period, typically 24 to 72 hours. Such an investigation normally involves interviews with the alleged victim and offender and an evaluation of risk factors in the child's environment. Where appropriate, the child may be removed from the home to safeguard his or her welfare.

If the allegations of abuse or neglect are found to be true, caseworkers from children and family services are involved in assisting the children involved in court proceedings and in formulating plans to provide services or treatment to both the children and the families involved (Sedlak et al., 2005). In cases where abuse or neglect occurs in institutional settings, the institutions involved, if allowed to remain open, are monitored by children and family services.

Court-Appointed Special Advocates

Court-appointed special advocates (CASAs) work closely with departments of children and family services on abuse and neglect cases (Siegel et al., 2003). CASA volunteers are trained citizen volunteers who are appointed by the court to give advice in the best interests of children who are victims of abuse or neglect. The volunteers are ordinary people, usually without legal expertise, who care about what happens to children who have been victimized by abusive or neglectful parents. The juvenile court rarely appoints CASAs in delinquency cases (this may happen only if the delinquent child has an extensive history of abuse or neglect that may be influencing his or her delinquent behavior).

In jurisdictions with CASA programs, CASA volunteers are assigned to one case at a time by the juvenile court judge. They are responsible for researching the background of the case, reviewing court documents, and interviewing everyone involved in the case, including the child. CASA volunteers also prepare a report for the court discussing what they believe is in the best interests of the child based on the evidence they have reviewed (Siegel et al., 2003). The judge may use this report when deciding on a disposition for the child. Once the judge has decided on the case, the CASA volunteers continue monitoring the case to ensure that the child and/or family receives the services ordered by the court.

Training and Competence of Juvenile Court Personnel

If the goals of the juvenile justice system are to be achieved, the system needs to be staffed by well-trained, competent practitioners. Unfortunately, a number of circumstances have prevented total success in this area. Prosecutors and defense attorneys who handle juvenile court cases generally have little to gain by large investments of time and money. Few defense attorneys have gained national renown as the result of their efforts in juvenile court. Few prosecutors can count on being reelected on the basis of successful prosecutions in juvenile court. In addition, in many locales the juvenile court is regarded as something less than a real court of law where technical proficiency in law is necessary. Prosecutors often assign inexperienced assistants to handle juvenile court cases, and few defense attorneys specialize in the practice of juvenile law. As a result, many cases presented in juvenile court are poorly prepared by both sides. Some prosecutors are not thoroughly familiar with the juvenile code governing their jurisdiction. Similarly, defense attorneys will at times accept hearsay evidence, fail to present witnesses for the defense, and fail to object to procedural violations that might result in the dismissal of the petitions concerning their clients. In short, although the frequency of legal representation for both the state and defense has increased considerably during the past decade, the quality of such representation often leaves something to be desired.

Many judges handle juvenile cases as a part-time assignment. Although many clearly have the best interests of juveniles at heart, far too many show the same unfamiliarity with juvenile codes that characterizes many attorneys appearing before them. In fact, as we have observed, some appear to disregard juvenile codes altogether and rule their jurisdictions as dictators whose decisions on the bench are law.

A particularly disturbing example of judicial lack of familiarity with juvenile law was a case in which a part-time juvenile court judge sentenced a 14-year-old truant (minor requiring authoritative intervention [MRAI]) to the department of corrections. This clearly violates the juvenile code prohibiting status offenders from being transferred to that department. Intervention by the prosecutor and probation officer prevented this illegal act, which otherwise might have gone unchallenged until the department of corrections refused to accept the juvenile.

It should not be too much to ask that attorneys and judges practicing in juvenile court read and become familiar with applicable juvenile codes. If they do not, none of the constitutional guarantees or court decisions regarding due process in juvenile cases will have any impact. Treating juvenile court cases as if they did not involve the real practice of law has made practice before the juvenile court unattractive to many lawyers and judges and will continue to do so in the future. Fortunately, there is some evidence that a corps of better-informed sincere lawyers and judges is beginning to emerge. To encourage the growth of such a corps, proper recognition and rewards must be forthcoming.

Many jurisdictions require a bachelor's degree for employment in probation and social service positions, and a number of practitioners in these positions have master's degrees. The typical juvenile probation officer, for example, is a college-educated white male earning between $20,000 and $39,000 annually with a caseload of 41 juveniles (Torbet, 1996, p. 1). The OJJDP (1996) reported that more than three-quarters of all probation officers responding to their survey earned less than $40,000 a year, and 30% of these do not receive yearly pay increases. In some states, probation officers' salaries, typically paid by the county, are subsidized by state funds in an attempt to alleviate this problem (Torbet, 1996, p. 2).

In *Standards for the Administration of Juvenile Justice*, published by the Institute of Justice Administration and the American Bar Association (1980), various sections address the issue of training for juvenile court personnel. For example, one recommendation states,

> Family court judges should be provided with preservice training on the law and procedures governing subject matter by the family (juvenile) court, the causes of delinquency and family conflict, [and] a thorough understanding of agencies responsible for intake and protective services. In addition, inservice education programs should be provided to judges to assure they are aware of changes in law, policy, and programs. (sec. 1.4220)

Other recommendations (secs. 1.423, 1.424, and 1.425) address similar issues of **preservice/inservice training** in juvenile matters with prosecutors, public defenders, and other court personnel and their staffs. Today there is a good deal of inservice training available to juvenile justice court personnel. The National Council of Juvenile and Family Court Judges, for example, sponsors training programs for court personnel on a continuing basis and publishes the *Juvenile and Family Court Journal* to keep practitioners informed of the latest happenings in juvenile justice.

Career Opportunity— Youth Services Coordinator

Job Description: Responsible for coordinating the treatment and rehabilitation services of juvenile offenders. Provide for the assessment, classification, procurement, coordination, and evaluation of services for juvenile offenders incarcerated in state correctional and residential facilities. Required to work with families, governmental agencies, local courts, schools, and service agencies to create and provide comprehensive treatment programs for troubled youth. May provide counseling to youth.

Employment Requirements: Usually required to have 1 year of professional experience in the juvenile justice field. Required to have knowledge, experience, and an understanding of group and individual counseling, interactional strategies, and child development and behavior. College education needed in the areas of criminal justice, psychology, sociology, social work, education, and other closely related fields. Must complete an oral interview process before being hired.

Salary and Benefits: Benefits are provided by the state and include health and life insurance, paid vacations and holidays, and retirement plans. Salaries vary depending on the geographical location of the position but can range from $26,000 to $38,000.

SUMMARY

Key figures in juvenile court proceedings include attorneys for the state and for the defendant, the judge, representatives from the department of children and family services, and the probation officer. Although the frequency of legal representation in juvenile court is increasing, the quality of this representation needs to be improved. The practice of juvenile law must be taken more seriously if we do not want to deal with juveniles who repeat their offenses and eventually come before adult courts.

Competent lawyers and judges need to be rewarded for their performances in juvenile court proceedings. Whenever possible, juvenile court judges should be assigned exclusively to juvenile court for a period of time. Judges who combine the best elements of the parent figure and lawgiver roles are a definite asset to the juvenile justice system. Probation officers and department of children and family services personnel are crucial if juvenile justice philosophy is to be implemented. Their services to the court and to juveniles with problems complement the roles of the other juvenile court personnel. Although the overall quality of juvenile court personnel is improving, there is still considerable variance. Continued emphasis on training and competence at all levels is essential.

Note: Please see the Companion Study Site for Internet exercises and Web resources. Go to www.sagepub.com/juvenilejustice6study.

Critical Thinking Questions

1. Discuss the roles of the prosecutor and defense counsel in juvenile court. Why is the presence of legal representatives for both sides crucial in contemporary juvenile court? Discuss the relationship between the prosecutor and defense counsel.
2. Why is the judge such a powerful figure in juvenile court? What are the advantages and disadvantages of the judge as lawgiver and parent figure? How well trained are juvenile court judges?
3. In what sense is the role of juvenile probation officer ambiguous? What are the consequences of this ambiguity? How important is the probation officer in juvenile court proceedings?
4. What role do representatives from the department of children and family services play in juvenile court proceedings? Why are court-appointed special advocates important in juvenile court proceedings?

Suggested Readings

Berlow, A. (2000, June 5). Requiem for a public defender. *American Prospect, 11,* 28–32.

Bishop, S. J., Murphy, M. J., & Hicks, R. (2000). What progress has been made in meeting the needs of seriously maltreated children? The course of 200 cases through the Boston Juvenile Court. *Child Abuse and Neglect, 24,* 599–610.

Bridges, G. S., & Steen, S. (1998). Racial disparities in official assessments of juveniles: Attributional stereotypes as mediating mechanisms. *American Sociological Review, 63,* 554–570.

Fox, R. W., Kanitz, H. M., & Folger, W. A. (1991). Basic counseling skills training program for juvenile court workers. *Journal of Addictions and Offender Counseling, 11*(2), 34–41.

Gahr, E. (2001, June). Judging juveniles. *American Enterprise,* pp. 26–28.

Payne, J. W. (1999, January). Our children's destiny. *Trial, 35,* 83–85.

Reddington, F. P., & Kreisel, B. W. (2000). Training juvenile probation officers: National trends and patterns. *Federal Probation, 64*(2), 28–32.

Rubin, H. T. (1980). The emerging prosecutor dominance of the juvenile court intake process. *Crime & Delinquency, 6,* 229–318.

Rush, J. P. (1992). Juvenile probation officer cynicism. *American Journal of Criminal Justice, 16*(2), 1–16.

Siegel, L. J., Welsh, B. C., & Senna, J. J. (2003). *Juvenile delinquency: Theory, practice, and law* (8th ed.). Belmont, CA: Wadsworth/Thomson Learning.

Too poor to be defended [editorial]. (1998, April 9). *The Economist,* pp. 21–22.

Torbet, P. M. (1996). *Juvenile probation: The workhorse of the juvenile justice system.* Washington, DC: Office of Juvenile Justice and Delinquency Prevention.

Viljoen, J. L., Klaver, J., & Roesch, R. (2005). Legal decisions of preadolescent and adolescent defendants: Predictors of confessions, pleas, communication with attorneys, and appeals. *Law and Human Behavior, 29,* 253–277.

PREVENTION AND DIVERSION PROGRAMS

CHAPTER LEARNING OBJECTIVES

On completion of this chapter, students should be able to

- *Discuss the advantages and disadvantages of prevention and diversion programs*
- *Describe three major types of prevention*
- *List and discuss several specific prevention and diversion programs*
- *Discuss the concept of restorative justice*
- *Refer to services provided by children and family service agencies*
- *Critique prevention and diversion programs*

KEY TERMS

Preadjudication intervention

Postadjudication intervention

Primary prevention

Secondary prevention

Tertiary prevention

Functionally related agencies

Diversion

Pure diversion

Secondary diversion

Territorial jealousy

Social promotions

Wilderness programs

Restorative justice

Faith-based initiatives

Head Start

Follow Through

JUMP

Teen courts

Drug courts

Scared Straight

Radical nonintervention

▲ A juvenile offender is given a tour of a maximum-security prison, the Ferguson Unit, by other inmates in one variation of programs intended to frighten juveniles and so discourage them from becoming involved in a life of crime. Programs based on the Scared Straight model allow juvenile offenders to hear about the realities of prison life directly from the older prisoners.

Our society spends millions of dollars annually attempting to apprehend, prosecute, and correct/rehabilitate delinquents and child abusers. Although some of these attempts prove to be more or less successful with some offenders, the results are not particularly impressive on the whole. It would seem logical, therefore, to explore the possibilities of concentrating our resources on programs that might provide better returns. Many authorities have come to believe that most of our money is spent at the wrong end of the juvenile justice process. As Mays and Winfree (2000) indicated,

> Delinquency prevention is an attractive idea—in the abstract. Preventing delinquency means stopping undesired juvenile conduct in its tracks, before it can become delinquent, and before adolescents come to the juvenile justice system's attention. If preventive efforts were perfect, there would be no need for a separate juvenile justice system and, in all likelihood, far less adult crime. (p. 341)

But to reduce the problems, we need to address the root causes, not just disagree with the behaviors (Friedenberg, 1965).

In most cases, we wait until a juvenile comes into official contact with the system before an attempt is made to modify the behavior that has, by the time contact becomes official, become more or less ingrained. Our legal system generally prevents intervention by justice authorities without probable cause, and we would have it no other way. Still, this makes it more difficult for corrections personnel, or personnel in related agencies, to modify offensive behavior after the fact either by intervening prior to adjudication (**preadjudication intervention**) or by intervening after the juvenile has been adjudicated (**postadjudication intervention**). A recent study by the Office of Juvenile Justice and Delinquency Prevention (2003, p. 9) pointed out that the earlier intervention can be introduced, the better the opportunity to change the behavior. It would seem most logical, then, to bring as many resources as possible to bear so as to prevent the offender from engaging in illegal behavior in the first place (predelinquent intervention) or to try to divert the juvenile, as early as possible, when he or she does encounter the justice system (Lundman, 1993).

For example, consider the difficulty of trying to rehabilitate a juvenile addicted to heroin. By the time the juvenile is addicted, apprehended, and processed, he or she has probably developed problems in the family, problems in school, and delinquent habits oriented toward ensuring his or her supply of heroin (e.g., burglary, mugging, pushing drugs). To rehabilitate the juvenile, we need to deal with all of these problems. If, however, we had effective programs to detect and help resolve problems that are likely to lead to heroin use (or child abuse), the necessity for solving all of these complicated related problems would be eliminated. Suppose, for example, that we found the juvenile in question to be dissatisfied with traditional education but interested in pursuing a specific vocation. Suppose that we were to provide an alternative education that enabled the juvenile to pursue that vocation and heightened his or her interest in success within the system. We might, then, prevent the juvenile from dropping out of school, joining a heroin-abusing gang, and developing the undesirable behavior patterns just mentioned. Or suppose we found that child abuse was common among single teenage unwed mothers primarily because they had no conception of child rearing and all it entails. Could we prevent a good deal of the abuse by providing parenting skills to such mothers?

Prevention

There are three major types of prevention programs. **Primary prevention** is directed at preventing illegal acts among the juvenile population as a whole before they occur by alleviating social conditions related to the offenders. **Secondary prevention** seeks to identify juveniles who appear to be at high risk for delinquency and/or abuse and to intervene in their lives early. **Tertiary prevention** attempts to prevent further illegal acts among offenders once such acts have been committed (Office of Juvenile Justice and Delinquency Prevention, 2000). None of these programs is a cure-all, and there are a number of difficulties in attempting to develop and operate such programs (Mays & Winfree, 2000, p. 324). Nonetheless, it may well be that our resources could be employed more effectively in prevention rather than in correction of offensive behavior.

During the 1930s, several projects addressed the issue of delinquency prevention. The Chicago Area Project involved churches, social clubs, and community committees that sponsored recreation programs for juveniles; addressed problems associated with law enforcement, health services, and education; targeted local gangs; and helped to reintegrate juveniles who had been adjudicated delinquent. In spite of these efforts, no solid evidence that delinquency was prevented or reduced resulted from the project (Lundman, 1993).

During the late 1960s, the President's Commission on Law Enforcement and Administration of Justice (1967) recommended the establishment of alternatives to the juvenile justice system. According to the report, service agencies capable of dealing with certain categories of juveniles should have these juveniles diverted to them. The report further recommended the following (pp. 19–45):

1. The formal sanctioning system and pronouncement of delinquency should be used only as a last resort.
2. Instead of the formal system, dispositional alternatives to adjudication must be developed for dealing with juveniles, including agencies to provide and coordinate services and procedures to achieve necessary control without unnecessary stigma. Alternatives already available, such as those related to court intake, should be more fully exploited.
3. The range of conduct for which court intervention is authorized should be narrowed, with greater emphasis on consensual and informal means of meeting the problems of difficult children.

During the early 1970s, the National Advisory Commission on Criminal Justice Standards and Goals (1973) stated that "the highest attention must be given to preventing juvenile delinquency, minimizing the involvement of young offenders in the juvenile and criminal justice system, and reintegrating them into the community" (p. 36). The commission further recommended minimizing the involvement of the offender in the system. This does not mean that we should coddle the offender. It recognizes that the further the offender penetrates into the system, the more difficult it becomes to divert him or her from a criminal career. Minimizing a child's involvement with the juvenile justice system does not mean abandoning the use of confinement for certain individuals or failing to protect victims of abuse or neglect. Until more effective means of treatment are found, chronic and dangerous delinquents should be incarcerated to protect society, and abused children must be made wards of the court and removed from unsafe conditions. However, the juvenile justice system must search for beneficial programs outside institutions for juveniles who do not need confinement or sheltered care.

Both labeling and learning theories stress the desirability of prevention rather than correction. The basic premise of labeling theory is that juveniles find it difficult to escape the stigmatization of being known as delinquents or abuse victims. Once labeled, juveniles are often forced out of normal interaction patterns and forced into associations with others who have been labeled. From this perspective, the agencies of the juvenile justice system that are established to correct delinquent behavior often contribute to its occurrence even as they try to cope with it. Learning theory holds that individuals engage in delinquent behavior because they experience an overabundance of interactions, associations, and reinforcements with definitions favorable to delinquency. Therefore, if agencies cast potential or first-time delinquents into interaction with more experienced delinquents, the process of learning delinquent behavior is enhanced greatly. Alternatively, concentration on the problems of youth that tend to lead to delinquent behavior and abuse or neglect not only may result in preventing some juveniles from becoming involved in progressively more serious offenses but also might allow the justice network to concentrate efforts on hardcore delinquents and abusers whose labels and stigmatization have been earned.

Because delinquency and abuse or neglect are complex problems, no single program is likely to emerge as being effective in preventing all such behaviors. Delinquency prevention, for example, involves many variables, and no one program is likely to be foolproof. Inherent in the multifaceted problems of delinquency and child abuse prevention is the fact that these behaviors have roots in the basic social conditions of our society. Increasing urbanization with accompanying problems of poverty, inferior education, poor housing, health and sanitary problems, and unemployment are but a few social conditions that seem to be related to delinquency and abuse or neglect. Therefore, we should focus our attention on these problems if preventive efforts are to have a chance of success (Johnson, 1998; Kowaleski-Jones, 2000; Lane & Turner, 1999; Liddle & Hogue, 2000; Yoshikawa, 1994). Although a number of programs are important for the prevention of delinquency and child abuse, we would be remiss if we focused only on programs directed specifically at preventing such behaviors and ignored these underlying conditions. Large-scale social change directed at the areas just discussed is clearly an important preventive measure and would enable more people to achieve culturally approved goals without needing to resort to illegal means.

In June 1970, a group was invited by the Youth Development and Delinquency Prevention Administration of the Department of Health, Education, and Welfare to meet in Scituate, Massachusetts, to consider the problem of youth development and delinquency prevention. The document produced at that meeting stated, "We believe that our social institutions [school, family, church, etc.] are programmed in such a way as to deny large numbers of young people socially acceptable, responsible, and personally gratifying roles. These institutions should seek ways of becoming more responsible to youth needs" (Youth Development and Delinquency Prevention Administration, 1971, p. 2). The group further stated that any strategy for youth development and delinquency prevention should give priority to "programs which assist institutions to change in ways that provide young people with socially acceptable, responsible, personally gratifying roles and assist young people to assume such roles" (p. 2).

It follows from this premise that the development of viable strategies for the prevention and reduction of delinquency and abuse or neglect rests on the identification, assessment, and alteration of those features of institutional functioning that impede development of juveniles, particularly those whose social situations make them most prone to developing delinquent careers, becoming victims of abusive behavior, or participating in collective forms of withdrawal and deviancy. This approach does not deny the occurrence of individual deviance, but it does assert that in many cases the deviance is traceable to the damaging experiences of juveniles in institutional encounters.

As Katkin, Hyman, and Kramer (1976) pointed out some time ago,

> It is social institutions in the broader community—families, churches, schools, social welfare agencies, etc.—which have the primary mandate to control and care for young people who commit delinquent acts. It is only when individuals or institutions in the community fail to divert (or decide not to divert) that the formal processes of the juvenile justice system are called into action. (p. 404)

In this respect, Yoshikawa (1994) found that comprehensive family support combined with early childhood education may well be successful in bringing about long-term prevention. Similarly, Johnson (1998) noted that the actions of parents and teachers may reduce juvenile crime more effectively than do those of the police. Lane and Turner (1999) discussed the importance of interagency cooperation in preventing delinquency, and Liddle and Hogue (2000) found that family-based intervention in the form of the multidimensional family prevention model can help to build resilient family ties and strong connections with prosocial agencies among adolescents. Kowaleski-Jones (2000) found that residential stability and schools perceived as high quality by mothers were factors related to preventing juveniles from getting into trouble. Finally, Mihalic, Fagan, Irwin, Ballard, and Elliot (2004) stated that limiting opportunities for bullying and other school victimization reduces delinquency in schools (and perhaps in communities as well).

The responsibility for dealing with juveniles who have problems has been placed too frequently solely on juvenile justice practitioners. The public has been more than willing to place the blame for failures in preventing delinquency and abuse or neglect on these practitioners and has been quick to criticize their efforts. These practitioners are often faced with the task of attempting to modify undesirable behavior that has become habitual and deep-rooted and that a variety of other agencies have failed to modify. In addition, the time period available for rehabilitation is usually short.

In our society, there are a number of agencies with which juveniles come into contact earlier, more consistently, and with less stigmatization than the juvenile justice system. Some of these agencies or institutions are functionally related to the juvenile justice system. The term **functionally related agencies** is used to describe those agencies having goals similar to those of the juvenile justice system—improving the quality of life for juveniles by preventing offensive behavior, providing opportunities for success, and correcting undesirable behavior. Functionally related agencies may or may not work in conjunction with the juvenile justice system.

Diversion Programs

One form of prevention is **diversion,** which has carried many different, and sometimes conflicting, meanings. Diversion is often used to describe prejuvenile justice, as well as postjuvenile justice, activities. Some diversion programs are designed to suspend or terminate juvenile justice processing of juveniles in favor of release or referral to alternate services. Likewise, some diversionary activities involve referrals to programs outside the justice system prior to juveniles entering the system. The latter category is often referred to as **pure diversion,** and programs of this type are not as numerous as those in the former, or **secondary diversion,** category. In Practice 9.1 gives an example of one of the types of diversionary programs that have been initiated throughout the United States. Children involved in this program meet with their victims, community members, and other interested parties to resolve the issues raised by the deviant behavior. Yet these children do not need to face a juvenile court judge and are not involved in the justice system.

Diversion is not without pitfalls. It sometimes permits intervention into juveniles' lives and their families with little or no formal processes and inadequate safeguards of individual liberties. One of the major concerns with diversion programs is that they result in "net widening," or bringing to the attention of juvenile authorities children who otherwise would not be labeled, thereby increasing rather than decreasing stigmatization.

Before discussing specific diversion programs, a major problem in coordinating such programs and the agencies sponsoring them should be mentioned. The problem is one of **territorial jealousy,** which refers to a belief commonly held by agency personnel that attempts to coordinate efforts are actually attempts to invade the territory they have staked out for themselves. Agency staff members have a tendency to view themselves as experts in their particular field, to resent suggestions for change made by outsiders, and to fear that they will be found

In Practice 9.1

Want to Resolve a Dispute? Try Mediation

What Is Mediation?

In a process called mediation, a person trained as a mediator helps two (or more) people resolve a conflict or disagreement. The conflict being resolved might be as simple as who should pay for a damaged locker. Or it might be as complex as which parent should receive custody of a child in the case of a divorce. In either situation, mediation involves solving the dispute through peaceful means. The mediator, however, does not simply listen to the conflict and draw up the terms of a solution; the people with the conflict (the participants or disputants) do that. In addition, it is the participants, not the mediator, who enforce the agreed-upon solution.

The mediator plays a special role. He or she doesn't decide what is right or wrong or find people guilty or innocent as a judge would in a courtroom. Instead, the mediator tries to help the disputants find and agree upon a peaceful way to resolve their conflict. . . .

How Does Mediation Prevent or Reduce Crime?

Conflicts are not always minor and harmless. Assaults or threatened assaults often happen between people who know each other and, in many of these cases, start off with small arguments or disagreements. The mediation process provides a way for these people to resolve their disagreements before either party resorts to violence. It also helps people reach agreements without feeling they have had to "give in." In this way, both sides in mediation come out winners! . . .

Peer mediators help the disputants rechannel anger and reach peaceful agreements. When a disagreement or conflict arises, a teacher, an administrator, a concerned student, or the fighting students themselves can refer the issue to peer mediation. A peer mediator is quickly assigned, and the mediation process begins, resolving the issue and preventing further discord. Playground mediators in elementary schools similarly help prevent fights and resolve disagreements between much younger students.

Mediation programs run for and by youth have enjoyed great success across the nation. Students in Buncombe County, N.C., for example, have conducted more than 1,100 mediation hearings at middle schools and high schools. The disputes were handled by more than 330 student mediators, all of whom received training and technical assistance from the Mediation Center of Asheville, N.C. According to Buncombe County school officials, the mediation sessions were a huge success. They eliminated 742 days of in-school suspension and 1,220 days sof out-of-school suspension. The school system also reported reduced violence in the schools as a result of the mediation program.

Warning

If you are planning a mediation program, be sure to work closely with your local community mediation program, police department, sheriff's office, or school administration. Some disputes are too complex or potentially violent to be solved by peer mediation alone.

SOURCE: National Criminal Justice Reference Service. (2000, March). Want to resolve a dispute? Try mediation. *Youth in Action.* Available: www.ncjrs.gov/html/ojjdp/youthbulletin/2000_03_1/contents.html

to be lacking in competence. As a result, these staff members tend to keep agency operations secret and reject attempts by personnel from other agencies to provide services or suggest improvements.

Perhaps an example will help to clarify the concept of territorial jealousy. An attempt was made by youth services agency personnel to provide services to a school district with one of the highest dropout rates in the state. The services offered included educational and vocational counseling, alternative educational programs, and some immediate employment opportunities. When contacted by the director of the youth services agency, the principal of the local high school indicated that the school system "really had no dropout problem," that school counselors handled any existing problems, and that he would initiate contact when he needed help from a social work agency.

As the example clearly indicates, the consequences of territorial jealousy can be extremely serious for both juveniles and taxpayers. Duplication of services is a costly enterprise in a time of budgetary cutbacks and financial restraints; however, denial of available services to juveniles with problems can be disastrous. Lack of cooperation, understanding, and confidence among agency personnel greatly hampers attempts to provide for the welfare of juveniles.

Some Examples of Prevention and Diversion Programs

School Programs

In Chapter 3, the importance of school personnel in shaping the behavior of children was discussed. No other institution in our society, with the possible exception of the family, has as much opportunity to observe, mold, and modify youthful behavior as does the school. Early detection of problems frequently leads to their solutions before they become serious. The importance of education as a stepping-stone to future opportunities for success cannot be stressed too much. Stephens and Arnette (2000) pointed out, "Many at-risk young people make the disastrous choice of dropping out of school or behaving in ways that cause them to be abandoned by or pushed out of the school setting. Next to the family, school is the most formative influence in a child's life" (p. 2). The provision of meaningful educational opportunities for children who have been labeled as delinquent or in need of supervision is of great importance in attempts to reintegrate these children into society.

Although it was once possible, and fairly common, for educators to deal with "problem youth" by pushing them out of the educational system, recent court decisions indicate that all children have the right to an education. Therefore, children who have been found delinquent and status offenders can no longer be dismissed from school legally without due process. School counselors who formerly concerned themselves with academic and career counseling, advising, and scheduling also face the reality of coping with behavioral and emotional problems. It is hoped that teachers who in the past simply passed juveniles with such problems on to their colleagues by refusing to fail problem youth (giving them **social promotions**) will begin to seek other more desirable alternatives. Illinois, for example, has created the Regional Safe Schools Program, which targets sixth through twelfth graders. This program allows children who were traditionally expelled or suspended to transfer into an alternative learning environment to continue with academic work, counseling, community service, and vocational activities.

Moreover, educational personnel play an important role in preventing and correcting delinquent behavior by providing personal counseling and appropriate referrals.

There are numerous school programs designed to prevent juveniles from engaging in delinquent activities or to divert them from such activities once they become involved. In Chapter 7, we mentioned the police officer school liaison programs that have come into existence during recent years. To prevent the distraction of delinquency activities from the educational environment, school districts such as that in Camden County, New Jersey, have hired police officers to handle the public safety issues on campus. Camden County hired police officers to work 10 months while school is in session and to spend their time responding to criminal acts on campus, investigating them, and making arrests when necessary (Dunn, 1999). The officers also carry out community policing activities in the school, provide drug education programs, and serve as liaisons between the school and the police department. A peer mediation program coordinated by the police officers is used to resolve issues between students. Early evaluation of the program showed a reduction in fighting and drug-related problems among students as well as an increase in support from parents who had previously viewed the district as dangerous for their children (Dunn, 1999). This program is not uncommon, with school resource officers being a mainstay in most schools in the United States since the school shootings of the 1990s (Garry, 1996).

Another program presented in the schools by police officers is the DARE (Drug Abuse Resistance Education) program. Originally developed in California in 1983, the program spread rapidly to other states. The goal of the semester-long program aimed at fifth and sixth graders is to equip juveniles with the skills to resist peer pressure to use drugs. Trained police officers present the program as a part of the regular school curriculum in an attempt to provide accurate information about drugs and alcohol, teach students decision-making skills, help students to resist peer pressure, and provide alternatives to drug abuse.

Unfortunately, research has shown that participation in the DARE program during elementary school has no effect on later alcohol use, cigarette smoking, or marijuana use in twelfth grade, although it may deter a small amount of use of illegal and more deviant drugs such as inhalants, cocaine, and LSD among teenage males (Dukes, Stein, & Ullman, 1997). Other research has shown that the impact of DARE on drug-related behavior of children who have been through the program is minimal (Cauchon, 1993; Walker, 1998, p. 275). After reviewing various studies on DARE programs during the 1990s, Kanof (2003) reported that DARE had no statistically significant effect on long-term drug use (p. 2). Proponents argue, perhaps with some justification, that at a minimum the programs introduce police officers and children to one another as real people at an early age and that the effects of classroom interaction may have beneficial effects for both.

In response to the criticisms of the DARE program, a new research-based curriculum that focused on prevention science and drug use was created and implemented in 2001. The University of Akron and the DARE program combined to execute the program in six U.S. cities. (Carnevale Associates, 2006). The new program targets children in seventh grade with a 10-week curriculum and provides another short program to those same children as they enter the ninth grade. "To date, University of Akron researchers have preliminary results that the Take Charge of Your Life program may be effective in reaching those adolescents who are at elevated risk for substance abuse" (p. 4). These results seem promising, although the study has had a number of methodological problems (e.g., Hurricane Katrina causing the closure of some test schools, declining rates of drug use in the general population, implementation of similar programs in the schools).

▲ Danielle Aihini, a sixth grader at Copeland Middle School in Rockaway, New Jersey, discusses gang awareness with Raymond Vonderheide, a member of the New Jersey State Patrol Gang Unit. Students are participating in the GREAT program. Over the course of 4 weeks, students meet with a police officer to discuss gangs and violence, how to communicate, how to develop empathy for others' anger, and how to resolve conflict with others.

Yet another program involving police and school cooperation is the GREAT (Gang Resistance Education and Training) program. Unlike DARE, the GREAT program is a 13-week curriculum offered in middle schools, elementary schools, during the summer, and to families (GREAT, 2006). Its focus is on reducing the number of children joining gangs. It is taught by uniformed police officers and was recently evaluated. The research found that students who had completed the program gained more education on the "consequences of gang involvement, and they develop[ed] favorable attitudes toward the police. . . . However, the program did not reduce gang membership or future delinquent behavior" (Esbensen, 2004, p. 4). It was suggested that other strategies to reduce gang involvement and delinquency should be used:

1. Alternative education programs have also become more widespread. Such programs include enhanced skills training, community internship programs, and more general attempts to integrate the schools and the community in the interests of serving the needs of marginal students and those who do not anticipate attending college. Life Skills Training Programs, implemented by teachers in the classroom, are directed at sixth and seventh graders and are designed to prevent or alleviate tobacco, alcohol, and marijuana use. (Mihalic, Irwin, Elliot, Fagan, & Hansen, 2001)

Character education programs that put forth values that are generally accepted by all (e.g., honesty, integrity, responsibility, tolerance) are also being offered in schools with high numbers of at-risk youth (Stephens, 1998). Improved school counseling services and programs designed to encourage student and parent participation in combating violence and vandalism in the schools are indicative of other attempts to expand and improve the role of school personnel in preventing delinquency and diverting delinquents from further inappropriate actions. The Center for the Study and Prevention of Violence has identified a number of other programs that have been shown to be effective in reducing youth violence, including the following (Mihalic et al., 2001):

2. The Bullying Prevention Program, implemented primarily by school staff members, involves school-based intervention for the reduction of bullying. Research has shown that the program results in significant reductions in reports of students being bullied or bullying others and of antisocial behaviors. It also shows improvements in the overall "social climate" of the school (Clemson University, 2002, p. 2).
3. Big Brothers/Big Sisters of America (BBBSA) provides adult mentoring programs for children. Research has shown that children with mentors are 46% less likely to initiate drug use, 27% less likely to initiate alcohol use, and 32% less likely to commit assault.
4. The PATHS (Promoting Alternative Thinking Strategies) curriculum is primarily school based but also includes activities for parents. It is aimed at promoting emotional and social competencies.

Perhaps the most successful school program to prevent delinquency is the High/Scope Perry Preschool Project in Ypsilanti, Michigan. This program has focused on a group of 123 African American 3- and 4-year-olds identified as being at risk for school failure in 1962. Of these children, 58 were assigned to the preschool program, and 65 were assigned to a control group. Data concerning these children were collected periodically for some 40 years. The results of the research showed that children who had participated in the preschool program for 2.5 hours per day, Monday through Friday, for 2 years (a) had lower rates of delinquency than did the control group, (b) had lower teen pregnancy rates than did the control group, (c) were less likely to be dependent on welfare than were the control group members, and (d) were more likely to graduate from high school than were the control group members (High/Scope Educational Research Foundation, 2002).

Wilderness Programs

Wilderness programs had their origins in the forestry camps of the 1930s. These programs involve small, closely supervised groups of juveniles who are confronted with difficult physical challenges that require teamwork and cooperation to overcome them. The intent of the programs is to improve the self-esteem of the juveniles involved while teaching them the value of cooperative interaction. Wilderness programs, which last from roughly 1 month to 1 year or longer, do not typically accept violent juveniles. Some provide counseling and follow-up services, whereas others do not. Juveniles may be sent directly to these programs as an alternative to detention or may participate in the programs after more traditional dispositions have been imposed.

Evaluations of wilderness programs have been fraught with methodological difficulties. Much of the information concerning the success of the programs has been provided by program developers and staff and is anecdotal in nature. Most of the evidence provided under these less-than-ideal conditions shows that the programs lead to somewhat lower recidivism rates than no programs at all, but the effectiveness of the programs is in question.

Restorative Justice Programs

The philosophy of **restorative justice** centers on the assertion that crime and delinquency affect persons instead of the traditional assertion that crime affects the state. In fact, restorative justice defines a "crime [as] an offense against human relationships" (National Victims Center, 1998, p. 52). Howard Zehr, viewed as the leading visionary in restorative justice, defined restorative justice as "a process to involve, to the extent possible, those who have a stake in a specific offense and to collectively identify and address harms, needs, and obligations in order to heal and put things as right as possible" (Zehr, 2002, pp. 19–20). "It is a future-focused model that emphasizes problematic problem solving instead of just deserts" (Wilkinson, 1997, p. 6). "It is grounded in the belief that those most affected by crime should have the opportunity to become actively involved in resolving the conflict" (Umbreit, Vos, Coates, & Lightfoot, 2005, p. 255).

Restorative justice advocates programs such as victim–offender mediation, victim impact panels, community service, and community sentencing. Such programs, aimed at creating or enticing emotions within criminal offenders, appear to be gaining momentum in the criminal and juvenile justice networks in the United States.

Restorative justice programs have been variously called community conferencing, family group conferencing, community justice, community corrections, balanced and restorative justice, victim–offender dialogue, and real justice, depending on the agency applying the concepts. Although the name may change, the definition and core concepts of restorative justice—accountability, competency, and public safety—remain the same in all applications. Crime victims, offenders, and the community are the keys to success in restorative justice.

The question is, how well does restorative justice work in practice? Most studies on restorative justice have focused on satisfaction rates of the participants, not recidivism. Feelings of satisfaction and fairness have consistently been high among the participants (Bradshaw & Roseborough, 2005). But what about recidivism? Walker (1998) indicated, "Evaluations of experimental programs have tended to find slightly lower recidivism rates for offenders receiving restorative justice than for those given traditional sentences of prison or probation. The differences are not always consistent, however, and many questions remain regarding the implementation and outcomes of such programs" (p. 224). Bradshaw and Roseborough (2005) found in their study that victim–offender mediation and family group conferencing were two restorative programs that showed the most promise in reducing recidivism.

Although it is likely that the informal sanctions imposed by family and community are more effective than threats of formal punishment, what happens when there is no sense of family or community? The sense of family and community may well be lacking in drug-ridden and economically ravaged neighborhoods. The concept of restorative justice may make sense for young middle-class offenders involved in minor first offenses but may be totally irrelevant for those living in the most crime-ridden areas of America (Walker, 1998, p. 225). In fact, the failure of these institutions is largely responsible for the development of our current criminal justice system, and there is little evidence of the rebirth or strengthening of these institutions today (Cox & Wade, 1996, pp. 48–50; Walker, 1998).

Other **faith-based initiatives** are also becoming popular with the federal government (see In Practice 9.2). President George W. Bush established an Office of Faith-Based and Community Initiatives tasked with strengthening and expanding social services offered to the needy (in particular those in poverty). Through this office, faith-based organizations can apply for grants and funding to support programs aimed at drug prevention, violence prevention, at-risk youth, gang behaviors, and so on. Because this is a fairly new endeavor, research is scarce; however, Ericson (2001) reported that faith-based groups are usually open to developing relationships with other agencies, have poor administrative or organizational structures (e.g., hiring practices, bookkeeping skills), and avoid preaching to the children about religion. Instead, they rely on relationship building and support. This is one area that will be worth watching in future studies.

In Practice 9.2

A Different Kind of Justice: Peoria Crime Victim Appeals to Court for Chance to Mentor Teen Who Robbed Him

"Be joyful in hope, patient in affliction, faithful in prayer."

As Montelle Talley recites his favorite Scripture from memory, Dwight Winnett reaches for the Bible. The two sit in plain office chairs, facing each other in Winnett's cluttered office at United Disciples Christian Church in West Peoria.

Winnett, a minister, thumbs through pages to the Book of Romans, Chapter 12, Verse 12.

"Yep, it's just what you said, joyful in hope."

Talley's mother used to send him Bible verses while he was in jail, awaiting sentencing on armed robbery charges. "And that's just what I did," Talley continues, "be faithful in prayer."

He smiles, and his voice flows from the stop-start rhythms of rap and hip-hop to the low rolling timbre of a Sunday morning sermon. But it is a Thursday afternoon a few weeks ago.

"See, the Bible will help you understand your situation," Talley says. "Things seemed bad; they were talking about eight years."

Talley's prospects for an eight-year prison sentence dried up considerably one day last February when Winnett stood in a Peoria County courtroom telling a judge Talley could have shot him, Talley could have killed him, but if the court would allow, he would rather mentor Talley than see him to go prison.

The language of criminal justice doesn't fit Talley and Winnett.

A crime victim and a criminal?

A guilty robber and an innocent victim?

(Continued)

(Continued)

Two men—one 19, the other 54; one a minister, the other a rapper in the Dirty Boys Click—sharing Bible verses in a cramped church office?

In a criminal justice system built on adversary [proceedings] and punishment, there are no words for what they are becoming.

Restorative Justice

Largely because of Winnett's intervention, Tally got probation instead of prison that day in court. Since then, he and Winnett have been meeting fairly regularly. Talley and his mother visited Winnett's church. Through a friend, Winnett helped Talley find a part-time job. Mainly, they have gotten to know each other outside the narrow boxes marked victim and assailant.

"This is restorative justice," says Sally Wolf, consultant and trainer for Illinois Balanced and Restorative Justice Initiative, a not-for-profit coalition based in Paxton. She has introduced the restorative concepts and practices to juvenile court programs, probation departments, police departments, and schools, including, recently, Manual and Woodruff high schools.

"At its purest, restorative justice just happens: It comes up from the people, the community, not from bringing in people like me to try and start it," says Wolf.

Restorative justice, also known as balanced and restorative justice, is not a program. By law, balanced and restorative justice is supposed to be the philosophical underpinning of all juvenile court systems in Illinois.

In reality, its implementation is spotty, but growing, in juvenile court systems in Illinois and virtually unheard of in adult criminal court, where Talley and Winnett's case was decided.

"It's a hard sell," Wolf says, "because we have been brainwashed into one way of thinking for so long."

Restorative justice is a philosophy, a set of practices easiest defined by what it is not—which is the dominant law of the land, a system of adversarial, retributive, or punitive justice.

"It's almost a religious embrace that the pathway to progress is through incarceration," Bart Lubow, a national expert on alternatives to incarceration, said during a recent conference on juvenile detention in Peoria. "We are wedded to the idea that we can achieve a more civil society through incarceration."

But Peoria County State's Attorney Kevin Lyons has qualms about a concept like restorative justice.

"It seems so darn nebulous," he says. "I can't get my arms around it."

However, he admits he misjudged recent judicial innovations, including drug court, which combines elements of the restorative justice philosophy with substance abuse treatment.

"I used to say there's no way you're going to get me to approach domestic violence cases in a different manner or try drug court," he says.

"I have been dragged into those things, kicking and screaming, but I was wrong."

How It Works

Adversarial justice systems start with what law was broken, who broke it, and how to punish the offender.

Restorative justice policies ask what harm was done, how to repair the harm, and who's responsible for repairing that harm.

By the short definition, restorative justice aims to achieve a more civil society through building, or repairing, relationships between or among people.

A broad variety of practices can fall under its umbrella, some of them familiar, such as victim's restitution or meaningful community service.

(Continued)

(Continued)

Deeply restorative concepts, in one form or another, have been applied to efforts to reconcile racial groups in post-apartheid South Africa, end gang wars in South Central Los Angeles, resolve neighborhood disputes, and cut truancy rates and fighting in schools.

Under restorative justice principles, crime victims don't have to develop relationships with their assailants, as Winnett and Talley have. Victims don't have to help them find jobs, as Winnett has for Talley. Victims are free to participate as much, or as little, as they wish in the process of repairing the harm.

But the mending between Talley and Winnett highlights the guiding principles behind restorative justice.

"A Little Stop"

Their first encounter was early, about 9:30 a.m. in March 2005. Winnett was carrying groceries into his mother-in-law's house along West Malone Street. He saw Talley and a boy walking down the street, but he didn't think anything of it.

Talley, walking with a younger friend, was on his way home. It was a weekday. He had been suspended from school after spending the night at another friend's house. He had just started getting into guns.

"I was on my way home," Talley repeats several times, as if to emphasize he hadn't started out to rob Winnett. But there he was, standing in a secluded driveway. And here was Talley, with a gun tucked in his waist.

The thought of robbing someone frightened Talley a little, but then again, it might be easy. He'd just "make a little stop, keep on going."

Winnett doesn't say a word as Talley speaks.

"But it's almost as if I'm the objective outsider, I hadn't known what was going through his head," Winnett says.

When Winnett speaks, however, Talley almost relives the emotions swirling through his head that day—and the emotions he felt when he, himself, was a victim of an armed robbery.

At one point, Winnett mentions that he didn't see Talley in the driveway. Talley realized Winnett didn't see him.

"That's the part where my heart dropped," Talley interrupts. "I still had time to change my mind."

From memory, Winnett recites exactly what Talley ordered him to do. Close the trunk of the car, come here, put your hands on your head.

Talley interrupts again. "So much was going through my mind, I didn't hear nothing he was saying."

Hearing Winnett, Talley continually flashes back to a night several years earlier, when a man stuck a gun in his chest demanding money and a cell phone. Talley's assailant ordered him to turn around and walk away, but Talley refused. "If he was going to shoot me, he was going to have to look at me." Talley says.

Talley negotiated his way out of the robbery and walked away, never calling police. But he didn't forget how it made him feel. That's why he still can't understand what made him do the same thing to Winnett.

Repairing the Harm

From Winnett's vantage point, the two robberies are slightly different.

"When I saw his eyes, I was convinced this young man wants to rob me, he doesn't want to shoot me," Winnett says.

(Continued)

(Continued)

"You expect, when you're getting robbed, the stereotypical gangster-rap kind of stuff. He didn't treat me like that at all. He was very quiet, very polite, and respectful. I really felt safe, even though he had a gun in his hand."

And, Winnett noticed, Talley never pointed the gun directly at him. In fact, he wondered if he could get the gun from him.

Talley laughs. "You probably could have."

Instead, Talley took Winnett's wallet and, along with his friend, ran. Winnett called police. The two were arrested almost immediately.

Restorative justice holds offenders accountable in a different kind of way, Wolf says.

"When people sit down and really listen to how someone has been harmed, when they realize what they've done, it's amazing," she says.

In a jurisdiction with formalized restorative justice policies, Talley, Winnett, and others affected by the crime, including Talley's family and Winnett's family, might have had a sit-down meeting with a judge, attorneys, police officers, probation officers, even community residents to try to reach an agreement on how to repair the harm.

As it was, Winnett did much of the legwork himself, meeting Talley's mother [and] talking to Talley's minister, while Talley sat in Peoria County Jail for almost a year.

During his time in jail, Talley earned a general education diploma. Indirectly, he got word to Winnett, apologizing for what he had done.

Finally, Winnett talked to a judge and attorneys, who agreed to let him and Talley try to repair the harm on their own.

Good for Both

Their work is not finished. It has not been easy or perfect. Talley must make it through four years of probation, including a year of intensive probation. Plus, he was nervous and a little leery of meeting Winnett.

"On our first meeting, we went to Steak 'N Shake," Talley recalls. "He told me he forgave me. That made it easier for me to talk to him."

After initially meeting weekly for about four months, Winnett's workload at the church got heavier. They met and talked sporadically during the summer. But neither plans to stop the repair job they've started.

Winnett says this is an opportunity for him as well as Talley.

"All of this is not just working for Montelle's good, it's for my own good."

SOURCE: Adams, Pam. (2006, August 27). A different kind of justice: Peoria crime victim appeals to court for chance to mentor teen who robbed him. *The Peoria Journal Star,* Record Number: 0000776060.

Children and Family Services

As noted in Chapter 7, child protective services (children and family services) agencies have goals similar to those of the juvenile justice system. These agencies provide, among other services, day care programs, foster care programs, youth advocacy programs, and advice to

▲ Early childhood education programs may help to prevent later delinquency.

unwed mothers. In addition, they investigate reported cases of child abuse and neglect. Children and family services agencies deal with all categories of juveniles covered by most juvenile court acts, provide individual and family counseling services, and are empowered to refer suitable cases to appropriate private agencies. In addition, they can provide financial aid to children and families in need.

Like most state agencies, children and family services agencies are often caught up in political change. Although many of these agencies require a bachelor's or master's degree for employment and emphasize the need for professionalism among staff members, skillful and competent administrative personnel are often replaced when the political party in power changes. As a result, the continuity of policies implemented by these agencies frequently leaves much to be desired. Nonetheless, state agencies concerned with providing services to children and families often have considerable power and, when administered appropriately, can provide multiple services to children in trouble. When not administered in the proper way, results can be disastrous.

Federal Programs

The federal government has sponsored many programs that, although not designed specifically as delinquency prevention programs, did encourage children to accept and attain lawful objectives through institutionalized means of education and employment. Examples of some of the varied federal programs provide some insight into the value of these programs in preventing delinquency and crime, illustrate the focal points of these programs, and show how they attempt to improve the social ills that result in delinquency (Yablonsky & Haskell, 1988).

There have been a number of federally funded programs aimed at improving educational and occupational opportunities for disadvantaged children. A secondary benefit of many of these programs was believed to be a decrease in the likelihood of delinquency among the children involved. The projects **Head Start** and **Follow Through** were designed to help culturally deprived children catch up or keep pace during their preschool and early school years. Previously, many children from culturally and/or economically deprived parents lagged behind other children in verbal and reading skills. Starting far behind in basic skills, many of these children never caught up and school too often became an experience characterized by failure and rejection. As a result, many dropped out of school as soon as possible, often during their first or second year of high school. Of those who did drop out, many went on to become delinquent. Head Start and Follow Through have shown that children who are socioeconomically disadvantaged can, and do, make progress when parents, teachers, and volunteers focus their efforts on these children (Eitzen & Zinn, 1992, p. 399). The High/Scope Perry Preschool Project discussed earlier in this chapter is another example of a federally funded program. Demonstrating a trickle-down effect, states have picked up on this trend by funding prekindergarten programs in public schools. "As of the 2002–03 school year, over 750,000 children were in these programs" (Clifford, Bryant, & Early, 2005, p. 50).

Along slightly different lines, a number of federal laws providing assistance to the hardcore unemployed were passed. For example, the Manpower Development and Training Act, the Vocational Education Act, the Economic Opportunity Act, the Rehabilitation Program for Selective Service Rejectees, the Comprehensive Employment and Training Act, the Job Training Partnership Act, and the President's Youth Opportunities Campaign had the major objective of aiding young people in finding employment by helping them to become more readily employable. The basic assumption underlying these programs has been that employment is an important key to solving the problems of many young people.

The emphasis of youth opportunity centers is to increase employability through counseling or to provide vocational and prevocational training and work training programs. This approach recognizes that if young people, handicapped by inadequate education and lack of occupational skills, are to become employable, they must somehow be provided with additional training. It is hoped that these young people will then be absorbed into the labor market once their performance capabilities are improved.

Similarly, the Job Corps program was directed at individuals between 16 and 21 years of age with the principal objective of providing training in basic skills and a constructive work experience. The Job Training Partnership Act of 1981 also promised new hope for young people seeking their first jobs when it replaced the scandal-ridden Comprehensive Employment and Training Act.

All of these programs have been geared toward providing youth with employment opportunities that, it is hoped, will lead them to a better life. The basic underlying assumption seems to be that young people employed in jobs for which they are suited are less likely to engage in delinquent or criminal activity than are young people who are not employed and have little hope of finding any worthwhile employment.

During recent years, the federal government has given attention to the concept of mentoring. In 1992, Congress amended the Juvenile Justice and Delinquency Prevention Act of 1974 to include the Juvenile Mentoring Program (**JUMP**) because of a growing belief that positive bonds between children and adults can forge actions or behaviors essential to a healthy life (Bilchik, 1998b). According to the 1998 Juvenile Mentoring Program (JUMP) report to Congress,

> Historically, the notion of one individual providing caring support and guidance to another individual has been reflected in a variety of arenas. In the clinical mental health field, we talk about bonding and the importance of a child feeling connected to a nurturing adult in the early years of life. In the adoption field, we talk about the need for attachment. In schools, tutors help support successful educational experiences. In juvenile and family court, Court Appointed Special Advocates (CASAs) provide support and advocacy for children in need of assistance. In the substance abuse field, we make use of sponsors to support sobriety. In the business field, we create teams to ensure that new employees have the support they need to be successful in the corporate organizational system. Currently, there are many types of formal mentoring programs generally distinguishable by the goals of their sponsoring organization. Most youth oriented programs recognize the importance of ensuring that each child they serve has at least one significant adult in his/her own life that can be friend, role model, guide, and teacher of values. If that person is not available in the child's family, mentors can help fill the critical gap. (p. 5)

By using JUMP, the federal government is hoping to modify behaviors committed by children that can lead to juvenile delinquency, gang participation, and increased school dropout rates and to enhance the academic performance of the children participating in the program. All JUMP programs have been sponsored by local community organizations with the help of federal grants during recent years. Current research on the program is still ongoing, although early findings indicate that both children and their mentors found the relationship to be rewarding (Novotney, Mertinko, Lange, & Baker, 2000, p. 5). Research on how the program meets or accomplishes its goals is still not available.

The role of the federal government in programs designed specifically to prevent delinquency has been somewhat limited as a result of the belief that the primary responsibility for these programs rests with the states. Although there have been scattered efforts in the field of juvenile justice by the federal government (e.g., the development of the Children's Bureau in 1912, the development of various federal commissions and programs in 1948, 1950, and 1961), the ones most relevant to prevention occurred in 1968 with the Juvenile Delinquency Prevention and Control Act and in 1974 with the Juvenile Justice and Delinquency Prevention Act. The Juvenile Delinquency Prevention and Control Act permits allocation of federal funds to the states for delinquency prevention programs, and the Juvenile Justice and Delinquency Prevention Act attempts to create a coordinated national program to prevent and control delinquency (Office of Juvenile Justice and Delinquency Prevention, 1979). The Juvenile Justice and Delinquency Control Act also called for an evaluation of all federally assisted delinquency programs, a centralized research effort on problems of juvenile delinquency, and training programs for persons who work with delinquents. This law directs spending of funds on diverting juveniles from the juvenile justice system through the use of community-based programs such as group homes, foster care, and homemaker services. In addition, community-based programs and services that work with parents and other family members to maintain and strengthen the family unit are recommended.

The Juvenile Justice Amendments of 1977 made it clear that, in the opinion of Congress, the evolution of juvenile justice in the United States had resulted in excessive and abusive use of incarceration under the rubric of "in the best interests of the child" and that the prohibitions of contact with adult offenders and incarceration of status offenders and nonoffenders (e.g., dependent or neglected children) were to be taken seriously (Office of Juvenile Justice and Delinquency Prevention, 1980).

A wide variety of community and state agencies have become involved in delinquency prevention. Most efforts have been independent and uncoordinated. By the 1950s, the delinquency prevention effort in virtually every state and large city was like a jigsaw puzzle of services operating independently. The agencies concerned with delinquency prevention included the schools, recreation departments, public housing authorities, public welfare departments, private social agencies, health departments, and medical facilities. Davidson, Redner, and Amdur (1990) came to the conclusion that although diversion programs can provide positive results, territorial jealousies remain difficult to overcome.

Other Diversion and Prevention Programs

Although it would be impossible to list and discuss all prevention and diversion programs, we mention a few more here. Recreational and activity programs conducted by local police, civic, and religious groups are often aimed at preventing delinquency. The concept of **teen courts** has originated as a way to keep first-time juvenile offenders who commit minor offenses and are willing to admit guilt from being processed in the formal juvenile justice system. Local civic agencies or schools, in conjunction with the police department and the juvenile court, sponsor most of these programs. The courts are made up of teens under 17 years of age who process the cases by acting as prosecutor, defense counsel, bailiff, and clerk and who determine the punishment for the cases by acting as the jury. An adult attorney acts as the judge to ensure the fairness and legality of the sentencing. The offender is required to complete the sentence handed down by the teen jury. If the offender does not abide by the sentencing guidelines, he or she is referred to the juvenile court for formal processing. The goal of these programs is to hold the juveniles accountable for their actions, but not to stigmatize the juveniles by formally processing them in the juvenile justice system, while attempting to divert them from further delinquency.

Drug courts are another attempt to prevent children and adults from continuing deviant behaviors. Drug courts aim to stop the abuse of alcohol and other drugs through the use of intensive therapeutic supervision. "As of 2006, there were 1,557 drug courts in operation in the United States" and more than 300 more in the planning phase (National Institute on Drug Abuse, n.d., p. 98). Research on rates of recidivism after drug court attendance is virtually nonexistent at this point. The research available remains focused on training, development, implementation, and expected outcomes. As suggested by the Bureau of Justice Assistance (2003), drug courts are in dire need of rigorous evaluation standards and reporting guidelines.

In addition to the prevention and diversion programs already mentioned, there have been a number of attempts to scare juveniles away from delinquent behavior. The best known, although not the earliest, of these programs was publicized nationally through a television film call **Scared Straight.** The film recorded a confrontation between juveniles brought into Rahway State Prison in New Jersey and inmates housed in the prison. Such confrontation was based on the theory that inmates could frighten juveniles to the extent that they would be deterred from committing further delinquent acts. Scared Straight reported that of the 8,000 juveniles participating in such sessions through 1978, 90% had not been in trouble with the law again. Nationwide attention was focused on attempts to frighten juveniles out of delinquency, and such programs were viewed by some as a panacea for delinquency problems (Finkenauer, 1982). However, more objective evaluations of this and other such programs have yielded, at best, mixed results. It is certain that such programs are not a panacea for delinquency, and some appear to increase rather than decrease the frequency of recidivism. In fact, Lundman (1993) recommended the permanent abandonment of efforts to scare and inform juveniles "straight."

Yet another attempt at preventing delinquency and diverting delinquent children involves the use of community policing models oriented toward juveniles. These programs operate on the assumption that community policing officers are more likely to favor problem-solving and peacekeeping roles with children than are their traditional counterparts (see Chapter 7). Officers who view their roles in these terms may be more likely to try to help children before they get into trouble or to divert them away from the juvenile justice network (Bazemore & Senjo, 1997; Belknap, Morash, & Trojanowicz, 1987).

Community-based programs in Harrisburg, Pennsylvania, sponsored by the juvenile justice system of that state, attempted to reduce minority overrepresentation in the juvenile justice system by reducing rates of arrest and rearrest for clients and by reducing educational failure, dropout, and truancy. A 2-year follow-up evaluation of the results of the programs showed that they were somewhat successful in reducing recidivism but had little effect on school failure, dropout, or truancy (Welsh, Jenkins, & Harris, 1999). Santa Cruz, California, also addressed issues in minority representation recently. City officials changed policies, procedures, practices, and programs that they identified as barriers to services for minorities. In particular, they implemented objective decision-making techniques, implemented cultural sensitivity practices, and decreased unnecessary delays in detention releases and court appearances (Dighton, 2003).

There are a host of other agencies providing services that complement those of the juvenile justice system. These include YMCAs and YWCAs, both of which often provide counseling and recreation programs. One alarming trend among these agencies is that membership fees have tended to eliminate the opportunity for some children to use the services available. Some YMCA and YWCA programs seem to discourage rather than encourage the participation of children who have little interaction with adults and have few resources.

In many areas, community mental health clinics provide services based on a sliding fee scale. Other agencies, such as Catholic Social Services, Vocational Rehabilitation Services, and the Boy and Girl Scouts of America, also use a sliding scale to determine fees for counseling, membership, testing, and employment referrals. Still other agencies provide essentially the same services free of charge. These agencies typically include community centers, Big Brother/Big Sister Volunteer Programs, alcohol and drug clinics, and hotline programs. In addition, many colleges and universities offer counseling services free of charge or based on a sliding scale.

Not all diversion and prevention programs originate in urban areas. There are many programs that have been implemented effectively in rural areas as well. One example is mentoring programs that highlight the importance of a caring adult in helping to prevent delinquency, increasing the likelihood of school success, and developing relationship skills. Mentors provide tutoring, job skills, and informal counseling and also make referrals to agencies as necessary. Big Brother/Big Sister programs are among the best-known mentoring programs, and they have branches in rural areas as well as urban areas around the country. Teen and drug courts are also spreading to small towns and rural areas, as is the notion of restorative justice in general.

Some Criticisms

As indicated previously, delinquency prevention programs usually employ one of two strategies: either reform of society or individual treatment. Both strategies, as generally employed, have had difficulties. Programs oriented toward reforming society have been quite costly in terms of the results produced, depending on whether results are measured in terms of alleviating educational, occupational, and economic difficulties or in terms of reducing delinquency. Lack of coordination among various programs, interprogram jealousy, considerable duplication, and mismanagement have seriously hampered the effectiveness of these programs. As a result, much of the money intended for juveniles with problems ends up in staff salaries, and many of the personnel hired to help supervise, train, and educate these juveniles are tied up in dealing with administrative red tape. In addition, programs attempting to improve societal conditions may take a long time to show results. The extent to which any results can be attributed to a specific program is extremely difficult to measure. As a result, the public is frequently hesitant to finance prevention programs because they have no immediately visible payoffs. In fact, it may be that diversion programs simply do not work either because the concept is flawed or because the current system does not provide an opportunity for them to work. Some see diversion as an interesting concept with "unanticipated negative consequences" (Mays & Winfree, 2000, p. 116).

There are two basic types of prevention programs directed at providing individual treatment. The first deals with children who have already come into contact with the juvenile justice system and attempts to prevent further contact. As noted previously, there are inherent difficulties in attempting to reform or rehabilitate juveniles after they have become delinquent. Many of the basic assumptions about programs directed at preventing future delinquent acts by those already labeled as delinquent are highly questionable. For example, it is doubtful whether individual therapy will be successful if the juvenile's problems involve family, school, and/or peers. Similarly, the belief that recreational or activity programs, in and of themselves, are beneficial in reducing delinquency seems to be more a matter of faith than a matter of fact at this time.

The second type of individual treatment program attempts to identify juveniles who are likely to become delinquent before a delinquent act is committed. These programs may be called early identification programs or predelinquency detection programs. Although these programs are clearly intended to nip the problem in the bud, they may be criticized for creating the very delinquency they propose to reduce; that is, identifying a juvenile as predelinquent focuses attention on the juvenile as a potential problem child and, therefore, labels him or her in much the same way as official juvenile justice agencies label juveniles as delinquent. In one sense, then, the juvenile is being treated (and sometimes punished) for something that he or she has not yet done. Programs directed toward pure prevention may, unintentionally, lead juveniles to be labeled earlier by identifying them at an earlier stage. This phenomenon is often referred to as net widening (discussed earlier in this chapter).

Some time ago, Edwin M. Schur encouraged the development of an approach to delinquency prevention. We believe (and have suggested at other points in this book) that it has considerable merit. His approach is called **radical nonintervention.** According to Schur (1973), "The primary target for delinquency policy should be neither the individual nor the local community setting, but rather the delinquency-defining processes themselves" (p. 154). Rather than consistently increasing the number of behaviors society refuses to tolerate, we should develop policies that encourage society to tolerate the "widest possible diversity of behaviors and attitudes" (p. 154). Much of the behavior currently considered delinquent is characteristic of adolescence, is nonpredatory in nature, and is offensive only because it is engaged in by juveniles. Because in one sense it is rules that produce delinquents, it might make more sense to change the rules (as we have done at the adult level in terms of alcohol consumption, abortion, and homosexuality) than to attempt to change juveniles or the entire society overnight. One approach, then, would be to make fewer activities delinquent and to concentrate on enforcing rules for violations that may actually be harmful to the juvenile, society, or both. In other cases, our best strategy may be simply to "leave kids alone wherever possible" (p. 154).

Supporting Schur's (1973) contention is the fact that the Office of Juvenile Justice and Delinquency Prevention (1979) found that a number of programs have no defensible basis whatsoever (e.g., those based on presumed personality differences or biological differences), others are poorly implemented (e.g., behavior modification programs in treatment settings without community follow-up), and still others of questionable merit are based only on preliminary evidence (e.g., most predelinquency identification programs).

Finally, Sherman and colleagues (1997), in a systematic review of literature on crime prevention, concluded that some programs seem to work but that many do not. More important, perhaps, the authors concluded that the programs that work best are those in communities that need them least and that true prevention probably lies outside the realm of criminal justice. Indeed, many programs seem to work where schools and families are stable, but few appear to be successful where schools and families are torn apart by drugs, crime, and violence. Walker (1998), in reviewing this and other studies, concluded,

> We found that most current crime [delinquency] policies and proposed alternatives are not effective. We found that both conservatives and liberals are guilty of peddling nonsense with respect to crime policy. . . . The truth about crime policy seems to be that most criminal justice-related policies will not make any significant reduction in crime. (p. 279)

Thus, as we indicated earlier in the book, if we wish to prevent at least some crime and delinquency, we must seek solutions in the broader social structure by focusing on unemployment and discrimination, maintaining stable families (whatever their structure), and providing meaningful education for all children.

Child Abuse and Neglect Prevention Programs

Many of the specific programs we have discussed are oriented toward diverting delinquents and preventing delinquency. There are also numerous programs aimed at preventing child abuse and neglect. We examine a few of these programs here.

Illinois instituted a program to provide in-home assistance to families in which children have been mildly abused so as to prevent foster home placement. Mild abuse has been defined as including things such as overspanking and bruising of the child that does not result in serious injury to the child. Under the plan, when a report of abuse is filed, investigators from the Department of Children and Family Services talk with the family to determine whether the child is safe in the home and whether the family is likely to benefit from the plan (Perkiss, 1989).

In the state of Minnesota, politicians have made childhood assistance a clear priority in their campaigns. Among the programs already under way is the Child Care Fund, which helps to subsidize the cost of day care. The Children's Health Plan provides benefits to poor children under the 8 years of age and may be expanded to help poor children up to 18 years of age (Madigan, 1989).

The Child Abuse Prevention Program in New York uses puppets to provide information to school children on physical and sexual abuse. The program has been in operation since 1988 and has been presented to more than 280,000 children (Child Abuse Prevention Program, 2006). The program addresses safe, unsafe, and confusing touches; identifying, resisting, and reporting abuse; and the differences between abuse and discipline (Child Abuse Prevention Program, 2006).

Operation KID (Kids Identification), initiated by the Edmond Police Department and Edmond Memorial Hospital in Edmond, Oklahoma, is a program addressing the issue of missing children. The program provides tips for preventing abduction of children, provides fingerprints and dental records for parents to be used in the event a missing child who cannot be identified is found, and in such cases also provides a profile sheet of information about the child (Russi, 1984).

St. Luke's Hospital in Cedar Rapids, Iowa, opened a Child Protection Center in 1987. The center provides centralized access to services for sexually and physically abused children and coordinates efforts to identify, treat, and prevent child abuse. Specially trained personnel conduct examinations of abused children and collect evidence for prosecution based on briefings provided by police investigators (Hinzman & Blome, 1991).

These are only a few of the many programs currently being used to prevent abuse and neglect. As is the case with delinquency prevention programs, none of them is foolproof—none is a panacea.

Career Opportunity—Big Brother/Big Sister Program Director

Job Description: Work with children and their families interested in forming relationships with adult mentors. Work with adult volunteers who choose to develop mentoring relationships with children. Interview, assess, and train volunteers, children, and families. Supervise and monitor adult-to-child mentoring matches. Facilitate support groups. Be responsible for recruiting children and adult mentors from the community.

Employment Requirements: Must have a minimum of a bachelor's degree in social work, counseling, guidance, psychology, or a related field. Must have experience in working directly with children within a social service agency or similar surroundings, excellent oral and written communication skills, assessment and counseling skills, problem-solving skills, experience with diverse populations, and (usually) some experience in public speaking. Required to work some evenings and, on occasion, weekends.

Salary and Benefits: Salary ranges from $20,000 to $40,000, depending on the location of the position. Benefits vary by geographical location but typically include paid medical, dental, and vacation as well as a 401(k) plan.

SUMMARY

All practitioners interested in the welfare of juveniles with problems should be familiar with the wide range of programs available in most communities. Teachers should not hesitate to consult personnel from children and family services agencies or law enforcement officials when appropriate or to enter into long-term agreements about sharing information in the interests of intervening appropriately with children in trouble. It is important to remember that the goal of each of these agencies is the same—to provide for the best interests of children. Territorial jealousy must be eliminated, and practitioners must learn to share their expertise with those outside their agency. It is not a sign of failure or weakness to recognize and admit that a particular problem could be dealt with more beneficially by personnel from an agency other than one's own. Concerned practitioners should provide direct services when it is possible and should not hesitate to make referrals when doing so is necessary or desirable.

Probably the best way to combat delinquency and child abuse is to prevent them from occurring in the first place. There are at least three ways to accomplish some form of prevention: changing juvenile behavior, changing the rules governing that behavior, and changing societal conditions leading to that behavior. Although the last named probably holds the most promise for success, it is also the least likely to occur.

By establishing good working relationships among schools, families, and juvenile justice practitioners, early detection of serious juvenile problems may be facilitated, and proper referrals may be made. Clearly, if the old adage that "an ounce of prevention is worth a pound of cure" is true, early detection and the support of the family as the primary institution influencing juvenile behavior are crucial to prevention programs. It is true that educational and vocational projects, community treatment programs, and the use of volunteers and nonprofessionals show some effectiveness. Recreation, individual and group counseling, social casework, and the use of detached workers (gang workers) may also be effective under some conditions.

At the same time, it is clear that many juvenile offenses are of a nonserious nature and that the statutes creating these offenses could be changed. We need to assess the necessity or desirability of many statutes and move to change those that serve no useful purpose and those that do more harm than good.

Practitioners are also in an excellent position to detect and report types of behavior that, in their experience, frequently lead to the commission of serious delinquent acts. Use of their experiences in combination with well-designed research projects will, it is hoped, lead to modified, more satisfactory theories of causation. Recognizing the variety of factors involved, the range of alternative programs available, and the strengths and weaknesses of prevention programs should lead to greater success in dealing with both categories of juveniles.

Preventing delinquency and child abuse is more desirable than attempting to rehabilitate delinquents or salvage battered and neglected juveniles from an economic viewpoint, from the viewpoint of the juveniles involved, and from society's viewpoint. It is hoped that commitment by both government and the private sector will facilitate more effective prevention and lead to the abandonment of ineffective programs. Examination of some of the basic assumptions of current prevention programs is essential.

There are a number of agencies operating programs that complement or supplement juvenile justice programs. Coordinating and organizing these programs to eliminate duplication and increase efficiency has been shown to be difficult as the result of territorial jealousy. Nonetheless, the best way to ensure the welfare of juveniles with problems is to share knowledge through interagency cooperation and referral, and budgetary restraints are currently dictating that this be accomplished.

Note: Please see the Companion Study Site for Internet exercises and Web resources. Go to www.sagepub.com/juvenilejustice6study.

Critical Thinking Questions

1. What are the major approaches to delinquency prevention? What are the strengths and weaknesses of each? Discuss some contemporary attempts to prevent delinquency or divert delinquents and tell why you believe they are effective or ineffective.
2. List some of the assumptions you believe are basic to delinquency prevention and diversion programs. To what extent do you feel each of these assumptions is justified? Why is the public often unwilling to finance prevention programs, and what are the consequences of this unwillingness?
3. What is territorial jealousy? Why does it occur, and what are some of its consequences?
4. Discuss at least two agencies or programs with goals similar to those of the juvenile justice system. In your opinion, how successful are these agencies in achieving their goals?
5. Discuss some of the attempts currently being made to prevent child abuse and neglect. Are such programs operating in your community?

Suggested Readings

Carnevale Associates. (2006). *A longitudinal evaluation of the new curricula for the D.A.R.E. middle (7th grade) and high school (9th grade) programs: Take Charge of Your Life—Year Four progress report.* Available: www.dare.com/home/Resources/documents/DAREMarch06ProgressReport.pdf

Esbensen, F-A. (2004). *Evaluating G.R.E.A.T.: A school based gang prevention program* (Research for Policy). Washington, DC: National Institute of Justice.

Flores, R. J. (2003). *Child delinquency: Early intervention and prevention.* Washington, DC: U.S. Department of Justice, Office of Juvenile Justice and Delinquency Prevention.

Gang Resistance Education and Training. (2006). *Welcome to the G.R.E.A.T. Web site.* Available: www.greatonline.org

Garry, E. (1996). *Truancy: First step to a lifetime of problems.* Washington, DC: U.S. Department of Justice, Office of Juvenile Justice and Delinquency Prevention.

Konaf, M. (2003). *Youth illicit drug use prevention: DARE long term evaluations and federal efforts to identify effective programs.* Washington, DC: U.S. General Accounting Office. Available: www.eric.ed.gov/ERICDocs/data/ericdocs2/content_storage_01/0000000b/80/27/f7/98.pdf

Mihalic, S., Fagan, A., Irwin, K., Ballard, D., & Elliot, D. (2004). *Blueprints for violence prevention.* Washington, DC: U.S. Department of Justice, Office of Juvenile Justice and Delinquency Prevention.

National Conference of State Legislators. (1999). *Comprehensive juvenile justice: A legislator's guide.* Available: www.ncsl.org/programs/cj/jjgover.htm

Stephens, G. (1998, May). Saving the nation's most precious resources: Our children. *USA Today Magazine,* pp. 54–57.

Stephens, R. D., & Arnette, J. L. (2000). *From the schoolhouse to the courthouse: Making successful transitions.* Washington, DC: U.S. Department of Justice, Office of Juvenile Justice and Delinquency Prevention.

Zehr, H. (2002). *The little book of restorative justice.* Intercourse, PA: Goodbooks.

DISPOSITIONAL ALTERNATIVES

CHAPTER LEARNING OBJECTIVES

On completion of this chapter, students should be able to

- *List and describe dispositional alternatives*
- *Discuss the dispositional phase of the juvenile justice process*
- *Discuss probation, conditions of probation, and revocation*
- *Discuss the relationship between probation and restorative justice*
- *List advantages and disadvantages of foster homes*
- *List advantages and disadvantages of treatment centers*
- *Discuss juvenile corrections, dilemmas, and consequences*
- *Present arguments for and against capital punishment for juveniles*
- *Address some possible solutions to the effects of incarceration*

KEY TERMS

Probation

John Augustus

National Probation Act

Revocation of probation

Technical violation

Probation as conditional release

Labeling process

Intensive supervision

Victim–offender reconciliation program

Community Response to Crime program

Victims of crime impact panels

Victim–offender mediation programs

Foster homes

Away syndrome

Private detention facilities

Public detention facilities

Capital punishment

Shock intervention

Boot camp

Positive peer culture

When attempts to divert a child from the juvenile justice network fail, an adjudicatory hearing is held to determine whether the juvenile should be dismissed or categorized as a delinquent, as a minor in need of supervision (or authoritative intervention), or as an abused, neglected, or dependent child. After adjudication, the judge must make a decision concerning appropriate disposition. The judge uses his or her own expertise and experience, the social background investigation report, and sometimes the probation officer's or caseworker's recommendation in arriving at a decision.

Many states use a bifurcated hearing process, so that the adjudicatory and dispositional hearings are held at different times. This is often preferred because different evidentiary rules apply at the two hearings. Whereas only evidence bearing on the allegations contained in the petition is admitted at the adjudicatory hearing, the totality of the juvenile's circumstances may be heard at the dispositional hearing.

▲ While the impact of shock incarceration is still being evaluated, boot camps have frequently been criticized for tolerating physical violence and inhumane conditions.

The alternatives available to the judge differ depending on the category in which the juvenile has been placed, but in general they range from incarceration to treatment, to foster home placement, to probation. In the *Gault* case, the U.S. Supreme Court specifically declined to comment on the applicability of due process requirements during the dispositional phase of juvenile court proceedings. Thus, we must turn to state statutes or lower court decisions in analyzing this process. Keep in mind that the purpose of the dispositional hearing is to determine the best way to correct or treat the juvenile in question while protecting society. To accomplish these goals, the court must have available as much information as possible about the juvenile, his or her background (e.g., family, education, legal history), and available alternatives. Evidence pertaining to the welfare of the juvenile is generally admissible at this stage of the proceedings, and the juvenile should be represented by counsel.

Although some nondelinquent juveniles, typically those found to be in need of supervision, may be confined temporarily in specifically designated facilities, the trend had been toward diverting them to other types of programs. In some cases, the child is permitted to remain with the family under the supervision of the court; in others, custody reverts to the state, with placement in a foster or adoptive home. The extent of state intervention has been a subject of considerable controversy, but when the welfare of the child is involved, termination of parental rights may be the only way to provide adequate protection.

Delinquent conduct always involves violation of law, unlike some of the other conduct dealt with by the juvenile court. There are numerous available dispositions for juveniles in this category, including probation (release after trial with court supervision) under conditions prescribed by the court, placement in a restrictive/secure facility not operated by the department of corrections, and commitment to a public correctional facility. The latter disposition is generally used as a last resort but may be necessary to protect society. In some cases, restitution is used in addition to probation or as a disposition in and of itself. In other cases, weekend incarceration or community-based correctional programs are used. These programs allow juveniles to remain in the community, where they may attend school, work part-time, and participate in supervised activities. The effectiveness of such programs is an empirical question, and many of these programs are not adequately evaluated.

Probation

A juvenile delinquent on **probation** is released into the community with the understanding that his or her continued freedom depends on good behavior and compliance with the conditions established by his or her probation officer and/or the judge. Probation, then, gives the delinquent a second chance to demonstrate that he or she can function in the community. The history of probation goes back to the 14th century, when offenders could be entrusted to the custody of willing citizens to perform a variety of tasks. The founding father of probation is said to be **John Augustus,** who attended criminal court proceedings during the 1850s and took selected offenders into his home so that they might avoid prison. The city of Boston had hired a probation officer by 1878, other cities and states followed suit, and all states had adopted probation legislation by 1925. The **National Probation Act,** passed in 1925, authorized federal district court judges to hire probation officers as well (Cromwell, Killinger, Kerper, & Walker, 1985).

A major finding of past presidential commissions has been that the earlier and deeper an offender goes into the juvenile justice system, the more difficult it is for him or her to get out successfully. Unnecessary commitments to correctional institutions often result in "criminalized" juveniles. The revolving door of delinquency and criminality is perpetuated as a result. The fact that there may be a short-term benefit from temporarily removing some juveniles from society should be tempered with the realization that, once released, some juveniles are more likely to jeopardize the community than if they had been processed under adequate probation services in the community where they must eventually prove themselves anyway. Because the goal of the juvenile court is therapeutic rather than punitive, probation is clearly in accord with the philosophy of the court. When circumstances warrant probation, when the juveniles for whom probation is a viable alternative are carefully selected, and when adequate supervision by probation officers is available, probation seems to have potential for success. Failure to take proper precautions in any of these areas, however, jeopardizes chances of success and adds to the criticism of probation as an alternative that coddles delinquents.

Probation is clearly the most frequent disposition handed down by juvenile court judges, accounting for more than 90% of all dispositions in some jurisdictions. Despite pressures exerted by the mass media (in the form of coverage of some exceptionally disturbing offenses committed by probationers), juvenile court judges have generally adopted the philosophy that a juvenile delinquent will usually benefit more from remaining with his or her family or under the custody of other designated persons in the community than from being incarcerated.

In making a disposition, the juvenile court judge traditionally places heavy emphasis on the current offense, the preferences of the complainant, and the juvenile's prior legal history, family background, personal history, peer associates, school record, home, and neighborhood. In addition, consideration is given to whether justice would be best served by granting probation or whether incarceration is necessary for the protection of the public. There are a multitude of other factors considered by judges, including the juvenile's attitude toward the offense and whether the juvenile participated in the offense in a principal or secondary capacity. The degree of aggravation and premeditation, as well as mitigating circumstances, is also considered. All of this information is provided to the judge in the social background investigation.

Once probation has been granted, certain terms and conditions are imposed on the probationer. Within broad limits, these terms and conditions are left to the discretion of the judge and/or probation officer. The requirements that the probationer obey all laws of the land, attend school on a regular basis, avoid associating with criminals and other persons of ill repute, remain within the jurisdiction, and report regularly to the probation officer for counseling and supervision are general terms and conditions usually imposed by statutory decree. Other requirements that the court may impose include curfews, drug testing, counseling, community service, and restorative justice programming. Although the court has broad discretion in imposing terms and conditions of probation, these terms and conditions must be reasonable and relevant to the offense for which probation is being granted. For example, in *People v. Dominguez* (1967), a condition that the female defendant could not become pregnant while unmarried was not considered to be related to the robbery for which she was adjudicated delinquent. The appellate court reasoned that a possible pregnancy had no reasonable relationship to future criminality. In *Jones v. Commonwealth* (1946), an order of a juvenile court requiring regular attendance at Sunday school and church was held to be unconstitutional because "no civil authority has the right to require anyone to accept or reject any religious belief or to contribute any support thereto." However, a condition of probation that requires a defendant to pay costs or make restitution is generally upheld provided that the amounts ordered to be paid are not excessive in view of the financial condition of the defendant. Any condition that cannot reasonably be fulfilled within the period fixed by the court is not likely to be upheld.

The importance of adhering to the terms and conditions of probation is stressed because violations constitute a basis for **revocation of probation** and the imposition or execution of the sentence that could have been given originally by the judge. There are generally three types of violations: technical, rearrest for a new crime or act of delinquency, and absconding or fleeing jurisdiction. A **technical violation** is usually characterized by the probationer flagrantly ignoring the terms or conditions of probation but not actually committing a new act of delinquency. For example, deliberately associating with delinquent peers might lead to revocation if such behavior was prohibited as a condition of probation. Typically, technical violations include minor infractions on behalf of the probationer. Technical violations are generally worked out between the probationer and probation officer, and they usually do not result in revocation action unless the probationer develops a complete disregard for the terms or conditions of probation. A rearrest or new custody action due to a new act of delinquency is obviously a serious breach of probation. The seriousness of the new act of delinquency is important in determining whether revocation proceedings will be initiated. Most rearrests are viewed by probation officers as serious and usually result in the revocation of probation, although there is some room for discretion. Although absconding or fleeing the juvenile court's jurisdiction may be considered a technical violation, it is generally considered separately and may result in revocation action.

Release on probation is conditional (i.e., "**probation as conditional release**"); that is, the liberty of the probationer is not absolute but rather subject to the terms and conditions being met. Although the probation officer may seek a revocation of probation, the court will ultimately determine whether to revoke probation. When juveniles violate the conditions of supervised release and face revocation of probation, issues of due process with respect to right to counsel and standard of proof arise. In *Morrissey v. Brewer* (1972), the U.S. Supreme Court held that although a parole revocation proceeding is not a part of the criminal prosecution, the potential loss of liberty involved is nevertheless significant enough to entitle the parolee to due process of law. First, the Court held that the parolee is entitled to a preliminary hearing to determine whether there is probable cause to believe that a violation of a condition has occurred. Second, an impartial examiner will conduct the hearing. Finally, notice of the alleged violation, purpose of the hearing, disclosure of evidence to be used against the parolee, opportunity to present evidence on the parolee's own behalf, and limited right to cross-examination are allowed under due process. Subsequently, in *Gagnon v. Scarpelli* (1973), concerning the issue of probation revocation proceedings, the Court held that a probationer was entitled to the same procedural safeguards announced in *Morrissey v. Brewer*, including requested counsel. Previously, in *Mempa v. Rhay* (1967), the Court had held that when the petitioner had been placed on probation and his sentence deferred, he was entitled by due process of law to the right to counsel in a subsequent revocation proceeding because the revocation proceeding was a continuation of the sentencing process and, therefore, the criminal prosecution itself. Most courts, in the absence of statute, have held that the probation violation need be established only by a preponderance of the evidence even if the violation is itself an offense.

There are several dispositions available in revocation hearings. If the charges are vacated, the probationer may be restored to probation or the conditions may be altered, may be amended, or may even remain the same. The revocation may be granted with a new disposition generally resulting in commitment to a juvenile correctional institution. The juvenile may also be sentenced to a treatment center if the revocation was due to behavior requiring treatment such as drug or alcohol abuse.

Although the length of probation varies among states, the maximum term of probation for the juvenile is usually not beyond the maximum jurisdiction of the juvenile court. Most terms of juvenile probation are between 6 months and 1 year, with possible extensions in most states. Probation dispositions are usually indeterminate, leaving the release date up to the discretion of the probation officer. On successful completion of the probation period, or on the recommendation of the probation officer for early discharge, termination of probation releases the juvenile from the court's jurisdiction.

Although probation serves the purpose of keeping the juvenile in the community while rehabilitation attempts are being made, there are some potential dangers built into this disposition. Learning and labeling theories indicate that proper supervision of probationers is essential if rehabilitation is to occur. Otherwise, the juvenile placed on probation may immediately return to the "old gang" or behavior patterns that initially led to his or her adjudication as delinquent.

Similarly, the juvenile placed on probation, while remaining with his or her family, may end up in the same negative circumstances that initially led to delinquent behavior except that he or she has now been labeled and is, more or less, "expected" to misbehave. The **labeling process** may exaggerate problems in family, school, and peer relations, and the juvenile may find it difficult to meet the expectations established for him or her. In many cases, the only positive role model available is the probation officer, whose caseload may preclude seeing the juvenile for more than a few minutes a week.

In an attempt to remedy the problems of limited probation officer time and lack of sufficient supervision of the probationer, several strategies are being employed. The first of these is electronic monitoring, which uses technology to track the whereabouts of the probationer. A bracelet is placed on the wrist or ankle of the juvenile in question, and his or her whereabouts can be determined by signals transmitted and picked up by a receiver maintained by the probation officer. In some cases, the juvenile is placed under house arrest for a specified period; in other cases, the juvenile may be allowed to go to school or work but must be home during certain hours. A second strategy involves **intensive supervision,** which is usually reserved for juveniles facing their last chance before incarceration. Probation officers working in intensive supervision programs have limited caseloads (usually no more than 15 to 20 probationers), make frequent contacts with their charges, make contacts with the families of the probationers, contact school authorities and/or employers periodically, work with clients at times other than normal working hours, and keep extensive records of their contacts. They typically review the conditions of probation regularly and adjust them as needed. Intensive supervision usually lasts 6 to 12 months. The assumption on which these programs are based is that the probation officer as role model, supervisor, and disciplinarian will be more effective if he or she spends more time with each client. Empirical testing of the programs is ongoing. Another attempt at providing better probationary services for delinquents involves contracting with private agencies. The Office of Juvenile Justice and Delinquency Prevention and other state and local agencies have contracted with private concerns to provide services (e.g., counseling, job readiness skills, structured and wilderness programs) for probationers to supplement the public services provided. The American Correctional Association (2005) supports the use of private services in its policies, maintaining that the government is ultimately responsible for corrections and should use all resources available to accomplish the goals of corrections. In 1999, the American Correctional Association conducted a survey concerning private-sector involvement in juvenile corrections. The survey revealed that 46 jurisdictions indicated they had at least one active private-sector contract. The main reason given for such a contract was that private-sector vendors could provide services and expertise that were lacking in the jurisdictions in question (Levinson & Chase, 2000).

Restorative justice (see Chapter 9) is a new philosophy that is being adopted quickly by juvenile courts as a supplement to probation services. The roots of restorative justice can be traced to 1974 in Ontario, Canada. The Mennonite Central Committee, through the help of a probation officer, created the first mediation program involving the basic principles of restorative justice. This program, called a **victim–offender reconciliation program,** used the payment of restitution directly to the victim by the offender as its core. Traditionally, payment of restitution to the victim was handled directly by the probation office in an impersonal manner. By forcing the offender to pay the restitution directly to the victim, the process was construed as a repayment for loss and damages to an individual rather than a state-mandated court fine for a harm done to the state. The success of this program initiated interest in restorative justice in the United States and in other parts of Canada.

Elkhart, Indiana, was the first U.S. city to initiate a victim–offender mediation program during the late 1970s. As the philosophy grew, a nonprofit organization called the Center for Community Justice, based on the restorative justice philosophy, was created in 1979. Since the 1980s, restorative justice has been called by a variety of different names depending on the agency applying its concepts. Although the name may change, the definition and core concepts of restorative justice—accountability, competency, and public safety—remain the same in all programs.

First, accountability in restorative justice is used to explain how offenders are to respond to the harm they have caused to victims and the community. Accountability requires that offenders take personal responsibility for their actions, face those they have harmed, and take steps to repair harm by making amends. Much of the literature regarding restorative justice calls this process "making things right" or "repairing the harm" (Center for Restorative Justice and Mediation, 1996a; Restorative Justice for Illinois, 1999). "Accountability is more than the 'guilty party' stating they committed a wrongful act" (Restorative Justice for Illinois, 1999, p. 1). The state of Illinois adopted the restorative justice philosophy as the basis of its juvenile court in 1999 and has been implementing restorative justice programming across the state. Many of the programs focus on accountability. The state of Minnesota has implemented the accountability concept in a **Community Response to Crime program.** This program uses a community intervention team that meets with the offender to let him or her know how the behavior affected the community, how the community expects the offender to make amends, and how the community is willing to support the offender while he or she makes amends (Center for Restorative Justice and Mediation, 1996b, p. 2). In stories published by the Minnesota Department of Corrections, the Community Response to Crime panel is very involved in all aspects of the juvenile's life. In one publication on the program, the Minnesota Department of Corrections (n.d.) stated,

> Several volunteer community members of a Community Response to Crime panel gave their home phone numbers to a juvenile, suggesting that he call them if he has a problem. An eighty year old victim of an attempted burglary, disappointed that the offender re-offended after promising never to do it again, met with him and asked insistently over and over, "How am I going to know you won't do this again?" She calls him regularly to make sure he stays out of trouble. A victim of juvenile vandalism participated in the circle process. Shortly after the case was resolved, he was diagnosed with terminal cancer. A circle member who had worked closely with him throughout the case visited him in the hospital, taking him homemade soup and flowers. Another circle member played the violin at his funeral. As a result of the interventions used in these cases, informal support systems were created which did not depend upon formal system services. (para. 8)

Second, restorative justice requires competency on behalf of offenders. Competency is not the mere absence of bad behavior; it is provision of resources for persons to make measurable gains in educational, vocational, social, civic, and other abilities that enhance their capacity to function as productive citizens (Bazemore & Day, 1996; Restorative Justice for Illinois, 1999). Restorative justice suggests that programs be designed to promote empathy in offenders, to teach effective communication skills to offenders, and to develop conflict resolution skills in offenders. Programs such as **victims of crime impact panels** (VCIPs), **victim–offender mediation programs,** and programs sponsored by community-run self-help groups such as Mothers Against Drunk Driving (MADD) strive to teach competency to offenders. One competency program is being used in southeast Missouri for juvenile offenders. This program uses a VCIP to increase empathy levels in juvenile offenders by asking victims of crime to tell offenders how the crimes have affected their lives. MADD offers a similar program. MADD also offers victims panels on a nationwide basis for empathy development in offenders of drunk driving. In Dakota County, Minnesota is using a "crime repair crew" that is "made up of offenders who are called to scenes of property crimes to fix and clean up the damage. The crime repair crew gives offenders the opportunity to 'give back' to the community while learning skills in construction and painting" (Center for Restorative Justice and Mediation, 1996b, p. 2).

Public safety is the third area of restorative justice. "Public safety is a balanced strategy that cultivates new relationships with schools, employers, community groups, and social agencies" (Restorative Justice for Illinois, 1999, p. 1). Public safety also facilitates new relationships with victims. "The balanced strategy of restorative justice invests heavily in strengthening a community's capacity to prevent and control crime" (Bazemore & Day, 1996, p. 7). The concept of public safety relies heavily on the community. The community, according to restorative justice, should make sure that "the laws which guide citizens' behaviors are carried out in ways which are responsive to our different cultures and backgrounds—whether racial, ethnic, geographic, religious, economic, age, abilities, family status, sexual orientation, and other backgrounds—and all are given equal protection and due process" (Center for Restorative Justice and Mediation, 1985, p. 1). Restorative justice also proclaims that crime control is not the sole responsibility of the criminal justice system but rather is the responsibility of the members of the community. Sentencing circles, reparative boards, and citizens councils are examples of the public safety concept in application. A program in northern Canada uses sentencing circles,

> which are groups of community members who decide how a crime will be resolved. Originally used in Aboriginal or Native communities . . . , the circles seek to get everyone involved and the perspective of the victim is valued. They often include "law and order" [i.e., criminal justice personnel] participants, and the plan for repairing the harm nearly always includes a community-based solution. (Center for Restorative Justice and Mediation, 1996a, p. 2)

Reducing recidivism is typically the baseline for showing that a program is effective. In a review of restorative justice programs, Umbreit, Vos, and Coates (2006) found that reports examining recidivism rates after victim–offender mediation and group circles have been mixed. Hayes and Daly (2003) reported that those juveniles who believe that the reparation plan is arrived at in consensus instead of being forced on them are less likely to offend than are those who do not. Peacemaking circles do not try to reduce recidivism, so research on them in this regard is scarce.

The participants in a balanced and restorative justice system are crime victims, offenders, and the community. Crime victims are essential to the success of the restorative justice process because they are involved in the healing and reintegration of the offenders and themselves. Crime victims receive support, assistance, compensation, and restitution. The offenders participating in restorative justice programs provide repayment to their communities and are provided with work experience and social skills necessary to improve decision making and citizen productivity. The community is involved by providing support to both the offenders and crime victims. The community provides individuals, besides criminal justice personnel, to act as mentors to the offenders and also provides employment opportunities for the offenders.

Juveniles placed on probation with families or support persons who are concerned and cooperative may benefit far more from this disposition than from placement in a correctional facility. In an attempt to provide this solid foundation for juveniles whose own families are unconcerned, uncooperative, or the source of the delinquent activity or abuse or neglect in question, the juvenile court judge may place the juvenile on probation in a foster home.

Foster Homes

When maintenance of the family unit is clearly not in the juvenile's best interests (or in the family's best interests, for that matter), the judge may place a juvenile in a foster home. Typically, **foster homes** are reserved for children who are victims of abuse or neglect. Delinquent children may spend a short time in a foster home, but these children seem more suited for treatment facilities and the services they offer. Ideally, foster homes are carefully selected through state and local inspection and are to provide a concerned comfortable setting in which the juvenile's behavior may be modified or in which the abused or neglected child can be nurtured in safety.

Foster parents provide the supervision and care that are often missing in the juvenile's own family and provide a more constant source of supervision and support than does the probation officer. As a result, the juvenile's routine contacts should provide a more positive environment for change than would be the case if the juvenile were free to associate with former delinquent companions or unconcerned, abusive, or criminal parents. Foster homes are frequently used as viable alternatives for minors who have been abused or neglected, or who are dependent or in need of supervision, because many of these children are caught up in dangerous situations at home. It is often clearly in their best interests to be removed from their natural families.

Foster homes clearly have a number of advantages for children who are wards of the court provided that the selection process for both foster parents and the children placed with them is adequate. Unfortunately, some couples apply for foster parent status in the belief that the money paid by the state or county for housing such juveniles will supplement their incomes. If this added income is the basic interest of potential foster parents, only limited guidance and assistance for foster children can be expected. In addition, many of these couples soon find that the money paid per foster child is barely adequate to feed and clothe the child and, therefore, does not enhance their incomes. Thus, careful selection of foster parents is imperative; foster parents who may injure or kill children in their care are unsuitable and must be weeded out during the process (see In Practice 10.1). The number of foster children is growing more rapidly than the number of foster parents willing to take on the responsibility.

No matter how careful the juvenile court judge is in selecting children for foster home placement, some placements are likely to involve children whose behavior is difficult to control. Raising any juvenile presents problems, and caring for delinquent and abused juveniles frequently adds to these problems. As a result, the number of couples willing to provide foster care for delinquent and abused juveniles is never as great as the need. Foster families must be carefully screened through on-site visitations and interviews and must possess those physical and emotional attributes that will be supportive for any child placed with them. Assuming responsibility for a delinquent, abused, or neglected juvenile placed in one's home

In Practice 10.1

Execution Proposed for Foster Deaths: Child Welfare Chiefs Want Stiffer Penalty for Parents Who Kill

The state's child welfare directors want Ohio lawmakers to automatically make it possible for foster parents who kill a child in their care to be sentenced to death.

Although Franklin County Children Services has never had such a case go to court, Director John Saros is among those leading the charge based on the recent death of a 3-year-old developmentally disabled boy in Clermont County.

"There is nothing more egregious than for a person who has come forward saying, 'You can trust me,' to turn around and kill a defenseless child who has been removed from their home because of abuse, neglect, or another troubling circumstance," Saros said. "I view it as an aggravating circumstance that shouldn't be treated any differently than someone who murders a police officer or firefighter."

That and other proposals for reform will be delivered to the legislature soon, officials said.

Saros said Children Services also will be checking up on the 1,570 foster children it has in private care after a state report blasted Lifeway for Youth, the New Carlisle group charged to care for Marcus Fiesel.

In its report, the Ohio Department of Job and Family Services faulted Lifeway, which has 523 foster homes in Ohio and operates in six states, for not watching carefully over the Middletown boy.

Marcus died in August after being left alone in a closet for two days, wrapped in a blanket and packing tape, while his foster parents went to a family reunion in Kentucky.

Liz and David Carroll, Jr., have been charged with murder, kidnapping, felonious assault, and child endangering.

The state report cites Lifeway for 15 violations, including failing to conduct a complete home study, not visiting the home frequently enough, allowing a relative to serve as a reference, lying about the amount of training the couple received, and overbilling the state for training.

Lifeway Executive Director Michael Berner did not return phone calls yesterday.

Children Services stopped sending youths to Lifeway's Cincinnati office after Marcus's death and sent caseworkers to visit the more than 200 children who the Franklin County agency had in Lifeway homes at the time. The agency has sent 372 children to Lifeway so far this year, [and] 198 remain in the group's care, Saros said.

Instead of limiting its scrutiny to Lifeway, Children Services will examine all of its 42 private foster care companies as a precaution.

Children Services will ask the private groups in a few weeks for electronic copies of criminal checks, details of parents' backgrounds, home studies, licenses, references, and other materials for all the foster parents caring for Franklin County children. Private agencies place 81 percent of the agency's foster children.

Although the effort will stretch Children Services' capabilities and funding, it is necessary, Saros said.

"Ninety-nine [percent] of our foster parents are wonderful, caring people worthy of our trust," he said. "But then you have the people who are duplicitous and are willing to lie and misrepresent themselves who will always be difficult to catch."

Several providers yesterday said they understand the need for additional inspection.

(Continued)

(Continued)

"When something as incredibly terrible as this happens and you're in the people's business, you do everything possible to prevent further deaths," said Nicholas Rees, Buckeye Ranch's vice president of development.

Others said they hoped the increased scrutiny will be short term.

"I really understand Children Services' need for this," said Robert J. Marx, executive director of the Rosemont Center. "But I really worry that if you ask for every home study, training record, and piece of paper in a foster parent's file, no one will have time to do anything else."

Sometimes, he said, "bad people will do bad things," no matter the safeguards.

State officials said they applaud efforts by individual child welfare agencies to protect the children in their care.

But the state is focused on getting the 54 recommendations in its report adopted, said Dennis Evans, spokesman for the Department of Job and Family Services. The reforms include toughening foster care licensing and screening standards.

The agency also is reviewing Lifeway's operation to decide whether to recertify the group when its license expires Jan. 18. The Public Children Services Association of Ohio, which represents the state's child welfare agencies, is drafting the death penalty proposal and three other foster care measures directors say are needed:

- Matching children who have severe emotional, mental, and physical disabilities with people trained to care for their needs
- Creating a new category of foster care providers to make it easier for people who want to help a particular child or siblings
- Changing how funding works so that agencies also would be paid for helping families keep their children instead of simply providing funding for foster care

The association's executive director, Crystal Ward Allen, supports the state proposals but said she worries they could have a chilling effect.

"Becoming a foster parent is already a daunting process, and we're about to make it even more daunting," she said.

"We need to better support foster parents, not overload them."

Source: Pyle, Encarnacion. (2006, December 22). Execution proposed for foster deaths: Child welfare chiefs want stiffer penalty for parents who kill. *The Columbus Dispatch,* p. 1A.

requires a great deal of commitment, and many juveniles who might benefit from this type of setting cannot be placed due to the lack of available families. Alternatives available to the judge in such cases include placement in a treatment center, placement in a group home, and incarceration in a juvenile correctional or shelter care facility.

Treatment Centers

Throughout this book, it has been indicated that juveniles should be diverted from the juvenile justice system when the offenses involved are not serious and when viable alternatives are available. Status offenders and abused, neglected, or dependent juveniles clearly should not be incarcerated. There may, of course, be times when the only option available to the court is to provide temporary placement in shelter care facilities, foster homes, or group homes when conditions preclude a return to the family. In cases where the juvenile in question may present a danger to himself or herself or to others, or where the juvenile may flee, temporary placement may be necessary.

Placement may also be necessary in cases where the juvenile's family is completely negligent or incapable of providing appropriate care and/or control. Temporary custody of dependent, neglected, and in need of supervision juveniles, as well as nonserious delinquents, should be in an environment conducive to normal relations and contact with the community. Numerous private and public programs directed at such juveniles have emerged during the past decade.

Sentencing a child to a treatment center is often used in conjunction with probation but can be used alone. Children are sent to treatment programs for a variety of reasons, including chemical dependency, behavioral or emotional problems, sexual assault counseling, problems resulting from previous abuse or neglect, and attitudinal or empathy therapy. Facilities such as Boys Town of America specialize in treating children with behavioral problems. This facility uses small family-oriented cottages focused on behavior modification to teach delinquent children how to control impulsive behaviors that may lead to criminal acts. Other treatment centers use positive peer culture treatment programs, play therapy programs, anger management therapy, conflict resolution programs, and life skills programs, to name only a few. Summit Academy, for example, uses a prep school culture to reform children living in the facility (see In Practice 10.2). Treatment centers are rarely administered by the state, so the juvenile court contracts with private institutions to provide these services. Most delinquent children sentenced to terms in treatment centers are one step away from being sentenced to a correctional institution. Thus successful completion of the treatment program determines whether the delinquent children will return to society or go to a correctional institution.

In Practice 10.2

Summit: A Reform School With Prep Culture

While getting ready for school each day at Summit Academy, LaMont Valentine can hardly believe how nice he looks wearing a dress coat and tie.

"I picture myself like this when I get older," said the 16-year-old senior. "I want to be a businessman."

A year ago, Mr. Valentine was on the streets of Philadelphia dealing drugs, running with the wrong crowd, always looking over his shoulder, and trying to stay one step ahead of the police.

People who knew him then wouldn't recognize him now. Although he's doing time in juvenile detention, his jail uniform happens to be khaki pants, penny loafers, a tie, a dress shirt, and a blue blazer with the school crest emblazoned on the left breast pocket, resembling what is worn at some elite boarding schools.

Prep culture is the whole idea behind Summit Academy, a reform school for delinquent teenage boys located in the boondocks of Butler County, far from the temptations of urban street life and the bad influence of old friends who would steer them to a life of crime.

(Continued)

(Continued)

"We show them this is how you comport yourself," said Sam Costanzo, the school's chief executive officer and founder, known as Mr. C to the students and staff. "We tell them this is how you dress and cut your hair. And it becomes a habit."

Only boys ages 14 to 18 who have been sentenced to juvenile detention can attend Summit Academy, but no one is guaranteed admission even if he's been referred there by a judge.

Because it is a private high school, administrators can reject any kid they feel is unlikely to benefit from the experience, including [boys] with a history of homicide, sex crimes, arson, mental illness, or running away from another institution.

"If you run, you're done. That's my motto," Mr. Costanzo said.

Unlike a traditional lockup facility or jail where bad behavior results in a longer sentence, an extended stay at Summit Academy is a privilege that comes with good behavior after [boys] have been admitted. All 38 of June's graduates were accepted to a college or technical school. And every student who graduates from Summit receives a $2,000 scholarship.

The 270 students currently attending Summit Academy have committed all kinds of juvenile and adult criminal offenses, but the most common by far is drug dealing.

About 100 are from Philadelphia; about 40 are from Allegheny County, and the rest come from 50 or so counties spread over six states. Counties that refer boys to Summit Academy pay $90 a day to house each one for the length of his sentence.

"I think it's a good program," said Allegheny County Judge Kim Berkeley Clark. "We've used their program for as long as I've been on the bench, [seven years]. It provides a wide range of exposure for the kids, and I think they can only benefit from that."

One of the things Summit Academy prides itself on is its food.

A typical breakfast includes pancakes, sausage patties, eggs, and fresh fruit. Shrimp creole, chicken cordon bleu, pulled pork, or crab cakes might be served for lunch, and dinner includes dishes like beef burgundy, Yankee pot roast, [and] Mexican lasagna.

Three meals and a late night snack of assorted deli sandwiches are planned each day by a certified chef, Chris Graczk, a former juvenile delinquent who credits Mr. Costanzo with saving his life.

"When I was 16, I didn't have a lot of positive in my life. My family situation was bad," said Mr. Graczk, who is now 40 and has been the chef for 10 years. "I was involved in burglary. It wasn't until I got admitted to the Academy that I had something positive to grab onto."

Summit Academy, with an annual budget of $12 million, is part of the Academy System, which has been in existence for 30 years in Pittsburgh. The Butler County facility was founded in 1996 with 10 boys from Allegheny County.

Located on 125 acres of rolling hills in the rural community of Herman in Butler County, the imposing red brick Summit Academy building has at various times been used as a monastery, a vacation resort, a band camp, and a police academy.

Summit Academy might be the first place where these young men have ever learned how to adjust a necktie, speak proper English, or been required to wear their trousers securely fastened at the waist.

Their education at Summit also includes some knowledge of visual and performing arts, which is reinforced with field trips to Pittsburgh, where students attend musicals and the symphony.

"I was forced to come here, but I'm OK with it now," said Andrew Scott, 17, a sophomore from Penn Hills. "I think everybody should take advantage of this. But some don't appreciate it as much."

(Continued)

(Continued)

John Payne, 19, of Philadelphia, will enroll at Indiana University of Pennsylvania in January to major in business and pre-law. He graduated from Summit Academy in June but has been allowed to stay on campus until he enrolls in college.

"I talked to my judge and asked him not to let me go back to Philly," said Mr. Payne, a former drug dealer and user who came to Summit as an eleventh-grader with an adult criminal record. "My family is there, but it's in my best interest to stay away because of all the violence.

"I look at myself in the mirror and think I'm a whole new person. I like myself in a suit. Here I'm in classes with kids who want to make a change, and they are college-minded too."

About 70 percent of the student body at Summit Academy is black, 20 percent is white, and 10 percent is made up of other races and ethnic groups.

Every student is required to take 10 mathematics classes each week. The GED exam is offered each month. Fifty-six students have passed it so far this year, and some seniors who participate in the College Within a High School program leave Summit Academy with up to six college credits they earn from Butler County Community College.

"We tell them they must have a plan when they leave here," said Steve Sherer, the school director and head football coach at Summit Academy. "They must have some kind of employment or be going to trade school or college. We try to get them to put value on an education."

About 80 percent of students who enter Summit Academy successfully complete the program. They enter a reintegration program when they leave the facility, which requires them to work with a counselor back home who helps keep them on track.

Mr. Sherer and John McCloud, an administrator and the head wrestling coach, say one of the reasons for their high success rate is the athletic program. They use athletics at Summit to reinforce positive behavior.

Students compete in the Tri-County North Conference with school districts such as Mars, Union, Cornell, and Riverside. The football program was 0–33 when it started seven years ago but has come close to making the playoffs the past two years. The team finished this season with an overall 4–5 record.

"We were approaching the longest losing streak in high school football," Mr. Costanzo said. "Our objective is to build character and make gentlemen of these young men. But our concern was they'd be identified as losers.

"We compete in WPIAL, and we want our kids to feel we are at the same level as everyone else," Mr. Costanzo said. "They go on the basketball and football playing field with brand new sneakers."

New students arrive at Summit Academy year round, so that presents some challenges to the school culture.

Leave is determined by the courts. Some students don't get to go home for three months. But school administrators assess them after 60 days for a home pass. Most of them will leave for Christmas for five days. Thanksgiving, spring, and summer breaks also are five days each.

In between those breaks, they earn home passes Friday through Sunday.

"It breaks my heart I can't send every kid home for Christmas and Thanksgiving," Mr. Costanzo said.

He recalls one boy from Pittsburgh who was preparing to go home for the weekend who respectfully asked if it was all right if he took his Summit uniform home so he could wear it to church.

"He told me his grandmother had never seen him in a jacket and tie. He wanted to know if he could please take it home and make her proud," he said. "I just hugged him."

SOURCE: Grant, Tim. (2006, November 26). A reform school with prep culture. *The Pittsburgh Post-Gazette*, p. B1.

▲ The most severe dispositional alternative for juveniles is commitment to a correctional facility.

Juvenile Corrections

The most severe dispositional alternative available to the juvenile court judge considering a case of delinquency is commitment to a correctional facility. There are clearly some juveniles whose actions cannot be tolerated by the community. Those who commit predatory offenses or whose illegal behavior becomes progressively more serious might need to be institutionalized for the good of society. For these delinquent juveniles, alternative options may have already been exhausted, and the only remedy available to ensure protection of society may be incarceration. Because juvenile institutions are often very similar to adult prison institutions, incarceration is a serious business with a number of negative consequences for both juveniles and society that must be considered prior to placement.

Although incarcerating juveniles for the protection of society is clearly necessary in some cases, correctional institutions frequently serve as a gateway to careers in crime and delinquency. The notion that sending juveniles to correctional facilities will result in rehabilitation has proved to be inaccurate in most cases. In 1974, Robert M. Martinson completed a comprehensive review of rehabilitation efforts and provided a critical summary of all studies published since 1945. He concluded that there was "pitifully little evidence existing that any prevailing mode of correctional treatment had an appreciable effect on recidivism" (Martinson, 1974, p. 54). Bernard (1992) arrived at the same conclusion two decades later (p. 587). In spite of the fact that most of the research on the effects of juvenile correctional facilities substantiates the conclusions of these authors, we have developed and frequently implement what may be termed an **away**

syndrome. When confronted with a juvenile who has committed a delinquent act, we all too frequently ask, "Where can we send him [or her]?" This away syndrome represents part of a more general approach to deviant behavior that has prevailed for many years in America. The away syndrome applies not only to juveniles but also to the mentally ill, the retarded, the aged, and the adult criminal. This approach frequently discourages attempts to find alternatives to incarceration, arises frequently when we become frustrated by unsuccessful attempts at rehabilitation, and is frequently accompanied by an "out-of-sight, out-of-mind" attitude. Our hope seems to be that if we simply send deviants far enough away so that they become invisible, the juveniles and their problems will disappear. However, walls do not successfully hide such problems, nor will they simply go away. Not only do "graduates" from correctional institutions reappear, but also their experiences while incarcerated often seem to solidify delinquent or criminal attitudes and behavior. Most studies of recidivism among institutionalized delinquents lead to the conclusion that although some programs may work for some offenders some of the time, most institutional programs produce no better results than does the simple passage of time.

There are a number of alternative forms of incarceration available. For juveniles whose period of incarceration is to be relatively brief, there are many public and private detention facilities available. Treatment programs and security measures vary widely among these institutions. Both need to be considered when deciding where to place a juvenile. In general, **private detention facilities** house fewer delinquents and are less oriented toward strict custody than are facilities operated by the state department of corrections. Many of these private facilities provide treatment programs aimed at modifying undesirable behavior as quickly as possible to facilitate an early release and to minimize the effects of isolation. The cost of maintaining a delinquent in an institution of this type may be quite high, and not every community has access to such a facility.

Public detention facilities frequently are located near larger urban centers and often house large numbers of delinquents in a cottage- or dormitory-type setting. As a rule, these institutions are used only when all other alternatives have been exhausted or when the offenses involved are quite serious. As a result, most of the more serious delinquents are sent to these facilities. In these institutions, concern with custody frequently outweighs concern with rehabilitation. A number of changes have occurred in juvenile correctional facilities during the past half-century. Cottage-type facilities have been replaced by institutional-type settings, and the number of juveniles incarcerated has increased, as has the severity of their offenses (Gluck, 1997). Typically, we see fences, razor wire, and guards at these facilities more often than we see treatment providers.

As the discussion of learning and labeling theories indicates, current correctional environments are not the best places to mold juvenile delinquents into useful law-abiding citizens. Sending a delinquent to a correctional facility to learn responsible, law-abiding behavior is like sending a person to the desert to learn how to swim. If our specific intent is to demand revenge of youthful offenders through physical and emotional punishment and isolation, current correctional facilities will suffice. If we would rather have those incarcerated juveniles return to society rehabilitated, a number of changes must be made.

First, we need to be continually aware of the negative effects resulting from isolating juveniles from the larger society, especially for long periods of time. This isolation, although clearly necessary in certain cases, makes reintegration into society difficult. The transition from a controlled correctional environment to the relative freedom of society is not an easy one to make for those who have been labeled as delinquent. This was demonstrated by Krisberg, Austin, and Steele (1989), who found recidivism rates of 55% to 75% among juvenile parolees (and these figures seem to remain fairly accurate today).

Second, it is essential to be aware of the continual intense pressure to conform to institutional standards that characterizes life in most correctional facilities. Although some juvenile institutions provide environments conducive to treatment and rehabilitation, many are warehouses concerned only with custody, control, and order maintenance. Correctional personnel frequently deceive the public, both intentionally and unintentionally, about what takes place in their institutions by providing tours that emphasize orderliness, cleanliness, and treatment orientation. Too often, we fail to see or consider the harsh discipline, solitary confinement, and dehumanizing aspects of correctional facilities. We often fail to realize that the skills needed to survive in these institutions may be learned very well, but these are not the same skills needed to lead a productive life on the outside. It has been recommended that concerned citizens, prosecutors, public defenders, and juvenile court judges spend a few days in correctional facilities to see whether the state is really acting in the best interests of juveniles who are sent there.

Third, the effects of peer group pressure in juvenile correctional facilities must be considered. There is little doubt that behavior modification will occur, but it will not necessarily result in the creation of a law-abiding citizen. The learning of delinquent behavior may be enhanced if the amount of contact with those holding favorable attitudes toward law violation is increased. Juvenile correctional facilities are typically characterized by the existence of a delinquent subculture that enhances the opportunity for dominance of the strong over the weak and gives impetus to the exploitation of the unsophisticated by the more knowledgeable.

Into this quagmire we sometimes thrust delinquents who become involved in forced homosexual activities, who learn to settle disputes with physical violence or weapons, who learn the meaning of shakedowns and "the hole," and who discover how to "score" for narcotics and other contraband. Juvenile institutions have long been cited in cases of brutal beatings and other inhumane practices between residents (inmates) and between staff and residents. We are then surprised when juveniles leave these institutions with more problems than they had prior to incarceration.

It is clearly counterproductive to send juveniles to educational or vocational training 6 to 8 hours a day, only to return them to a cottage or dormitory where "anything goes" except escape. Juveniles who are physically assaulted or gang raped in their cottage at night are seldom concerned about success in the classroom the next day. The delinquent subculture existing in juvenile correctional facilities is based on toughness and the ability to manipulate others. Status is determined largely by position within this delinquent subculture, which often offsets the efforts of correctional staff to effect positive attitudinal and/or behavioral change. Because, as we saw earlier, the behavior demanded within the delinquent subculture is frequently contrary to behavior acceptable to the larger society, techniques for minimizing the negative impact of that subculture must be found.

A fourth problem frequently encountered in juvenile correctional facilities is the assignment to cottages and/or existing programs based on vacancies rather than on the benefit to the particular juvenile. Juveniles who need remedial education may end up in vocational training. Any benefits to be derived from treatment programs, therefore, are minimized.

A fifth problem involves mutual suspicion and distrust among staff members who see themselves as either rehabilitators or custodians. Rehabilitators often believe that custodians have little interest and expertise in treatment, whereas custodians often believe that rehabilitators are "too liberal" and fail to appreciate the responsibilities of custody. The debate between these factions frequently makes it difficult to establish a cooperative treatment program. In addition, juveniles frequently try to use one staff group against the other. For example, they may tell the social worker that they have been unable to benefit from treatment efforts because the guards harass them physically and psychologically, keeping them constantly upset. This kind of report often contributes to the feud between guards and caseworkers, who occasionally become so concerned with staff differences that the juveniles are left to do mostly as they please.

Finally, the development of good working relationships between correctional staff members and incarcerated juveniles is difficult. The delinquent subculture, the age difference, and the relative power positions of the two groups work against developing rapport in most institutions. Frankly, there is often little contact between treatment personnel and their clients. It is very difficult for the caseworker, who sees each of his or her clients 30 minutes a week, to significantly influence juveniles, who spend the remainder of the week in the company of their delinquent peers and the custodial staff. Because the custodial staff members enforce institutional rules, there is a built-in mistrust between the staff members and their charges. Nonetheless, guards deal with the day-to-day problems of incarcerated juveniles most frequently, even though the guards are generally not regarded by caseworkers as particularly competent.

Under these circumstances, it is not difficult to see why rehabilitative efforts often end in failure. Finding solutions to these problems is imperative if we are to improve the chances of rehabilitating juveniles who must be incarcerated.

Capital Punishment and Youthful Offenders

Clearly, there are some juveniles who are extremely dangerous to others and who do not appear to be amenable to rehabilitation. Thus, all states have established mechanisms for transferring or waiving jurisdiction to adult court in such cases (as we indicated in Chapter 5). Once this transfer occurs, the accused loses all special rights and immunities and is subject to the full range of penalties for criminal behavior. This includes, in some jurisdictions, "absolute" sentences such as life in prison without parole (Cothern, 2000, p. 1; Dorne & Gewerth, 1998, p. 203). In yet another deeply divided decision, in 2005, the U.S. Supreme Court held in *Roper v. Simmons* that states cannot execute offenders (i.e., **capital punishment**) who are under the age of 18 years. According to the ruling, juveniles are not as culpable as typical criminals, and executing juveniles would violate both the Eighth Amendment and the Fourteenth Amendment (Death Penalty Information Center, n.d.). The Court's ruling affected 72 juveniles in 12 states.

The first recorded juvenile execution in America occurred in 1642. Since that time, 361 individuals have been executed for crimes they committed as juveniles (Cothern, 2000, p. 3; Streib, 2000). The first case that the U.S. Supreme Court heard on the death penalty for juveniles was *Eddings v. Oklahoma* (1982). In this case, the Court did not rule on the constitutionality of the death penalty for minors, but it did hold that the age of the minor is a mitigating factor to be considered at sentencing. In *Thompson v. Oklahoma* (1988), the Court found that the death penalty did not constitute cruel and unusual punishment in that particular case, although the justices were deeply divided over the issue. The Court reaffirmed this ruling in *Stanford v. Kentucky* (1989) and *Wilkins v. Missouri* (1989). The decision in *Roper* overturns these prior decisions.

Some Possible Solutions

All rehabilitative programs are based on some theoretical orientation to human behavior, running the gamut from individual to group approaches and from nature to nurture. Knowledge of these various approaches is critical for all staff members working in juvenile correctional facilities. Nearly all juvenile institutions use some form of treatment program for the juveniles in custody—counseling on an individual or group basis, vocational and educational training, various types of therapy, recreational programs, and religious counseling. In addition, they provide medical and dental programs of some kind as well as occasional legal service programs. The purpose of these various programs is to rehabilitate the juveniles within the institutions—to turn them into better-adjusted individuals and send them back into the community as productive citizens. Despite generally good intentions, however, the goal of rehabilitation has been elusive, and it may be argued that it is better attained outside the walls of institutions.

Solving the problems created by the effects of isolation on incarcerated juveniles is a difficult task. We need to be certain that all available alternatives to incarceration have been explored. We must remember that virtually all juveniles placed in institutions will eventually be released into society. If those juveniles are to be released with positive attitudes toward reintegration, we must orient institutional treatment programs toward that goal. This can be accomplished through educational and vocational programs brought into the institutions from the outside and through work or educational release programs for appropriate juveniles. In addition, attempts to facilitate reintegration through the use of halfway houses or prerelease guidance centers seem to be somewhat successful.

Unfortunately, in many instances correctional staff members begin to see isolation as an end in itself. As a result, attempts at treatment are often oriented toward helping the juveniles adapt to institutional life rather than preparing them for reintegration. Ignoring life on the outside and failing to deal with problems that will be confronted on release simply add to the problem. Provision of relevant educational and vocational programs, employment opportunities on release, and programs provided by interested civic groups should take precedence over concentrating on strict schedules, mass movements, and punishment. The out-of-sight, out-of-mind attitude should be eliminated through the use of programs designed to increase community contact as soon as possible. This is not meant to belittle the importance of institutional educational, vocational, and recreational programs for the juvenile delinquents. However, such programs will fail unless they are supported by an intensive continual orientation to success outside the walls of the institution. This will require both correctional personnel and concerned citizens to pull their heads out of the sand in a cooperative effort to serve the best interests of both the incarcerated juveniles and society.

One example of a program aimed at rehabilitating juveniles is the Texas Youth Commission. The commission has made incarceration a tough and demanding experience for the chronic and violent offenders with whom it deals, thereby attempting to convince youthful offenders that a return to incarceration is undesirable. At the same time, dedicated employees are implementing and perfecting resocialization programs to make the transition from prison to the larger society more successful (Robinson, 1999).

In some instances, it appears that no matter what correctional officials have attempted in traditional programs, some juveniles just do not get the message. In an attempt to get the attention of such juveniles, programs using **shock intervention** and/or **boot camp** principles have been introduced. These programs are usually relatively short in duration (3–6 months) with an emphasis on military drill, physical training, and hard labor coupled with drug treatment and/or academic work (Inciardi, Horowitz, & Pottieger, 1993; Klein-Saffran, Chapman, & Jeffers, 1993). The juveniles sent to boot camps may have started to use illegal substances or have minor legal problems that include nonviolent offenses (Boot Camps for Teens, n.d.). Drill-sergeant-like supervisors scream orders at the juveniles, demand strict obedience to all rules, and otherwise try to shock young offenders out of crime while imposing order and discipline. Although these programs have received a good deal of media attention, there is some doubt about their overall effectiveness. Whereas some maintain that the programs build self-esteem and teach discipline, others argue that serious delinquents are unlikely to change their behavior as the result of marching, physical exertion, and shock tactics (MacKenzie & Souryal, 1991). Boot camps are not suitable for all children. As Abundant Life Academy (2006) suggested, "There are children that need a more clinical setting than a boot camp. If a child has suicidal issues, is severely depressed, [is] self-mutilating, or has a serious psychiatric diagnosis, they would be better served in a therapeutic boarding school, residential treatment center, or in some cases even a psychiatric hospital" (para. 7).

Changes are needed in rehabilitation and treatment programs within the walls of the institution as well. Some programs are based on faulty assumptions. Others fail to consider the problems arising from the transition between the institution and the community on release. Some further examples should help to illustrate the advantages and disadvantages of different types of treatment programs.

Many institutions rely on individual counseling and psychotherapy as treatment modalities. Treatment of this type is quite costly, and contact with the therapist is generally quite limited. In addition, treatment programs of this type rest on two highly questionable assumptions: that the delinquents involved suffer from emotional or psychological disorders and that psychotherapy is an effective means of relieving such disorders. Most delinquents have not been shown to suffer from such disorders. Whether those who do are suffering from some underlying emotional difficulty or from the trauma of being apprehended, prosecuted, adjudicated, disposed of, and placed in an institution is not clear. Finally, whether psychotherapeutic techniques are effective in relieving emotional or psychological problems when they do exist is a matter of considerable disagreement.

Another type of program involves the use of behavior modification techniques. In programs of this type, the delinquent is rewarded for appropriate behavior and punished for inappropriate behavior. Rewards may be given by the staff, by peers, or by both, with rewards given by both showing the best results. Research on behavior modification programs has shown encouraging results. It is reasonable to assume that most delinquent behavior can be modified under strictly controlled conditions. Although it is possible to control many conditions within the walls of the institution, such controls cannot be applied to the same degree following release. In addition, as indicated earlier, behavior that is punished within the institution may be rewarded on the outside and vice versa. Again, transition from the institutional setting to the community is crucial. There are also ethical issues to consider that concern granting institutional staff members the power to modify behavior while still protecting the rights of the juveniles.

Other treatment techniques frequently employed in juvenile detention facilities center on change within the group. These include the use of reality therapy, group counseling sessions, psychodrama or role-playing sessions, transactional analysis, activity therapy, guided group interaction, and self-government programs. All of these techniques are aimed at getting the juveniles to talk through their problems, to take the roles of other people so as to better understand why others react as they do, and to assume part of the responsibility for solving their own problems. All of these seem to be important given that lack of communication, lack of understanding other people's views, and failure to assume responsibility for their own actions characterize many delinquents. Continuing access to behavior modification programs after release could provide valuable help during and after the period of reintegration.

Assuming that we have worthwhile rehabilitation programs in juvenile institutions, serious attempts should be made to match juveniles with appropriate programs and to stop convenience assignments such as those based on program vacancies and ease of transfer. It is important to classify offenders into treatment-relevant types based on juveniles' current behavior, self-evaluations, and past histories. Assignment of youthful offenders to specific programs and living areas based on these categories must be associated with specific types of treatment and training programs. Treatment programs will vary according to the juveniles' behavioral characteristics, maturity levels, and psychological orientations. Whereas one behavioral type may benefit from behavior modification based on immediate reinforcement (positive/negative), another behavioral type may benefit more through increasing levels of awareness and understanding. Inappropriate behavior will result in a loss of privileges or points toward a specific goal. Although it may be risky to assume that there are clearly delineated behavioral categories with accompanying treatment for each category, systematic attempts along these lines would appear to be a step in the right direction (Harris & Jones, 1999).

Because the peer group plays such an important role in correctional facilities, some way must be found to use its influence in a positive manner. Some institutions have adopted a **positive peer culture** orientation in which peers are encouraged to reward one another for appropriate behavior and to help one another eliminate inappropriate behavior. Although correctional staff members frequently believe that these programs are highly successful, in many cases juveniles simply learn to play the game; that is, they make appropriate responses when being observed by staff members but revert to undesirable behavior patterns on their return to the dorm or cottage. This frequently happens because correctional personnel get taken in by their own institutional babble. They sometimes begin to believe that the peer culture they see is positive when it is actually mostly negative. One way to avert this problem is to view rehabilitation as more than an "8 to 5" job. Unfortunately, the problems that confront incarcerated juveniles do not always arise at convenient times for staff members. Assistance in solving these problems should be available when it is needed.

Another beneficial step taken in some institutions has been to move away from the dormitory or large cottage concept to rooms occupied by two or three juveniles. These juveniles are carefully screened for the particular group in which they are included in terms of seriousness of offense, type of offense, past history of offenses, and so forth. This move holds some promise of success because "rule by the toughest" may be averted for most inmates. In this way, nonviolent offenders, such as auto thieves and burglars, run less risk of being "contaminated" by their more dangerous peers, for example, those who committed offenses involving homicide, battery, or armed robbery.

Finally, relationships between therapeutic and custodial staff members, and between all staff members and inmates, need to be improved. The solution is obvious. All staff members in juvenile correctional facilities should be employed on the basis of their sincere concern with preparing inmates for their eventual release and reintegration into society. Distinctions between custodial and treatment staff members should be eliminated, rehabilitation should be the goal of every staff member, and every staff member should be concerned about custody when necessary. Training and educational opportunities should be available to help staff members keep up with new techniques and research.

Providing concerned and well-trained correctional personnel will not guarantee better relationships with all incarcerated juveniles, but it should improve the overall quality of relationships considerably. Although initial costs of employment may be somewhat higher, the overall costs will not exceed those now incurred by taxpayers who often pay to have the same juveniles rehabilitated time and time again. According to Hibbler (1999), the National Juvenile Corrections and Detention Forum addressed this issue, recognizing that new laws dealing with juveniles have often led to a distancing from the use of appropriate intervention techniques that might help juveniles to grow into responsible adults. Forum participants concluded that incarcerated juveniles should be taught to understand and respect societal rules, that vocational training should be included in their correctional programs, and that bridge programs should be developed to help incarcerated juveniles to complete the transition to society (Hibbler, 1999).

We have focused, for the most part, on dispositional alternatives available to delinquents. There are other types of alternatives available to dependent, addicted, abused, and neglected minors as well. In addition to foster home placement, these include placement of juveniles in their own homes under court supervision (protective supervision); use of orders of protection that detail when, where, and under what circumstances parents or guardians may interact with the juveniles in question; and commitment to drug rehabilitation or mental health programs.

Career Opportunity—Recreation Officer I or II

Job Description: Responsible for facilitating and implementing planned recreational activities for incarcerated juveniles. Facilitate indoor and outdoor supervised sports, conduct group games, organize field trips, and facilitate other recreational programs that meet the varied interests, abilities, and needs of the juveniles. Maintain facility policies and enforce behavior management strategies in the course of recreational programs.

Employment Requirements: Requires a 4-year degree with a specialization in recreation, physical education, leisure management, or a closely related field. If without a college education, must possess 4 years of diversified experience in the field of group recreation or physical education, must have graduated from high school, and must have experience in organizing, implementing, scheduling, and overseeing recreation activities. General college education may be substituted for up to 2 years of experience.

Salary and Benefits: Salary ranges from $22,000 to $36,000. Benefits are provided according to the state benefits program, which usually includes health and life insurance, paid vacations and holidays, and a retirement program.

SUMMARY

It is clear that careful consideration should be given to available alternatives to incarceration of juveniles. Probation, whether within the juvenile's own family or in a foster home, has the advantage of maintaining ties between the juvenile and the community. Proper supervision and careful selection procedures to determine whether a juvenile can benefit from probation are essential. When incarceration is necessary to protect society, programs directed toward the eventual return of the juvenile to society should be stressed.

Changes are required in society's belief that juveniles who are "out of sight" will automatically remain "out of mind." Nearly all of these juveniles will eventually return to society, and efforts must be made to ensure that time spent in institutions produces beneficial results, not negative results. Thus, juveniles should not be randomly assigned to correctional treatment programs, nor can the negative effects of the delinquent subculture that develops in most institutions be ignored. All programs should be routinely evaluated to determine whether they are meeting their goals and the more general goals of rehabilitating juveniles while protecting society.

Note: Please see the Companion Study Site for Internet exercises and Web resources. Go to www.sagepub.com/juvenilejustice6study.

Critical Thinking Questions

1. What are some of the possible negative consequences of placing juveniles in correctional facilities? In your opinion, what circumstances would warrant such placement? Why?

2. What are the major advantages and disadvantages of probation as a disposition? How are these advantages and disadvantages modified by foster home placement? Why has so much criticism been aimed at probation as a disposition?

3. What is restorative justice? What are the three primary concepts used in restorative justice? Who is involved in the implementation of restorative justice programs?

4. If you were superintendent of a juvenile correctional facility today, what steps would you take to ensure that juveniles would be better prepared for their return to society? Why would you take these steps?

5. What are intermediate sanctions? What is shock intervention? How likely do you think the latter is to help rehabilitate serious delinquents?

Suggested Readings

Center for Restorative Justice and Mediation. (1985). *Principles of restorative justice.* St. Paul: University of Minnesota, School of Social Work.

Center for Restorative Justice and Mediation. (1996). *Restorative justice: For victims, communities, and offenders—What is the community's part in restorative justice?* St. Paul: University of Minnesota, School of Social Work.

Hayes, H. (2004). Assessing reoffending in restorative justice conferences. *Australian and New Zealand Journal of Criminology, 38*(1), 77–101.

Hayes, H., & Daly, K. (2003). Youth justice conferencing and reoffending. *Justice Quarterly, 20,* 725–764.

Kilgore, D. (2004). Look what boot camp's done for me: Teaching and learning at Lakeview Academy. *Journal of Correctional Education, 55,* 170–185.

Levinson, R. B., & Chase, R. (2000). Private sector involvement in juvenile justice. *Corrections Today, 62*(2), 156–159.

Patterson, E. (2004). Homosexuals' rights and the placement of children. *Policy and Practice of Public Human Services, 62*(1), 28.

Wilkinson, R. A. (1997, December). Back to basics. *Corrections Today, 59,* 6–7.

Zehr, H. (1990). *Changing lenses.* Scottsdale, PA: Herald Press.

CHILD ABUSE AND NEGLECT

CHAPTER LEARNING OBJECTIVES

On completion of this chapter, students should be able to

- *Discuss domestic violence*
- *Define and discuss physical abuse of juveniles*
- *Discuss the importance of mandated reporting*
- *Define and discuss child neglect*
- *Discuss the vicious cycle of child abuse*
- *Enumerate the consequences of psychological/emotional abuse of juveniles*
- *Define and discuss sexual abuse of juveniles*
- *Discuss intervention strategies*

KEY TERMS

Domestic violence

Abused and Neglected Child Reporting Act

Child neglect

Types of neglect

Emotional or psychological abuse

Sexual abuse

Criminal sexual abuse

Criminal sexual assault

Sexual exploitation

Pedophile

Internet exploitation

Intervention

White v. Illinois

Munchausen's syndrome by proxy

Child death review teams

Over the past four decades, studies of the family have touched on an aspect of the family that was rarely discussed before—**domestic violence** (spousal and/or child abuse). The family, which had traditionally been viewed as an institution characterized by love, compassion, tenderness, and concern, proved to be an institution in which members are at considerable risk due to increasing reported episodes of physical abuse and violence among members. In fact, the Bureau of Justice Statistics (2005a) reported that from 1998 to 2002, roughly 3.5 million violent crimes were committed by one family member against another. Of these, 49% involved crimes against a spouse, 11% involved crimes against a son or daughter, and 41% involved crimes committed against other family members (e.g., grandparents, parents, grandchildren). The most frequently reported crime was simple assault. Roughly 75% of these crimes occurred in or near the victims' residences, and 40% of the victims were injured. Roughly three-fourths of the victims of family violence were females, and approximately three-fourths of the offenders were male. In 2004, 62.4% of child victims experienced neglect, 17.5% were physically abused, 9.7% were sexually abused, 7.0% were psychologically maltreated, and 2.1% were medically neglected. In addition, "14.5 percent of victims experienced such 'other' types of maltreatment as 'abandonment,' 'threats of harm to the child,' or 'congenital drug addiction.' . . . These maltreatment type percentages total more than 100 percent because children who were victims of more than one type of maltreatment were counted for each maltreatment" (U.S. Department of Health and Human Services [DHHS], 2006). Overall, the rate of maltreatment declined slightly between 2001 and 2004 (DHHS, 2006), but maltreatment remains a very serious issue. According to the most recent report of data from the National Child Abuse and Neglect Data System, approximately 872,000 children were found to be victims of child abuse or neglect in calendar year 2004. The maltreatment rate was 11.9 per 1,000 children in 2004 (Child Welfare Information Gateway, 2006).

Although the privacy of the home and family has made research on this topic difficult, there is now little doubt that the seeds of violence are frequently sown in this setting or that one cause of violence among juveniles is that of being reared in a violent family.

> Nearly 84 percent (83.4%) of victims were abused by a parent acting alone or with another person. Approximately two-fifths (38.8%) of child victims were maltreated by their mothers acting alone; another 18.3 percent were maltreated by their fathers acting alone; 18.3 percent were abused by both parents. Victims abused by such non-parental perpetrators as an unmarried partner of parent, legal guardian, or foster parent accounted for 10.1 percent of the total. (DHHS, 2006)

The privacy of the home and the fear of retaliation and/or exposure make identifying and helping maltreated juveniles (including abused and neglected juveniles) extremely difficult. And some juvenile court judges who hear cases of suspected child abuse are hesitant to break up the family by removing the child to other circumstances, a trait not difficult to understand in light of the emphasis of most juvenile court acts on preserving the integrity of the family. It may be, however, that preserving the family also preserves child abuse and perpetuates violence on the part of some abused children as they grow into adulthood.

High divorce rates, increasing numbers of stepparents, increasing numbers of children reported as abused, the development of coalitions against domestic violence, and changes in state statutes dealing with domestic violence all indicate that family life is often problematic and

▲ Nixzmary Brown's body lies in a casket at her wake at the Ortiz Funeral Home in New York on January 16, 2006. The abuse case was another in a string of incidents that forced Mayor Michael Bloomberg to call for investigations and legislative reforms.

sometimes violent. Furthermore, child abuse is typically not a one-time event, as data for 2004 clearly indicate. Children who had been prior victims of maltreatment were 84% more likely to experience a recurrence than were those who had not been prior victims (DHHS, 2006).

Figure 11.1 indicates the reported distribution of child abuse by age for 2004. The graph breaks the victim population into age groups as follows: 0–3, 4–7, 8–11, 12–15, and 16–17 years. According to this chart, the most frequently victimized age group is the youngest one, with a rate of 16.1 per 1,000 children. The oldest children were victimized the least frequently.

Physical Abuse

In 2004, an estimated 1,490 children died from abuse or neglect (an average of roughly 4 children per day). These deaths are simply the tip of the iceberg. Children who are brain damaged or maimed are less visible but far more frequent. Research by Finkelhor and Ormrod (2001) indicates that homicides of young children may be seriously undercounted (see In Practice 11.1). According to the U.S. Department of Health and Human Services (2006), in 2004, the three main categories of maltreatment related to fatalities were physical abuse (28.3%), neglect (35.5%), and various combinations of maltreatments (30.2%).

As a result of the frequency of occurrence of child abuse, legislatures in all 50 states have enacted child abuse reporting laws. In Illinois, the **Abused and Neglected Child Reporting Act**

Figure 11.1 Child Maltreatment Victimization Rates by Age Group, 2004

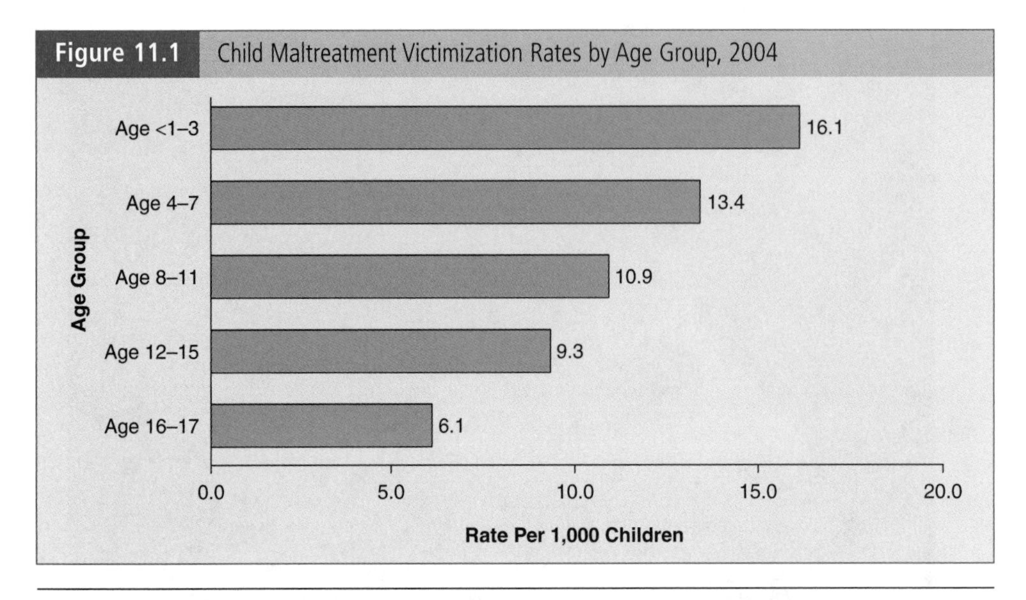

Source: Adapted from U.S. Department of Health and Human Services. (2006). *Child maltreatment, 2004.* Figure S-1. Available: www.acf.dhhs.gov/programs/cb/pubs/cm04/index.htm

In Practice 11.1

Review of Fulton DFCS Files Launched: After the Mishandled Case Involving a 2-Year-Old's Death, State Official Fears Other Kids May Be in Danger

Fearing that a 2-year-old Atlanta girl's death could be a sign that other children are in peril, a state official Wednesday announced an immediate audit of child welfare cases handled by the Fulton County Department of Family and Children Services.

DFCS supervisors overruled a front-line caseworker's recommendation, sending the child, Nateyonna Banks, to be placed with her mother—who is now charged in the toddler's beating death.

The case was handled so badly, and so many warning signs were apparent, that other children under the agency's watch could be in jeopardy, said Dee Simms, head of the Georgia Office of the Child Advocate.

"Our initial findings leave me very, very concerned," Simms said. "If this is common practice in Fulton County, then other kids may be in danger. ... There may be immediate decisions that need to be made."

Simms is performing a separate investigation into the death of Nateyonna, who police say died Nov. 9 after a beating from her mother.

This summer, Fulton County DFCS moved the girl from foster care to the care of her mother, Shandrell Banks, a 29-year-old woman with a history of cocaine possession, child sexual abuse, and emotional problems, according to court and criminal records.

Simms said Wednesday her preliminary review indicated that the child should not have been placed with the mother and that once she was there, problems arose that should have prompted the Fulton County agency to remove the child.

(Continued)

(Continued)

For example, Simms said the child suffered an eye injury in October and that the mother never adequately explained how it happened, changing her story several times.

"There Were Red Flags"

Simms, having reviewed the case file, said the mother had suicidal thoughts. Simms also said Banks was overwhelmed with the care of Nateyonna and wanted DFCS to remove the child.

"There were red flags," she said. "This was almost a recipe for disaster."

Simms' office was created in 2000 by then-Gov. Roy Barnes, who gave her the responsibility of ensuring the safety of children in the state child welfare system. After years of problems, including the deaths of several children, DFCS has said it was improving. But some child welfare advocates worry that Nateyonna's death may reflect larger problems.

Simms said her office will review hundreds of cases to make sure the agency is ensuring that children are kept safe. Many of these children are victims of abuse and neglect, but they have not been removed from their homes. Instead, they remained there while DFCS monitored the family and provided services intended to strengthen the family.

In recent years, DFCS officials have stressed the desire to keep children with their parents, and Simms said she will examine whether Fulton [County] workers were overzealous about that goal in this case.

The child's death prompted three Fulton County supervisors to resign and two to be placed on paid administrative leave pending an internal review of the case.

DFCS officials said they invited Simms to perform the review of case files. Simms said she will look at cases over the past several months, examining whether caseworkers visited the child, provided the right services, and whether the parents improved in their care of the child.

Family members of Nateyonna have also criticized Fulton County DFCS's handling of the case.

Nateyonna had been raised since she was a few days old by an aunt, Carolyn Banks of College Park, who criticized the agency's handling of the child's case.

"It was just gross negligence," Banks said. "They are supposed to protect children."

She said she contacted DFCS this summer for financial help and had no idea that her call would lead to the agency's taking away the child and placing her with Shandrell Banks. She said Banks was unfit to care for the child. "I'm the only mother she ever knew," said Carolyn Banks, 49. "Shandrell had mental issues."

Lack of Trust at DFCS

[Carolyn Banks] said Nateyonna was a loving child who loved to eat chicken and whose favorite toy was a Winnie-the-Pooh doll that said "I love you" and "Can you pick me up?"

She added, of the events leading up to Nateyonna's death, "I wouldn't have imagined [it] in my wildest imagination."

Shandrell Banks was accused by two other daughters of sexual abuse in 2004, according to a report by the state Department of Human Resources filed in Fulton County Juvenile Court. Those two children were removed from the home and remain in foster care, officials said.

Meanwhile, the head of the local union for DFCS caseworkers, Ralph Williams, said he was upset that the decision to place Nateyonna with her mother was made by DFCS supervisors who overruled a caseworker's recommendation to place the child with her aunt.

Williams said the case illustrates a lack of trust between caseworkers and supervisors at the agency.

"Too many caseworkers in my organization have no trust in DFCS management because too often it works against them," said Williams.

Source: Schneider, Craig. (2006, December 7). Review of Fulton DFCS files launched: After the mishandled case involving a 2-year-old's death, state official fears other kids may be in danger. *The Atlanta Journal-Constitution*, A1.

(*Illinois Compiled Statutes* [*ILCS*], ch. 325, art. 5, sec. 5/1–11, 2006) not only designates the state agency for investigating reports made under the act but also lists persons mandated to report such acts. Other states have similar acts mandating reporting for medical, social service, school, and law enforcement personnel. Civil immunity for persons reporting in good faith as well as waiver of the spousal and physician–patient privilege is typically spelled out in these acts.

In general, child abuse occurs when a child under a specific age (typically 18 years) is maltreated by a parent, another immediate family member, or any person responsible for the child's welfare. Child maltreatment includes physical, sexual, and emotional abuse as well as physical, emotional, and educational neglect by a caretaker. Although legal definitions of physical abuse are quite specific, it is important to realize that, in practical terms, what constitutes abuse differs considerably depending on time, place, and audience and that the line between abuse and discipline is often vague. Does spanking a 4-year-old with an open hand on the buttocks constitute child abuse? What if the child is 2 years of age? Suppose that a belt is used instead of the hand? Suppose that the child is struck on the torso instead of the buttocks? On the head? What kind of behavior are we talking about here? Most people in a given society can agree that certain behaviors are unreasonable—kicking biting, cutting, burning, strangling, shooting, and so on—when it comes to dealing with children. Cases involving these behaviors are relatively clear-cut and, although not problem free, present the lowest degree of difficulty for intervening authorities. It is the more frequent, less clear-cut cases that are most difficult to resolve.

Physical abuse can be defined as any physical acts that cause or can cause physical injury to a child (Snyder & Sickmund, 1999). Physical abuse can be described as a vicious cycle involving parents with unrealistic expectations for their children, perhaps prior experiences as victims of abuse themselves, and often feelings of insecurity. The result is conflict between the two parties or perhaps parentally perceived conflict in the case of infants. For example, the parent wants a young child to eat nicely in the presence of guests. As is often the case with young children, the child does not eat, plays with his food, and eventually ends up wearing a good deal of it. The parent may regard this as a direct reflection of her child-rearing abilities and may discipline the unruly child as a result. The extent of discipline depends on the extent of anger and frustration present in the parent, the level of parenting skills involved, the age of the child, the nature of the audience, and so on. With older children who have clearly defined goals in the interactive process, the conflict may be more intentional, for example, when an adolescent chooses to go out with friends rather than respect his parents' wishes to stay home and clean his room. In the negotiations that follow, physical abuse is one of several options available to the parent and is more likely under certain circumstances. These circumstances often exist in situations where a teenage single parent attempts to raise a child (or children) in conditions bordering on poverty. The young parent might not have learned how to care for an infant, might not know what realistic expectations are for the child, and might be frustrated by needing to raise a dependent child alone, thereby reducing his or her own life chances. If the child fails to meet the expectations that the parent has established (or that have been mutually established in the case of older children), disappointment results. When this is expressed by the parent, it may lead to lower self-esteem on behalf of the child and then to underachievement and further failure to meet parental expectations. The parent, disappointed and fearing that he or she may be perceived as a failure, responds with emotional and/or physical abuse and the cycle may begin again (Crosson-Tower, 1999; Cunningham, 2003; DePaul & Domenech, 2000; DiLillo, Tremblay, & Peterson, 2000; Pears & Capaldi, 2001). Goldman, Salus, Wolcott, and Kennedy (2003) concluded,

> While the estimated number varies, child maltreatment literature indicates that some maltreating parents or caregivers were victims of child abuse and neglect themselves. Research suggests that about one-third of all individuals who are maltreated as children will subject their children to maltreatment, further contributing to the cycle of abuse. Children who either experience maltreatment or witness violence between their parents or caregivers may learn violent behavior and may also learn to justify that behavior. (para. 8)

It is important to point out that child abuse occurs among all social classes and racial and ethnic groups. Still, some researchers have found relationships between child abuse and factors such as age of mother and socioeconomic, educational, and employment factors (Brown, Cohen, Johnson, & Salzinger, 1998; Cadzow, Armstrong, & Fraser, 1999; Paxson & Waldfogel, 1999). The disproportionate number of official cases involving children in lower socioeconomic circumstances may be a result of differing abilities to pay for medical services. But according to the Children's Defense Fund, poverty is the largest predictor of child abuse and neglect. Children in families with annual incomes of less than $15,000 are 22 times more likely to be abused or neglected than are children in families with annual incomes of $30,000 or more (Children's Defense Fund, 2006).

As just indicated, available data indicate that abusers are more likely to be women, perhaps because they spend more time with children (DHHS, 2006). At least three of four abusers are parents of the children involved, with other family members constituting the next largest group. Babysitters, friends of the family, and others account for a small proportion of child abuse (DHHS, 2006).

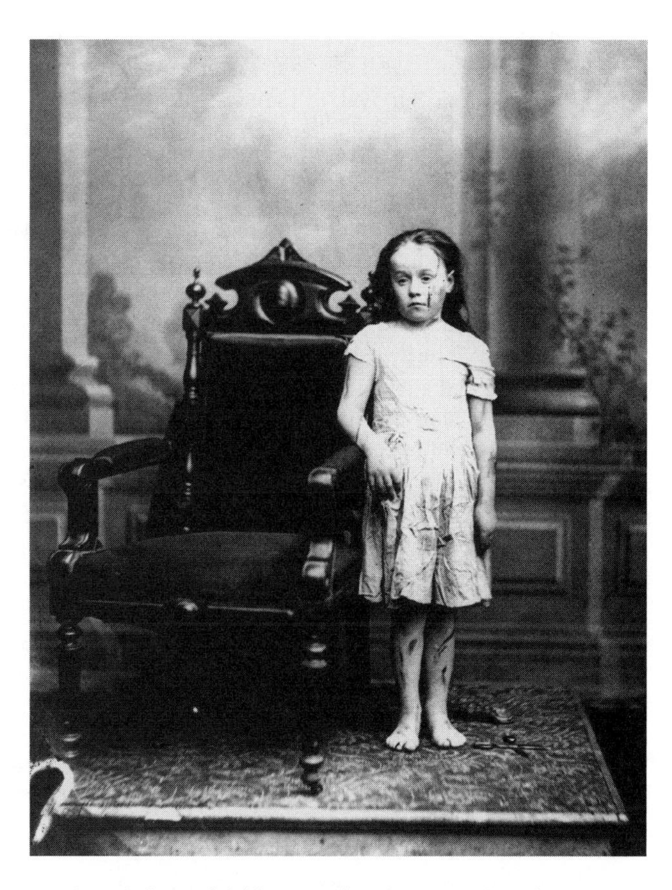

▲ Physical abuse of children must be taken seriously in the interests of both current and future generations.

Child Neglect

Child neglect generally involves an individual under the age of 18 years whose parent, or another person responsible for the child's welfare, does not provide the proper or necessary support, education as required by law, or medical or other remedial care recognized under state law as necessary, including adequate food, shelter, and clothing, or who is abandoned (see

Chapter 5). There are three **types of neglect:** physical, emotional, and educational. Physical neglect includes abandonment, expulsion from the home ("throwaway child"), failure to seek medical help for the child, delay in seeking medical care, inadequate supervision, and inadequate food, clothing, and shelter. Emotional neglect includes inadequate nurturing or affection, permitting maladaptive behavior such as illegal drug or alcohol use, and inattention to emotional and developmental needs. Educational neglect happens when a parent or caretaker permits chronic truancy or ignores educational or special needs (Snyder & Sickmund, 1999). Although the impact of neglect may be less obvious than that of abuse, the long-term consequences for the child may be equally harmful (Brown et al., 1998). Emotional, behavioral, and physical development may be impaired, the juvenile may drop out of school, medical problems may ensue, and encounters with the juvenile and/or criminal justice system(s) may result.

Emotional Abuse of Children

Emotional abuse occurs in families where the children's opinions do not count or where they are never sought. It occurs in families where the adult members fail to spend quality time with their children and where children's requests are met with responses such as "not right now," "maybe later," "we'll see," and "after a while." This type of abuse occurs in families fighting for economic survival, in families where drugs and their pursuit are more important than children, and in dual-career families where there just never seems to be enough time to do things with the children, where the dvd player is a constant built-in babysitter, and where giving the latest toy or electronic game takes the place of giving time. Some estimates are that roughly one-third of all school-age children, an estimated 5 million between the ages of 5 and 13 years, are so-called latchkey children—kids who care for themselves while their parents are at work. In Phoenix schools, for example, roughly 50% of third and fourth graders are latchkey kids, and one-third of all complaints to child welfare agencies involve latchkey children. "The two biggest fears facing a young latchkey child are an encounter with an intruder and a parent who doesn't come home on time" (City of Phoenix, 2006).

Some states include **emotional or psychological abuse** within the general definition of harm to a child that includes "mental injury." This term refers to "an injury to the intellectual or psychological capacity of a child as evidenced by a discernible and substantial impairment in the ability to function within the normal range of performance and behavior" (*Florida Statutes Annotated,* 415, 503(9)(a), (12) Supp., 2005). Wyoming's statute, for example, defines mental injury as "an injury to the psychological capacity or emotional stability of a child as evidenced by an observable or substantial impairment in the ability to function within a normal range of performance and behavior with due regard to their culture" (*Wyoming Statutes Annotated,* sec. 14–3-202(a)(ii), 2001). Emotional abuse can more generally be defined as "an act (including verbal or emotional assault) or omissions that caused or could have caused conduct, cognitive, affective, or other mental disorders" (Snyder & Sickmund, 1999, p. 40). Ambiguous definitions of emotional abuse often preclude protective agencies from intervening in suspected emotional abuse cases (Hamarman & Bernet, 2000).

Sexual Abuse of Children

Sexual abuse of a child is "involvement of the child in sexual activity to provide sexual gratification or financial benefit to the perpetrator, including contacts for sexual purposes, prostitution, pornography, or other sexually exploitative activities" (Snyder & Sickmund, 1999, p. 41). Under most statutes, it includes incest (sexual relations with family members), criminal sexual abuse, and criminal sexual assault. In general, **criminal sexual abuse** involves the intentional fondling of the genitals, anus, or breasts, or any other part of the body, through the use of force or threat of force of a victim (child) unable to understand the nature of the act, for the purpose of sexual gratification. **Criminal sexual assault** involves contact with or intrusion into the sex organs, anus, mouth, or other body part by the sex organ of another, or some other object wielded by another, with accompanying force or threat of force or with a victim (child) unable to understand the nature of the act.

Some states include **sexual exploitation** in an expanded statutory definition of sexual abuse. Such statutes typically include references to exploitation for pornographic purposes and to prostitution (*Florida Statutes Annotated,* ch. 827.071, 2006; *Maryland Annotated Code,* art. 27, sec. 35A9a)(2), 2001).

Despite increasing numbers of child maltreatment reports during the 1990s, the percentage of reports that were child sexual abuse allegations decreased. Sexual abuse reports dropped from 16% of all child maltreatment reports in 1986 to an average of 8% of reports from 1996 to 1998 (Office of Juvenile Justice and Delinquency Prevention, 2001). Victims typically range in age from 3 to 17 years, although there are numerous cases of infant abuse reported as well. Indications are that sexual abuse of children by perpetrators known to them involves multiple incidents of abuse over a relatively long period of time (up to 6 or 8 years at least). Furthermore, the vast majority of children (75%–80%) who are sexually abused are abused by a parent or stepparent. In most cases, the abuser is an adult male and the victim is a female child, but all other combinations are reported as well. Perhaps as a consequence, it has been reported that girls in the juvenile justice system have high rates of past sexual abuse. Some support for this relationship was indicated in a study by Goodkind, Ng, and Sarri (2006) that found that girls who had experienced sexual abuse had more negative mental health, school, substance use, risky sexual behavior, and delinquency outcomes. Whatever the gender of the child victim, when the offender is a parent the nonoffending spouse is sometimes aware of the sexual abuse but does little to prevent it. For a variety of reasons, the nonoffending spouse may even take the side of the perpetrator, possibly defending the perpetrator's innocence in court.

In some instances, the offender is a **pedophile** who seeks out children for purposes of sexual gratification. Such individuals may be relatives, friends, or strangers, and they often victimize a number of children. Although they frequently fail to develop meaningful sexual relationships with adults, pedophiles can be skillful predators when it comes to children (see In Practice 11.2).

Detection of child sexual abuse is difficult for a variety of reasons. First and foremost, sexual interaction with children is a very complex phenomenon. A great deal of ambiguity exists about what is and what is not appropriate behavior, especially in the mind of the child. In determining inappropriate behavior, we must ask the question: At what point do touching, fondling, kissing, and stroking become sexual? Cases that might appear clear-cut to an adult are often far less so to a child, particularly when the adult involved is an authority figure (parent) who assures the child that the behavior is okay if it is kept secret. Second, once the child begins to question the appropriateness of the sexual behavior, several difficult alternatives emerge. Does the child tell the nonoffending parent? Will that parent or any other adult believe the child? What will the adult to whom the report is made think of the child? What will happen to the child if the offending adult is arrested or, perhaps worse, confronted but not arrested? How important is the love of the offending adult to the child? Is the child in some

In Practice 11.2

Accused Pedophile Refereed: Police Fear Farmington Hills Man, Charged in Girl's Assault, Had Inappropriate Contact With Others

A suspected pedophile arrested Tuesday was a longtime local youth sports referee who was known to warn parents and coaches of the dangers of child predators, police have learned.

That's among the background uncovered by investigators as they began reviewing potential evidence, including child pornography, found inside the 46-year-old's home and car.

Richard R. Gerard is being held in the Oakland County Jail in lieu of $2 million bond and is scheduled for a preliminary exam today in Berkley District Court for 12 felony-related counts of child sexual abuse involving an 11-year-old girl in 1998. If convicted, he could face up to life in prison.

"We've already recovered some child pornography from inside his car," said Farmington Hills Police Chief William Dwyer. "The real job is going to be reviewing all these tapes for possible victims. That could take weeks."

Dwyer said more than a half-dozen detectives will sort through nearly 400 videotapes, some similar to the one that led to criminal charges against Gerard, that were seized Tuesday night during a police search of his residence on Country Bluff in the Crosswinds condominium complex at 14 Mile and Haggerty roads.

One of Gerard's ex-wives—he's been married and divorced five times—went to police this week after finding a videotape in her basement of Gerard performing sexual acts on a sleeping girl. The graphic videotape was allegedly filmed over a two- to three-year period beginning in 1998, when the child was 11, investigators believe. In all incidents, the girl appears unconscious.

Dwyer said one of the ex-wife's sons—also Gerard's son—identified the girl to investigators. The girl—now 19 years old—told investigators she recalled being given drugs and alcohol by Gerard when he used to babysit for her at her Berkley home while her mother was at work.

Police fear Gerard may have had inappropriate contact with other children. Court records indicate he has fathered three sons, now all in their 20s, by one former wife and a daughter, now 7 years old, by another ex-wife.

But police are also concerned because Gerard, a Southgate mortgage loan officer, has lived throughout Metro Detroit over the past 25 years, including in Warren, Ferndale, Beverly Hills, and Royal Oak.

Gerard, a referee for 25 years, was currently director of referees for Oakland Macomb Youth Football Association, according to Jackie Kage, vice chairwoman of the league, which supervises football and cheerleading for boys and girls, ages 6 to 13 years old, in 10 Oakland and Macomb county suburbs.

"We were shocked to hear this," she said. "We closely background everyone here, and there were times when he brought in articles about predators as examples of why we all had to be very careful about kids' safety. He seemed as concerned as anyone."

Kage said board members and parents were always present during games between August and November. She stressed no adult is ever alone with any of the 2,000 children who participate in the weekend games. She said Gerard also refereed at high school sports events.

(Continued)

(Continued)

"He was very good and kind to the kids," she said.

"This is the typical 'You never think he would do something like that,'" said Kage, who said members would vote to sever Gerard's relationship with the league at their Monday meeting.

"There was never one complaint about him," she said. "I would know about it."

Anyone with information about Gerard is asked to call detectives at (248) 871-2779 or 2780.

SOURCE: Martindale, Mike. (2006, November 17). Accused pedophile refereed: Police fear Farmington Hills man, charged in girl's assault, had inappropriate contact with others. *The Detroit News*, 1B.

way responsible for what has happened? These and other questions make it difficult for the child to disclose sexual behavior considered inappropriate and, therefore, make child sexual abuse difficult to detect. Clearly, many of these questions are related to the possibility of creating conflict within the interactive patterns of the family and a desire to avoid doing so.

Sometimes children do not perceive sexual acts by a member of the family or a family friend as abuse because children might not think that such a person can abuse them. At least two studies have found that the identity of the perpetrator makes a difference in whether sexual abuse is reported, finding that cases involving strangers are more likely to be reported than are those involving adults known the children (Hanson, Resnick, & Saunders, 1999; Stroud, Martens, & Barker, 2000). Public exposure of such abuse may help to prevent further abuse, yet publicity, medical evaluation, court appearances, interviews by investigators, and unnecessary visibility might not be in the best interests of the children. Very young children do not know that incest is bad or wrong at the same level as do older children and adolescents (Yates & Comerci, 1985). A sense of guilt may develop, be buried in the subconscious of the child, and then surface during later years, often in the teens when the child is approaching adulthood. Acting-out behavior—running away, attempting suicide, engaging in self-mutilation, becoming sexually promiscuous, being on drugs, and having a high level of apathy—tend to occur more often among older children (Cyr, McDuff, Wright, Theriault, & Cinq-Mars, 2005; "Memory Loss," 1991).

Internet Exploitation

No discussion of child abuse today is complete without considering **Internet exploitation.** In announcing a new assistance program for state and local law enforcement investigations of predators that use the Internet to locate their victims, U.S. Attorney General Alberto Gonzales noted that "we are in the midst of an epidemic of sexual abuse and exploitation of our children" (U.S. Department of Justice, 2006, para. 2). "Despite highly publicized arrests, law enforcement officials say that the sexual exploitation of children on the Internet is growing dramatically. Over the past four years, the number of reports of child pornography sites to the National Center for Missing & Exploited Children (NCMEC) has grown by almost 400 percent" (Scherer, 2005). The Internet provides child sexual predators with a means of communicating with potential victims. Because of its anonymity, rapid transmission, and often unsupervised nature, the Internet has become a medium of choice that predators use to contact juveniles and transmit and/or receive child pornography. Cyberspace provides child sexual predators with the opportunity to engage children in exchanges that can lead to personal questions designed to lure children into sexual conversations and sexual contact. It should be noted that not all children who are victimized via the Internet are innocents. Some are curious, rebellious, or troubled adolescents who are seeking sexual information or contact and are easily seduced and manipulated because they fail to fully understand or recognize the possible consequences of their actions (Armagh, 1998).

Intervention

Child abuse cases have typically been difficult to litigate for several reasons, including the following:

1. Establishing the competency and credibility of the child victim/witness
2. Questions concerning the admissibility of the child's out-of-court statements
3. Questions concerning the applicability of husband–wife, physician–patient, and clergy–penitent privileges
4. The use of character witness evidence in the form of either evidence of prior acts of abuse or expert testimony on the "battered child" or "battering parent" syndrome
5. The difficulty of the child victim confronting the alleged perpetrator in court

Intervention begins with someone reporting the abuse or suspected abuse, moves into the investigatory stage that typically involves a home visit and interviews with the parties involved, and then moves to risk assessment and a decision concerning what type of action to take. When charges appear to be substantiated, the question of removing the child from the home must be considered, as must the propriety of arresting the suspect(s). Sometimes a medical team becomes involved either as the result of emergency needs on behalf of the child or to attempt to determine whether abuse has occurred. If it is determined that abuse has occurred, the police and the investigator for the child protection agency involved take up the case and present it to the prosecutor for further action. Along the way, educators and mental health professionals often become involved as well in an attempt to ensure the well-being of the child (Crosson-Tower, 1999).

In 1990, the U.S. Supreme Court gave tacit approval to procedures designed to protect the child victim in abuse cases. These procedures include the use of videotaped testimony, testimony by one-way closed-circuit television, and testimony by doctors and other experts in child abuse (Carelli, 1990).

In spite of court decisions and other attempts to improve the way we deal with abused children, a study in Boston that compared court data from the mid-1980s with data from 1994 found few differences; children were still in the protective system an average of 5 years, and court cases still required approximately 1.6 years to complete. Half of those children permanently removed from parental custody were still in so-called temporary foster care 4 years later (Bishop, Murphy, & Hicks, 2000).

In 1992, the U.S. Supreme Court took another step in facilitating the intervention process. In ***White v. Illinois*** (1992), the Court affirmed the use of hearsay statements in child sexual abuse cases. In this case, the 4-year-old child victim did not testify, but others (her mother, a doctor, a nurse, and a police officer) to whom she had talked about the assault were allowed to testify ("Statistical Base," 1992, p. 3).

More and more police agencies are recognizing the importance of using specially trained investigators to conduct initial interviews with victims of abuse. Many investigations are now carried out jointly between the police and child protective agencies (Heck, 1999). Investigators have been briefed concerning the needs of all the various agencies involved in such cases, eliminating the need for repeated interviews that sometimes result in conflicting testimony due to fear of the interviews themselves, poor recall, new settings for the interviews, and/or different responses to questions that are worded differently even though they are addressing the same issue. The Child Development–Community Policing (CDCP) program developed by the New Haven Police Department and the Yale University Child Study Center in New Haven, Connecticut, is one example of such interagency cooperation. Teams of police officers and child development specialists work together to recognize and minimize the impact of violent crime scenes on physically or emotionally maltreated child witnesses (Krisberg, 2005, p. 139).

Special investigative techniques have been used by police in the investigation of a form of child abuse known as **Munchausen's syndrome by proxy** (MSBP). In this form of abuse, the abuser fabricates (or sometimes creates) an illness in the child victim. The child is then taken to a physician, usually by the mother, who knows that hospitalization for tests and observation is likely to be recommended because the symptoms are described as severe but no apparent cause exists. Tests, especially those that may be painful for the child, are welcomed by the apparently concerned parent. In addition, the parent may inject foreign substances (e.g., feces) into the hospitalized child, and there are documented reports of attempts by the parent to suffocate the hospitalized child. Perpetrators have been apprehended with the cooperation of medical staff members and the use of hidden video cameras, and prevention has been accomplished through placing the child in an open ward where medical staff members are in constant attendance. The former alternative is desirable because if the perpetrator is not arrested, she or he may relocate and further injure or kill the child (McMahon, 2002).

In an attempt to curb Internet crimes against children, U.S. attorneys are partnering with Internet Crimes Against Children task forces in implementing "Project Safe Childhood" to help law enforcement and community leaders prevent Internet crimes and investigate and prosecute sexual predators targeting children ("U.S. Offers Funds," 2006). Obviously, however, those in the best position to intervene in Internet-related activities are the parents of the children involved. By monitoring the Internet activities of their children, parents can identify and report suspicious contacts and can counsel young users as to the dangers inherent in certain types of Web sites and contacts.

Other prevention initiatives include **child death review teams** (consisting of experts with medical, social services, and/or law enforcement backgrounds), established in most states to review suspicious deaths of children, and statutory changes facilitating the prosecution of those involved in child maltreatment (Finkelhor & Ormrod, 2001, p. 11).

Finally, legal alternatives are available at state and federal levels to prosecute those involved in child maltreatment. There are, for example, a number of federal statutes dealing with child abuse and exploitation. To mention just a few, 18 U.S.C. § 2422(b) prohibits engaging in any type of enticement or coercion of individuals under 18 years of age to engage in prostitution or other criminal sexual activity, 18 U.S.C. § 2251(a) prohibits employing or using minors to engage in sexually explicit conduct for the purpose of creating a visual depiction of that conduct, and 18 U.S.C. § 2251A(a) prohibits parents or guardians from selling or transferring custody, or offering to do so, of minors knowing that the minors will be portrayed in a visual depiction of sexually explicit conduct.

It would be inappropriate to conclude our discussion of violence against juveniles without emphasizing the difficulties involved in dealing with them as victims. Effective advocacy for such juveniles is imperative for a variety of reasons. First, as we have seen, they are often ashamed, unable, or afraid to tell anyone about their plights. In many cases, even though they are being regularly and severely abused, children will not tell others because of the fear (sometimes instilled by their abusers) that their parents will be taken away from them if they do seek help. For many young children, this prospect is more frightening than their fear of continued abuse. Second, in many cases that do reach the courts, children are unable to testify effectively due to fear and/or inability to express themselves adequately. There are now adequate means available to deal with this problem, but these means are of little value unless they are recognized and used. Third, even when children are able to express themselves adequately, perhaps as a result of the hesitancy to break up families (discussed earlier), judges or children and family services officials might not remove juveniles from their homes. Even where there is evidence of abuse based on the testimony of teachers, caseworkers, and physicians, children have been returned to the homes in which they were being abused. In several such cases the abuse continued, and in some cases the children involved were killed by the abusing parents following the decision not to remove them from their homes (recall In Practice 11.1). To avoid such occurrences, it is crucial that the rights of child abuse victims be ensured by making certain that the children have proper representation and counseling and that their testimony is taken seriously. A study in Denver revealed that less than 3% of the allegations of sexual abuse made by children were demonstrated to be erroneous. The conclusion of the authors was that erroneous concerns about sexual abuse by children are rare (Oates et al., 2000).

If some parents are bent on destroying their own children, it is imperative that the state exercise the right of *parens patriae* to protect such children. Last but not least, the state should proceed as rigorously as possible in the prosecution of abusers, if only to prevent them from abusing their own spouses or children again. "The extent and complexity of the problem, and the fact that the juvenile homicide rate in the United States continues to be substantially higher than in other modern democracies, suggest that much remains to be done" (Finkelhor & Ormrod, 2001, p. 11).

SUMMARY

With respect to violence committed against juveniles, we must be aware that the incidence of such violence is great even by the most conservative estimates. All suspected cases should be treated as serious and given immediate attention so as to protect the juveniles involved, to prevent the juveniles from learning violent behavior that they may duplicate later in life, and to attempt to seek treatment or prosecution of the offenders.

Although family integrity is important, maintaining such integrity in cases of domestic violence or child abuse may be less important than saving lives or limbs (of either children or parents). Most states now have in place legislation that enables the state to protect children from abuse, but many practitioners remain hesitant to take official action that would break up the involved families. One need only read any newspaper with a large circulation to note the sometimes deadly consequences of failure to remove abused children from the homes of abusers. The failure to remove abused children from the homes in which they are abused is sometimes rationalized by pointing to the uncertainty of appropriate foster home or shelter care placement. Although it is true that such placement is sometimes problematic, leaving a child who has been, or is being, physically or sexually abused in the home of the abuser is unconscionable.

Violence against juveniles has received considerable attention over the past decade or so. Even though violence committed against juveniles appears to be declining in the United States, it is still considered an epidemic by some. Clearly, child abuse in its various forms is a relatively common occurrence, although it is likely that only a small proportion of abuse cases are reported or discovered. Child abuse is particularly alarming because of the physical and psychological damage done to children, because most research indicates that at least some parents who were abused as children go on to abuse their own children, and because, in spite of numerous programs designed to help prevent or halt child abuse, child abuse is by nature difficult to detect and control.

Note: Please see the Companion Study Site for Internet exercises and Web resources. Go to www.sagepub.com/juvenilejustice6study.

Critical Thinking Questions

1. It is often said that child abuse is intergenerational. Does available evidence support this claim? Why or why not? Why is child abuse so difficult to deal with? What, in your opinion, would be required for us to deal more effectively with such abuse?
2. How much progress have we made in dealing with those who abuse children? In protecting children from abuse? How has the Internet affected our ability to protect children from sexual exploitation?
3. What is emotional abuse? Does it always occur when physical abuse occurs? Does it occur independent of physical abuse? Can you provide some examples of emotional abuse without accompanying physical abuse?

Suggested Readings

Bishop, S. J., Murphy, J. M., & Hicks, R. (2000). What progress has been made in meeting the needs of seriously maltreated children? The course of 200 cases through the Boston Juvenile Court. *Child Abuse & Neglect, 24,* 599–610.

Crosson-Tower, C. (1999). *Understanding child abuse and neglect* (4th ed.). Boston: Allyn & Bacon.

Ernst, J. S. (2000). Mapping child maltreatment: Looking at neighborhoods in a suburban county. *Child Welfare, 79,* 555–572.

Finkelhor, D., & Ormrod, R. (2001, October). Homicides of youth and children. *Juvenile Justice Bulletin.* (Washington, DC: Office of Juvenile Justice and Delinquency Prevention)

Goldman, J., Salus, M. K., Wolcott, D., & Kennedy, K. Y. (2003). *A coordinated response to child abuse and neglect: The foundation for practice.* Washington, DC: U.S. Department of Health and Human Services. Available: www.childwelfare.gov/pubs/usermanuals/foundation/foundatione.cfm

Goodkind, S., Ng, I., & Sarri, R. C. (2006). The impact of sexual abuse in the lives of young women involved or at risk of involvement with the juvenile justice system. *Violence Against Women, 12,* 456–477.

Hannon, K. A. (1991, December). Child abuse: Munchausen's syndrome by proxy. *FBI Law Enforcement Bulletin,* pp. 8–11.

Hanson, R. F., Resnick, H. S., & Saunders, B. E. (1999). Factors related to the reporting of childhood rape. *Child Abuse & Neglect, 23,* 559–569.

McGowan, B. G., & Walsh, E. M. (2000). Policy challenges for child welfare in the new century. *Child Welfare, 79,* 11–27.

Pears, K. C., & Capaldi, D. M. (2001). Intergenerational transmission of abuse: A two-generational prospective study of an at-risk sample. *Child Abuse & Neglect, 25,* 1439–1461.

Scherer, R. (2005, August 18). Child porn rising on the Web. *Christian Science Monitor.* Available: www.csmonitor.com/2005/0818/p01s01-stct.html

Smart, C. (2000). Reconsidering the recent history of child sexual abuse, 1910–1960. *Journal of Social Policy, 29*(1), 55–71.

12

VIOLENT JUVENILES AND GANGS

CHAPTER LEARNING OBJECTIVES

On completion of this chapter, students should be able to

- *Discuss the history and current status of gangs in the United States*
- *Assess the various theories of gang development/membership*
- *Recognize the relationships among gangs, violence, and drugs*
- *Understand the role of firearms in youth violence*
- *Understand the relationship between delinquency and gang membership*
- *Discuss the characteristics of gang members in terms of age, race/ethnicity, gender, monikers, jargon, and graffiti*
- *Discuss a variety of public and private responses to gang activities*
- *Assess various alternatives to incarceration for violent juveniles*

KEY TERMS

Delinquent subculture
Sociological factors
Psychological factors
Social disorganization
Violent Crime Index offenses
Neo-Nazism
Politicized gangs
Street gangs
El Rukn
Bloods and Crips
Blackstone Rangers
Wilding
Narcotics trafficking
Black Disciples
Folks
Vice Lords
Latin Kings
People
Simon City Royals
Wannabes
Street corner family
Vice Queens
Monikers
Graffiti
Jargon
Spergel model
Alternatives to incarceration

Although there are a number of theoretical attempts to explain why juveniles engage in antisocial conduct, it is well known that many delinquent acts are committed in the company of others. As a result, much attention has been given to the role of the gang. Research has focused basically on two areas: the factors that direct or encourage a juvenile to seek gang membership and the effects of the gang on the behavior of its membership.

Cohen (1955) concluded that much delinquent behavior stems from attempts by lower-class youth to resolve status problems resulting from trying to live up to middle-class norms encountered in the educational system. Juveniles who determine they cannot achieve in this system often seek out others like themselves and form what Cohen called a **delinquent subculture.** According to Cohen, it is in the company of these "mutually converted" associates that a great deal of delinquency occurs.

According to many scholars (Bursik & Grasmick, 1995; Klein, 1967; Wooden & Blazak, 2001), there are numerous factors that bring juveniles into gangs. **Sociological factors** and physical factors include place of residence, school attended, location of parks and hangouts, age, race, and nationality. **Psychological factors** include dependency needs, family rejection, and impulse control. Still other factors are related to the structure and cohesiveness of the gang and peer group pressure. In the early appraisals of gangs, causation was tied to the theories of the slum community and its inherent attributes of **social disorganization** (Klein, 2005).

During the 1920s and 1930s, a group of sociologists at the University of Chicago, including Frederick Thrasher, Frank Tannenbaum, Henry McKay, Clifford Shaw, and William Whyte, conducted a number of studies of gangs in Chicago. According to Thrasher (1927), the gang is an important contributing factor facilitating the commission of crime and delinquency. The organization of the gang and the protection it affords make it a superior instrument for execution of criminal enterprises.

Interest in the relationship between gangs and delinquency waned during the 1960s and 1970s but increased during the late 1980s and early 1990s with reports of gang activities among minority groups both in and out of correctional facilities (Klein, 2005). Chicano gang activity on the West Coast and Chinese and Vietnamese gang activity have received media attention recently as well as attention from the National Institute of Justice (Hanser, 2007a; Valdez, 2005). It has been suggested that the amount of attention gangs receive is directly related to the ideology of the political party in power, to economic concerns of citizens, and to fear of victimization (Bookin & Horowitz, 1983; McGloin, 2005). Fear of victimization in the form of random and drive-by shootings has captured the attention of the media, the public, and Congress. The gangs of the late 1980s and 1990s appear to be better armed, more violent, and more mobile than their predecessors, and research on gangs is once again in vogue (Bilchik, 1999c; Mays, 1997; Wooden & Blazak, 2001).

Over the past two decades, violent crimes committed by juveniles have again received a great deal of attention, much of which has focused on juvenile gangs and the crimes of violence they perpetrate (see In Practice 12.1). "Although the violent juvenile crime rate has been decreasing dramatically since 1994, high-profile incidents such as school shootings serve to keep the problem of juvenile violence at the forefront of national attention" (Bilchik, 1999c, p. iii). As we noted in earlier chapters, one result of this emphasis on juvenile violence is that all states now have laws making it easier to try violent juveniles in adult courts and making it possible to prescribe more severe penalties for such juveniles. To what extent are the concerns about growing violence by juveniles based in fact? What, if anything, can be done to effectively reduce violence by juveniles? In the following sections, we examine these and other questions concerning the involvement of juveniles in violent activities.

In Practice 12.1

Police Tie Jump in Crime to Juveniles

Police in cities across the USA are linking the recent jump in the nation's violent crime rate to an increasing number of juveniles involved in armed robberies, assaults, and other incidents.

In Minneapolis, Milwaukee, Washington, Boston, and elsewhere, police are reporting spikes in juvenile crime as a surge in violence involving gangs and weapons has raised crime rates from historical lows early this decade. The rising concern about juveniles comes a month after the FBI said the nation's rate for violent crimes—murder, rape, robbery, and aggravated assault—rose in 2005, the first time in five years.

Minneapolis police estimate that this year, juveniles will account for 63% of all suspects in violent and property offenses there, up from 45% in 2002.

In Washington and Boston, police say there have been alarming increases in robberies by juveniles. This year, 42% of all robbery suspects in Washington have been juveniles, up from 25% in 2004, the police department says. A series of homicides—14 in July—has led D.C. Police Chief Charles Ramsey to declare an emergency that allows him to put more cops in troubled areas. Four suspects have been arrested in the slaying of a British man in the upscale Georgetown area Sunday; they include a 15-year-old.

In Boston, juvenile arrests for robbery rose 54% in 2005; weapons arrests involving youths rose 103%. "Kids are jumping into this violence," police Superintendent Paul Joyce says. "We're very concerned."

The forces behind rising juvenile crime vary by city, but officials cite some common factors. Among them:

- *Reduced funding for police and community programs.* Localities often complain they don't have enough money; now the chorus is getting louder. Tight budgets and an emphasis on terrorism have shifted federal and state money from police and programs for youths. "It should be no surprise that the streets are more violent," Minneapolis Mayor R. T. Rybak says. Since 2003, he says, Minneapolis has lost at least $35 million a year in state funding for city programs.
- *A changing social climate.* In Boston and other cities, gang leaders imprisoned a decade or longer ago are being released and are reclaiming their turf. Joyce says they're recruiting—or forcing—youths to carry guns or deliver drugs to shield older gang members from additional charges. The weapons can turn disputes among teens into violent confrontations, he says.

"Every 10 years, we seem to go through a cycle of violence," says Tom Cochran of the U.S. Conference of Mayors. "Everybody is trying."

SOURCE: Johnson, K. (2006, July 13). Police tie jump in crime to juveniles. *USA Today.* Available: www.usatoday.com/news/nation/2006-07-12-juveniles-cover_x.htm

Violent Juveniles

Reports of violent juveniles are widespread and are regular parts of newspaper/magazine headlines, television specials on youthful violence, comments on behalf of political officials promising a "get tough" approach (increasing the severity of punishment) to young offenders, and citizen action groups concerned about juvenile violence. Many of the newspaper articles and comments are based on analyses of official statistics that, as we pointed out earlier in the book, can be highly misleading or misinterpreted when it comes to assessing juvenile delinquency.

But are these statistics currently being misinterpreted? If we consider violent juvenile offenders to be those who commit (and, in terms of official statistics, are arrested for) criminal homicide, rape, robbery, or aggravated assault/battery, what has been the trend during recent years? In its report, "Juvenile Arrests 1999," the Office of Juvenile Justice and Delinquency Prevention indicated that, based on Federal Bureau of Investigation (FBI) statistics, juveniles accounted for 16% of all arrests for violent crimes in 1999 (Snyder, 2000).

> The substantial growth in juvenile violent crime arrests that began in the late 1980's peaked in 1994. In 1999, for the fifth consecutive year, the rate of juvenile arrests for **Violent Crime Index offenses**—murder, forcible rape, robbery, and aggravated assault—declined. Specifically, between 1994 and 1999, the juvenile arrest rate for Violent Crime Index offenses fell 36%. As a result, the juvenile violent crime arrest rate in 1999 was the lowest in a decade. The juvenile murder arrest rate fell 68% from its peak in 1993 to 1999, when it reached its lowest level since the 1960's. (p. 1, boldface added)

Prior to these recent reports, the picture of juvenile violence looked considerably different. Between 1965 and 1990, the overall murder arrest rate for juveniles increased 332%, accompanied by a 79% increase in the number of juveniles who committed murder with guns. Juvenile arrests for murder also showed an increase between 1987 and 1993, when 3,800 juveniles were arrested for murder. By 1999, the number of juvenile arrests for murder had declined to 1,400 (Snyder, 2000, p. 1). All in all, roughly one-third of 1% of 10- to 17-year-olds were arrested for violent crimes in 1999 (p. 6).

The forcible rape juvenile arrest rates increased by 44% between 1980 and 1991 but declined between 1991 and 1999 when compared with the 1980 rate. Aggravated assault arrest rates more than doubled between 1983 and 1994 and then declined to 24% below the 1994 peak but remained 69% above its 1983 low point. And the juvenile arrest rate for robbery increased 70% between 1988 and 1994 but then declined in 1999 to its lowest levels since at least 1980 (Snyder, 2000, p. 8). Even with the clear decline in official violence by juveniles, fear of such violence persists and is fed by events such as school shootings.

Although this trend toward lower juvenile crime has been observed during the past few years, more recent data from the first 6 months of 2006 (released while this chapter was being written) demonstrate that the decline in overall violent criminal activity has been reversed, with a 3.7% increase in the violent crime category. According to the FBI (2006), the following have been observed:

- Nationwide, robbery offenses nationwide increased 9.7%, murder offenses increased 1.4%, and aggravated assault offenses increased 1.2%. Forcible rape offenses decreased less than 0.1%.
- The volume of reported robbery offenses for the first 6 months of 2006 was up in all of the nation's city population groupings when compared with the 2005 reported data. The largest increase, 12.8%, occurred in cities with populations of 10,000 to 24,999.
- Reported robbery offenses also increased in the nation's metropolitan counties, up 8.4%.
- Cities with populations of 500,000 to 999,999 had the most marked increase in reported murder offenses, up 8.4%.

▲ Police stand outside the east entrance of Columbine High School as bomb squads and SWAT teams secure students on April 20, 1999, in Littleton, Colorado, after two masked teens on a "suicide mission" stormed the school and blasted fellow students with guns and explosives before turning the weapons on themselves. Such school-related incidents have continued to occur during recent years.

In addition, there are indications that violent juvenile crime is now back on the rise as many police departments are noting that the increase in robberies seems to include more juvenile offenders. Although the full analysis of national data is not yet complete, it is apparent that violent juvenile crime is back on the rise in many metropolitan areas of the United States, demonstrating the continually changing nature of juvenile crime.

Firearms and Juvenile Violence

American researchers consistently find that the most common weapons used in juvenile homicides are firearms. Sickmund, Snyder, and Poe-Yamagata (1997) found that 79% of victims of juvenile homicide offenders were killed with firearms. Lizotte, Tesorio, Thornberry, and Krohn (1994) identified two types of juvenile gun owners: low risk and high risk. High-risk juveniles who owned guns were more likely to carry guns regularly, own guns for protection, seek respect from others by carrying guns, own handguns and sawed-off long guns, and associate with peers who owned guns for protection. Furthermore, high-risk gun owners reported higher rates of antisocial behavior and bullying than did low-risk gun owners.

However, the prediction of gun violence is typically considered problematic by social scientists due to the inherent difficulty in predicting low base rate behavior. In other words, because juvenile gun violence is not a routinely occurring behavior, there are limited cases to analyze as a means of deriving conclusions. Furthermore, the risk of gun violence changes with the specific motivation for the juvenile's use of a firearm. For example, motivations might be due to any of the following: membership in a gang, victimization and bullying from other juveniles, a desire to mimic others observed, and the intent to harm parents, teachers, or others disliked by the juvenile. All of these examples are considerably different from one another, and this demonstrates the contextual and subjective element of risk prediction associated with gun violence. Despite the difficulty in prediction, Kaser-Boyd (2002) provided five indicators that serve as potential warning signs of potential child/juvenile gun violence (p. 198):

1. *Exposure to violence, either in the home or in the community.* Although exposure to television violence is not commonly cited, it is a factor in a number of homicides. Furthermore, the preoccupation with violent images is a definite warning sign. This preoccupation is often stimulated by media exposure to violent acts.
2. *A lack of success with the normal tasks of adolescence* (e.g., failing in school, having no extracurricular involvement)
3. *Social rejection and poor social supports.* Alienation and lack of empathy develop in large part from social deprivation.
4. *Intense anger that has built up from previous events*
5. *An inability to express or resolve intense feelings in adaptive ways and a proclivity for externalizing defenses or acting out*

It should be noted that young children may provide a variety of clues or indicators of future gun violence in artwork or other similar school assignments (Hanser, 2007b). However, teachers, parents, and other child-care workers may find it quite difficult to distinguish usual fantasy drawings from those that are true warning signs (Hanser, 2007b). But as children grow older, drawings are not as likely to correlate with future gun violence (Hanser, 2007b); rather, many older juveniles will provide verbal references or warnings that are often dismissed by others who otherwise would be able to prevent the escalation of violence (Hanser, 2007b; Kaser-Boyd, 2002). In either event, most variables are not clearly discernible in many cases, with the possible exception of membership in a gang that has used violence and/or firearms in past gang-related incidents (Hanser, 2007b). Even with this information, most experts agree that there are no truly predictive variables that are particularly effective in distinguishing school firearms offenders from other students (Kaser-Boyd, 2002).

"The kids who get picked on now too have a weapon: fear. Goths, punks, geeks, drama clubbers, bandies, and the math club all have a tool against the popular cliques in the threat of gun violence" (Wooden & Blazak, 2001, p.114). This naturally brings to mind incidents such as the Columbine High School massacre and other such acts of violent retaliation among juveniles (Hanser, 2007b). Still, the *1998 Annual Report on School Safety* showed a decline in the number of high school students who brought weapons to school over the 1993–1997 period from 8% to 6% (U.S. Department of Justice, 1998). The percentage of juveniles who reported bringing a gun to school within the 4 weeks prior to the survey remained constant at 3% (U.S. Department of Justice, 1998).

A USA Today/CNN/Gallup Poll conducted in 1993 found that respondents believed that male teenagers were the people most likely to commit crimes, and in high-crime areas 71% of the respondents indicated that male teens were most likely to commit crimes (Meddis, 1993). Bilchik (1999c) concluded,

> Twenty years of research repeatedly has shown that in any city or neighborhood a small percentage of offenders are responsible for committing a large proportion of the crime that occurs there. . . . Overall, juvenile violence is committed primarily by males and often intraracially among minority males. While some younger adolescents do commit violent offenses, the majority of juvenile offenders and victims are sixteen- and seventeen-year-olds. An examination of neighborhood factors indicates that many violent juvenile offenders live in disruptive and disorganized families and communities. However, as the surveys with the children living in high-risk neighborhoods show, the majority of youth who live in such environments are not involved in serious delinquency. (pp. 8–9)

The preceding comments should not be interpreted to mean that juvenile females are not involved in violent offenses. While crime by male offenders tended to drop during the period between 2000 and 2004, female crime remained stable or actually increased (FBI, 2006; Valdez, 2005). In addition, the types of offenses for which females are arrested and incarcerated have changed (Valdez, 2005). In some areas of the United States, female gangs have actually assumed a role that is quite distinct and independent from that of local male gangs, and some have even excelled in criminal enterprise ventures such as drug trafficking (Valdez, 2005).

The heightened involvement of females in violent offenses has been attributed to prior victimization of females (Bloom, Owen, & Covington, 2003). This has been referred to as the "victim-turned-offender hypothesis" by some researchers (Hanser, 2007b). From an abundance of literature, it is clear that female offenders are often victims of sexual abuse and physical abuse during early childhood and that they also tend to be victims of sexual assault and/or domestic abuse during adulthood (Bloom et al., 2003; Hanser, 2007b). The suggestion is that females become violent perpetrators in response to their own victimization, although substance abuse, economic conditions, and dysfunctional family lives have also been linked to violent offending by females (Acoca, 1998; Bloom et al., 2003; Peters & Peters, 1998).

Some violent crimes have been attributed to the rebirth of **neo-Nazism** in the form of "skinhead" groups such as the Nazi Low Riders, White Aryan Resistance, Hammer Skins, World Church of the Creator, and other groups whose members perpetrate hate crimes (Hamm, 1993; Wooden & Blazak, 2001, p. 131). Whether or not these groups can be legitimately defined as gangs is a matter of perspective, but they are often excluded from the traditional definition because they are typically organized around the overt ideology of racism rather than as the result of shared culture and experiences (Hamm, 1993). Still, they may be considered as **politicized gangs** because they use violence for the purpose of promoting political change by instilling fear in innocent people. Perhaps they are best considered as a terrorist youth subculture, as Hamm (1993) suggested.

The debate rages as to whether violent crimes are related to violence in the media (Centerwood, 1992; Kaser-Boyd, 2002; Wooden & Blazak, 2001, pp. 113–114). As a result of the latter, there has been considerable pressure and some success in getting television networks to tone down violence and to label programs with violent content as such. Many of the violent crimes attributed to juveniles are drug-related crimes involving gang disputes over drug territories or attempts to steal to get money to buy drugs (McGloin, 2005). And of course, a great deal of violence is associated with traditional **street gangs** as well as with skinheads.

▲ Gang signs and symbols may be included in graffiti that may also identify a gang's turf.

Gangs

> Gangs pose a serious social problem in the United States. It is no secret that in U.S. communities, large and small, the fear wrought by teen gangs has spread rapidly. With gang victimizations reported daily, many people have become virtual prisoners in their own homes. The trepidation caused by gang drive-by shootings causes schools to practice "ducking drills" and people to huddle in their darkened homes, hide their children in bathtubs, and be afraid to let their children play outside. (Peak, 1999, p. 51)

It should be noted that although we once could have referred to gangs as "juvenile" gangs, such a distinction is no longer totally appropriate because many gangs now include older adults among their memberships. Indeed, some researchers and practitioners have noted that gangs can be intergenerational in nature (Valdez, 2005). In communities that have intergenerational membership, it is often found that multiple family members—all of different age ranges—may be current or past gang members. In such families, children are essentially socialized into the gang subculture through their families and through the surrounding community. In fact, the majority of citizen members in these communities may be gang members or at least gang sympathizers, making it very difficult for police or community supervision personnel to combat gang activity in these areas (Valdez, 2005).

The "turf" gangs of yesteryear have been replaced in many instances by sophisticated criminal organizations involved in drug trafficking, extortion, murder, and other illegal activities (Klein, 2005; McGloin, 2005). These street gangs destroy entire neighborhoods, maiming and killing their residents. They destroy family life, render school and social programs ineffective, deface property, and terrify decent citizens. Last but not least, they have grown into national organizations that support and encourage criminal activities not only in local neighborhoods but also across the country and internationally. Some 800,000 gang members were thought to be active in more than 30,000 gangs across the United States in 1997. Every city with a population of 250,000 or more reported the presence of gangs (Bilchik, 1999c; McGloin, 2005). The number of gang members in rural counties increased by 43% between 1996 and 1998 (Wilson, 2000). Nationally, in 1998, 46% of all gang members were thought to be Hispanic, 34% were thought to be African American, 12% were thought to be Caucasian, and 8% were thought to be Asian or "other." More than one-third of all youth gangs were thought to have memberships including members of two or more racial groups (McGloin, 2005). The largest proportion of gang members involved in burglary or breaking and entering was reported in rural counties (Wilson, 2000, p. xv).

A Brief History of Gangs

Thrasher's classic study of juvenile gangs was published in 1927 and included information based on more than 1,300 gangs in the Chicago area, including fraternities, play groups, and street corner gangs. His study was the first to emphasize the organized purposeful behavior of youth gangs. He found that gangs emerged from the interstitial areas as a result of social and economic conditions, became integrated through conflict, gradually developed an esprit de corps or solidarity, and protected their territory against outsiders much like today's gangs.

According to Thrasher (1927), gangs originate naturally during the adolescent years from spontaneous play groups, which eventually find themselves in conflict with other groups. As a result of this conflict, it becomes mutually beneficial for individuals to band together in a gang to protect their rights and satisfy needs that their environment and families cannot provide. By middle adolescence, a gang has distinctive characteristics that include a name, a geographical territory, a mode of operation, and usually an ethnic or racial distinction. Thrasher not only analyzed gang behavior and activity but also was concerned about the effect of the local community on gangs. He found that if the environment is permissive and lacks control, gang activity will be facilitated. If there is a high presence of adult crime, a form of hero worshiping occurs with high status given to adult criminals. This type of environment is conducive to, and supportive of, gang behavior. Although Thrasher did his analysis more than 75 years ago, his conclusions appear to be borne out in terms of contemporary gang and neighborhood activities. Gangs clearly flourish where control of streets and children has been lost and under circumstances where role models are basically older males involved in criminal activity.

The Chicago school spurred other studies of gangs that generally supported the earlier images of that school. Shaw and McKay (1942) found that most offenses were committed in association with others in gangs and that most boys were socialized into criminal careers by other offenders in the neighborhood.

As indicated earlier, Cohen (1955), in his book *Delinquent Boys,* emphasized that gang juveniles have a negative value system by middle-class standards. This results in a "status frustration" that is acted out in a "nonutilitarian, negativistic" fashion through the vehicle of the gang.

Yablonsky (1962) indicated that violent gangs, at least, are not as well organized and highly structured as some theorists had supposed. In addition, Yablonsky indicated that the police, the public, and the press may help to create and unify gangs by attributing to gangs numerous acts that gangs did not commit. Strong support for Yablonsky's conclusions can be found in Dawley's (1973) book, *A Nation of Lords,* which is an autobiography of sorts about the Vice Lords in Chicago. Over the past decade, however, violent gangs have unquestionably become more organized, with gangs such as **El Rukn** and the **Bloods and Crips** illustrating that such organizations extend even into prisons. Other gangs are also demonstrating the ability to organize mobile units and maintain branch units in scattered cities, largely in response to drug markets.

Cohen, Miller, and others (see Chapter 4) further developed theories of gang delinquency during the 1950s and 1960s. During this period, delinquency came to be regarded as a product of social forces rather than individual deviance. Gang members were viewed as basically normal juveniles who, under difficult circumstances, adopted a gang subculture to deal with their disadvantaged socioeconomic positions. Gangs attracted a good deal of attention as a result of their apparent opposition to conventional norms and sometimes were romanticized, as in the popular *West Side Story.*

By the late 1960s and early 1970s, the United States was in a period of social upheaval marked by civil disturbances, racial protests, antiwar demonstrations, and student protests. Gangs were largely forgotten by the media and sociologists. Definitions of crime and delinquency came into question. Political liberals focused on abuse of power and crimes by the wealthy. Labeling theory came into vogue, postulating that members of the lower social class are more likely to be labeled as deviant than are those in the middle and upper classes as a result of the balance of power resting with the latter. Gangs were viewed as a response to injustice and oppression. Conservatives, however, viewed crime and delinquency as products of immorality, poor socialization, and/or lack of sufficient deterrence. Control theory was popular with this political faction because it postulated that delinquency was largely an individual matter, developing early in life and occurring due to a lack of internal and external controls. Failure of institutions such as the family, police, and corrections became the focal point of those representing the conservative viewpoint, obviating the need to deal with the social structure and conditions that were the focus of the liberal camp. The latter group continued to view gang members as juveniles in need of help rather than punishment. While these groups argued over the source of responsibility for crime and delinquency in general, developments that would soon lead society to take another look at the gang phenomenon were occurring.

During the 1960s, a Chicago gang known as the **Blackstone Rangers** (later called the Black P Stone Nation and still later known as El Rukn) emerged as a group characterized by a high degree of organization and considerable influence. The Blackstone Rangers sought and were granted federal funds, as well as funds from private enterprises, to support their activities. This funding gave the gang an appearance of political and social respectability. Street gangs in America were becoming politicized. As Miller (1974) stated, the notion of

> transforming gangs by diverting their energies from traditional forms of gang activities—particularly illegal forms—and channeling them into "constructive" activities is probably as old, in the United States, as gangs themselves. Thus, in the 1960s, when a series of social movements aimed at elevating the lot of the poor through ideologically oriented, citizen-executed political activism became widely current, it was perhaps inevitable that the idea be applied to gangs. (p. 410)

Jacobs (1977, p. 145) offered three explanations as to why Chicago street gangs, as well as those in many other urban areas, became politicized during the 1960s:

1. Street gangs adopted a radical ideology from the militant civil rights movement.
2. Street gangs became committed to social change for their community as a whole.
3. Street gangs became politically sophisticated, realizing that the political system could be used to further their own needs—money, power, and organized growth.

Jacobs maintained that the third explanation is applicable to the Blackstone Rangers and many other large gangs in metropolitan areas. The leadership learned how to use the system to provide capital for the gangs' illegal activities. Gangs showed increased sophistication in organizing their activities along the lines of organized crime. Individual felonies were replaced by major criminal activity involving drugs, weapons, extortion, prostitution, and gambling. Fistfights were replaced by violent acts involving the use of weapons.

During the 1980s, society became increasingly concerned with violence and prescriptions for crime control, and this concern carried over into the 1990s and the new century (Bilchik, 1999c). Attention has once again focused on crime and delinquency resulting from failures of social institutions, inadequate deterrence, and insufficient incapacitation. Deterrence research has become popular, focusing on police, probation, and corrections activities rather than on gang dynamics. Current emphasis is on preventing juveniles from joining gangs through community education and involvement and on bringing to a halt the violent activities of gangs through stricter laws, better prosecution, more severe sanctions, and negotiated peace agreements between feuding gangs.

"The last quarter of the 20th century was marked by significant growth in youth gang problems across the United States. In the 1970's, less than half of the states reported youth gang problems, but by the late 1990's, every state and the District of Columbia reported youth gang activity" (Miller, 2001, p. iii). The states with the largest numbers of gang problem cities in 1998 were California, Illinois, Texas, Florida, and Ohio (Klein, 2005; McGloin, 2005).

Although gang problems continue to be a big-city problem, as we enter the 21st century, they are by no means confined to such cities. In fact, the number of cities with populations between 1,000 and 5,000 reporting gang activities has increased 27 times, and the number of cities with populations between 5,000 and 10,000 reporting gang activities has increased more than 32 times (Miller, 2001, p. x).

The Nature of Street Gangs

In general, street gangs may be identified by the following characteristics (Illinois Department of Corrections, 1985):

1. They are organized groups with recognized leaders who command the less powerful.
2. They are unified at peace and at war.
3. They demonstrate unity in obvious recognizable ways (i.e., wearing of colors, certain types of graffiti, use of gang signs).
4. They claim a geographic area and an economic and/or criminal enterprise (i.e., turf, drugs).
5. They engage in activities that are delinquent and/or criminal or are somehow threatening to the larger society.

Short and Strodtbeck (1965) developed a somewhat similar set of criteria for gang members:

1. Recurrent congregation outside the home
2. Self-defined inclusion/exclusion criteria
3. A territorial basis consisting of customary hanging and ranging areas, including self-defined use and occupancy rights
4. A versatile activity repertoire
5. Organizational differentiation (e.g., by authority, roles, prestige, or friendship)

Delinquent and Criminal Gang Activities

Antisocial and criminal conduct by members of juvenile gangs is not a new phenomenon. Early immigrant groups arriving in this country frequently found themselves located in the worst slums of urban areas, and gangs soon emerged. Among the earliest juvenile gangs were those of Irish background, followed later by Italian and Jewish gangs and eventually by gangs of virtually all ethnic and racial backgrounds. Typically, members of these gangs left gang activities behind as they grew older, married, found employment, and raised families (Huff, 1996; Siegel, Welsh, & Senna, 2003). Some, however, gravitated to adult gangs and into organized criminal activities (Abadinsky, 2003). The path from juvenile gang membership to adult crime seems to have broadened during recent years, so although it is true that some street gangs are still little more than collections of neighborhood juveniles with penchants for macho posturing, many are emerging as drug–terrorism gangs that terrify residents of inner-city neighborhoods. In fact, the National Criminal Justice Reference Service (NCJRS) provides data that demonstrate this observation. For instance, in 1996, 50% of gang members were juveniles (i.e., under 18 years) and 50% were adults (i.e., 18 years or over). However, just 3 years later (in 1999), it was found that roughly 37% were juveniles and 63% were adult offenders (NCJRS, 2005). Thus, it is clear that adults have the greater influence over gang activity. These adult members likewise tend to form the leadership for most street gangs (Valdez, 2005). However, there are indications that many juveniles that do engage in gang activity may be more willing to engage in violence (see In Practice 12.2).

Street gangs violate civilized rules of behavior, engaging in murder, rape, robbery, intimidation, extortion, burglary, prostitution, drug trafficking, and (starting in the late 1980s) a phenomenon known as **wilding** where gang members attack individuals at random, committing any of the offenses just mentioned. During the 1990s, drive-by shootings became another tool of gang members who were seeking retribution but were unconcerned about the lives of innocent bystanders often shot in the process. The activities of gangs have become increasingly serious, more sophisticated, and more violent, and they are more likely to involve the use of weapons (Siegel et al., 2003; Valdez, 2005). Gangs have become problematic in California, where the legislature passed a law making it a felony to belong to a gang known to engage in criminal activities, and other jurisdictions are in the process of establishing similar legislation. Although the constitutionality of these laws has yet to be established, their mere existence indicates how serious the gang problem is perceived to be. During the late 1980s, *Newsweek* magazine described the drug gangs found in urban ghettos and barrios as consisting of young men whose poverty and deprivation have immunized them to both hope and fear. The result is a casual acceptance of—and sometimes enthusiasm for—torture and murder, drive-by shootings, and public mayhem. "If they don't kill you, they'll kill your mother" ("Crack, Hour by Hour," 1988).

In Practice 12.2

Gangs Getting Younger, Bolder: Probe Leads to 40 arrests, Half Kids

Antelope Valley gangs are becoming younger and more brazen.

That's the conclusion of a Palmdale law enforcement official who helped lead one of the biggest Antelope Valley gang-specific investigations of its kind.

Sheriff's Detective Key Budge said the three-month investigation culminated with the arrests of approximately 40 gang members, half of whom were juveniles.

Most of the arrests were on suspicion of residential burglaries, street robberies, and car thefts.

Law enforcement officials seized tens of thousands of dollars of stolen property and cash from the gang that Budge said represents just a fraction of merchandise stolen over the period of the investigation.

Budge said nearly all the adults charged in the case have reached plea agreements, but many of the cases involving juveniles have not yet been resolved.

Most of the juveniles are repeat offenders trained by adults to commit the crimes because minors receive minimal penalties, Budge said.

Budge was startled by what he called the arrested juveniles' arrogance, noting that they believed their experience with the criminal justice system boosted their gang stature.

"It's something that is new to us," Budge said. "We're seeing more and more violent crimes being committed by [juveniles]."

Budge's troublesome conclusions follow a statewide trend. Assemblyman Mervyn Dymally, D-Compton, soon will introduce legislation proposing pilot gang prevention programs for children as young as 9 in Compton, Inglewood, and Oakland schools, his spokeswoman said.

Budge wouldn't specify the ages of the juveniles involved in the gang because they are being prosecuted, but he said he's learned from the investigation that the gang recruits children in fifth grade.

Budge said several prominent members of the gang are from Los Angeles, but he thinks the gang functions independently of larger L.A. gangs.

"It doesn't appear that they take their marching orders from L.A. gangs," he said.

Budge said the sting involved a collaboration between Lancaster and Palmdale sheriff's stations. He said four detectives heading the investigation summoned approximately 200 sheriff's deputies for the Nov. 2 sweep.

The gang was primarily involved in crime activity described as somewhat crude—strong-armed retail and home invasion burglaries and street robberies using real weapons or simulated weapons—on a larger scale than local law enforcement officials are accustomed to seeing.

The investigation's focus has since shifted to uncovering layers of the gang suspected of selling the stolen property. Budge said adult family members sold much of the merchandise to pawn shops and on the streets.

Budge expressed frustration with an overburdened juvenile justice system and laws governing juvenile crime that he says have created a legal loophole gangs are exploiting.

"The laws that pertain to juveniles are outdated," Budge said. "If there was a true penalty, I think that might be a true deterrent for these kids."

Source: Rubin, Gideon. (2007, January 18). Gang members getting younger, bolder: Probe leads to 40 arrests, half kids. Retrieved from: http://www.dailynews.com/news/ci_5042384

The days when rival gangs fought each other over "turf" and "colors" are fading fast. Today, gang conflicts are more of the form of urban guerrilla warfare over drug trafficking (Hanser, 2007a). Gang turf is now drug sales territory. Informers, welchers, and competitors are ruthlessly punished or assassinated. Street warfare and the bloody rampage of gang violence are the norm in many inner cities. As an illustration of these points, it is estimated that 3,340 member-based gang homicides were committed in 1997 (Bilchik, 1999c, p. 15; Valdez, 2005). Gang members commit a disproportionate share of crime for their numbers (Klein, 2005; McGloin, 2005). Statistical data indicate that although gang membership in a given jurisdiction might not be high and gang members constitute only a small percentage of all criminals, they typically commit more offenses than their non-gang member criminal counterparts (NCJRS, 2005). In 1997, youth gangs were involved in 18% of homicides nationwide. Gang violence is increasing in intensity and is spreading throughout the country. In large metropolitan areas such as Chicago, New York, Miami, and Los Angeles, gang-related homicides number in the hundreds annually (Bilchik, 1999c, pp. 15, 20; Valdez, 2005). Many of these homicides result from gang wars and retaliations, and often the victims are innocent bystanders or those unable to defend themselves (Hughes, 2005). The macho image of gang members confronting each other in open warfare is largely a creation of the media. More often, gang killings occur on the streets, in the dark, as a result of gang members in a speeding vehicle firing shots at their intended victim(s).

Major gangs have made **narcotics trafficking** (sale and distribution of drugs) an important source of income, and activities in this area have become even more lucrative with the advent of a street market for a variety of drugs. For instance, Latino street gangs are highly active in the trafficking of heroin and cocaine that originate in Mexico and South America, respectively (Hanser, 2007a; Thompson, 2004). Some of the increase in gang violence is a result of competition over turf ownership related to the sales of these products. Gangs involved in profit-oriented schemes frequently resort to violence to protect their illicit businesses (Hanser, 2007a; Thompson, 2004). With this shift to more business-oriented activities, some gangs have gone underground; members no longer openly display colors or graffiti, sometimes leading to the mistaken assumption that gang activity in a particular area has ceased (Hughes, 2005; Klein, 2005; McGloin, 2005).

On other occasions, even when officials know that gangs are behind a good deal of the illegal activity occurring in their jurisdictions, they deny the importance of gangs for political reasons (Valdez, 2005). Political officials, including some police chiefs, would prefer not to make themselves look bad by admitting that gang activity in their areas is uncontrollable. It is estimated that 42% of youth gangs are involved in the street sale of drugs for the purpose of making a profit for their gangs and that 33% of youth gangs in the United States are involved in the distribution of drugs. Crack cocaine and marijuana are thought by law enforcement agencies to lead the list of drugs sold (Bilchik, 1999c, pp. 22, 25, 28).

An illustration of the drug gang problem can be seen in South-Central Los Angeles, where the Bloods and Crips reign (Huff, 1996; Valdez, 2005). These gangs consist of confederations of neighborhood gangs, each with a relatively small number of members. The gangs have traditionally been involved in robbery, home invasion, burglary, and homicide, but they became involved in drug trafficking as a major enterprise with the production of crack cocaine. The Bloods (whose color is red) and the Crips (whose color is blue) date back to the late 1960s and early 1970s and consist of "rollers" or "gang-bangers" who are in their 20s and 30s—gang veterans who have made it big in the drug trade (Huff, 1996; Valdez, 2005). These veterans supervise and control the activities of younger members who are involved in drug trafficking and operating and supplying crack houses, activities that bring in millions of dollars a week. The drug trafficking of the Bloods and Crips spread into other cities, including Seattle, Portland, Denver, Kansas City, Des Moines, and even Honolulu and Anchorage (Hieb, 1992). Police sources there say that between January and November 2000, there were some 64 gang-related shootings and 34 gang-motivated gun battles between the Bloods and the Crips (McPhee, 2000). In 2004, more than half of the almost 1000 homicides committed in Los Angeles and Chicago were considered to be gang related (OJJDP Fact Sheet, 2006).

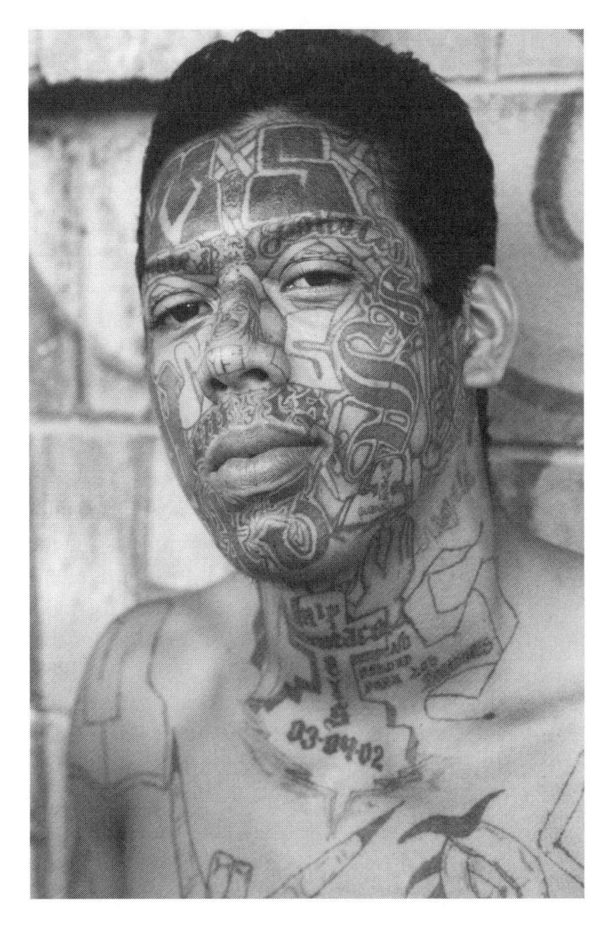

▲ Tattoos may be used to signify gang affiliation as well as to express individuality.

Similar gangs exist in other major cities. In Chicago, for example, major street gangs include the **Black Disciples**/Black Gangster Disciple Nation/Brothers of the Struggle (BOS) who are collectively known as **folks.** They are in competition with the **Vice Lords,** also known as the Conservative Vice Lord Nation, who have aligned themselves with other groups such as the **Latin Kings** (Main, 2001; Main & Spielman, 2000). These gangs are collectively referred to as **people** (Babicky, 1993a, 1993b; Dart, 1992). Splinter gangs of "people" and "folks" have been found in Minneapolis, Des Moines, Green Bay, and the Quad Cities area of Illinois. Activities include extortion, drug trafficking, and violent crime in the form of homicide, robbery, drive-by shooting, and battery.

Although street gangs are predominantly black and Hispanic, the **Simon City Royals** ("folks") are a white gang that originated in Chicago during the early 1960s. Originally formed to stop the "invasion" of Hispanic gangs into their area of the city, the Royals became actively involved in burglaries and home invasions during the 1980s. As leaders of the gang were imprisoned, they formed alliances with the Black Gangster Disciples for protection, and the alliances continue today (Babicky, 1993a, 1993b).

Latino gangs exist in every major urban area and many suburban areas as well. Indeed, Latino gangs have spread to various areas of the United States, including cities in the Southwest, Florida, New York, and the city of Chicago. In fact, the NCJRS (2005) estimated that nearly 47% of all gang members are Latino in racial orientation. Furthermore, in states such as Arizona, it is reported that fully 62% of gang members are Latino in demographic orientation. Other areas of the Southwest report similar levels of Latino gang activity (NCJRS, 2005).

Among gangs in general, including Latino gangs, there has been an emergence of gang nations (Attorney General of Texas, 2002). The term *gang nation* is an informal name for a very large gang that may have several affiliated subsets of members, sometimes stretching across vast areas of the United States (Attorney General of Texas, 2002). Increasingly, law enforcement officials are considering the Surenos and Nortenos in California, as well as the Mexican Mafia in California and Texas, to be gang nations (Hanser, 2007a). These two Latino gang nations are made up of smaller "sets" that share certain symbols and loyalties. These gang nations have become so large that different sets of the same gang might not even know each other except by recognition of some common sign or insignia (Attorney General of Texas, 2002). In fact, these sets, although belonging to the same nation, may develop rivalries among themselves while also rallying against a common enemy (Attorney General of Texas, 2002; Hanser, 2007a).

In the United States, there has recently been an emergence of Latino gangs whose memberships come largely from nations in Central America (Hanser, 2007a; Thompson, 2004; Valdez, 2005). During the past decade, gangs have migrated across Central America and into Mexico, ultimately to settle in the United States. This migration has occurred due to military targeting of these gangs in their native nations of El Salvador, Guatemala, and Honduras (Thompson, 2004). The arrival of these new immigrants has generated an unending stream of gang-related violence in many Latino neighborhoods in various areas of California (Hanser, 2007a; Thompson, 2004).

The two largest Central American gangs in the United States are the Mara Salvatruchas (known as MS-13) and the Mara 18, and both of these gangs originated on the streets of Los Angeles (Hanser, 2007a; Thompson, 2004). The Salvatruchas gang was started by the children of refugees from various war-torn countries in Central America (Thompson, 2004). Conversely, the 18th Street gang was started by Mexican immigrants who arrived during the 1970s. The 18th Street gang and the Mara Salvatruchas are polar enemies of one another (Hanser, 2007a; Thompson, 2004; Valdez, 2005). Both gangs have experienced tremendous growth, although the MS-13 has gained the greatest national notoriety. Given the sheer numbers of illegal immigrants entering the United States from Mexico, prevention of gang immigrants seems to be virtually impossible for law enforcement agencies in the southwestern region of the United States. As a result, these gangs have become two of the fastest-growing and most notorious Latino gangs in the United States to date (Thompson, 2004).

Although Asian gangs exist, they are much less numerous than those consisting of African American and/or Latino American members. Asian gangs are located in Chinese, Filipino, Vietnamese, and Cambodian communities and are involved in a variety of illegal activities such as extortion, protection of illegal enterprises, and summary executions of members of rival gangs (Peak, 1999, pp. 53–54). Often, these gang members victimize their own ethnic groups and/or communities (Abadinsky, 2003; Valdez, 2005).

Gangs engage in a wide variety of activities, including the following:

1. *Vandalism*—graffiti and wanton destruction
2. *Harassment and intimidation*—to recruit members, to exact revenge on those who report their activities, and so forth
3. *Armed robbery and burglary*—the elderly and, more recently, suburban communities targeted

4. *Extortion of the following:*
   a. *Students in schools*—protection money
   b. *Businesses*—protection money to avoid burglaries, fires, vandalism, and general destruction
   c. *Narcotics dealers*—protection money to operate in a specific geographic area and a percentage of the "take"
   d. *Neighborhood residents*—who pay for the ability to come and go without being harassed and for the "privilege" of not having their property destroyed

Gang crime continues to grow in smaller suburban and even rural communities, which are frequently perceived as easy marks for theft, burglary, robbery, and shoplifting, among other crimes. In other cases, gangs migrate to these areas to avoid the intense competition of drug trafficking in the cities (see In Practice 12.2).

Nearly every community has experienced **wannabes** or juveniles who may wear gang colors and post graffiti in an attempt to emulate big-city gang members. Yet in many communities, gangs have been largely ignored. Part of the reason for the continued expansion of gang activities is the view that such activity is not our problem. Because many street gangs are ethnically oriented, it is easy to perceive the problem as affecting only certain groups or neighborhoods. Because members of traditional gangs are predominantly from the lower social class, gangs are perceived as problematic basically in lower-class areas. However, if gang activity is not dealt with quickly, the consequences soon spread to the larger community, including middle-class neighborhoods, and the problem may become unmanageable (Pattillo, 1998). Osgood and Chambers (2000) analyzed juvenile arrest rates for 264 nonmetropolitan counties in four states. They concluded that juvenile violence in nonmetropolitan areas is associated with residential instability, family disruption, ethnic heterogeneity, and population size. Evans, Fitzgerald, and Weigel (1999) found no significant differences in gang membership or pressure to join gangs between rural and urban juveniles. Urban juveniles, however, were more likely to report having friends in gangs and being threatened by gangs than were their rural counterparts.

Gang Membership

Among the generally accepted reasons for gang development and membership are the following:

1. The gang provides peer support during the transition from adolescence to adulthood.
2. The gang results from a lower-class cultural reaction to the values or goals of the dominant society.
3. The gang provides the opportunity to attain goals adopted by the larger society, but through alternative (illicit) means.
4. The gang provides self-esteem, economic opportunities, a sense of belonging, affection, and so on, all of which are missing from the lives of many juveniles.

Furthermore, individual motivations for joining gangs may be categorized as follows:

1. *Identity/Recognition.* This allows members to achieve a status they believe is impossible to achieve outside the gang. Many are failures in legitimate endeavors such as academics and athletics.
2. *Protection/Survival.* Many believe that it is impossible to survive living in a gang-dominated area without becoming a member.
3. *Intimidation.* Juveniles are told and shown (e.g., through beatings) by gang members that membership is essential.
4. *Fellowship and brotherhood.* The gang offers psychological support to its members and provides the companionship that may be lacking in the home environment.

As Jacobs (1977) explained, "Time and again gang members explained that, whether on the street or in prison, the gang allows you to feel like a man; it is a family with which you can identify. Many times young members have soberly stated that the organization is something, the only thing, they would die for" (p. 150). Bloch and Neiderhoffer (1958) believed that gangs were cohesive and organized and, therefore, satisfied deep-seated needs of adolescents. Whyte (1943) portrayed the gang as a **street corner family.** "Gangs are residual social subsystems often characterized by competition for status and, more recently, income opportunity through drug sales" (Curry & Spergel, 1988).

Characteristics

In general, gang membership can be divided into three categories: leaders, hardcore members, and marginal members. The leaders within gangs usually acquire their positions of power through one of two methods: either by being the "baddest"/meanest member or by possessing charisma and leadership abilities. In addition, the leaders tend to be older members who have built up seniority. Hardcore members are those whose lives center on the totality of gang activity. They are generally the most violent, streetwise, and knowledgeable in legal matters. Marginal or fringe members (sometimes referred to as "juniors" or "peewees") drift in and out of gang activity. They are attached to the gang but have not developed a real commitment to the gang lifestyle. They associate with gang members for status and recognition and tend to gravitate toward hardcore membership if no intervention from outside sources occurs (Babicky, 1993a, 1993b; Mays & Winfree, 2000, p. 302).

Age

Gang members ranging from 8 to 55 years of age have been detected. For many younger juveniles, gang members serve as role models whose behavior is to be emulated as soon as possible to become full-fledged gang-bangers. Consequently, these children are often exploited by members of gangs and manipulated into committing offenses such as theft and burglary to benefit the gang as part of their initiation or rite of passage. As gang crimes become more profitable, as in the case of drug franchises, the membership tends to be older. As members become older, they move away from street crime and move up in stature within the gang hierarchy. The younger members maintain the turf-oriented activities, and the adults move into more organized and sophisticated activities.

Gender

Street gangs are predominantly male. Although girls have been a part of gangs since the earliest accounts from New York during the early 1800s, their role has traditionally been viewed as peripheral. Thrasher (1927), in his classic study of gangs, discussed two female gangs in Chicago. Females have been described in the literature primarily as sex partners for male gang members or as members of auxiliaries to male gangs (Campbell, 1995). Yet as we have indicated throughout this book, involvement in crime, including violent crime, among juvenile females has clearly increased over the past decade. In fact, a study by Bjerregaard and Smith (1993) found that rates of participation in gangs are similar for males and females.

With respect to female gang membership, Campbell (1997) noted,

> Although there is evidence that young women have participated in urban street gangs since the mid-nineteenth century, it is only recently that they have received attention as a topic of study in their own right. . . . Previous work on female gang members has placed considerable emphasis upon their sexuality either as [an] area for reform through social work, as a symptom of their rejection of middle-class values, or as the single most important impression management problem they face. . . . These young women are stigmatized by ethnicity and poverty as well as gender. (p. 129)

Clearly, a good deal more research is required to understand female participation in gang behavior.

The research that does exist focuses largely on Hispanic girls in gangs, with less attention given to black, white, and Asian female gang members (Hanser, 2007a; Valdez, 2005). An exception to this trend is Fishman's (1995) ethnographic study of the black **Vice Queens** in Chicago. Fishman noted that female gangs are moving closer to being independent, violence-oriented groups. The Vice Queens were routinely involved in strong-arming and auto theft, and they frequently engaged in aggressive and violent behavior, activities viewed as traditionally performed almost exclusively by males. Esbensen, Deschenes, and Winfree (1999) also found that gang girls are involved in a full range of illegal activities, although perhaps not as frequently as are gang males. Harper and Robinson (1999) found that involvement of black juvenile females in sexual activity, substance abuse, and violence was clearly related to membership in gangs. Thus, the relationship between gang membership and a variety of delinquent and criminal behaviors appears to be similar for girls and boys.

Campbell (1995) argued that, to account for female gang membership, we must consider the community and class context within which the girls involved live, the problems that face poverty-class girls, and the problems for which they seek answers in gangs. Among these problems are the following:

1. A future of meaningless domestic labor with little possibility of escape
2. Subordination to the man of the house
3. Sole responsibility for children
4. The powerlessness of underclass membership

Furthermore, many of these girls will be arrested, "and the vast majority will be dependent on welfare. The attraction of the gang is no mystery in the context of the isolation and poverty that awaits them" (Campbell, 1995, pp. 75–76). "Without the opportunity to fulfill themselves in mainstream jobs beyond the ghetto, their sense of self must be won from others in the immediate environment. . . . Their association with the gang is a public proclamation of their rejection of the lifestyle which the community expects from them" (Campbell, 1997, p. 146). Curry (1998) agreed, finding that gang membership in these circumstances can be a form of liberation. Finally, Esbensen and colleagues (1999) found that female gang members have lower levels of self-esteem than do male gang members, lending further support to Campbell's conclusions that the gang provides a source of support and feelings of self-worth. Of course, not all female gang members are found in urban areas, as we saw in Chapter 3. Gangs have begun to appear in both suburban and rural areas previously thought to be immune from gang activity (Bjerregaard & Smith, 1993).

Monikers

Many gang members have **monikers** or nicknames that are different from their given names. In general, these monikers reflect physical or personality characteristics or connote something bold or daring. Often gang members do not know the true identities of other members, making detection and apprehension difficult for law enforcement officials.

Graffiti

Street gang **graffiti** is unique in its significance and symbolism. Graffiti serves several functions: it is used to delineate gang turf as well as turf in dispute, it proclaims who the top-ranking gang members are, it issues challenges, and it proclaims the gang's philosophy. Placing graffiti in the area of a rival gang is considered an insult and a challenge to the rival gang, which inevitably responds. The response may involve anything from crossing out the rival graffiti to committing drive-by shootings or engaging in other forms of violence. The correct interpretation of graffiti by the police can offer valuable information as to gang activities. Symbols that represent the various gangs are typically included in graffiti and are known as "identifiers."

Jargon

Gang members frequently use **jargon** to exchange information. In fact, understanding gang jargon may be critical to obtaining convictions for gang-related crimes. In 1986, Jeff Fort and several high-ranking members of the Chicago-based El Rukn were charged with plotting acts of terrorism in the United States for a sum of $2.5 million from Libya. The FBI recorded 3,500 hours of telephone conversations, most of which were in code, in this case. A former high-ranking member of El Rukn, who became a prosecution witness, translated portions of the confusing conversations for jurors. Fort was convicted and sentenced to an extended prison term.

Recruitment

Gangs continue to recruit new members to defend their turf and expand criminal activities so as to increase profits (Hughes, 2005). There is often intense competition among new young gang members to prove themselves to the hardcore membership. Brutal initiation rituals in which recruits are severely beaten with fists, feet, and other objects are not uncommon (Hughes, 2005; Mays & Winfree, 2000; Valdez, 2005). This competition results in younger gang members (10–13 years old) being very dangerous. Gang members know that juveniles in this age group are not likely to be prosecuted in adult court, making them particularly valuable in the commission of serious offenses (Hughes, 2005; McGloin, 2005; Szymanski, 2005). Because the young members want the approval of the older gang members, they are highly motivated to prove themselves and are likely to do whatever they are told to do. Recruitment of these new members occurs anywhere juveniles gather—shopping malls, bowling alleys, skating rinks, public parks, neighborhoods, and schools (Hieb, 1992; Huff, 1996; Siegel et al., 2003). There has been an increase in the recruitment of members outside major urban centers and in the number of middle-class juveniles approached as potential members.

Response of Justice Network to Gangs

The problems presented by juvenile gangs are not easily addressed. In large part, the origins of these problems are inherent in the social and economic conditions of inner-city neighborhoods across the United States, and the issue is complicated by the continued existence of racial and ethnic discrimination in the educational and social arenas (Walker, Spohn, & DeLone, 2004). These conditions are largely beyond the control of justice officials, whose efforts are hampered greatly as a result. Because we appear to be unwilling to confront the basic socioeconomic factors underlying gang involvement, our options are limited largely to responding to the actions of gang members after they have occurred (McGloin, 2005; Walker et al., 2004). In general, the response in these terms has been to propose, and often pass, legislation creating more severe penalties for the offenses typically involved, specifically, drug-related and weapons-related offenses (Finkelhor & Ormrod, 2001; Siegel et al., 2003). Recognizing the fact that gangs are now more mobile and that splinter gangs exist in numerous communities, law enforcement officials have attempted to respond by establishing cooperative task forces of combined federal, state, and local authorities who share information and other resources to combat gang-related activities.

At the federal level, several past presidents have called for a war on drug trafficking and appointed a "drug czar" to oversee efforts in this area. President Bill Clinton appointed Lee Brown, a former Houston police chief and New York City police commissioner, to the post. Prosecutors at the federal and state levels have become involved in extremely complex expensive cases to incarcerate known gang leaders and to send a message to gangs that their behavior is not to be tolerated. Gang crimes units and specialists have emerged in most large urban, and some medium-sized city, police departments. School liaison programs (discussed in Chapter 7) have been implemented in the hope of reducing gang influence in the schools (Brown, 2006; Ervin & Swilaski, 2004; Valdez, 2005). Parent groups have mobilized to combat the influence of gangs on children, and media attention has focused on the consequences of ignoring gang-related crimes (Ervin & Swilaski, 2004; Hanser, 2007b). Forfeiture laws have been passed, making it possible for government agencies to seize and sell or use cars, boats, planes, and homes in the pursuit of illegal activities (Abadinsky, 2003). In short, there is now considerable effort directed toward controlling gang activities. Whether such effort is properly organized, coordinated, and directed and whether the effort will have the desired consequences remain empirical questions.

In 1995, the Office of Juvenile Justice and Delinquency Prevention awarded grants to five communities to implement and test the **Spergel model** intended to reduce gang crime and violence. This model involves developing a coordinated team approach to delivering services and solving problems (Burch & Kane, 1999). Strategies involved include mobilization of community leaders and residents; use of outreach workers to engage juveniles in gangs; access to academic, economic, and social opportunities; and gang suppression activities. Such collaborative efforts that use resources from multiple agencies in the community have been widely adopted throughout the nation (Clear & Cole, 2003; Ervin & Swilaski, 2004; Hanser, 2007b; Hughes, 2005).

Public, Legislative, and Judicial Reaction

There is little doubt that some violent juveniles must be dealt with harshly and incarcerated for the protection of society in spite of the fact that processing these juveniles as adults clearly violates the philosophy of the juvenile court network by labeling them as criminals at an early age and by placing them in incarceration with automatic transfer to adult facilities at the age of majority (McGloin, 2005; Moffitt & Caspi, 2001; Szymanski, 2005). As we have seen, public perception that there was a dramatic increase in violent and serious crime by juveniles during the 1980s and 1990s resulted in considerable pressure on legislators to pass new, more stringent laws relating to the prosecution and incarceration of violent juvenile offenders (Siegel et al., 2003; Szymanski, 2005). As noted in Chapter 5, the juvenile court acts of many states have been amended to remove juveniles charged with criminal homicide, rape, or armed robbery from the jurisdiction of the juvenile court if they were over a certain age at the time they committed the offense.

However, legislative attempts to solve crime problems by passing tougher laws, such as mandatory sentencing laws (e.g., for drug crimes) and "three strikes" laws (after the third offense, they "throw away the key"), have resulted in less than desirable outcomes (Clear & Cole, 2003; Walker et al., 2004). Walker (1998), for example, concluded that "three strikes and you're out laws are a terrible crime policy" because there is no evidence that they reduce serious crime and considerable evidence that they lead to the incarceration of many people who would not commit other crimes anyway (p. 140; see also Clear & Cole, 2003).

Gang suppression, gang sweeps, zero tolerance policies, and loitering ordinances are tactics employed by the police to minimize gang activity (Huff, 1996). Attention has been focused on developing partnerships between the police and other community agencies in an attempt to intervene in gang activities. In the city of Reno, Nevada, for example, police decided to focus on the top 5% of gang members using a repeat offender program to target them. At the same time, the police coordinate efforts to deal with the 80% of gang members who are not considered hardcore. A community action team (CAT) was formed to accomplish these tasks, interview and develop intelligence on gang members, deal with the families of gang wannabes, develop neighborhood advisory groups (NAGs), and implement an improved media policy. This collaborative effort has resulted in increasingly positive evaluations of police performance and reduced amounts of gang violence (Weston, 1993).

Alternatives to Incarceration for Violent Juveniles

Although numerous causes of youth violence have been posited, solutions have been less apparent. Should we censor the media? Are more prisons or longer sentences the answer?

Would gun control help? Is the juvenile justice network, as currently conceived, out of date and ill-equipped to handle violent juveniles?

In his study of juvenile homicide, Sorrells (1980) found that a disproportionate number of juveniles who commit this offense come from communities with a high incidence of poverty and infant mortality. He also noted that such offenders are products of "violent chaotic families." He concluded that juveniles who kill are likely to fall into one of three categories (p. 152):

1. Youngsters who lack the capacity to identify with other human beings
2. Prepsychotic juveniles who kill as an expression of intense emotional conflicts and who are also high suicide risks
3. Neurologically fearful youngsters who kill in overreacting to a genuinely threatening situation

As **alternatives to incarceration,** Sorrells suggested identifying high-risk communities and pooling agency resources to combat specific problems characterizing each community, screening violent juveniles for emotional problems, developing treatment programs focusing on resolving such emotional problems, and removing children from violent chaotic families where possible.

Other researchers have provided general support for Sorrells's (1980) findings and made similar recommendations (Moffitt, 1993; Moffitt & Caspi, 2001). One study of recidivism among juvenile offenders, for example, found that 100% of those recidivists had arrest records prior to the arrests on which the recidivism was based, 88% had unstable home lives, 86% were unemployed, and more than 90% had school problems (Ariessohn, 1981). Other researchers have also found that when early intervention is not effective, somewhere between two-thirds and three-fourths of violent juvenile offenders on probation will recidivate, committing essentially the same type of offense within a few months (Buikhuisen & Jongman, 1970; Moffitt, 1993; Moffitt & Caspi, 2001). Clearly, parents can play a major role with respect to violent juveniles (Hanser, 2007b; Moffitt, 1993; Siegel et al., 2003). It is important to note, however, that they can serve as negative role models as well as positive ones (Hanser, 2007a; Moffitt, 1993; Siegel et al., 2003).

Schools and teachers can also play important roles in preventing violence and gang membership (Kaser-Boyd, 2002). The GREAT (Gang Resistance Education and Awareness Training) program (discussed in Chapter 7) taught by police in the schools may help juveniles to resist gangs. Parents and teachers can work together to improve the interpersonal, cognitive, problem-solving skills of juveniles. This approach focuses on modifying thinking processes rather than behaviors themselves. "I Can Problem Solve" helps children learn to solve interpersonal problems by focusing on means–ends thinking (a step-by-step approach to pursuing goals), weighing pros and cons (of carrying out behaviors), alternative solution thinking, and consequential thinking (the ability to think about different outcomes from a specific behavior). Although more research needs to be done, there is some evidence that teaching children such skills at very young ages can help them to develop prosocial attitudes and behaviors that last well into adulthood (Hanser, 2007b; Moffitt & Caspi, 2001).

There has been an abundance of research that has sought to differentiate between long-term and repeat juvenile offenders and those who are likely to eventually desist from further crime (Moffitt, 1993; Moffitt & Caspi, 2001; Szymanski, 2005). Because some juveniles persist in violent behavior even after conventional interventions are used, some researchers have concluded that it may be necessary to deal with juveniles who engage in progressively more serious assaultive behavior by commitment to a detention facility for a 3- to 6-month period (thereby preventing recidivism in the community during this very high-risk period), during which time the juvenile's behavior is stabilized and brought under control (Moffitt, 1993; Moffitt & Caspi, 2001; Szymanski, 2005). Many researchers and authors maintain that effectively targeting and detaining select juveniles most likely to recidivate can produce maximal results in recidivism outcomes (Hanser, 2007b; Moffitt, 1993; Moffitt & Caspi, 2001; Szymanski, 2005). In an effort to explore the viability of such an approach, the Office of Juvenile Justice and Delinquency Prevention sponsored a 2-year program to identify, select, prosecute, and enhance treatment for serious habitual juvenile offenders. Analysis showed that such programs result in more findings of guilt and more correctional commitments as well as that linking such efforts with special correctional treatment programs for juveniles is highly problematic due to the necessity of subcontractual relationships between prosecutors and service providers and the unavailability of special correctional programs to meet the diverse needs of serious habitual offenders (National Institute of Justice, 1988).

Because of past research on juvenile detention and juvenile treatment outcomes, many researchers and advocates now call for a comprehensive strategy aimed at eliminating all risk factors for further delinquency, including gang involvement (Ervin & Swilaski, 2004; Evans & Sawdon, 2004; Hanser, 2007b; Valdez, 2005). This strategy called for a wide spectrum of services and sanctions to be used to protect potential and current delinquents from the womb to school and beyond (Clear & Cole, 2003; Ervin & Swilaski, 2004; Evans & Sawdon, 2004; Hanser, 2007b; National Youth Gang Center, 1997).

Unfortunately, getting to and treating potentially violent juveniles is not an easy task. Although the proportion of juveniles who commit violent crimes is relatively small, this group commits a sizable number of offenses (Moffitt, 1993). Protecting the best interests of children is clearly an important goal of the juvenile justice network, but so is protection of society (Hanser, 2007b; Siegel et al., 2003). It may well be that police, prosecutors, and judges need to deal with violent juveniles earlier and more severely than they have done in the past (Szymanski, 2005). Although giving juveniles the benefit of the doubt in early encounters with the police and courts may be well intentioned, in the case of violent offenders at least, it is also dangerous to others (Hanser, 2007b; Moffitt, 1993). It is clear that violence becomes a pattern of behavior when intervention either does not occur or is not effective (Moffitt, 1993). In the interests of protecting society and attempting to rehabilitate violent juveniles by delivering the best programs available as early as possible, violent juveniles must be identified, apprehended, and evaluated or judged as soon as possible (Moffitt, 1993; Seigel et al., 2003).

It seems that many of the factors involved in improving opportunities for juveniles while reducing delinquency lie outside the scope of the traditional juvenile justice system. It is because of this that anti-gang programs must incorporate comprehensive intervention that assists these juveniles in separating themselves from their gang affiliation. This can be particularly challenging when juveniles come from neighborhoods where gang activity is quite prevalent. In response to the challenges in reforming youth gang members, many areas throughout the United States and Canada have developed gang exit programs for youthful gang members. We now turn our focus to these programs.

Establishing a Juvenile Gang Exit Program

Effective gang intervention programs must include community advocacy that can facilitate a sense of cohesion among neighborhoods so that these communities will not be intimidated by gangs and gang activity. One such program in Toronto, Canada, addresses this issue through a regimen that has three specific components: (a) assessment and intake, (b) intensive training and personal development, and (c) case management (Evans & Sawdon, 2005; Hanser, 2007b).

The assessment and intake phase examines the interest and motivation of the individual gang member, the amount of gang involvement by that member, and the gang member's family and social history (Evans & Sawdon, 2005). The next phase is referred to as the gang member intensive training and personal development phase and consists of two separate curricula—one for male gang members and the other for female gang members—that address different aspects relevant to both genders (Evans & Sawdon, 2005; Hanser, 2007b). Female gang members have special issues that are not relevant to male gang offenders as frequently, including child-rearing concerns, the fear of sexual victimization, and issues related to childhood molestation (Hanser, 2007b). Likewise, male gang members may have issues related to the definitions of masculinity, lack of respect for the female population, and other problems that are not usually pertinent to female offenders (Hanser, 2007b). Both curricula include 60 hours of intensive training and interactional exercises addressing typical topics such as anger management, racism, and communication skills. The last phase of this intervention is the gang member case management phase and includes individualized therapeutic sessions as well as ongoing group meetings for ex-gang members. This phase is designed to be a relapse or recidivism prevention mechanism for prior gang members.

One aspect of this intervention program that is particularly effective is the use of group facilitators who are also prior gang members (Evans & Sawdon, 2005; Hanser, 2007b). These staff members maintain community connections by visiting local community centers and other youth services and/or recreational establishments to provide information pertaining to the program. Staff members are likewise given leadership skills training, empathy building, counseling, and the development of their own personal stories that help the juveniles to identify with staff members during their outreach role as well as their role as facilitator during group sessions (Evans & Sawdon, 2005; Hanser, 2007b).

Overall, this program has been found to be quite effective in terms of both treatment and outreach to gang members in the community. In addition to the initial treatment regimen and the relapse sessions, this program attempts to separate juveniles from their gang-oriented surroundings. This is often necessary because prior gang members must otherwise combat members from their prior gangs. Furthermore, the tug of the gang subculture tends to be powerful, and this makes the juveniles more likely to recidivate (Hanser, 2007b). Thus, the gang exit strategy will employ a combination of individualized and group interventions while relocating prior gang members to an area that is not as likely to pull the youthful offenders back into gang activity (Evans & Sawdon, 2005; Hanser, 2007b).

Career Opportunity—DARE Officer

Job Description: Work full-time as certified police officer teaching the DARE program curricula to local school children from fifth grade through high school. Act as role model in the classroom, drawing attention to the hazards of drug and alcohol use, peer pressure, and violence as well as the issues involved in racial and gender stereotyping. Trained in the DARE curricula at a training center sponsored by the national DARE program.

Employment Requirements: Required to be full-time uniformed police officer with at least 2 years of policing experience. Must have met at least the minimum training standards set by the local policing agency, and must go through an oral interview during the DARE screening process. In addition, the officer and his or her agency must have an agreement with the local school district to teach the DARE curricula at the school. Other qualities important to the position include a demonstrated ability to interact and relate well with children, good oral and written communication skills, organization skills, promptness, the ability to develop personal relationships, and flexibility and ability to handle unexpected situations, statements, and actions. Must successfully complete 80 hours of training to be certified to teach the core curriculum and the K–4 program. If teaching at the high school level, in the DARE parent program, or in special education classes, must complete an additional 40 hours of training and teach the core curriculum for at least two semesters prior to the additional training.

Salary and Benefits: Salary varies from jurisdiction to jurisdiction. Good benefits and retirement packages are typically included.

SUMMARY

Violence by and against juveniles has received considerable attention over the past decade or so. Stories appearing in the mass media have led many to believe that violence committed by juveniles is an epidemic, and current official statistics and other sources of information indicate that violent acts committed by juveniles have indeed increased during recent years. There is little doubt that those juveniles who commit violent offenses deserve our immediate attention because research indicates that they are likely to continue to commit such acts unless early effective intervention occurs.

Even though there have been past declines in violent crimes by juveniles, a substantial proportion of violent crime is still attributed to them. Furthermore, there is reason to believe that juvenile violent crime is now increasing. For those juveniles who do commit violent offenses, incarceration or effective intervention very early in the offending career may well be the best means of protecting society. Evidence indicates that unless one of these two alternatives is employed, recidivism is very likely.

Careful screening of juveniles to ensure that only those who actually commit violent acts are processed according to laws intended to deal with such offenders is imperative in terms of costs to both the juveniles involved and society.

Our society is confronted by a multitude of problems relating to gangs. Preventing juveniles from becoming involved in gang activities, particularly in inner-city neighborhoods, is extremely difficult if not impossible. Juveniles who do not join gangs voluntarily risk their lives as well as the lives of their family members. Thus, early identification of new recruits and comprehensive knowledge concerning the membership and actions of existing gangs are essential. Identification of juveniles who are in the process of becoming gang members may be accomplished through a variety of means. Sudden changes in friendships; minor but chronic problems with police, school, and family; wearing the same color patterns daily (although colors now appear to be diminishing in importance as a symbol of gang membership); discovery of strange logos/insignia on juveniles' bodies, notebooks, or clothing; use of new nicknames (monikers); flashing of hand signs; and unexplained money may be signs of impending or actual gang involvement (Hieb, 1992). Spotting the signs of pre- or early gang activity is, of course, largely up to parents and teachers, who then need to take appropriate action to address the issue (Corbitt, 2000).

Even early intervention does not ensure that gang influence will be reduced given that the juveniles in question are most likely to be returned to the neighborhoods in which the gangs operate or, in some cases, to a correctional facility that is also largely controlled by gangs. Incarceration of adult gang leaders may have some impact, but evidence indicates that these leaders often continue to control gang activities on the outside while they are in prison and frequently control the gangs within the prisons themselves.

The best available strategy is to identify the signs of gang activities as early as possible and prosecute gang members to the full extent of the law so as to send gang leaders the message that their actions will not be tolerated by the community. These programs are often referred to as "zero tolerance" programs, meaning that no amount of gang activity will be accepted. Such action by the community and the justice network may persuade gang leaders looking to expand their spheres of influence to move elsewhere. Where gangs are already clearly established, as in most metropolitan areas, a massive coordinated effort addressing socioeconomic conditions as well as criminal behavior will be required if gang behavior is to be brought under some degree of control. Some such efforts are now being made, and careful evaluation of their impact is crucial.

Gang activities have a long history in the United States, but attention has been redirected toward gangs recently as a result of their involvement with drug trafficking and gunrunning, which are multimillion-dollar enterprises. The complexion of gangs has changed somewhat over the years, and referring to gangs as "juvenile" gangs is not totally appropriate at this time due to the strong influence of adult gang leaders who supervise, organize, and control gang activities.

Juveniles continue to join gangs to attain status and prestige lacking in the domestic and educational arenas. Gang members continue to fight territorial wars, wear colors, extort protection money, and exclude from membership those from different racial or ethnic groups. Gangs exist in all urban areas, have extensive organizations in most prisons, and are spreading out to medium-sized and even smaller cities.

Gang involvement in violent activities—sometimes random and sometimes carefully planned—has received a good deal of attention from both the media and justice officials. The latter are organizing to better combat gang activities, but their success has yet to be carefully evaluated. Similarly, "get tough" legislation has been passed at all levels, but the impact of such legislative action remains in question.

Note: Please see the Companion Study Site for Internet exercises and Web resources. Go to www.sagepub.com/juvenilejustice6study.

Critical Thinking Questions

1. Is violence committed by juveniles on the increase in the United States? Support your answer. In your opinion, are adults in the United States afraid of juveniles? Should they be?
2. Is probation likely to be effective in deterring violent juveniles from recidivating? Why or why not? Are there more effective programs for deterring violent juveniles?
3. What are the relationships among guns, drugs, and violence? Would gun control cut down on the number of violent crimes committed by juveniles? Against juveniles?
4. Describe the conditions under which gang membership is most likely to be attractive to juveniles. What kinds of responses do we, as a society, need to make to help control gangs?
5. Are there major differences in reasons for joining gangs and behaviors engaged in while in gangs between male and female gang members? Are female gang members more similar to or different from their male counterparts today regarding criminal activity?

Suggested Readings

Bilchik, S. (1999, December). *1997 National Youth Gang Survey.* Washington, DC: U. S. Department of Justice.

Burch, J., & Kane, C. (1999, July). *Implementing the OJJDP comprehensive gang model* (OJJDP Fact Sheet No. 112). Washington, DC: U.S. Department of Justice.

Campbell, A. (1991). *The girls in the gang* (2nd ed.). Cambridge, MA: Basil Blackwell.

Corbitt, W. A. (2000). Violent crimes among juveniles: Behavioral aspects. *FBI Law Enforcement Bulletin, 69,* 18–21.

Curry, G. D., & Spergel, I. A. (1988). Gang homicide, delinquency, and community. *Criminology, 26,* 381–405.

Davis, N. (1999). *Youth crisis: Growing up in a high-risk society.* Westport, CT: Praeger.

Dawley, D. (1973). *A nation of lords: A history of the Vice Lords.* Garden City, NY: Anchor Books.

Esbensen, F., Deschenes, E. P., & Winfree, L. T. (1999). Differences between gang girls and gang boys: Results from a multisite survey. *Youth & Society, 31,* 27–53.

Evans, D. G., & Sawdon, J. (2004, October). The development of a gang exit strategy: The Youth Ambassador's Leadership and Employment Project. *Corrections Today.* Available: www.cantraining.org/BTC/docs/Sawdon%20Evans%20CT%20Article.pdf

Evans, W. P., Fitzgerald, C., & Weigel, D. (1999). Are rural gang members similar to their urban peers? Implications for rural communities. *Youth & Society, 30,* 267–282.

Hanser, R. D. (2007). *Special needs offenders in the community.* Upper Saddle River, NJ: Prentice Hall.

Hunter, J. A., Hazelwood, R. R., & Slesinger, D. (2000). Juvenile sexual homicide. *FBI Law Enforcement Bulletin, 69,* 1–9.

Mays, G. L. (1997). *Gangs and gang behavior.* Chicago: Nelson–Hall.

Miller, W. B. (2001, April). *The growth of gangs in the United States: 1970–1998* (OJJDP Report). Washington, DC: U.S. Department of Justice.

National Criminal Justice Reference Service. (2005). *Gangs: Facts and figures.* Washington, DC: U.S. Department of Justice. Available: www.ncjrs.org/spotlight/gangs/facts.html

Osgood, D. W., & Chambers, J. M. (2000). Social disorganization outside the metropolis: An analysis of rural youth violence. *Criminology, 38*, 81–115.

Pattillo, M. E. (1998). Sweet mothers and gangbangers: Managing crime in a black middle-class neighborhood. *Social Forces, 76*, 747–774.

Shure, M. B. (1999, April). Preventing violence the problem-solving way. *Juvenile Justice Bulletin.* (Washington, DC: U.S. Department of Justice)

Wilson, J. J. (2000, November). *1998 National Youth Gang Survey* (OJJDP Summary). Washington, DC: U.S. Department of Justice.

Valdez, A. (2005). *Gangs: A guide to understanding street gangs.* San Clemente, CA: LawTech Publishing.

Yablonsky, L. (1970). *The violent gang.* Baltimore, MD: Penguin Books.

THE FUTURE OF JUVENILE JUSTICE

CHAPTER LEARNING OBJECTIVES

On completion of this chapter, students should be able to

- *Evaluate the extent to which the goals of the juvenile justice network have been met as we enter the 21st century*
- *Suggest alternatives for the future of the juvenile justice network*
- *Explain the restorative justice and "get tough" approaches to juvenile justice*
- *Discuss the possible demise of the juvenile justice network*
- *Discuss ways of improving upon the current juvenile justice network*

KEY TERMS

Hybrid sentencing

Restorative or balanced justice

"Get tough" or "just deserts" approach

Revitalized juvenile justice

Throughout this book, we have discussed in varying detail the philosophies of the juvenile justice network, the procedural requirements of that network, and some of the major problems with the network as it now operates. We have seen that the juvenile justice system is subject to numerous stresses and strains from within. The good intentions are sometimes met with less than desirable results; thus, change and trial and error are imperative. We have arrived at a number of conclusions, some of which are supported by empirical evidence and others of which are more or less speculative, based on our observations and those of concerned practitioners and citizens.

The initial underlying assumption of the juvenile justice network is that juveniles with problems should be treated and/or educated rather than punished. Adult and juvenile justice networks in the United States were separated because of the belief that courts should act in the best interests of juveniles and because of the belief that association with adult offenders would increase the possibility that juveniles would become involved in criminal careers. The extent to which we have achieved the goals of the juvenile justice network continues to be debated. Where do we go from here? Although it is always risky to speculate, it appears to us that there are four more or less distinct possibilities for the future of the juvenile justice network.

Possibility number one is that the juvenile justice network will cease to exist as a separate entity. Possibility number two is that the juvenile system will begin to rely heavily on **hybrid sentencing** with regard to violent juveniles. Possibility number three is that the **restorative or balanced justice** movement (see Chapter 10) will triumph and we will return to a more caring personal approach to juvenile justice. Possibility number four is that those favoring a "get tough" approach will be victorious and the goals of the juvenile justice network will change dramatically. As Lindner (2004) pointed out, the juvenile court today is more focused on accountability and punishment than ever before. Thus, possibility number four is that the **"get tough" or "just deserts" approach** supporters will reform the juvenile justice system so that an increasing number of juveniles are dealt with in the adult justice network. "A popular view [among policymakers and supporters of "get tough"] is that, to protect the public safety, youth who commit serious or violent acts should be subject to the same punishments as apply to adults" (Bishop, 2004, p. 635). Although many states have adopted this approach and are now operating in full swing with transfers to adult court, this philosophy is contradictory to the purpose of juvenile court.

> There is little argument that the current juvenile justice system is indeed in turmoil and lacks the foresight and preventive measures required for lasting reform. . . . The challenge before us is to move from the rhetoric to the reality of what we are going to do to save their [juveniles'] lives and our collective futures. (Hatchette, 1998, pp. 83–84)

If this sounds familiar, it may be because it has been the theme throughout this text.

As Ohlin (1998) noted, "Confidence in the ability of our institutional system to control juvenile delinquency has been steadily eroding. Public insecurity, fear, and anxiety about youth crime are now intense and widespread, despite the juvenile court and probation system and the training schools that have evolved over the past century" (p. 143). Lindner (2004) echoed this statement by pointing out that public perception has been that juvenile crimes are becoming more and more serious. The public has also perceived that the juvenile justice system is too lenient on offenders. Public outcry has led to more punitive laws, increased waivers to adult court, and a shift from the best interests of the child to punishment. In actuality, however, "the rate of juvenile violent crime arrests has consistently decreased since 1994, falling to a level not seen since at least the 1970s" (Snyder & Sickmund, 2006, p. 2). Unfortunately, "get tough" proponents are not choosing to advertise the decreased crime rates; instead, they have focused on the upswing in juvenile violence over the past year.

Uniformity of juvenile law has yet to be achieved. Many citizens still adopt an "out-of-sight, out-of-mind" attitude toward juveniles with problems, and both citizens and practitioners are often frustrated by our supposed failure to curb delinquency and abuse or neglect in spite of the millions of dollars invested in the enterprise. We continue to refuse to address the larger societal issues of race, class, and gender as they relate to crime and delinquency (Ohlin, 1998, p. 152). Susman (2002) discussed racism in sentencing, legislation, and the focus on crime and minority offenders (see In Practice 13.1).

▲ In El Salvador, members of the Mara Salvatrucha in the Tonacatepeque penitentiary for underage prisoners are shown.

There is undoubtedly room for a great deal of improvement in juvenile justice. To some extent, such improvement depends on changes in societal conditions such as poverty, unemployment, and discrimination. Changes in the family and in the educational network that improve our ability to meet the needs of juveniles are also crucial. Changes in the rules that govern juveniles may be appropriate in some instances. Making better use of the information made available to us by researchers and practitioners is yet another way to improve the network. In the end, taking a rational, calculated approach to delinquency and abuse or neglect will pay better dividends than will adhering to policies developed and implemented as a result of fear and misunderstanding.

Bilchik (1998a) concluded, "A revitalized juvenile justice system needs to be put into place and brought to scale that will ensure immediate and appropriate sanctions, provide effective treatment, reverse trends in juvenile violence, and rebuild public confidence in and support for the juvenile justice system" (p. 89). Such a **revitalized juvenile justice** system would include swift intervention with early offenders, an individualized comprehensive needs assessment, transfer of serious or chronic offenders, and intensive aftercare. This system would require the coordinated efforts of law enforcement, treatment, correctional, judicial, and social service personnel. This approach to delinquency control represents a form of community programming that might help to reintegrate troubled youth into mainstream society rather than further isolate and alienate them (Bazemore & Washington, 1995; Zaslaw & Balance, 1996). To accomplish this goal, the resistance to change that characterizes most institutions must be overcome.

(Text continues on page 316)

In Practice 13.1

Doubting the System: Laws on Juveniles Stir Debate Over Punishment and Racism

Maybe if she had been stricter with him. Maybe if she hadn't married an abusive man. Or maybe if she hadn't let him live with a dad who was a drug addict and an uncle who she says beat him up. Maybe if, as a 16-year-old girl growing up in Roosevelt, Long Island, she had stayed in school and not dropped out to have a baby.

Kim Williams recites the "maybes" of her life like a wistful mantra as she discusses her son, Glenn Sims, who was 16 in November 1998 when he was arrested in an Atlanta suburb and charged with holding up a convenience store. Williams was stunned. Sims was no angel, but Williams said she'd never expected him to be accused of such a thing.

"He knew better because we never raised him that way," said Williams, sitting in the small, one-bedroom apartment she was sharing with her 13-year-old daughter until money worries led them to move in with her parents in May. Williams assumed that Sims' young age would bring time in a juvenile facility or perhaps house arrest and probation. She urged him to confess, thinking honesty would be rewarded with lighter treatment. "I thought there was some kind of a law for first-time offenders," she said.

What Williams didn't know, and what she and thousands of other parents have come to find out, is that under a law passed with little fanfare in 1994 under the benign name of the "School Safety and Juvenile Justice Reform Act," Sims was an adult in the eyes of Georgia's penal code. His guilty plea brought a mandatory 10-year prison term with no chance of parole.

"He was scared, and I'll tell you, when the judge said 10 years, the tears fell out of his eyes," Williams said. "I cried, I guess, for about two years," she added, only half-jokingly.

And like most youths caught up in the law, charged and convicted of one of the "seven deadly sins"—murder, voluntary manslaughter, rape, armed robbery, aggravated sodomy, aggravated child molestation, and aggravated sexual battery—that automatically turn 13- to 16-year-olds into adults in Georgia's courts, Sims is black. Since Senate Bill 440 was passed in May 1994, 76 percent of the more than 3,800 teenagers arrested for SB440 offenses have been black, although [blacks] comprise 34 percent of the state's teenage population, according to the Georgia Indigent Defense Council, a state agency that tracks SB440 cases.

Proponents of keeping juveniles in juvenile courts say the nationwide trend toward pushing them into the adult system is tainted by racism—sometimes blatant, sometimes unintentional—because the laws are passed by mainly white legislatures and enforced by judges and prosecutors, the overwhelming majority of whom are white.

"There really is in operation an unspoken sense that these are throwaway kids. If they didn't think they were throwaway kids, they wouldn't treat them that way," said Malcolm Young, executive director of the Sentencing Project, a Washington, D.C., think tank that studies criminal justice issues.

Proponents of tougher treatment for juveniles deny institutional racism and say black teens simply commit more, and more serious, crimes. "If you took all the black males between the ages of 16 and 25 and put them on an island in the Pacific, crime would drop 80 percent overnight," said Brian Silverman, the former chief of the juvenile division at the Cook County, Ill., public defender's office, who acknowledges

(Continued)

(Continued)

the racial disparities in prisons. "I'm not convinced it's caused by racism. If it is, that's bad and we should do something to change it. But if the cause is something other than racism, then it's a problem maybe society can't handle."

Researchers say that both sides are looking at it too simplistically, that factors such as policing methods and the prior records juveniles bring into court must be considered when weighing the level of racial bias.

Nobody denies the numbers, though, and even if blacks do account for a higher number of teens arrested for violent crime, figures from the nation's 75 largest counties in 1998 indicated a difference in the treatment of them. In the counties surveyed, whites were 48 percent of the youths charged as juveniles with violent crimes—exactly the same as the figure for blacks. They exceeded blacks charged with murder in juvenile court, 59 percent to 36 percent.

But in those same counties, whites accounted for 25 percent of juveniles charged as adults with violent crimes, compared with 73 percent for blacks.

For nonviolent drug and public order offenses, the disparities were greater. Building Blocks for Youth, a Washington, D.C.-based organization that studies juvenile justice issues, surveyed 18 counties in 1998, including Queens, Bronx, Kings, and New York in New York State, and found that while blacks were 64 percent of youths arrested for felony drug offenses, they represented 76 percent of the drug offenses handled in adult courts. They were 68 percent of those arrested for public order crimes such as gun possession and rioting, but accounted for 76 percent of youths charged as adults with such offenses.

In many states, the population of black youths in adult prisons is several times that of white youths. In Georgia, blacks account for 83 percent of the minors in adult prisons and whites 17 percent. In Alabama, Mississippi, South Carolina, and Virginia, blacks account for about 75 percent of the minors in adult prisons, while whites are about 20 percent. Studies show that minors housed in adult prisons are several times more likely to be physically assaulted and to attempt suicide and are more inclined to commit crimes again once they have been released.

Yet throughout the 1990s, state after state passed laws making it easier to try young offenders as adults, spurred in part by criminologists' thunderous warnings that social and demographic changes had combined to create a new breed of child and teenage "superpredator."

In fact, juvenile crime has dropped since the theory swept the country, following an increase in the 1980s and early 1990s that was blamed on factors ranging from the crack cocaine epidemic to the economy. According to the FBI [Federal Bureau of Investigation], violent crime arrest rates for youths ages 15 to 17 declined 44 percent from 1994 to 2000. Criminologists' ominous words, however, had struck a chord with legislators who had witnessed rising juvenile crime rates, including Georgia's then-Gov. Zell Miller. His state saw juvenile arrests for violent crime jump from about 1,900 in 1990 to 3,000 in 1993.

"Kids who commit adult crimes ought to be tried and sentenced as adults," Miller, now a U.S. senator, said after the bill passed in 1994. Among other things, he said the law would let teachers focus on teaching rather than "worrying about whether little Johnny is packing a gun."

Critics of such laws say that they were thrown together haphazardly by politicians more concerned about public support than public safety and that there was no thought of what sort of people the jails would turn loose on society once the young prisoners' sentences were fulfilled. But for legislators pressured by public fears of crime, repealing such laws, even those that have proved to be the most racially skewed, is out of the question, leaving thousands of young black men such as Sims to spend their developmental years behind bars.

(Continued)

(Continued)

"The unfortunate thing is that it's across-the-board mandatory. Some child who might benefit from counseling or treatment won't get it. Juvenile facilities offer schools, mental help, [and] counseling to a much greater degree than an adult would get," said Susan Teaster of the Georgia Indigent Defense Council, who was a public defender when SB440 passed. "They weren't thinking about the consequences 10 years down the road when they passed this law. You tell a 13- or 14-year-old he's getting 10 or 12 years in jail, and you're taking all hope away from him. We'll be looking at people who've been incarcerated all the years when they're supposed to learn to be productive members of society. They'll have been surrounded by other people convicted of crimes. They'll have no probation, no parole, and they'll have no idea how to function in society."

The Office of Juvenile Justice and Delinquency Prevention, part of the Department of Justice, said a host of unforeseen problems had arisen as more states treated juvenile offenders as adults. These included the issue of how to obtain quick medical treatment in prison for minors needing parental consent for surgery or medication, the added workload on criminal courts forced to handle formerly "juvenile" offenders, and the addition of thousands of new and immature inmates to already crowded prison systems. It also noted the overrepresentation of minorities among juveniles charged as adults, and it questioned the deterrent effect of the laws, saying few states had taken steps to educate juveniles of the sanctions they might face from breaking the law.

Even get-tough advocates who favor trying some juveniles as adults, such as Silverman and Peter Reinharz, a former prosecutor in Manhattan who was chief of the city's family court, say that laws that treat all juveniles the same without considering an individual's circumstances, and that impose minimum sentences without allowing judges to consider alternative punishments, are misguided.

"Automatically treating them as adults in every situation is ludicrous," said Reinharz, now the managing attorney for Nassau County. "If someone is 14 and doing an armed robbery, I certainly don't want to be in the community with him. Something has to be done. On the other hand, the idea of saddling this guy with a felony conviction for the rest of his life is really stupid. He's going to wind up not being able to get a job and having no other choices down the road than a life of crime. There has to be some common sense kind of plan, and unfortunately there isn't a lot of common sense type of planning with these laws."

Such arguments haven't convinced Miller, who said Georgia's drop in juvenile arrests since 1994 proves the law is deterring young offenders. "The juvenile justice system we had was not adequate to handle the violence of today's young criminals," Miller said in a written reply to questions. "These are not the Cleaver kids soaping up some windows. These are middle school kids conspiring to hurt their teachers, teenagers shooting people and committing rapes, young thugs running drug gangs and terrorizing neighborhoods."

As for the racial disparity in SB440 offenses, Miller said there will always be critics who oppose all punishment for young offenders, black or white. "I trust the prosecutors and judges in my state and believe they have exercised the proper discretion in enforcing this law."

Not everyone believes this, least of all parents of black teenagers jailed in Georgia. They share the belief that because of race, their sons were caught in a system that equates young black men with thuggery and assumes black teens arrested for serious crimes ended up in jail because they came from ghetto neighborhoods without adult supervision. And while they all acknowledge that their sons are not blameless, they agree with the experts who say putting them in prison for at least 10 years for crimes committed as teenagers will do more harm than good.

(Continued)

(Continued)

"When you're dealing with children, it's just so idiotic," said Billie Ross, the president of Mothers Advocating Juvenile Justice, a group of parents with children convicted under SB440. Ross is convinced that if her son were white, police and prosecutors would have considered his parents' professional backgrounds and their stable lifestyle when they arrested him in 1996 for armed robbery. Glenwood Ross III, better known as Trey, was a slightly built 15-year-old then. He was convicted and sentenced to the mandatory 10 years. His mother was spurred to get her master's degree in social work, in part, by Trey's arrest. His father, Glenwood Ross, is a professor of economics at Morehouse College in Atlanta.

"I think they just have this box that they put a lot of these black kids into, and he fell into it," Billie Ross said of Trey. "Let's face it, if you're a black person in America, particularly a young black man, you have to worry about this."

Few did, however, until they collided with laws such as SB440.

Williams never imagined she'd spend a decade of weekends and holidays crisscrossing Georgia's rolling, monotonous countryside to visit Sims in prison for a crime committed when he was 16. She never imagined he would be forced to fight off rapists in the first prison where he was held, the maximum-security Arrendale State Prison in Alto, Ga., where all boys convicted as adults are housed.

"He got cheated because I was young when I had him," said Williams, who was 16 when Sims was born. "All I knew was partying and running in the streets. I wasn't the best parent. I was too young to know what a parent was."

That's not to say Sims was trouble free or that Williams considers him without fault. The move from Long Island to Georgia when Sims was 10 didn't go well for the boy who preferred hanging out with his cousins and friends back in New York. He missed going out for Chinese food, ice skating, playing sports on his old school teams, and the snowy winters. He also missed his father, who never married his mother but who lived around the corner from them in Roosevelt. Life in Atlanta seemed dull by comparison.

"He came down here with that New York mentality, that no one could tell him what to do. And he found out that Georgia doesn't play that way," said Williams, whose cheerful nature belies the difficulties of her life. Now 36, she works as a secretary by day and a security guard on weekend nights to support herself and her daughter and to set aside money for both kids' futures. At least twice a month and on holidays, she makes the two-hour drive to Hancock State Prison in Sparta to spend a few hours with Sims, who was transferred there in October 2000 after nearly two years at Arrendale.

The visits aren't easy. "My mom works too hard to be coming up here every two weeks," said Sims, a tall, gregarious young man who wears the outfit issued to all inmates: white pants with a black stripe down the side and a short-sleeved white shirt with a stripe up the front. "She cries a lot. I tell her, 'look, I did what I did, and I'm gonna do my time for it.'"

For Williams, Sims' arrest was a crushing blow in a years-long battle to establish a stable life. Her first husband, to whom she was married for seven years, beat her so badly, she said, that she fled with the children to a shelter. Sims said he remembers his first stepfather pummeling his mother until she was bruised and bleeding, even going to her workplace to drag her out and hit her.

Thus began a series of moves, which ended in 1990 when Williams decided to follow other family members who had gone to Georgia. As Sims struggled with the upheaval, she granted his wish to return to Long Island to live with his biological father for a while. When his father was jailed for drugs, Sims' uncle was in charge. His answer to Sims' behavior problems was violence.

(Continued)

(Continued)

"His uncle wasn't into 'let's whip you with a belt.' His uncle was into 'let's beat you up with the fists,'" Williams said.

By the time Williams brought Sims back to Georgia, his path to disaster was set. "I kept hate in my heart. Hatred, hatred, hatred," he says now. "I don't know if it just came out in my teens or what." He returned to Atlanta more troubled than when he had left.

"He [Sims] wanted to stay out all night. I'd go check on him and find a bunch of clothes under the sheets to make it look like someone was in bed," Williams said, laughing at the things that angered her then but that now bring back memories of a time when Sims was home, not sleeping on a prison cot.

When the police arrived at Williams' home that night in November 1998 to question her son, Sims admitted his involvement in the robbery. Williams and her husband thought it best to come clean. They hoped a judge would take Sims' youth, turbulent past, and family ties into consideration. They didn't know about SB440.

"We really, really did not put up a fight," Williams said. "I don't know what I was thinking when I had him confess. Something in my mind told me they were not going to take that boy away. I knew he'd get time, but I didn't think it would be 10 years. Maybe five years tops."

Sims is scheduled for release in December 2008. He will be 26 years old. His grandmother has a room set aside for him in her house. His mother has been trying to gently nudge him to think about going to college when he gets out. "We're hoping to just shelter him with love," she said.

If studies of offenders are anything to go by, Sims will have a hard time avoiding trouble again. Research in New York, New Jersey, and Florida comparing re-offense rates among juveniles treated in adult versus juvenile courts shows that those sent to criminal court are more likely than others to become repeat offenders. "It's scandalous. We're sowing the seeds of having these kids grow up to be violent criminals," said Herschella Conyers, a public defender in Cook County, Ill., where a law mandating adult trials for youths accused of certain drug and weapons offenses has been criticized for producing racially skewed results. One portion of the law requires adult court for juveniles charged with drug and weapon offenses within 1,000 feet of public housing projects, areas that are predominantly black. More than 90 percent of the teenagers convicted as adults under the law are black.

"It's really incredible," Conyers said. "We *can* do better things than that with kids, even if they're guilty. It's just crazy. The consequences are both inhumane and disastrous."

Even the Georgia Department of Corrections, whose prison system has absorbed at least 574 minor inmates since SB440 took effect, acknowledges it was passed without much consideration for the pasts and futures of the youngsters caught up by it. Whereas lawmakers saw 13-, 14-, and 15-year-olds as street-smart and mean as adults, they didn't take into consideration other factors, said department spokesman Mike Light. "What we see is kids who've led very hard lives, for the most part, and who've made bad decisions. What drives these individuals to commit crimes so early? What are we going to do with these kids?" asked Light, who worries that unless such issues are addressed, prison could become the "grad school" for the next level of crime.

With a growing inmate population and limited staffing and resources, the special needs of children behind bars, such as mentoring, counseling, and education, simply cannot be adequately met, he said.

"I'm surprised he can read at all," Williams said of Sims shortly before his last birthday. "He's going to be 20 years old on Aug. 9 without an education, and that really hurts me."

(Continued)

(Continued)

"I wouldn't feel so bad if they prepared the kids for when they get out," Glenwood Ross, Trey's father, said bitterly. "But these kids are just sitting around doing nothing, so what's society going to get when they come out? A bunch of dummies who've been sitting around doing nothing."

Still, there appears to be little chance laws such as SB440 will be overturned. Child advocates concede that they are fighting an uphill battle. Families of children in prison represent a tiny constituency, and they are trying to persuade people who have no interest in juvenile justice issues to rally on behalf of juveniles accused of serious crimes.

"We're not saying don't punish the kids. We're saying use discretion, use judgment. But people don't want to hear that," said Billie Ross, who admits that until Trey was arrested, she had no interest in juvenile justice issues. "It wasn't anything I felt like I needed to be concerned with. Then all of a sudden this hit me. I guess I'm like a lot of other people—dumb and blind to a lot of what's going on until it strikes you directly."

Mothers Advocating Juvenile Justice is lobbying to introduce a parole possibility for juveniles convicted under SB440 but hasn't gotten far. The proposed amendment, House Bill 269, never made it out of the legislature's judiciary committee last year.

Percentages of Arrests by Race Under Georgia's SB440 Law for 1994–2002

	Percentage	
Year	**White**	**Black**
1994	16	83
1995	17	83
1996	19	78
1997	20	78
1998	23	75
1999	21	75
2000	25	68
2001	24	70
2002	23	70

SOURCE: Georgia Indigent Defense Council.

NOTE: Some juveniles may be arrested with more than one offense, so the percentages may exceed 100%.

SOURCE: Susman, Tina. (2002, August 21). Doubting the System: Laws on juveniles stir debate over punishment and racism. *Los Angeles Times*.

(Text continued from page 309)

Cohn (2004b) added to this by claiming that revitalizing the system includes focusing on "areas of attention and repair" (p. 43) such as the following:

1. Excessive caseloads that preclude meaningful interventions by an assumed well-trained staff
2. The failure to recognize that not the offender but the community is the real "customer" of services
3. Little understanding of the relationship between planning, change, and social policy
4. Political, hard-line rhetoric leading to inappropriate changes in the juvenile code, including automatic waivers to criminal courts
5. The changing character of youthful offenders, especially in terms of substance abuse and the use of weapons when committing offenses
6. Inadequate kinds and availability of treatment programs for detained youths and for probationers and those in after care and, for those that work, inadequate replication efforts
7. A stubborn refusal by some significant actors to work collaboratively in defining and resolving problems of a mutual nature such as diversion or graduated sanctions
8. The failure of the "leadership" to deal head-on with inappropriate and "wrong" changes in juvenile codes, especially in terms of waivers
9. The failure of judges to provide the leadership when it comes to advocacy
10. The lack of meaningful diversion programs, including the need for more informal processes for nonviolent offenders
11. The lack of meaningful programs that involve the families (parents) of offenders
12. The failure to engage in advocacy for needed programs in the community for youth and their families
13. Too much complacency (status quo), which results in lack of appropriate planning
14. The failure to involve critical stakeholders in the development of agency-based policies and procedures
15. Inadequate involvement of subordinate staff in identifying and implementing agency mission and goals
16. Inadequate staff development and training programs that are based on the identification of core competencies
17. The failure to recognize the need for and value of wraparound services and programs, resulting in poor case management by too many case managers in too many agencies in any given case situation (i.e., the failure to recognize that too few offenders and their families receive disproportionately high levels of human services)

18. Inadequate development of ongoing and meaningful communication with superordinates and appropriate stakeholders, which results in too many being unaware of "what works"
19. The failure to design and implement a total, systems-based information technology program that enhances data sharing
20. The ongoing failure to evaluate programs to determine worthiness that should lead to decisions about program continuation, expansion, or abandonment
21. The failure to develop and implement program and operational standards
22. The failure to think systemically
23. The problem of too much stupidity![1]

Basically, revitalizing the system means acknowledging that changes are needed within the structure of the system and within the individual agencies operating as part of the system. It does not mean focusing on harsher punishments or accountability. Revitalization puts juveniles and treatment first while fixing the broken aspects of juvenile justice. As Califano and Colson (2005) indicated, the goal of the juvenile justice system is protection and reform of young people who commit crimes, "but the reality is a grim, modern version of Charles Dickens' *Oliver Twist.* Instead of providing care and rehabilitation, many facilities are nothing more than colleges for criminality" (p. 34).

We believe that the first possibility, the complete demise of the juvenile justice network, is unlikely simply because the goals of the network are worthy and not likely to be abandoned completely. In addition, the separate juvenile justice system employs a large supporting staff that is unlikely to be summarily dismissed, especially in light of the restorative justice movement. Based on current trends, it appears likely that the juvenile justice network will continue to provide separate services for juveniles, but with an increase in the number of crimes for which waivers to adult court are used and with an increase in hybrid sentencing. Moore (2001/2002) suggested that the use of blended sentences is creating a third criminal justice system. As discussed previously, blended sentences allow juveniles to be sentenced to adult and juvenile dispositions by both adult court judges and juvenile court judges. This raises the question of "when is a juvenile offender a juvenile and when is he or she an adult?" Instead of classifying juvenile offenders as one or the other, hybrid sentences allow the convenience and benefits of both courts. Juveniles are given a second chance by being allowed to rely on the juvenile court for rehabilitation while knowing that the adult court sanctions apply if rehabilitation fails. "By focusing on blending both juvenile and adult criminal sentences, the courts are moving away from an all-or-nothing approach and recognizing the lack of a bright-line rule to define a juvenile versus an adult" (p. 138).

It is unlikely that restorative justice advocates will withdraw from the field of battle in the juvenile justice network. Along with faith-based initiatives, they have gained momentum as we enter the 21st century because they have focused on caring for victims too long forgotten by those in the juvenile justice network, and that is not likely to change in the near future. They have successfully infiltrated every type of agency working with juveniles (see In Practice 13.2), and in a way the policies they advocate fit rather well with the "just deserts" model because, in addition to being processed through the courts, offenders are forced to confront their victims

In Practice 13.2

Restoring Justice Where Gangs, Not Suspensions, Define Fear

It's 9:47 on a Tuesday morning at Fenger Academy, a sprawling three-story brick building that straddles two city blocks in the Far South Side neighborhood of Roseland.

Students are flooding out of their second-period classes when two groups of rival gang members start throwing punches in the second-floor hallway. Instantly, the swarm in front of the library swells dangerously as students start running toward the fight.

This is an all-too-typical scenario at city high schools such as Fenger, which last year had one of the highest rates of violent incidents and student arrests in Chicago Public Schools.

In the past, this kind of fight most likely would have turned bloody, emptied the school, and ended with carloads of children hauled off in handcuffs. That's because too many school administrators respond to student violence either by ignoring it or by cracking down with the harshest possible punishment.

This fight didn't end that way because of a new approach to dealing with violence, called "restorative justice," that emphasizes healing over punishment. After the district started leaning on those take-no-prisoners schools, rewrote its student discipline policy, and embraced this philosophy, student arrests dropped significantly. There were 7,400 arrests during the 2005–2006 school year, a decrease of 13 percent from the 8,500 arrests the previous school year and down nearly 20 percent overall from the high of 8,900 in 2003.

At Fenger, school leaders know it's flat-out foolish to expect violence to disappear. This is a tough place, where gang clashes are inevitable and students fear getting "punked" more than they do a suspension or bad grades.

So, administrators strike a middle ground between paralysis and rigidity and throw themselves into the fray.

A minute after the morning fight erupts, Fenger security guards and other staff tear through the crowd and start peeling the students apart, pinning their arms down and pushing them in the opposite direction.

Al Cruise, dean of students and head football coach, takes an errant punch on his left cheek when he steps between two flailing sophomores. Principal William Johnson runs up the stairs and plants himself at the edge of the chaos, ordering onlookers back to class with a loud voice and a few choice curse words.

One beefy security guard grabs three students and drags them down the stairs toward the discipline office. The fight is over, but students still are milling suspiciously in the hallway after the passing bell.

So, the principal orders a hallway sweep, giving every student one minute to get to class or face suspension. Teachers then lock their doors and stand at the entrance. No one gets in and no one leaves. Ten minutes after the fight erupted, the second-floor hallway is empty and quiet.

One student got knocked down and got kicked hard enough that his face showed the imprint of a sneaker. There are bruises and a few swollen eyes, but no blood was shed. No one ran down the hall to pull

(Continued)

(Continued)

a fire alarm, which would have emptied the school and triggered more fighting in the courtyard. This chain reaction was a common occurrence last year, when students set off about two dozen prank fire alarms.

With the fight over, administrators have some choices to make. The easiest would be to dump the problem on Chicago police [and] let them make the call about whether to haul the group off to jail on misdemeanor battery changes.

Johnson tries a different tack.

He and other school officials reach out to leaders of both gangs to understand what triggered the fight and to prevent it from spilling over into more dangerous retribution after school and in coming days. As it is, two of the teens targeted in the morning fight fled the discipline office and ran to the third floor to pick a fight with their rivals from Altgeld Gardens, a different South Side neighborhood that now feeds Fenger's population and is fighting to establish gang dominance.

Attendance Dean Reginald Holmes takes a casual walk down the hall with another neighborhood gang leader and learns that some students are talking about taking their beef to a nearby corner after school. The tip allows Holmes to alert police for the need for extra patrol cars. The dean sends the student back to class, tossing him a taffy apple. "Happy Halloween," Holmes says. The student grins. "Thanks, homey."

Some veteran educators would balk at this kind of conciliation, arguing that it offers legitimacy to gang members.

"We don't negotiate with gang leaders. That just validates their existence," said Susan Gross, assistant principal at Taft High, a diverse Northwest Side school in Norwood Park. Gross said gang fights are unusual at her school, but when they do occur, the students are arrested and referred for expulsion. "But I can only imagine what it's like to work in an environment where gangs are a major part of the culture. In that case, you have to find a way to deal with it."

The two teens who started the fight spent the day in handcuffs sitting in the disciplinary office. In the end, though, they weren't charged with a crime. They were suspended for 10 days, along with three others who also exchanged blows. When they return, they will have to bring a parent with them to help mediate some kind of truce.

The mother of one student—on his third suspension for fighting and facing almost certain failure for the semester—already has asked the principal for help. The principal tries to find the student a spot in an alternative school, while another administrator tries to set him straight.

The sophomore is unmoved. He fights, he says, because he doesn't have a choice. "If you ignore them, they are going to come at you even harder."

One fight. Hours of frustrating negotiations. Johnson knows it would have been so much easier to crack down and dump the students at the cops' doorstep. He just doesn't think this approach will make his school safer.

"You never give up your building . . . and you don't let it fester," Johnson said. "This is my reality, and I can't make it go away by hiding from it. We owe it to these children to do everything we can to help them."

Source: Dell'Angela, Tracy. (2006, November 19). Restoring Justice Where Gangs, Not Suspensions, Define Fear. *Chicago Tribune*.

in the interest of making both "whole" again. To the extent, if any, that such confrontations prove to be uncomfortable for offenders, this may be viewed as another form of punishment in addition to, or in the place of, judicial punishment.

We might close by looking at the future of gangs and efforts to deal with them as a way of summarizing the future of the juvenile justice network. The contemporary cycle of youth gang activities is likely to continue because members who are imprisoned manage to maintain some gang-related activities even while in prison and because recruitment, given the conditions outlined previously, is no problem. As Postman (1991) put it, we can ill afford to "hurtle into the future with our eyes fixed firmly on the rearview mirror" (cited in Osborne & Gaebler, 1992, p. 19). Or, as according to Papachristos (2005), "No amount of law enforcement will rid the world of gangs. Strategies at all levels must move beyond simple arrests and incarceration to consider the economic structures of the cities and neighborhoods that breed street gangs. Otherwise, there will be nothing there to greet them but the waiting and supportive arms of the gang" (p. 55). Community programs aimed at alleviating the causes of gang membership as well as providing opportunities for those inclined to gang membership will need to be forthcoming if we hope to confront the gang problem. Support for the police is essential, but so is support for myriad community programs directed at high-risk youth. The best efforts of school personnel, social services professionals, and the community at large will also be necessary. Only by producing our best efforts in this regard can we hope to maintain the integrity of the juvenile justice network while providing appropriate alternatives for juveniles who cannot, or will not, be helped through education, treatment, care, concern, and opportunity.

Note: Please see the Companion Study Site for Internet exercises and Web resources. Go to www.sagepub.com/juvenilejustice6study.

Note

1. This material taken from Cohn, A. W. (2004). Planning for the future of juvenile justice. *Federal Probation, 68*(3), 39–44. Reprinted with permission of the author.

Suggested Readings

Bilchik, S. (1998). A juvenile justice system for the 21st century. *Crime & Delinquency, 44*, 89–101.

Cohn, A. (1999). Juvenile justice in transition: Is there a future? *Federal Probation, 63*(2), 61–67.

Federle, K. H. (1999). Is there a jurisprudential future for the juvenile court? *Annals of the American Academy of Political & Social Science, 564*, 28–36.

Feld, B. C. (1999–2000). The juvenile court: Changes and challenges. *Update on Law-Related Education, 23*(2), 10–14.

Juvenile court: Today and Tomorrow. (1999–2000). *Update on Law-Related Education, 23*(2), 18–20.

Moon, M. M., Sundt, J. L., Cullen, F. T., & Wright, J. P. (2000). Is child saving dead? Public support for juvenile rehabilitation. *Crime & Delinquency, 46*, 38–60.

Morse, S. J. (1999). Delinquency and desert. *Annals of the American Academy of Political & Social Science, 564*, 56–80.

Umbreit, M., & Coates, R. B. (1999). Multicultural implications of restorative juvenile justice. *Federal Probation, 63*(2), 44–51.

Appendix
Uniform Juvenile Court Act

The Uniform Juvenile Court Act was drafted by the National Conference of Commissioners on Uniform State Laws and approved and recommended for enactment in all the states at its annual conference meeting in its seventy-seventh year, Philadelphia, Pennsylvania, July 22—August 1, 1968. Approved by the American Bar Association at its meeting at Philadelphia, Pennsylvania, August 7, 1968.

Section 1. [Interpretation.] This Act shall be construed to effectuate the following public purposes:

1. to provide for the care, protection, and wholesome moral, mental, and physical development of children coming within its provisions;
2. consistent with the protection of the public interest, to remove from children committing delinquent acts the taint of criminality and the consequences of criminal behavior and to substitute therefore a program of treatment, training, and rehabilitation;
3. to achieve the foregoing purposes in a family environment whenever possible, separating the child from his parents only when necessary for his welfare or in the interest of public safety;
4. to provide a simple judicial procedure through which this Act is executed and enforced and in which the parties are assured a fair hearing and their constitutional and other legal rights recognized and enforced; and
5. to provide simple interstate procedures which permit resort to cooperative measures among the juvenile courts of the several states when required to effectuate the purposes of this Act.

Section 2. [Definitions.] As used in this Act:

1. "child" means an individual who is:
    i. under the age of 18 years; or
    ii. under the age of 21 years who committed an act of delinquency before reaching the age of 18 years; [or]
    iii. under 21 years of age who committed an act of delinquency after becoming 18 years of age and is transferred to the juvenile court by another court having jurisdiction over him;]
2. "delinquent act" means an act designated a crime under the law, including local [ordinances] [or resolutions] of this state, or of another state, if the act occurred in that state, or under federal law, and the crime does not fall under paragraph (iii) of subsection (4) [and is not a juvenile traffic offense as defined in section 44] [and the crime is not a traffic offense as defined in Traffic Code of the State] other than [designate the more serious offenses which should be included in the jurisdiction of the juvenile court such as drunken driving, negligent homicide, etc.];
3. "delinquent child" means a child who has committed a delinquent act and is in need of treatment or rehabilitation;
4. "unruly child" means a child who:
    i. while subject to compulsory school attendance is habitually and without justification truant from school;
    ii. is habitually disobedient of the reasonable and lawful commands of his parent, guardian, or other custodian and is ungovernable; or
    iii. has committed an offense applicable only to a child; and
    iv. if any of the foregoing is in need of treatment or rehabilitation;
5. "deprived child" means a child who:
    i. is without proper parental care or control, subsistence, education as required by law, or other care or control necessary for his physical, mental, or emotional health, or morals, and the deprivation is not due primarily to the lack of financial means of his parents, guardian, or other custodian;
    ii. has been placed for care or adoption in violation of law; [or]
    iii. has been abandoned by his parents, guardian, or other custodian; [or]
    iv. is without a parent, guardian, or legal custodian;
6. "shelter care" means temporary care of a child in physically unrestricted facilities;
7. "protective supervision" means supervision ordered by the court of children found to be deprived or unruly;
8. "custodian" means a person, other than a parent or legal guardian, who stands in loco parentis to the child or a person to whom legal custody of the child has been given by order of a court;
9. "juvenile court" means the [here designate] court of this state.

Section 3. [Jurisdiction.]

a. The juvenile court has exclusive original jurisdiction of the following proceedings, which are governed by this Act:

 1. proceedings in which a child is alleged to be delinquent, unruly, or deprived [or to have committed a juvenile traffic offense as defined in section 44;]
 2. proceedings for the termination of parental rights except when a part of an adoption proceeding; and
 3. proceedings arising under section 39 through 42.

b. The juvenile court also has exclusive original jurisdiction of the following proceedings, which are governed by the laws relating thereto without regard to the other provisions of this Act:

 1. proceedings for the adoption of an individual of any age;]
 2. proceedings to obtain judicial consent to the marriage, employment, or enlistment in the armed services of a child, if consent is required by law;
 3. proceedings under the Interstate Compact of Juveniles; [and]
 4. proceedings under the Interstate Compact on the Placement of Children; [and]
 5. proceedings to determine the custody or appoint a guardian of the person of a child.]

Section 4. [Concurrent Jurisdiction.] The juvenile court has concurrent jurisdiction with [———] court of proceedings to treat or commit a mentally retarded or mentally ill child.]

Section 5. [Probation Services.]

a. [In [counties] of over————————population] the [————] court may appoint one or more probation officers who shall serve [at the pleasure of the court] [and are subject to removal under the civil service laws governing the county]. They have the powers and duties stated in section 6. Their salaries shall be fixed by the court with the approval of the [governing board of the county]. If more than one probation officer is appointed, one may be designated by the court as the chief probation officer or director of court services, who shall be responsible for the administration of the probation services under the direction of the court.]

b. In all other cases the [Department of Corrections] [state] [county] child welfare department] [or other appropriate state agency] shall provide suitable probation services to the juvenile court of each [county.] The cost thereof shall be paid out of the general revenue funds of the [state] [county]. The probation officer or other qualified person assigned to the court by the [Department of Corrections] [state [county] child welfare department] [or other appropriate state agency] has the powers and duties stated in section 6.]

Section 6. [Powers and Duties of Probation Officers.]

a. For the purpose of carrying out the objectives and purposes of this Act and subject to the limitations of this Act or imposed by the Court, a probation officer shall:
   1. make investigations, reports, and recommendations to the juvenile court;
   2. receive and examine complaints and charges of delinquency, unruly conduct or deprivation of a child for the purpose of considering the commencement of proceedings under this Act;
   3. supervise and assist a child placed on probation or in his protective supervision or care by order of the court or other authority of law;
   4. make appropriate referrals to other private or public agencies of the community if their assistance appears to be needed or desirable;
   5. take into custody and detain a child who is under his supervision or care as a delinquent, unruly or deprived child if the probation officer has reasonable cause to believe that the child's health or safety is in imminent danger, or that he may abscond or be removed from the jurisdiction of the court, or when ordered by the court pursuant to this Act. Except as provided by this Act a probation officer does not have the powers of a law enforcement officer. He may not conduct accusatory proceedings under this Act against a child who is or may be under his care or supervision; and
   6. perform all other functions designated by this Act or by order of the court pursuant thereto.

b. Any of the foregoing functions may be performed in another state if authorized by the court of this state and permitted by the laws of the other state.

Section 7. [Referees.]

a. The judge may appoint one or more persons to serve at the pleasure of the judge as referees on a full or part-time basis. A referee shall be a member of the bar [and shall qualify under the civil service regulations of the County.] His compensation shall be fixed by the judge [with the approval of the [governing board of the County] and paid out of [————].

b. The judge may direct that hearings in any case or class of cases be conducted in the first instance by the referee in the manner provided by this Act. Before commencing the hearing the referee shall inform the parties who have appeared that they are entitled to have the matter heard by the judge. If a party objects the hearing shall be conducted by the judge.

c. Upon the conclusion of a hearing before a referee he shall transmit written findings and recommendations for disposition to the judge. Prompt written notice and copies of the findings and recommendations shall be given to the parties to the proceeding. The written notice shall inform them of the right to a rehearing before the judge.

d. A rehearing may be ordered by the judge at any time and shall be ordered if a party files a written request therefore within 3 days after receiving the notice required in subsection (c).

e. Unless a rehearing is ordered the findings and recommendations become the findings and order of the court when confirmed in writing by the judge.]

Section 8. [Commencement of Proceedings.] A proceeding under this Act may be commenced:

1. by transfer of a case from another court as provided in section 9;
2. as provided in section 44 in a proceeding charging the violation of a traffic offense;] or
3. by the court accepting jurisdiction as provided in section 40 or accepting supervision of a child as provided in section 42; or
4. in other cases by the filing of a petition as provided in this Act. The petition and all other documents in the proceeding shall be entitled "In the interest of———, a [child] [minor] under [18] [21] years of age."

Section 9. [Transfer from other Courts.] If it appears to the court in a criminal proceeding that the defendant [is a child] [was under the age of 18 years at the time the offense charged was alleged to have been committed], the court shall forthwith transfer the case to the juvenile court together with a copy of the accusatory pleading and other papers, documents, and transcripts of testimony relating to the case. It shall order that the defendant be taken forthwith to the juvenile court or to a place of detention designated by the juvenile court, or release him to the custody of his parent, guardian, custodian, or other person legally responsible for him, to be brought before the juvenile court at a time designated by that court. The accusatory pleading may serve in lieu of a petition in the juvenile court unless that court directs the filing of a petition.

Section 10. [Informal Adjustment.]

a. Before a petition is filed, the probation officer or other officer of the court designated by it, subject to its direction, may give counsel and advice to the parties with a view to an informal adjustment if it appears;
    1. the admitted facts bring the case within the jurisdiction of the court;
    2. counsel and advice without an adjudication would be in the best interest of the public and the child; and
    3. the child and his parents, guardian or other custodian consent thereto with knowledge that consent is not obligatory.
b. The giving of counsel and advice cannot extend beyond 3 months from the day commenced unless extended by the court for an additional period not to exceed 3 months and does not authorize the detention of the child if not otherwise permitted by this Act.
c. An incriminating statement made by a participant to the person giving counsel or advice and in the discussions or conferences incident thereto shall not be used against the declarant over objection in any hearing except in a hearing on disposition in a juvenile court proceeding or in a criminal proceeding against him after conviction for the purpose of a pre-sentence investigation.

Section 11. [Venue.] A proceeding under this act may be commenced in the [county] in which the child resides. If delinquent or unruly conduct is alleged, the proceeding may be commenced in the [county] in which the acts constituting the alleged delinquent or unruly conduct occurred. If deprivation is alleged, the proceeding may be brought in the [county] in which the child is present when it is commenced.

Section 12. [Transfer to Another Juvenile Court Within the State.]

a. If the child resides in a [county] of the state and the proceeding is commenced in a court of another [county], the court, on motion of a party or on its own motion made prior to final disposition, may transfer the proceeding to the county of the child's residence for further action. Like transfer may be made if the residence of the child changes pending the proceeding. The proceeding shall be transferred if the child has been adjudicated delinquent or unruly and other proceedings involving the child are pending in the juvenile court of the [county] of his residence.

b. Certified copies of all legal and social documents and records pertaining to the case on file with the clerk of the court shall accompany the transfer.

Section 13. [Taking into Custody.]

a. A child may be taken into custody:
   1. pursuant to an order of the court under this Act;
   2. pursuant to the laws of arrest;
   3. by a law enforcement officer [or duly authorized officer of the court] if there are reasonable grounds to believe that the child is suffering from illness or injury or is in immediate danger from his surroundings, and that his removal is necessary; or
   4. by a law enforcement officer [or duly authorized officer of the court] if there are reasonable grounds to believe that the child has run away from his parents, guardian, or other custodian.

b. The taking of a child into custody is not an arrest, except for the purpose of determining its validity under the constitution of this State or of the United States.

Section 14. [Detention of Child.] A child taken into custody shall not be detained or placed in shelter care prior to the hearing on the petition unless his detention or care is required to protect the person or property of others or of the child because the child may abscond or be removed from the jurisdiction of the court or because he has no parent, guardian, or custodian or other person able to provide supervision and care for him and return him to the court when required, or an order for his detention or shelter care has been made by the court pursuant to this Act.

Section 15. [Release or Delivery to Court.]

a. A person taking a child into custody, with all reasonable speed and without first taking the child elsewhere, shall:
   1. release the child to his parents, guardian, or other custodian upon their promise to bring the child before the court when requested by the court, unless his detention or shelter care is warranted or required under section 14; or

2. bring the child before the court or deliver him to a detention or shelter care facility designated by the court or to a medical facility if the child is believed to suffer from a serious physical condition or illness which requires prompt treatment. He shall promptly give written notice thereof, together with a statement of the reason for taking the child into custody, to a parent, guardian, or other custodian and to the court. Any temporary detention or questioning of the child necessary to comply with this subsection shall conform to the procedures and conditions prescribed by this Act and rules of court.

b. If a parent, guardian, or other custodian, when requested, fails to bring the child before the court as provided in subsection (a) the court may issue its warrant directing that the child be taken into custody and brought before the court.

Section 16. [Place of Detention.]

a. A child alleged to be delinquent may be detained only in:
   1. a licensed foster home or a home approved by the court;
   2. a facility operated by a licensed child welfare agency;
   3. a detention home or center for delinquent children which is under the direction or supervision of the court or other public authority or of a private agency approved by the court; or
   4. any other suitable place or facility, designated or operated by the court. The child may be detained in a jail or other facility for the detention of adults only if the facility in paragraph (3) is not available, the detention is in a room separate and removed from those for adults, it appears to the satisfaction of the court that public safety and protection reasonably require detention, and it so orders.

b. The official in charge of a jail or other facility for the detention of adult offenders or persons charged with crime shall inform the court immediately if a person who is or appears to be under the age of 18 years is received at the facility and shall bring him before the court upon request or deliver him to a detention or shelter care facility designated by the court.

c. If a case is transferred to another court for criminal prosecution the child may be transferred to the appropriate officer or detention facility in accordance with the law governing the detention of persons charged with crime.

d. A child alleged to be deprived or unruly may be detained or placed in shelter care only in the facilities stated in paragraphs (1), (2), and (4) of subsection (a) and shall not be detained in a jail or other facility intended or used for the detention of adults charged with criminal offenses or of children alleged to be delinquent.

Section 17. [Release from Detention or Shelter Care—Hearing—Conditions of Release.]

a. If a child is brought before the court or delivered to a detention or shelter care facility designated by the court the intake or other authorized officer of the court shall immediately make an investigation and release the child unless it appears that his detention or shelter care is warranted or required under section 14.

b. If he is not so released, a petition under section 21 shall be promptly made and presented to the court. An informal detention hearing shall be held promptly and not later than 72 hours after he is placed in detention to determine whether his detention or shelter care is required under section 14. Reasonable notice thereof, either oral or written, stating the time, place, and purpose of the detention hearing shall be given to the child and if they can be found, to his parents, guardian, or other custodian. Prior to the commencement of the hearing, the court shall inform the parties of their right to counsel and to appointed counsel if they are needy persons, and of the child's right to remain silent with respect to any allegations of delinquency or unruly conduct.

c. If the child is not so released and a parent, guardian, or custodian has not been notified of the hearing, did not appear or waive appearance at the hearing, and files his affidavit showing these facts, the court shall rehear the matter without unnecessary delay and order his release unless it appears from the hearing that the child's detention or shelter care is required under section 14.

Section 18. [Subpoena.] Upon application of a party the court or the clerk of the court shall issue, or the court on its own motion may issue, subpoenas requiring attendance and testimony of witnesses and production of papers at any hearing under this Act.]

Section 19. [Petition—Preliminary Determination.] A petition under this Act shall not be filed unless the [probation officer,] the court, or other person authorized by the court has determined and endorsed upon the petition that the filing of the petition is in the best interest of the public and the child.

Section 20. [Petition—Who May Make.] Subject to section 19 the petition may be made by any person, including a law enforcement officer, who has knowledge of the facts alleged or is informed and believes that they are true.

Section 21. [Contents of Petition.] The petition shall be verified and may be on information and belief. It shall set forth plainly:

1. the facts which bring the child within the jurisdiction of the court, with a statement that it is in the best interest of the child and the public that the proceeding be brought and, if delinquency or unruly conduct is alleged, that the child is in need of treatment or rehabilitation;
2. the name, age, and residence address, if any, of the child on whose behalf the petition is brought;
3. the names and residence addresses, if known to petitioner, of the parents, guardian, or custodian of the child and of the child's spouse, if any. If none of his parents, guardian, or custodian resides or can be found within the state, or if their respective places of residence address are unknown, the name of any known adult relative residing within the [county], or, if there be none, the known adult relative residing nearest to the location of the court; and

4. if the child is in custody and, if so, the place of his detention and the time he was taken into custody.

Section 22. [Summons.]

a. After the petition has been filed the court shall fix a time for hearing thereon, which, if the child is in detention, shall not be later than 10 days after the filing of the petition. The court shall direct the issuance of a summons to the parents, guardian, or other custodian, a guardian ad litem, and any other persons as appear to the court to be proper or necessary parties to the proceeding, requiring them to appear before the court at the time fixed to answer the allegations of the petition. The summons shall also be directed to the child if he is 14 or more years of age or is alleged to be a delinquent or unruly child. A copy of the petition shall accompany the summons unless the summons is served by publication in which case the published summons shall indicate the general nature of the allegations and where a copy of the petition can be obtained.

b. The court may endorse upon the summons an order directing the parents, guardian or other custodian of the child to appear personally at the hearing and directing the person having the physical custody or control of the child to bring the child to the hearing.

c. If it appears from affidavit filed or from sworn testimony before the court that the conduct, condition, or surroundings of the child are endangering his health or welfare or those of others, or that he may abscond or be removed from the jurisdiction of the court or will not be brought before the court, notwithstanding the service of the summons, the court may endorse upon the summons an order that a law enforcement officer shall serve the summons and take the child into immediate custody and bring him forthwith before the court.

d. The summons shall state that a party is entitled to counsel in the proceedings and that the court will appoint counsel if the party is unable without undue financial hardship to employ counsel.

e. A party, other than the child, may waive service of summons by written stipulation or by voluntary appearance at the hearing. If the child is present at the hearing, his counsel, with the consent of the parent, guardian or other custodian, or guardian ad litem, may waive service of summons in his behalf.

Section 23. [Service of Summons.]

a. If a party to be served with a summons is within this State and can be found, the summons shall be served upon him personally at least 24 hours before the hearing. If he is within the State and cannot be found, but his address is known or can with reasonable diligence be ascertained, the summons may be served upon him by mailing a copy by registered or certified mail at least 5 days before the hearing. If he is without this State but he can be found or his address is known, or his whereabouts or address can with reasonable diligence be ascertained, service of the summons may be made either by delivering a copy to him personally or mailing a copy to him by registered or certified mail at least 5 days before the hearing.

b. If after reasonable effort he cannot be found or his post office address ascertained, whether he is within or without this State, the court may order service of the summons upon him by publication in accordance with [Rule] [Section]———[the general service by publication statutes.] The hearing shall not be earlier than 5 days after the date of the last publication.

c. Service of the summons may be made by any suitable person under the direction of the court.

d. The court may authorize the payment from [county funds] of the costs of service and of necessary travel expenses incurred by persons summoned or otherwise required to appear at the hearing.

Section 24. [Conduct of Hearings.]

a. Hearings under this Act shall be conducted by the court without a jury, in an informal but orderly manner, and separate from other proceedings not included in section 3.

b. The [prosecuting attorney] upon request of the court shall present the evidence in support of the petition and otherwise conduct the proceedings on behalf of the state.

c. If requested by a party or ordered by the court the proceedings shall be recorded by stenographic notes or by electronic, mechanical, or other appropriate means. If not so recorded full minutes of the proceedings shall be kept by the court.

d. Except in hearings to declare a person in contempt of court [and in hearings under section 44], the general public shall be excluded from hearings under this Act. Only the parties, their counsel, witnesses, and other persons accompanying a party for his assistance, and any other persons as the court finds have a proper interest in the proceeding or in the work of the court may be admitted by the court. The court may temporarily exclude the child from the hearing except while allegations of his delinquency or unruly conduct are being heard.

Section 25. [Service by Publication—Interlocutory Order of Disposition.]

a. If service of summons upon a party is made by publication the court may conduct a provisional hearing upon the allegations of the petition and enter an interlocutory order of disposition if:

   1. the petition alleges delinquency, unruly conduct, or deprivation of the child;
   2. the summons served upon any party (i) states that prior to the final hearing on the petition designated in the summons a provisional hearing thereon will be held at a specified time and place, (ii) requires the party who is served other than by publication to appear and answer the allegations of the petition at the provisional hearing, (iii) states further that findings of fact and orders of disposition made pursuant to the provisional hearing will become final at the final hearing unless the party served by publication appears at the final hearing, and (iv) otherwise conforms to section 22; and
   3. the child is personally before the court at the provisional hearing.

b. All provisions of this Act applicable to a hearing on a petition, to orders of disposition, and to other proceedings dependent thereon shall apply under this section, but findings of fact and orders of disposition have only interlocutory effect pending the final hearing on the petition. The rights and duties of the party served by publication are not affected except as provided in subsection (c).

c. If the party served by publication fails to appear at the final hearing on the petition the findings of fact and interlocutory orders made become final without further evidence and are governed by this Act as if made at the final hearing. If the party appears at the final hearing the findings and orders shall be vacated and disregarded and the hearing shall proceed upon the allegations of the petition without regard to this section.

Section 26. [Right to Counsel.]

a. Except as otherwise provided under this Act a party is entitled to representation by legal counsel at all stages of any proceedings under this Act and if as a needy person he is unable to employ counsel, to have the court provide counsel for him. If a party appears without the counsel the court shall ascertain whether he knows of his right thereto and to be provided with counsel by the court if he is a needy person. The court may continue the proceeding to enable a party to obtain counsel and shall provide counsel for an unrepresented needy person upon his request. Counsel must be provided for a child not represented by his parent, guardian, or custodian. If the interests of 2 or more parties conflict, separate counsel shall be provided for each of them.

b. A needy person is one who at the time of requesting counsel is unable without undue financial hardship to provide for full payment of legal counsel and all other necessary expenses for representation.

Section 27. [Other Basic Rights.]

a. A party is entitled to the opportunity to introduce evidence and otherwise be heard in his own behalf and to cross-examine adverse witnesses.

b. A child charged with a delinquent act need not be a witness against or otherwise incriminate himself. An extrajudicial statement, if obtained in the course of violation of this Act or which would be constitutionally inadmissible in a criminal proceeding, shall not be used-against him. Evidence illegally seized or obtained shall not be received over objection to establish the allegations made against him. A confession validly made by the child out of court is insufficient to support an adjudication of delinquency unless it is corroborated in whole or in part by other evidence.

Section 28. [Investigation and Report.]

a. If the allegations of a petition are admitted by a party or notice of a hearing under section 34 has been given the court, prior to the hearing on need for treatment or rehabilitation and disposition, may direct that a social study and report in writing to the court be made by the [probation officer] of the court, [Commissioner of the Court or other like officer] or other person designated by the court, concerning the child, his family, his environment, and other matters relevant to disposition of the case. If the allegations of the petition are not admitted and notice of a hearing under section 34 has not been given the court shall not direct the making of the study and report until after the court has heard the petition upon notice of hearing given pursuant to this Act and the court has found that the child committed a delinquent act or is an unruly or deprived child.

b. During the pendency of any proceeding the court may order the child to be examined at a suitable place by a physician or psychologist and may also order medical or surgical treatment of a child who is suffering from a serious physical condition or illness which in the opinion of a [licensed physician] requires prompt treatment, even if the parent, guardian, or other custodian has not been given notice of a hearing, is not available, or without good cause informs the court of his refusal to consent to the treatment.

Section 29. [Hearing—Findings—Dismissal.]

a. After hearing the evidence on the petition the court shall make and file its findings as to whether the child is a deprived child, or if the petition alleges that the child is delinquent or unruly, whether the acts ascribed to the child were committed by him. If the court finds that the child is not a deprived child or that the allegations of delinquency or unruly conduct have not been established it shall dismiss the petition and order the child discharged from any detention or other restriction theretofore ordered in the proceeding.

b. If the court finds on proof beyond a reasonable doubt that the child committed the acts by reason of which he is alleged to be delinquent or unruly it shall proceed immediately or at a postponed hearing to hear evidence as to whether the child is in need of treatment or rehabilitation and to make and file its findings thereon. In the absence of evidence to the contrary evidence of the commission of acts which constitute a felony is sufficient to sustain a finding that the child is in need of treatment or rehabilitation. If the court finds that the child is not in need of treatment or rehabilitation it shall dismiss the proceeding and discharge the child from any detention or other restriction theretofore ordered.

c. If the court finds from clear and convincing evidence that the child is deprived or that he is in need of treatment or rehabilitation as a delinquent or unruly child, the court shall proceed immediately or at a postponed hearing to make a proper disposition of the case.

d. In hearings under subsections (b) and (c) all evidence helpful in determining the questions presented, including oral and written reports, may be received by the court and relied upon to the extent of its probative value even though not otherwise competent in the hearing on the petition. The parties or their counsel shall be afforded an opportunity to examine and controvert written reports so received and to cross-examine individuals making the reports. Sources of confidential information need not be disclosed.

e. On its motion or that of a party the court may continue the hearings under this section for a reasonable period to receive reports and other evidence bearing on the disposition or the need for treatment or rehabilitation. In this event the court shall make an appropriate order for detention of the child or his release from detention subject to supervision of the court during the period of the continuance. In scheduling investigations and hearings the court shall give priority to proceedings in which a child is in detention or has otherwise been removed from his home before an order of disposition has been made.

Section 30. [Disposition of Deprived Child.]

a. If the child is found to be a deprived child the court may make any of the following orders of disposition best suited to the protection and physical, mental, and moral welfare of the child:
    1. permit the child to remain with his parents, guardian, or other custodian, subject to conditions and limitations as the court prescribes, including supervision as directed by the court for the protection of the child;
    2. subject to conditions and limitations as the court prescribes transfer temporary legal custody to any of the following:
        i. any individual who, after study by the probation officer or other person or agency designated by the court, is found by the court to be qualified to receive and care for the child;
        ii. an agency or other private organization licensed or otherwise authorized by law to receive and provide care for the child; or
        iii. the Child Welfare Department of the [county] [state,] [or other public agency authorized by law to receive and provide care for the child;]
        iv. an individual in another state with or without supervision by an appropriate officer under section 40; or
    3. without making any of the foregoing orders transfer custody of the child to the juvenile court of another state if authorized by and in accordance with section 39 if the child is or is about to become a resident of that state.
b. Unless a child found to be deprived is found also to be delinquent he shall not be committed to or confined in an institution or other facility designed or operated for the benefit of delinquent children.

Section 31. [Disposition of Delinquent Child.] If the child is found to be a delinquent child the court may make any of the following orders of disposition best suited to his treatment, rehabilitation, and welfare:

1. any order authorized by section 30 for the disposition of a deprived child;
2. placing the child on probation under the supervision of the probation officer of the court or the court of another state as provided in section 41, or [the Child Welfare Department operating within the county,] under conditions and limitations the court prescribes;
3. placing the child in an institution, camp, or other facility for delinquent children operated under the direction of the court [or other local public authority;] or

4. committing the child to [designate the state department to which commitments of delinquent children are made or, if there is no department, the appropriate state institution for delinquent children].

Section 32. [Disposition of Unruly Child.] If the child is found to be unruly the court may make any disposition authorized for a delinquent child except commitment to [the state department or state institution to which commitment of delinquent children may be made]. [If after making the disposition the court finds upon a further hearing that the child is not amenable to treatment or rehabilitation under the disposition made it may make a disposition otherwise authorized by section 31.]

Section 33. [Order of Adjudication—Non-Criminal.]

a. An order of disposition or other adjudication in a proceeding under this Act is not a conviction of crime and does not impose any civil disability ordinarily resulting from a conviction or operate to disqualify the child in any civil service application or appointment. A child shall not be committed or transferred to a penal institution or other facility used primarily for the execution of sentences of persons convicted of a crime.

b. The disposition of a child and evidence adduced in a hearing in juvenile court may not be used against him in any proceeding in any court other than a juvenile court, whether before or after reaching majority, except in dispositional proceedings after conviction of a felony for the purposes of a pre-sentence investigation and report.

Section 34. [Transfer to Other Courts.]

a. After a petition has been filed alleging delinquency based on conduct which is designated a crime or public offense under the laws, including local ordinances, [or resolutions] of this state, the court before hearing the petition on its merits may transfer the offense for prosecution to the appropriate court having jurisdiction of the offense if:
    1. the child was 16 or more years of age at the time of the alleged conduct;
    2. a hearing on whether the transfer should be made is held in conformity with sections 24, 26, and 27;
    3. notice in writing of the time, place, and purpose of the hearing is given to the child and his parents, guardian, or other custodian at least 3 days before the hearing.
    4. the court finds that there are reasonable grounds to believe that
        i. the child committed the delinquent act alleged;
        ii. the child is not amenable to treatment or rehabilitation as a juvenile through available facilities;
        iii. the child is not committable to an institution for the mentally retarded or mentally ill; and
        iv. the interests of the community require that the child be placed under legal restraint or discipline.

b. The transfer terminates the jurisdiction of the juvenile court over the child with respect to the delinquent acts alleged in the petition.

c. No child, either before or after reaching 18 years of age, shall be prosecuted for an offense previously committed unless the case has been transferred as provided in this section.

d. Statements made by the child after being taken into custody and prior to the service of notice under subsection (a) or at the hearing under this section are not admissible against him over objection in the criminal proceedings following the transfer.

e. If the case is not transferred the judge who conducted the hearing shall not over objection of an interested party preside at the hearing on the petition. If the case is transferred to a court of which the judge who conducted the hearing is also a judge he likewise is disqualified from presiding in the prosecution.

Section 35. [Disposition of Mentally Ill or Mentally Retarded Child.]

a. If, at a dispositional hearing of a child found to be a delinquent or unruly child or at a hearing to transfer a child to another court under section 34, the evidence indicates that the child may be suffering from mental retardation or mental illness the court before making a disposition shall commit the child for a period not exceeding 60 days to an appropriate institution, agency, or individual for study and report on the child's mental condition.

b. If it appears from the study and report that the child is committable under the laws of this state as a mentally retarded or mentally ill child the court shall order the child detained and direct that within 10 days after the order is made the appropriate authority initiate proceedings for the child's commitment.

c. If it does not appear, or proceedings are not promptly initiated, or the child is found not to be committable, the court shall proceed to the disposition or transfer of the child as otherwise provided by this Act.

Section 36. [Limitations of Time on Orders of Disposition.]

a. An order terminating parental rights is without limit as to duration.

b. An order of disposition committing a delinquent or unruly child to the [State Department of Corrections of designated institution for delinquent children,] continues in force for 2 years or until the child is sooner discharged by the [department or institution to which the child was committed]. The court which made the order may extend its duration for an additional 2 years, subject to like discharge, if:

   1. a hearing is held upon motion of the [department or institution to which the child was committed] prior to the expiration of the order;

   2. reasonable notice of the hearing and an opportunity to be heard is given to the child and the parent, guardian, or other custodian; and

3. the court finds that the extension is necessary for the treatment or rehabilitation of the child.

c. Any other order of disposition continues in force for not more than 2 years. The court may sooner terminate its order or extend its duration for further periods. An order of extension may be made if:

1. a hearing is held prior to the expiration of the order upon motion of a party or on the court's own motion;
2. reasonable notice of the hearing and opportunity to be heard are given to the parties affected;
3. the court finds that the extension is necessary to accomplish the purposes of the order extended; and
4. the extension does not exceed 2 years from the expiration of prior order.

d. Except as provided in subsection (b) the court may terminate an order of disposition or extension prior to its expiration, on or without an application of a party, if it appears to the court that the purposes of the order have been accomplished. If a party may be adversely affected by the order of termination the order may be made only after reasonable notice and opportunity to be heard have been given to him.

e. Except as provided in subsection (a) when the child reaches 21 years of age all orders affecting him then in force terminate and he is discharged from further obligation or control.

Section 37. [Modification or Vacation of Orders.]

a. An order of the court shall be set aside if (1) it appears that it was obtained by fraud or mistake sufficient therefore in a civil action, or (2) the court lacked jurisdiction over a necessary party or of the subject matter, or (3) newly discovered evidence so requires.

b. Except an order committing a delinquent child to the [State Department of Corrections or an institution for delinquent children,] an order terminating parental rights, or an order of dismissal, an order of the court may also be changed, modified, or vacated on the ground that changed circumstances so require in the best interest of the child. An order granting probation to a child found to be delinquent or unruly may be revoked on the ground that the conditions of probation have not been observed.

c. Any party to the proceeding, the probation officer or other person having supervision or legal custody of or an interest in the child may petition the court for the relief provided in this section. The petition shall set forth in concise language the grounds upon which the relief is requested.

d. After the petition is filed the court shall fix a time for hearing and cause notice to be served (as a summons is served under section 23) on the parties to the proceeding or affected by the relief sought. After the hearing, which may be informal, the court shall deny or grant relief as the evidence warrants.

Section 38. [Rights and Duties of Legal Custodian.] A custodian to whom legal custody has been given by the court under this Act has the right to the physical custody of the child, the right to determine the nature of the care and treatment of the child, including ordinary medical care and the right and duty to provide for the care, protection, training, and education, and the physical, mental, and moral welfare of the child, subject to the conditions and limitations of the order and to the remaining rights and duties of the child's parents or guardian.

Section 39. [Disposition of Non-Resident Child.]

a. If the court finds that child who has been adjudged to have committed a delinquent act or to be unruly or deprived is or is about to become a resident of another state which has adopted the Uniform Juvenile Court Act, or a substantially similar Act which includes provisions corresponding to sections 39 and 40, the court may defer hearing on need for treatment or rehabilitation and disposition and request by any appropriate means the juvenile court of the [county] of the child's residence or prospective residence to accept jurisdiction of the child.

b. If the child becomes a resident of another state while on probation or under protective supervision under order of a juvenile court of this State, the court may request the juvenile court of the [county] of the state in which the child has become a resident to accept jurisdiction of the child and to continue his probation or protective supervision.

c. Upon receipt and filing of an acceptance the court of this State shall transfer custody of the child to the accepting court and cause him to be delivered to the person designated by that court to receive his custody. It also shall provide that court with certified copies of the order adjudging the child to be a delinquent, unruly, or deprived child, of the order of transfer, and if the child is on probation or under protective supervision under order of the court, of the order of disposition. It also shall provide that court with a statement of the facts found by the court of this State and any recommendations and other information it considers of assistance to the accepting court in making a disposition of the case or in supervising the child on probation or otherwise.

d. Upon compliance with subsection (c) the jurisdiction of the court of this State over the child is terminated.

Section 40. [Disposition of Resident Child Received from Another State.]

a. If a juvenile court of another state which has adopted the Uniform Juvenile Court Act, or a substantially similar Act which includes provisions corresponding to sections 39 and 40, requests a juvenile court of this State to accept jurisdiction of a child found by the requesting court to have committed a delinquent act or to be an unruly or deprived child, and the court of this State finds, after investigation that the child is, or is about to become, a resident of the [county] in which the court presides, it shall promptly and not later than 14 days after receiving the request issue its acceptance in writing to the requesting court and direct its probation officer or other person designated by it to take physical custody of the child from the requesting court and bring him before the court of this State or make other appropriate provisions for his appearance before the court.

b. Upon the filing of certified copies of the orders of the requesting court (1) determining that the child committed a delinquent act or is an unruly or deprived child, and (2) committing the child to the jurisdiction of the juvenile court of this State, the court of this State shall immediately fix a time for a hearing on the need for treatment or rehabilitation and disposition of the child or on the continuance of any probation or protective supervision.

c. The hearing and notice thereof and all subsequent proceedings are governed by this Act. The court may make any order of disposition permitted by the facts and this Act. The orders of the requesting court are conclusive that the child committed the delinquent act or is an unruly or deprived child and of the facts found by the court in making the orders, subject only to section 37. If the requesting court has made an order placing the child on probation or under protective supervision, a like order shall be entered by the court of this State. The court may modify or vacate the order in accordance with section 37.

Section 41. [Ordering Out-of-State Supervision.]

a. Subject to the provisions of this Act governing dispositions and to the extent that funds of the [county] are available the court may place a child in the custody of a suitable person in another state. On obtaining the written consent of a juvenile court of another state which has adopted the Uniform Juvenile Court Act or a substantially similar Act which includes provisions corresponding to sections 41 and 42 the court of this State may order that the child be placed under the supervision of a probation officer or other appropriate official designated by the accepting court. One certified copy of the order shall be sent to the accepting court and another filed with the clerk of the [Board of County Commissioners] of the [county] of the requesting court of this State.

b. The reasonable cost of the supervision including the expenses of necessary travel shall be borne by the [county] of the requesting court of this State. Upon receiving a certified statement signed by the judge of the accepting court of the cost incurred by the supervision the court of this State shall certify if it so appears that the sum so stated was reasonably incurred and file it with [the appropriate officials] of the [county] [state] for payment. The [appropriate officials] shall thereupon issue a warrant for the sum stated payable to the [appropriate officials] of the [county] of the accepting court.

Section 42. [Supervision Under Out-of-State Order.]

a. Upon receiving a request of a juvenile court of another state which has adopted the Uniform Juvenile Court Act, or a substantially similar act which includes provisions corresponding to sections 41 and 42 to provide supervision of a child under the jurisdiction of that court, a court of this State may issue its written acceptance to the requesting court and designate its probation or other appropriate officer who is to provide supervision, stating the probable cost per day therefore.

b. Upon the receipt and filing of a certified copy of the order of the requesting court placing the child under the supervision of the officer so designated the officer shall arrange for the reception of the child from the requesting court, provide supervision pursuant to the order and this Act, and report thereon from time to time together with any recommendations he may have to the requesting court.

c. The court in this state from time to time shall certify to the requesting court the cost of supervision that has been incurred and request payment therefore from the appropriate officials of the [county] of the requesting court to the appropriate officials of the [county] of the accepting court.

d. The court of this State at any time may terminate supervision by notifying the requesting court. In that case, or if the supervision is terminated by the requesting court, the probation officer supervising the child shall return the child to a representative of the requesting court authorized to receive him.

Section 43. [Powers of Out-of-State Probation Officers.] If a child has been placed on probation or protective supervision by a juvenile court of another state which has adopted the Uniform Juvenile Court Act or a substantially similar act which includes provisions corresponding to this section, and the child is in this State with or without the permission of that court, the probation officer of that court or other person designated by that court to supervise or take custody of the child has all the powers and privileges in this State with respect to the child as given by this Act to like officers or persons of this State including the right of visitation, counseling, control, and direction, taking into custody, and returning to that state.

Section 44. [Juvenile Traffic Offenses.]

a. Definition. Except as provided in subsection (b), a juvenile traffic offense consists of a violation by a child of:

   1. a law or local ordinance [or resolution] governing the operation of a moving motor vehicle upon the streets or highways of this State, or the waterways within or adjoining this State; or

   2. any other motor vehicle traffic law or local ordinance [or resolution] of this State if the child is taken into custody and detained for the violation or is transferred to the juvenile court by the court hearing the charge.

b. A juvenile traffic offense is not an act of delinquency unless the case is transferred to the delinquency calendar as provided in subsection (g).

c. Exceptions. A juvenile traffic offense does not include a violation of: [Set forth the sections of state statutes violations of which are not to be included as traffic offenses, such as the so-called negligent homicide statute sometimes appearing in traffic codes, driving while intoxicated, driving without, or during suspension of, a driver's license, and the like].

d. Procedure. The [summons] [notice to appear] [or other designation of a ticket] accusing a child of committing a juvenile traffic offense constitutes the commencement of the proceedings in the juvenile court of the [county] in which the alleged violation occurred and serves in place of a summons and petition under this Act. These cases shall be filed and heard separately from other proceedings of the court. If the child is taken into custody on the charge, sections 14 to 17 apply. If the child is, or after commencement of the proceedings becomes, a resident of another [county] of this State, section 12 applies.

e. Hearing. The court shall fix a time for hearing and give reasonable notice thereof to the child, and if their address is known to the parents, guardian, or custodian. If the accusation made in the [summons] [notice to appear] [or other designation of a ticket] is denied an informal hearing shall be held at which the parties have the right to subpoena witnesses, present evidence, cross-examine witnesses, and appear by counsel. The hearing is open to the public.

f. Disposition. If the court finds on the admission of the child or upon the evidence that he committed the offense charged it may make one or more of the following orders:
   1. reprimand or counsel with the child and his parents;
   2. [suspend] [recommend to the [appropriate official having the authority] that he suspend] the child's privilege to drive under stated conditions and limitations for a period not to exceed that authorized for a like suspension of an adult's license for a like offense;
   3. require the child to attend a traffic school conducted by public authority for a reasonable period of time; or
   4. order the child to remit to the general fund of the [state] [county] [city] [municipality] a sum not exceeding the lesser of $50 or the maximum applicable to an adult for a like offense.

g. In lieu of the preceding orders, if the evidence indicates the advisability thereof, the court may transfer the case to the delinquency calendar of the court and direct the filing and service of a summons and petition in accordance with this Act. The judge so ordering is disqualified upon objection from acting further in the case prior to an adjudication that the child committed a delinquent act.]

Section 45. [Traffic Referee.]

a. The court may appoint one or more traffic referees who shall serve at the pleasure of the court. The referee's salary shall be fixed by the court [subject to the approval of the [Board of County Commissioners]].

b. The court may direct that any case or class of cases arising under section 44 shall be heard in the first instance by a traffic referee who shall conduct the hearing in accordance with section 44. Upon the conclusion of the hearing the traffic referee shall transmit written findings of fact and recommendations for disposition to the judge with a copy thereof to the child and other parties to the proceedings.

c. Within 3 days after receiving the copy the child may file a request for a rehearing before the judge of the court who shall thereupon rehear the case at a time fixed by him. Otherwise, the judge may confirm the findings and recommendations for disposition which then become the findings and order of disposition of the court.]

Section 46. [Juvenile Traffic Offenses—Suspension of Jurisdiction.]

a. The [Supreme] court, by order filed in the office of the [ ] of the [county,] may suspend the jurisdiction of the juvenile courts over juvenile traffic offenses or one or more classes effective and offenses committed thereafter shall be tried by the appropriate court in accordance with law without regard to this Act. The child shall not be detained or imprisoned in a jail or other facility for the detention of adults unless the facility conforms to subsection (a) of section 16.

b. The [Supreme] court at any time may restore the jurisdiction of the juvenile courts over these offenses or any portion thereof by like filing of its order of restoration. Offenses committed thereafter are governed by this Act.]

Section 47. [Termination of Parental Rights.]

a. The court by order may terminate the parental rights of a parent with respect to his child if:

    1. the parent had abandoned the child;
    2. the child is a deprived child and the court finds that the conditions and causes of the deprivation are likely to continue or will not be remedied and that by reason thereof the child is suffering or will probably suffer serious physical, mental, moral, or emotional harm; or
    3. the written consent of the parent acknowledged before the court has been given.

b. If the court does not make an order of termination of parental rights it may grant an order under section 30 if the court finds from clear and convincing evidence that the child is a deprived child.

Section 48. [Proceeding for Termination of Parental Rights.]

a. The petition shall comply with section 21 and state clearly that an order for termination of parental rights is requested and that the effect thereof will be as stated in the first sentence of Section 49.

b. If the paternity of a child born out of wedlock has been established prior to the filing of the petition, the father shall be served with summons as provided by this Act. He has the right to be heard unless he has relinquished all parental rights with reference to the child. The putative father of the child whose paternity has not been established, upon proof of his paternity of the child, may appear in the proceedings and be heard. He is not entitled to notice of hearing on the petition unless he has custody of the child.

Section 49. [Effect of Order Terminating Parental Rights.] An order terminating the parental rights of a parent terminates all his rights and obligations with respect to the child and of the child to him arising from the parental relationship. The parent is not thereafter entitled to notice of proceedings for the adoption of the child by another nor has he any right to object to the adoption or otherwise participate in the proceedings.

Section 50. [Commitment to Agency.]

a. If, upon entering an order terminating the parental rights of a parent, there is no parent having parental rights, the court shall commit the child to the custody of the [State County Child Welfare Department] or a licensed child-placing agency, willing to accept custody for the purpose of placing the child for adoption, or in the absence thereof in a foster home or take other suitable measures for the care and welfare of the child. The custodian has authority to consent to the adoption of the child, his marriage, his enlistment in the armed forces of the United States, and surgical and other medical treatment for the child.

b. If the child is not adopted within 2 years after the date of the order and a general guardian of the child has not been appointed by the [————] court, the child shall be returned to the court for entry of further orders for the care, custody, and control of the child.

Section 51. [Guardian ad litem.] The court at any stage of a proceeding under this Act, on application of a party or on its own motion, shall appoint a guardian ad litem for a child who is a party to the proceeding if he has no parent, guardian, or custodian appearing on his behalf or their interests conflict with his or in any other case in which the interests of the child require a guardian. A party to the proceeding or his employee or representative shall not be appointed.

Section 52. [Costs and Expenses for Care of Child.]

a. The following expenses shall be a charge upon the funds of the county upon certification thereof by the court:
   1. the cost of medical and other examinations and treatment of a child ordered by the court;
   2. the cost of care and support of a child committed by the court to the legal custody of a public agency other than an institution for delinquent children, or to a private agency or individual other than a parent;
   3. reasonable compensation for services and related expenses of counsel appointed by the court for a party;
   4. reasonable compensation for a guardian ad litem;
   5. the expense of service of summons, notices, subpoenas, travel expense of witnesses, transportation of the child, and other like expenses incurred in the proceedings under this Act.

b. If, after due notice to the parents or other persons legally obligated to care for and support the child, and after affording them an opportunity to be heard, the court finds that they are financially able to pay all or part of the costs and expenses stated in paragraphs (1), (2), (3), and (4) of subsection (a), the court may order them to pay the same and prescribe the manner of payment. Unless otherwise ordered payment shall be made to the clerk of the juvenile court for remittance to the person to whom compensation is due, or if the costs and expenses have been paid by the [county] to the [appropriate officer] of the [county].

Section 53. [Protective Order.] On application of a party or on the court's own motion the court may make an order restraining or otherwise controlling the conduct of a person if:

1. an order of disposition of a delinquent, unruly, or deprived child has been or is about to be made in a proceeding under this Act;
2. the court finds that the conduct (1) is or may be detrimental or harmful to the child and (2) will tend to defeat the execution of the order of disposition; and
3. due notice of the application or motion and the grounds therefore and an opportunity to be heard thereon have been given to the person against whom the order is directed.

Section 54. [Inspection of Court Files and Records.] [Except in cases arising under section 44] all files and records of the court in a proceeding under this Act are open to inspection only by:

1. the judge, officers, and professional staff of the court;
2. the parties to the proceeding and their counsel and representatives;
3. a public or private agency or institution providing supervision or having custody of the child under order of the court;
4. a court and its probation and other officials or professional staff and the attorney for the defendant for use in preparing a pre-sentence report in a criminal case in which the defendant is convicted and who prior thereto had been a party to the proceeding in juvenile court;
5. with leave of court any other person or agency or institution having a legitimate interest in the proceeding or in the work of the court.

Section 55. [Law Enforcement Records.] Law enforcement records and files concerning a child shall be kept separate from the records and files of arrests of adults. Unless a charge of delinquency is transferred for criminal prosecution under section 34, the interest of national security requires, or the court otherwise orders in the interest of the child, the records and files shall not be open to public inspection or their contents disclosed to the public; but inspection of the records and files is permitted by:

1. a juvenile court having the child before it in any proceeding;
2. counsel for a party to the proceeding;
3. the officers of public institutions or agencies to whom the child is committed;
4. law enforcement officers of other jurisdictions when necessary for the discharge of their official duties; and
5. a court in which he is convicted of a criminal offense for the purpose of a pre-sentence report or other dispositional proceeding, or by officials of penal institutions and other penal facilities to which he is committed, or by a [parole board] in considering his parole or discharge or in exercising supervision over him.

Section 56. [Children's Fingerprints, Photographs.]

a. No child under 14 years of age shall be fingerprinted in the investigation of a crime except as provided in this section. Fingerprints of a child 14 or more years of age who is referred to the court may be taken and filed by law enforcement officers in investigating the commission of the following crimes: [specifically such crimes as murder, non-negligent manslaughter, forcible rape, robbery, aggravated assault, burglary, housebreaking, purse snatching, and automobile theft].

b. Fingerprint files of children shall be kept separate from those of adults. Copies of fingerprints known to be those of a child shall be maintained on a local basis only and not sent to a central state or federal depository unless in the interest of national security.

c. Fingerprint files of children may be inspected by law enforcement officers when necessary for the discharge of their official duties. Other inspections may be authorized by the court in individual cases upon a showing that it is necessary in the public interest.

d. Fingerprints of a child shall be removed from the file and destroyed if:

   1. a petition alleging delinquency is not filed, or the proceedings are dismissed after either a petition if filed or the case is transferred to the juvenile court as provided in section 9, or the child is adjudicated not to be a delinquent child; or

   2. the child reaches 21 years of age and there is no record that he committed a criminal offense after reaching 16 years of age.

e. If latent fingerprints are found during the investigation of an offense and a law enforcement officer has probable cause to believe that they are those of a particular child he may fingerprint the child regardless of age or offense for purposes of immediate comparison with the latent fingerprints. If the comparison is negative the fingerprint card and other copies of the fingerprints taken shall be immediately destroyed. If the comparison is positive and the child is referred to the court, the fingerprint card and other copies of the fingerprints taken shall be delivered to the court for disposition. If the child is not referred to the court, the fingerprints shall be immediately destroyed.

f. Without the consent of the judge, a child shall not be photographed after he is taken into custody unless the case is transferred to another court for prosecution.

Section 57. [Sealing of Records.]

a. On application of a person who has been adjudicated delinquent or unruly or on the court's own motion, and after a hearing, the court shall order the sealing of the files and records in the proceeding, including those specified in sections 55 and 56, if the court finds:

   1. 2 years have elapsed since the final discharge of the person;

   2. since the final discharge he has not been convicted of a felony, or of a misdemeanor involving moral turpitude, or adjudicated a delinquent or unruly child and no proceeding is pending seeking conviction or adjudication; and

   3. he has been rehabilitated.

b. Reasonable notice of the hearing shall be given to:
   1. the [prosecuting attorney of the county];
   2. the authority granting the discharge if the final discharge was from an institution or from parole; and
   3. the law enforcement officers or department having custody of the files and records if the files and records specified in sections 55 and 56 are included in the application or motion.

c. Upon the entry of the order the proceeding shall be treated as if it never occurred. All index references shall be deleted and the person, the court, and law enforcement officers and departments shall properly reply that no record exists with respect to the person upon inquiry in any matter. Copies of the order shall be sent to each agency or official therein named. Inspection of the sealed files and records thereafter may be permitted by an order of the court upon petition by the person who is the subject of the records and only by those persons named in the order.

Section 58. [Contempt Powers.] The court may punish a person for contempt of court for disobeying an order of the court or for obstructing or interfering with the proceedings of the court or the enforcement of its orders subject to the laws relating to the procedures therefore and the limitations thereon.]

Section 59. [Appeals.]

a. An aggrieved party, including the state or a subdivision of the state, may appeal from a final order, judgment, or decree of the juvenile court to the [Supreme Court] [court of general jurisdiction] by filing written notice of appeal within 30 days after entry of the order, judgment, or decree, or within any further time the [Supreme Court] [court of general jurisdiction] grants, after entry of the order, judgment, or decree. [The appeal shall be heard by the [court of general jurisdiction] upon the files, records, and minutes or transcript of the evidence of the juvenile court, giving appreciable weight to the findings of the juvenile court.] The name of the child shall not appear on the record on appeal.

b. The appeal does not stay the order, judgment, or decree appealed from, but the [Supreme Court] [court of general jurisdiction] may otherwise order on application and hearing consistent with this Act if suitable provision is made for the care and custody of the child. If the order, judgment or decree appealed from grants the custody of the child to, or withholds it from, one or more of the parties to the appeal it shall be heard at the earliest practicable time.

Section 60. [Rules of Court.] The [Supreme] Court of this State may adopt rules of procedure not in conflict with this Act governing proceedings under it.

Section 61. [Uniformity of Interpretation.] This Act shall be so interpreted and construed as to effectuate its general purpose to make uniform the law of those states which enact it.

Section 62. [Short Title.] This Act may be cited as the Uniform Juvenile Court Act.

Section 63. [Repeal.] The following Acts and parts of Acts are repealed:

1.

2.

3.

Section 64. [Time of Taking Effect.] This Act shall take effect. . . .

Glossary

This glossary summarizes key terms that are used in this textbook as well as other juvenile justice terms that may or may not be included within this particular text. Readers should refer to this glossary to refresh their knowledge of discussions and case studies explored in chapters, tables, and In Practice articles.

Absolute deprivation. Lacking, or having only the minimal, means to survive within a society.

Abused children. Child victims of emotional, sexual, or physical assaults. These are victims of proactive behavior by an abuser.

Abused and Neglected Child Reporting Act. Legislation that designates the state agency for investigating reports of child abuse and neglect and also lists persons mandated to report such abuse and neglect.

Adjudicative facts. Facts of a juvenile adjudicatory hearing that represent the full scope of the facts alleged against the child named in the juvenile petition.

Adjudicatory hearing. Hearing similar to a trial in adult court. The merits of the case are heard and determined by the court.

Aftercare. Term used in approximately one-half of the states to describe parole for juveniles. It is essentially the juvenile justice system's equivalent to parole and is supervised early release of juveniles from a juvenile residential institution after a period of mandated confinement.

Age ambiguity. A problem due to the difficulty in setting agreed-on lower- and upper-limit legal age definitions, resulting in substantial ambiguity with respect to age definitions.

Age of majority. State and federal statutory designations that define individuals' status as juveniles by law. After offenders reach the age of majority, they are subject to criminal court jurisdiction.

Age of responsibility. Created nearly 2,000 years ago with origins in both Roman civil law and later canon (church) law. It made distinctions between juveniles and adults based on the notion of a minimum age of responsibility.

Alateen/Al-Anon. Alateen is a nationwide movement that is the teenage equivalent of Alcoholics Anonymous and is specifically organized to help teenage alcoholics. Al-Anon is also a nationwide movement that has been designed to assist those who are victimized or otherwise affected by the behavior of alcoholics who live in their households.

Alternatives to incarceration. Community-based programs that provide other sanctioning options besides a sentence to jail or prison. These community-based efforts can likewise be aimed at preventing juveniles from falling into crime within that community so as to avoid later potential incarceration.

AMBER alerts. Alerts that are broadcast when authorities suspect that a child has been abducted. A feature of the AMBER Plan, these alerts describe the victim, provide a profile of the suspected abductor, and describe any special circumstances of the case.

AMBER Plan. Voluntary partnership between law enforcement agencies and broadcasters to activate an urgent bulletin in the most serious child abduction cases. It is coordinated by state and local governments that ideally cooperate under a nationwide emergency alert umbrella when children are missing. AMBER is an acronym for America's Missing: Broadcast Emergency Response.

Anomalies. Individuals thought to be incapable of resisting the impulse to commit crimes except under very favorable circumstances.

Anomie theory. Sociological theory of criminal causation that generally refers to a state of "normlessness" in society.

Appeals. Requests to reexamine court rulings. Appeals from most juvenile courts are rare due to the absence of formality in the system.

Arrest data. Data that report the number of cases that were cleared by police agencies. Local law enforcement agencies report cleared cases to the Federal Bureau of Investigation, which publishes this information in its annual Uniform Crime Reports for regions across the nation.

At-risk juveniles. Juveniles who live in environments that are conducive to promoting deviant behavior. Also included are juveniles who exhibit antisocial behaviors that can lead to early deviance. In some states, *all* juveniles within defined age ranges are defined as at-risk juveniles.

Atavists. A biological theory of criminal causation developed during the 19th century by Cesare Lombroso. It argued that criminals are anthropological throwbacks to an undeveloped phase in human evolution and that this atavistic quality is indicated by physical abnormalities.

Attention deficit disorder. Behavioral disorder typified by deficient attention ability, chronic physical movements, and poor self-control.

Automatic waiver. Waiver that is automatically initiated to waive the exclusive jurisdiction of the juvenile court when specific offenses are allegedly committed by a juvenile.

Away syndrome. Approach that discourages attempts to find alternatives to incarceration, frequently arises in reaction to unsuccessful attempts at rehabilitation, and frequently is accompanied by an "out-of-sight, out-of-mind" attitude.

Bail. Release from custody pending trial after payment of a court-ordered sum.

Behavioral definitions. Definitions holding that those whose behavior violates statutes applicable to them are offenders whether or not they are officially labeled.

Behaviorists. Proponents of empirical objective approaches of observing and measuring behavior.

Bench trials. Resolution of legal and factual issues by a judge (the judicial "bench") without the participation of a jury. In these cases, the judge becomes the sole authority on the fate of accused offenders.

Beyond a reasonable doubt. Standard of proof applied for the resolution of delinquency cases. On a scale of 1% to 100%, proof beyond a reasonable doubt would typically be 95% to 100% certain.

Big Brothers/Big Sisters of America. Nationwide federation of youth service agencies that promote mentorship through individual relationships between adults and children. It promotes healthy interpersonal and socialization skills and behaviors in young people as they move through adolescence toward adulthood. Mentorship occurs through regular activities that many persons take for granted such as taking walks, playing outdoor games, going shopping, and bicycling.

Binding. Concept similar to probation that was established in England by the 14th century, when courts "bound over" offenders into the custody of members of the community for good behavior.

Biological theories. Theories that emphasize the belief that offenders differ from nonoffenders in some physiological way.

Biosocial criminology. A crime-study discipline that views behavior as the product of interaction between a physical environment and a physical organism and holds that contemporary criminology should represent a merger of biology, psychology, and sociology.

Black Disciples. Also known as the Black Gangster Disciple Nation/Brothers of the Struggle (BOS), a Chicago-based gang whose members are collectively known as "folks." They are in competition with the Vice Lords.

Blackstone Rangers. Original name of El Rukn, a Chicago gang that has emerged as a group characterized by a high degree of organization and considerable influence.

Bloods and Crips. Two gangs identified by the color red (Bloods) and the color blue (Crips). These gangs are enemies of one another, and each dates back to the late 1960s and early 1970s in South Central Los Angeles. They consist of "rollers" or "gang-bangers" who are in their 20s and 30s, gang veterans who have made it big in the drug trade. These veterans supervise and control the activities of younger members who are involved in drug trafficking and operating and supplying crack houses, activities that bring in millions of dollars a week. The drug trafficking of the Bloods and Crips spread into other cities, including Seattle, Portland, Denver, Kansas City, Des Moines, and even Honolulu and Anchorage.

Boot camp. Residential correctional facility where drill instructors replicate training techniques used in boot camps of the armed forces, adapting their methods to the special needs of juvenile offenders. The purpose is to "shock" residents into socially productive conformity over a period of 30 to 120 days.

Breed v. Jones. U.S. Supreme Court ruling that trying a juvenile as an adult in criminal court for the same crime that previously had been adjudicated in juvenile court violates the double jeopardy clause of the Fifth Amendment when the adjudication involves violation of a criminal statute.

Bridewell Workhouse. Institution established in 16th-century London to put offenders to work under strict discipline. Youthful offenders found begging or loitering were also sent to Bridewell.

Broken homes. Homes disrupted through divorce, separation, or desertion.

Capital punishment. Death penalty.

Career escalation. Progression from relatively minor status offenses toward more serious delinquent or criminal offenses.

Census of Juveniles in Residential Placement. National study conducted by the Office of Juvenile Justice and Delinquency Prevention that collects individual data on each juvenile held in a residential facility, including gender, date of birth, race, placement authority, most serious offense charged, court adjudication status, date of admission, and security status.

Certification. Procedural technicality whereby juveniles are legally certified as adults and are processed into the criminal justice system.

Chancery courts. Courts that operated under the guidance of the king's chancellor and were created to consider petitions of those who were in need of special aid or intervention, such as women and children left in need of protection and aid by reason of divorce, death of a spouse, or abandonment, and to grant relief to such persons. Through the chancery courts, the king exercised the right of *parens patriae.*

Child abuse reporting statutes. Laws mandating child abuse and neglect reporting procedures for certain institutions and professions. State agencies are often designated to investigate allegations of abuse. There may also be statutory reporting requirements for school employees, medical personnel, social services workers, and law enforcement officers.

Child death review teams. Teams composed of experts with medical, social services, and/or law enforcement backgrounds that review suspicious deaths of children and statutory changes facilitating the prosecution of those involved in child maltreatment.

Child neglect. Neglect that involves an individual under the age of 18 years whose parent or another person responsible for the child's welfare does not provide the proper or necessary support, education as required by law, or medical or other remedial care recognized under state law as necessary, including adequate food, shelter, clothing, or who is abandoned.

Child-saving movement. Mid-19th-century movement in the United States that sought to rescue children from unwholesome and dangerous environments. A fundamental tenet of the movement was that juveniles should receive treatment rather than punishment.

Children and family services. Personnel who do not actually work for the juvenile courts but still play a major role in the investigation, presentation of evidence, and dispositional recommendations in abuse and neglect cases.

Children in need. Juveniles who are not properly cared for by adults (e.g., abused, neglected, or abandoned children). Children in need are one population of juveniles served by the juvenile justice system.

Children in trouble. Juveniles who violate the law, either as juvenile delinquents or as status offenders. Some delinquents are waived into criminal court and tried as adults. Children in trouble are one population of juveniles served by the juvenile justice system.

CHINS. Children in need of supervision.

CHIPS. Children in need of protection or services.

Classical theory. Theoretical school holding that individuals are responsible for their deviant behavior. Punishments for these transgressors should always be proportional and never excessive. This theory was popular during the late 18th and early 19th centuries and was revived during the late 20th century in the United States.

Clear and convincing evidence. Standard of proof used in some circumstances that does not rise to the level of proof beyond a reasonable doubt but does exceed that which is based on a preponderance of the evidence. On a scale of 1% to 100%, clear and convincing evidence would typically be approximately 80% certainty.

Clearances. Arrests that have been made for reported crimes.

Common law. Law based on custom or use. Under common law, children under the age of 7 years were presumed to be incapable of forming criminal intent and, therefore, were not subject to criminal sanctions.

Community corrections. Nonresidential correctional alternatives that link juvenile offenders to the community and rely on community resources to provide correctional services. Probation, parole, and aftercare are examples of community corrections. They are also known as community-based corrections.

Community placement (benign). Less restrictive, and often nonrestrictive, community placements. Participants are permitted the same or similar freedoms of movement and association as are any other children. Foster homes are an example of this type of community placement.

Community placement (experiential). Controlled placements in rural communities. Participants are required to participate in a series of esteem- and character-building exercises over an established period of time. Wilderness programs are an example of this alternative.

Community placement (restrictive). Placements in the community under defined terms of probation or other diversionary agreements. The activities of participants are regulated, often quite intensively. Group homes are an example of this type of community placement.

Community policing. Era of policing beginning in the 1970s and lasting to the current time. This era's philosophy emphasized community political support, law, and professionalism. Law became the basic legitimation for police conduct. Community support and involvement were also sought and needed to solve problems. Crime control became just one goal of broader community problem solving. This type of policing requires development of intimate relationships between police and residents.

Community Resources Against Street Hoodlums. *See* CRASH.

Community Response to Crime program. Program that uses a community intervention team whose members meet with the offender to let him or her know how the behavior affected the community, how the community expects the offender to make amends, and how the community is willing to support the offender while he or she makes amends.

Concentric zone theory. Theory holding that zones of transition between residential and industrial neighborhoods consistently have the highest rates of crime and delinquency. It is commonly associated with the work of Earnest Burgess.

Conceptual schemes. Schemes that suggest relationships between variables but do not meet the requirements for theory.

Concurrent jurisdiction. Term indicating that two or more courts have jurisdiction over a potential case. For example, certain criminal acts may be concurrently under the jurisdiction of both the juvenile court and the criminal court.

Conditional probation. Probation programs that entail significant restrictions on movement and intensive communications with probation officers, usually involving personal meetings.

Conditioning. Process of rewarding for appropriate behavior and/or punishing for inappropriate behavior through which any type of social behavior can be taught.

Conditioning theory. Psychological theory of causation holding that environmental stimuli act as punishers or reinforcers for human behavior. Deviance can arise when criminal or delinquent behavior results in more pleasure than pain.

Conflict/Radical/Critical/Marxist theories. Theories that focus on whole political and economic systems and on class relations in those systems. Conflict theorists argue that conflict is inherent in all societies, not just capitalist societies, and focus on conflict resulting from gender, race, ethnicity, power, and other relationships.

Conflict theory of causation. Critical theory of causation hypothesizing that social tensions and conflicts often pit the "haves" against the "have-nots," with the latter being labeled as criminals or insurgents during these conflicts. Laws and rules are simply instruments of control used by ruling elites to maintain control of key institutions and thereby shut out others who might challenge the authority of the elites.

Continuance under supervision. Where the judge postpones adjudication and specifies a time period during which he or she (through court officers) will observe the juvenile.

Control theories. Theories assuming that all of us must be held in check or "controlled" if we are to resist the temptation to commit criminal or delinquent acts.

Cook County Juvenile Court. Court established under Illinois law in 1899. It was the first truly modern juvenile court given that it formulated concepts and procedures that continue to be applied in the current era.

Corrections-managed parole and aftercare. Systems where correctional administrators are given the authority to approve or deny early release. Senior officials in correctional facilities make recommendations on parole or aftercare for residents with the advice and counsel of professional social workers who are assigned to the facilities.

Cottage system. Model for juvenile correctional institutions in which 19th-century reformers designed rural reformatories as compounds with separate structures and dormitories. The belief was that residents who worked in industrial schools in the countryside would absorb wholesome rural values and virtues and thereby grow into responsible and productive adults.

Court-administered parole and aftercare. Parole and aftercare authority that has been exclusively granted to the court. The court has final oversight authority.

Court-appointed counsel. Attorneys who are either private attorneys appointed by the court or public defenders.

Court-appointed special advocates. Volunteers who work closely with the department of children and family services on abuse and neglect cases.

Crack. A cocaine-based stimulant drug that is inexpensive to manufacture and is much more affordable for drug users than is cocaine.

CRASH. Community Resources Against Street Hoodlums, a Los Angeles Police Department program created during the 1980s to aggressively confront gangs. Intimidation was expressly permitted as a matter of policy. The program was widely popular in the community, but it was disbanded after scandals involving illegal activities and behavior that violated the department's internal regulations.

Criminal deviance. Antisocial behavior by persons who violate laws prohibiting acts defined as criminal by city, county, and state lawmakers or the U.S. Congress. Both adults and juveniles (those waived into criminal courts) can be convicted of crimes.

Criminal gangs. Term commonly used for gangs that routinely engage in criminal behavior such as drug trafficking, burglary, vehicular theft, and assault.

Criminal sexual abuse. Abuse that involves the intentional fondling of the genitals, anus, or breasts, or any other part of the body, through the use of force or threat of force or of a victim (child) unable to understand the nature of the act, for the purpose of sexual gratification.

Criminal sexual assault. Assault that involves contact with or intrusion into the sex organs, anus, mouth, or other body part by the sex organ of another, or some other object wielded by another, with accompanying force or threat of force or of a victim (child) unable to understand the nature of the act.

Criminal subculture. Subculture in which juveniles are encouraged and supported by well-established conventional and criminal institutions.

Critical theories of causation. Theories of causation that broadly challenge the prevailing "orthodoxy" of criminology. These theories hypothesize that social tensions and conflicts are indelible features of society. Such tensions and conflicts are root causes of criminality and delinquency. Examples of critical theories in the United States are conflict theory and radical criminology.

Custody. Process of asserting physical control over juvenile offenders or victims. Custody is conceptually distinguishable from arrests and reflects the underlying duty of law enforcement agencies, as part of the juvenile justice system, to further the system's mission of protecting and rehabilitating children in trouble and children in need.

DARE. Drug Abuse Resistance Education, a program that emphasizes drug awareness education for elementary school students. Police officers from local station houses teach children about the different types of drugs, their effects, and how to recognize them. Officers also explain how to say "no" when drugs are offered.

Day treatment centers. Nonresidential institutions designed to provide treatment services to juvenile delinquents and other at-risk juveniles who might otherwise be placed in detention centers. Assignment of these centers can be done either as a condition of probation or as a diversionary alternative.

Defense counsel. Attorney for the juvenile who is processed through the juvenile court system.

Delinquency and drift. Theory that firm commitment to subcultural values is not necessarily a precursor of delinquent behavior, as argued by Sykes and Matza. Using techniques of neutralization, juveniles drift in an out of the delinquent subculture over time.

Delinquency, neglect, abuse, and dependency cases. Cases or occurrences where findings of delinquency, neglect, abuse, and dependency have been made.

Delinquent gangs. Term commonly used for gangs that routinely engage in criminal behavior such as drug trafficking, burglary, vehicular theft, and assault.

Delinquent subculture. Grouping of young people who "mutually convert" associates in the commission of delinquent activity.

Demeanor. People's appearance, particularly in relation to their mood or state of mind.

Demonology. Ancient belief holding that human deviance is the product of evil otherworldly forces such as demons and devils. Remedies included exorcism of the evil spirits, which often involved ordeals of pain.

Dependent children. Children who are abandoned by parents or guardians or who are otherwise uncared for.

Detention. Period between when a juvenile is taken into custody and prior to a hearing before the juvenile court. It is roughly comparable to the adult concept of being "jailed" because in both circumstances a suspect is temporarily housed in a detention facility pending further processing.

Detention hearing. Court hearing to determine whether detention is required.

Deterministic theories of criminal causation. Theories that eliminate individual responsibility for criminal behavior. These theories essentially conclude that other factors, such as biological and spiritual variables, explain an individual's lack of responsible self-control.

Deterrence theory. Punishment alternative based on the fear of swift and severe punishment. In theory, harsh punishment deters individuals who might otherwise be prone to break the law as well as those who have already violated the law.

Deviance. Behavior that is contrary to the standards of conduct or social expectation of a given group or society.

Differential association theory. Sociological theory of causation holding that criminals and law-abiding people learn their behavior from associations with others. People imitate or otherwise internalize the quality of these associations, so that criminality and delinquency are learned behaviors that are acquired from interacting with others.

Discretionary waiver. Waiver that may be left to the juvenile court judge to decide after a petition for a waiver has been filed and a hearing has been conducted on the advisability of granting the waiver. In general, the criteria used by juvenile court judges to determine the granting or denying of waivers of juveniles to criminal courts are rather vague and, for the most part, quite subjective.

Dispositional hearing. Hearing similar to the sentencing hearing in adult court. The court rules on what should become of the juvenile subsequent to the adjudication.

Diversion. Conceptual underpinning for community-based programs. This is a process that diverts juveniles away from the formal and authoritarian components of the juvenile justice system. Programs may be coordinated by youth agencies, juvenile courts, or the police.

Domestic violence. Violence involving one family or household member against another.

Double jeopardy. Criminal prosecution of a juvenile subsequent to proceedings in juvenile court involving the same act. In *Breed v. Jones,* the U.S. Supreme Court unanimously ruled that the Fifth Amendment's prohibition against double jeopardy precludes such criminal prosecution.

Drive-by shootings. Shootings by carloads of gun-wielding gang members. Occupants of the vehicles spray gunfire at members of rival gangs, firing wildly and often hitting innocent bystanders or residents inside their homes.

Dropouts. Juveniles who do not complete their secondary education. The term is also used to describe persons who do not participate throughout the full duration of a study.

Drug Abuse Resistance Education. *See* DARE.

Drug courts. Courts that attempt to prevent children and adults from continuing deviant drug-using behaviors. These courts aim to stop the abuse of alcohol and other drugs through the use of intensive therapeutic supervision.

Drug gangs. Street gangs that are organized around drugs as a criminal enterprise. They are fundamentally criminal organizations and are motivated by profit. They are usually composed of juveniles and young adults, with the latter being the leaders and "brains" of the operations.

Due process. Observing constitutional guarantees and rules of exclusion when processing court cases.

Ecological/Social disorganization approach. Approach that focuses on the geographic distribution of delinquency. It is commonly associated with the work of Shaw and McKay.

El Rukn. Chicago gang that was previously known as the Blackstone Rangers and later was called the Black P Stone Nation until it was called by the current name. This gang has emerged as a group characterized by a high degree of organization and considerable influence.

Electronic monitoring. Form of supervision that requires the juvenile to wear an anklet or other such device that allows the community supervision agency to electronically monitor the location of the juvenile on supervision.

Elmira Reformatory. Facility in New York where parole was first implemented in the United States in 1876. It was one of the first institutions in the United States to apply indeterminate sentencing with the possibility of early release. A new system of granting "good time" points, or credit, to inmates was designed to encourage proper behavior.

Emancipated juveniles. Juveniles who become emancipated from the control of their parents or the state under certain circumstances (e.g., marriage). This is predicated on a threshold age. They become *de jure* (legal) adults, which allows them to enter into contracts, own real estate, and accept responsibilities that would normally not be legally binding.

Emotional disturbance. Extremes in emotions exhibited by emotionally disturbed juveniles. Emotions can include impulsiveness, aggression, and polar (extreme) mood swings.

Emotional or psychological abuse. Injury to the intellectual or psychological capacity of a child, as evidenced by a discernible and substantial impairment in the ability to function within the normal range of performance and behavior.

Emotional neglect. Inadequate nurturance or affection given to children. It is basic inattention given to the emotional and developmental needs of juveniles.

Environmental factors. Effect of people, institutions, and other immediate associations on the personal development and behavioral traits of juveniles.

Era of socialized juvenile justice. The period between 1899 and 1967 in the United States. During this era, juveniles were considered not as miniature adults but rather as persons with less than fully developed morality and cognition.

Exclusive jurisdiction. Term indicating that the juvenile court will be the only tribunal legally empowered to proceed and that all other courts are deprived of jurisdiction.

Explorers programs. Programs that allow juveniles to learn about police work firsthand. Participants accompany officers on ride-alongs and other duties as a way to give positive experiences to juveniles and promote good relations.

Expungement. Destruction of files and records that document a particular juvenile's offense history. Expungement decrees generally originate with juvenile courts and are delivered to police departments.

Faith-based initiatives. Initiatives that are intended to strengthen and expand social services offered to the needy (in particular those in poverty). Faith-based organizations can apply for grants and funding to support programs aimed at drug prevention, violence prevention, at-risk youth, and gang behaviors.

Female associates (gangs). Girls and young women who belong to gangs but who are not exploited or subordinated by male members, in contrast to female auxiliaries. They are a "middle ground" between female auxiliaries and female gang-bangers.

Female auxiliaries (gangs). Girls and young women who play a subordinate role to male gang members. They do not share in leadership roles, they perform menial or otherwise secondary tasks, and they behave in a support role for the gang. Extreme exploitation and abuse of female members by male members may occur.

Female gang-bangers. Members of all-female gangs. They are completely independent from male gangs and act on their own volition without reference to, or permission from, male gang members. They are often formed as a reaction to the subordination and exploitation found in male gangs. They are a fairly rare phenomenon and are found in large cities.

Feminism. Approach to studying crime and delinquency that focuses on women's experiences, typically in the areas of victimization, gender differences in crime, and differential treatment of women by the justice network.

Finding of fact. Finding by a juvenile court judge concluding that a juvenile has committed a status offense, is delinquent, has been abused or neglected, or is dependent and in need of supervision.

FINS. Families in need of supervision.

"Folks." One of two alliances of gangs, or "sets," formed in Chicago. Along with the "people," the "folks" is an amalgamation of gangs that cuts across racial and ethnic identities. These alliances are not quite mega-gangs but rather groupings that identify fellow members as allies.

Follow Through. Similar to Head Start, program designed to help culturally deprived children catch up or keep pace during their preschool and early school years.

Forestry camps. Example of wilderness correctional programs. They are minimum-security residential corrections facilities, usually located on public lands such as state parks, where residents are required to perform forestry and land maintenance duties. Professional staff members manage the ranches and provide treatment and therapy.

Foster families. Families whose members care for children assigned to foster homes.

Foster homes. Nonsecure residential facilities that expand the concept of family replication by loosening the institutional restrictions found in group homes with a small number of supervisors and residents. Foster care is generally limited to victims of abuse and neglect rather than to lawbreakers. It attempts to substitute a foster family for a child's biological family.

Foster parents. Heads of group homes who, unlike their professional group home counterparts, are not professionally trained managers. Foster parents ideally nurture their residents and provide them with the guidance that was absent from the children's homes.

Fourth Amendment. Protection afforded against illegal search and seizure that extends to juveniles.

Free will approach. Crime control based on the belief that humans exercise free will and that human behavior results from rationally calculating rewards and costs in terms of pleasure and pain.

Free will theories of criminal causation. Category of theories that regard deviant behavior as a product of individual rational choice. Such rational choice is grounded in the human desire for pleasure and aversion to pain.

Functionally related agencies. Agencies that have goals similar to those of the juvenile justice system, namely, improving the quality of life for juveniles by preventing offensive behavior, providing opportunities for success, and correcting undesirable behavior.

Gang intervention. Approach to gangs where the philosophy mirrors that of gang prevention. The idea is to eliminate the conditions that lead to delinquency and criminal deviance. Thus, within the purview of prevention strategies, it is believed that delinquency, criminality, and gang creation can be "nipped in the bud" if certain identified conditions are corrected.

Gang prevention. Middle ground between gang intervention and gang suppression. Intervention policies require juvenile justice officials and the community to intervene in the lives of individuals who have crossed the line and adopted (or are about to adopt) the lifestyle of a gang-banger. Intervention specialists do not consider these individuals to be lost to society; rather, they see them as in need of being rehabilitated and reclaimed by the community.

Gang Resistance Education and Awareness Training. *See* GREAT.

Gang suppression. Policy to aggressively confront gangs as organizations and to confront gang members as individuals. The purpose of suppression is to either eliminate gangs or significantly repress their activity.

Gang-bangers. Slang term for members of youth gangs.

Gangs. Self-established, loosely organized, and antisocial groups of juveniles and/or young adults. Most gangs have adopted names, claimed territories, and devised symbols. They often have poor family and/or community relations and often present a significant threat of social disruption. Other terms for this phenomenon include street gangs, youth gangs, delinquent gangs, and criminal gangs.

***Gault* case.** U.S. Supreme Court ruling that in hearings that may result in institutional commitment, juveniles have the right to counsel, the right to confront and cross-examine witnesses, the right to remain silent, the right to a transcript of the proceedings, and the right to appeal.

"Get tough" or "just deserts" approach. Approach that would have an increasing number of juveniles processed within the adult justice network.

Graffiti. Various forms of painted and coded messages that are placed on buildings throughout the community as a means of communicating to allied and rival gang members. Street gang graffiti is unique in its significance and symbolism. Graffiti serves several functions; it is used to delineate gang turf as well as turf in dispute, it proclaims who the top-ranking gang members are, it issues challenges, and it proclaims the gang's philosophy.

Graffiti removal community service programs. Programs in some cities where taggers and gang members apprehended by the police are assigned to graffiti removal duty as part of their community service. Graffiti is often a self-described art form in which "taggers" draw or write messages on private property. It is also a means of communication among street gang members.

GREAT. Gang Resistance Education and Awareness Training, a program based on assumptions similar to, and also subject to criticisms similar to, those of the DARE program. These gang resistance programs train police officers to conduct comprehensive anti-gang education programs for juveniles who are not yet in high school.

Group homes. Nonsecure residential facilities managed by professional juvenile corrections workers who serve as administrators, teachers, and counselors. They try to replicate the values and social interactions of families; in practice, this means that most group homes are not large facilities. Group homes are integrated into neighborhoods.

Guardian ad litem. Court-appointed representative for juveniles who have been abused or neglected or who are dependent. He or she represents the personal interests of juveniles who are deemed unable to do so on their own behalf because of their immaturity.

Halfway house. Type of group home that houses up to 30 or 40 juveniles. Residents tend to be juveniles who are transitioning from correctional or substance abuse institutions to full release. Participants are sent to a halfway house as a way to prepare them for eventual reintegration into the community.

Head Start. Federal program that delivers preschool and early school educational services to children from poor or culturally marginalized environments. The program provides instruction in basic reading and verbal skills for children who might otherwise grow up with poor skills and thereby an increased possibility of future delinquency or criminality.

"Hidden" victimization. Unreported incidents of child abuse, neglect, exploitation, or other victimization. It often occurs in homes, schools, places of worship, or other insular settings.

Highfields Project. Common name used for the New Jersey Experimental Project for the Treatment of Youthful Offenders, an innovative program founded in New Jersey in 1950.

***Holmes* case.** U.S. Supreme Court ruling that because juvenile courts are not criminal courts, the constitutional rights guaranteed to accused adults do not apply to juveniles.

Home invasions. Violent trespass into homes by groups of juveniles who terrorize the residents. Crimes committed by invaders include assault, murder, rape, and robbery.

Homeward Bound. Variant of the Outward Bound model, established in 1970 in Massachusetts. This program developed personal growth through wilderness experiences. It also established a 6-week program that provided daytime challenges and nighttime education on survival skills and the natural environment.

House arrest. Confinement of juveniles to their residences under mandated terms and conditions. It is used in juvenile probation, parole, and aftercare programs.

Houses of refuge. Institutions founded during the 1820s to house vagrant and criminal juveniles. They are a typical example of Enlightenment-era juvenile justice reforms.

Hybrid sentencing. Means of matching a juvenile's sentence to the specific issues and circumstances associated with that offender. These sentences also allow juveniles to be sentenced to adult and juvenile dispositions by both adult court judges and juvenile court judges. Instead of classifying juveniles as one or the other, hybrid sentences allow the convenience and benefits of both courts. Juveniles are given a second chance by being allowed to rely on the juvenile court for rehabilitation while being aware that the adult court sanctions apply if rehabilitation fails.

Id, ego, and superego. Three components of the human conscious psyche according to psychoanalytic theory. The id involves instinctual and reactionary impulses, whereas the superego involves the moral sense of self. Both are balanced by the ego, a psychic component that mediates between the two extreme components of the human psyche.

Illegitimate opportunity structure. Scenario that where illegitimate opportunities are available, juveniles who are experiencing strain or anomie are attracted to that structure and are likely to become involved in delinquent activities.

In loco parentis. Latin term meaning "in the place of parents."

In need of supervision. Nondelinquent category to identify juveniles who commit status offenses but do not engage in aberrant behaviors that are serious enough to warrant a classification as delinquent.

Incapacitation. Punishment philosophy that imposes severe penalties, making offenders incapable of committing further crimes. Incapacitating judgments can include long-term imprisonment or execution.

Incorrigibles. Juveniles whose behavior is not controlled, or cannot be controlled, by authority figures. These are children who are in chronic conflict with authority figures.

Indeterminate sentencing. Sentencing philosophy that encourages rehabilitation by offering early release for behavioral conformity. The theory is that inmates will conform if sentences are flexible and indeterminate.

Initial contact. First occasion of police intervention when juveniles break the law or are victimized.

Integrated theories. Theories that attempt to combine two or more preexisting theories so as to provide more comprehensive explanations for criminal and delinquent behaviors.

Intensive aftercare supervision. Program that establishes guidelines for enhanced levels of postrelease supervision of juveniles that are more stringent and invasive than normal supervision.

Intensive supervision. Probation program mandated for persons who may pose a risk of flight or noncompliance with the terms of probation. Adult intensive supervision was originally designed during the 1980s to lower costs and reduce prison overcrowding. It is an intensified version of standard probation and emphasizes increased surveillance, more frequent contacts with probation officers, and enhanced control over participants. Electronic monitoring is commonplace.

Internet exploitation. Exploitation that occurs when child sexual predators use the Internet as a means of communicating with potential victims. Because of its anonymity, rapid transmission, and often unsupervised nature, the Internet has become a medium of choice that predators use to contact juveniles and transmit and/or receive child pornography.

Intervening institutions. Institutions whose officials are key personnel in the effort to help troubled and needy children. Examples include schools and law enforcement agencies.

Intervention. With respect to child abuse, a process that begins with someone reporting the abuse or suspected abuse, moves into the investigatory stage that typically involves a home visit and interviews with the parties involved, and then moves to risk assessment and a decision concerning what type of action, if any, to take.

Intermediate sanctions. Intensive alternatives to long-term institutionalization. These sanctions include house arrest, "shock" incarceration, electronic monitoring, and boot camps. Supervision of juveniles can be quite intensive.

Interrogation. Questioning by police. While in custody, juveniles have rights similar to those of adults with respect to interrogation.

Jails. Formerly the primary institution for holding juveniles, a practice that ended with the establishment of houses of refuge and later juvenile facilities. Although jails have long ceased to be the *principal* holding institution for juveniles, significant numbers of juveniles continue to be processed through jails.

Jargon. Terminology used by gang members. It can be very useful for practitioners to know and understand. Understanding gang jargon may be critical to obtaining convictions for gang-related crimes.

John Augustus. Founding father of probation.

Judicial reprieve. Early procedure similar to probation that permitted judges to suspend sentences for offenders pending appeal to the Crown for clemency. Judges also suspended unfair sentences when deemed necessary.

Judicial transfer. Systems that apply judicial waiver. Juvenile court judges have the authority to waive jurisdiction over juveniles, thereby sending them into the adult system. Variations of this model are most commonly found in the states.

JUMP. Juvenile Mentoring Program, a mentorship program administered by the Office of Juvenile Justice and Delinquency Prevention. Encounters consist of interpersonal mentorship between adults and children. The program is targeted to communities that potentially produce high rates of at-risk youth such as areas with a high incidence of crime, poverty, and/or school dropouts.

Jurisdiction (justice agencies). Lawful authority of a court or other justice agency. Jurisdictional power is delimited by territory, subject matter, or persons over which lawful authority is granted by statutes or constitutions.

Jurisdiction (personal). A court's power over certain persons, as granted by statutes or constitutions. It establishes judicial authority to render judgment in the interests of particular classes of people.

Jurisdiction (subject matter). Authority of a court to decide a particular class of cases.

Jurisdiction (territorial). Geographic scope of a court's authority. Unless otherwise specified by law, a court possesses the power to hear cases only within certain territorial limits defined by political boundaries such as specified municipalities or counties.

"Just Deserts" model. Model that applies the old adage of "an eye for an eye" when determining the fate of an offender. Conceptually, it advocates punishments that fit the crimes.

As a corresponding approach to the punishment model, juvenile offenders receive the punishments they deserve from juvenile court judges. Obviously, this is appropriate only for offenders, not victims.

Juvenile. Legal classification that is established within the parameters of culture and social custom. It fundamentally refers to those individuals who are below the age of adulthood, another classification. In the modern era, laws determine when a person is an adult, and juveniles are a defined class of nonadult persons who receive special treatment under the law.

Juvenile court (coordinated). Courts that are required to coordinate their duties with specialized courts such as family courts.

Juvenile court (designated). Juvenile systems that are subclassifications of larger county, municipal, or district systems. Juvenile courts function in accordance with their designations as divisions within these systems.

Juvenile court (independent). Juvenile systems that are completely independent from other courts. Although judges from other courts may be assigned to sit as juvenile court judges, the system itself is independent.

Juvenile court judge. Central participant in a juvenile court system who sits as the final arbiter for disputes and problems brought before the court. He or she is vested with the authority of the state, tangibly and personally representing *parens patriae.*

Juvenile court prosecutors. Attorneys representing the interests of the state in juvenile delinquency hearings. They retain the traditional exercise of prosecutorial discretion in deciding the question of whether to file petitions on behalf of juveniles and the charges to be brought against them.

Juvenile court statistics. Data on cases processed by juvenile courts. Research reports such as *Juvenile Court Statistics,* published by the Office of Juvenile Justice and Delinquency Prevention and the National Center for Juvenile Justice, contain statistics on juvenile court proceedings. These data include case statistics at the local jurisdictional level.

Juvenile criminals. Juveniles who officially cease to be juvenile delinquents and become defined as criminals.

Juvenile defense counsel. Attorneys for juveniles brought before the juvenile court. Juveniles accused of delinquent acts are entitled to legal representation under the same constitutional protections as their adult counterparts in the criminal justice system.

Juvenile delinquents. Individuals under the age of majority who commit offenses that would be classified as crimes in the adult criminal justice system. As delinquents, these offenders are processed through the juvenile justice system.

Juvenile detention centers. Residential correctional facilities for juveniles that are roughly comparable to adult jails. They are secure temporary preadjudication institutions that are used to house accused offenders until the final dispositions of their cases.

Juvenile deviance. Antisocial behavior by juveniles. It includes status offenses (violations of laws exclusively governing juvenile behavior) and delinquent acts (behavior that would be criminal if juveniles were tried as adults).

Juvenile intensive probation supervision. Closely monitored probation. It is provided to juveniles who may pose a risk of flight or noncompliance with the terms of probation. Electronic monitoring, using an electronic bracelet or some similar device, is commonplace.

Juvenile justice. Fair handling and treatment of juveniles under the law. It is a philosophy that recognizes the right of young people to due process and personal protections.

Juvenile justice process. Procedures established to ensure the fair administration of juvenile justice under the law. These procedures are carried out in accordance with institutions designed specifically for the administration of juvenile justice.

Juvenile justice system. Institutions that have been organized to manage established procedures as a way to achieve justice for juveniles. These institutions include the police, juvenile courts, juvenile corrections, and the community.

Juvenile Mentoring Program. *See* JUMP.

Juvenile petitions. Documents that request the court's intervention over matters involving juvenile delinquents, status offenders, children who have been abused or neglected, and dependent children. They are roughly equivalent to complaints or pressing charges in the adult system but are issued out of consideration for the best interests of the children.

Juvenile probation officer. Key figure at all levels of the juvenile justice system. He or she may arrange a preliminary conference among interested parties that may result in an out-of-court settlement between an alleged delinquent and the injured party or between parties in cases of abuse or neglect. After an adjudicatory hearing, the juvenile probation officer is often charged with conducting a social background investigation. This investigation is used to help the judge make a dispositional decision. This officer is also charged with supervising those juveniles who are placed on probation and released into the community as well as parents who have been deemed to have committed neglect or abuse.

Juvenile victimization. Ill treatment of children, including when juvenile delinquents victimize other children and when children suffer abuse or neglect from parents or other family members. Victimization may occur at home, in schools, or on the street.

***Kent* case.** U.S. Supreme Court ruling that during waiver hearings, juveniles are entitled to a hearing that includes the essentials of due process required by the Fourteenth Amendment.

Labeling process. Process whereby reactions of society, family, and the justice system can exaggerate problems in family, school, and peer relations, and the juvenile may find it difficult to meet the expectations established for him or her.

Labeling theory. Theory holding that society's reaction to deviant behavior is crucially important in understanding who becomes labeled as deviant.

Latchkey children. School-age children who return home from school to empty houses. Estimates indicate that there may be as many as 10 million children left unsupervised after school.

Latin Kings. Largely Hispanic gang whose members are aligned with the Vice Lords. This gang is known as "the people," and its members typically are not aligned with the Black Gangster Disciples or Simon City Royals.

"Lawgiver" judge. Judge who is primarily concerned that all procedural requirements are fulfilled in court proceedings.

Learning disabled. Possessing deficits in learning processes such as poor reading ability, lack of ability to memorize or follow directions, and incapability to distinguish or otherwise manage letters or numerals.

Learning theory. Theory holding that people learn behaviors through various forms of reinforcements, punishments, and social observations.

Legalistic approach. Formal and official approach with strict adherence to constitutional safeguards.

Legal definitions. Definitions holding that only those who have been officially labeled by the courts are offenders.

Legislative waiver. Procedure whereby the concepts of waiver and certification have been modified and codified into state legislation mandating the removal of specified offenders from juvenile court. This procedure, commonly known as statutorial exclusion, requires the assignment of juveniles suspected of murder or other serious offenses to the adult criminal system.

Long-term correctional facilities. One of two general types of juvenile correctional institutions. These facilities include training schools, youth ranches and forestry camps, and boot camps. The other general type of juvenile correctional institution is short-term temporary care facilities.

Mandated reporters. Reporters who are required to report suspected cases of abuse to the state.

McKeiver v. Pennsylvania. U.S. Supreme Court ruling that the due process clause of the Fourteenth Amendment does not require jury trials in juvenile court.

Mens rea. Latin term for "guilty mind." It centers on the question of when and under what circumstances children are capable of forming criminal intent.

Methamphetamine. Stimulant drug, also known as "meth," that is manufactured easily and sold inexpensively.

Monikers. Gang member nicknames that are different from members' given names. In general, these monikers reflect physical or personality characteristics or connote something bold or daring.

Munchausen's syndrome by proxy. Form of child abuse in which the abuser fabricates (or sometimes creates) an illness in the child victim. The child is then taken to a physician, usually by the mother, who knows that hospitalization for tests and observation is likely to be recommended because the symptoms are described as severe, but no apparent cause exists.

Narcotics trafficking. Illicit sale and distribution of narcotic drugs.

National Center on Child Abuse and Neglect. Organization that publishes data on abuse and neglect of children.

National Center for Juvenile Justice. Organization that collects and publishes information on the number of delinquency, neglect, and dependency cases processed by juvenile courts nationwide.

National Children's Advocacy Center. Organization that publishes data on abuse and neglect of children in tandem with a number of other agencies.

National Crime Victimization Survey. Data derived from annual household surveys conducted by the U.S. Department of Justice's Bureau of Justice Statistics in collaboration with the U.S. Census Bureau. It is usually made twice annually and asks approximately 40,000 to 50,000 respondents whether members of their households have been victims of crimes.

National Incident-Based Reporting System. System developed to collect information on each single crime occurrence. Under this reporting system, policing agencies report data on "offenses known to the police" (offenses reported to or observed by the police) instead of only those "offenses cleared by arrest," as was done in the original crime reporting process of the Uniform Crime Reports.

National Juvenile Court Data Archive. Extensive archive of automated records of juvenile court cases managed by the Office of Juvenile Justice and Delinquency Prevention. The archive was established to promote research on juvenile justice.

National Probation Act. Passed in 1925, legislation that authorized federal district court judges to hire probation officers as well.

National Youth Survey. Longitudinal study of a national sample of 1,725 juveniles involving the distribution of several series of questionnaires to the sample. It is an example of self-report data. This survey has generally reported that the incidence of violent delinquency among juveniles is significantly higher than that reported by official data such as the Uniform Crime Reports and the National Crime Victimization Survey.

Neglected children. Child victims who do not receive proper care from parents or other guardians. They include children who are unfed, poorly sheltered, left alone, not cleaned, or otherwise maltreated.

Neoclassical approach. Modification of the classical school that occurred during the early 19th century. It recognizes that juveniles and mentally ill adults do not make the same rational choices as do mature sane adults. Therefore, special consideration should be given to these classes of offenders.

Neo-Nazism. Variety of white supremacist gangs in the form of skinhead groups such as the Nazi Low Riders, White Aryan Resistance, Hammer Skins, World Church of the Creator, and other groups whose members perpetrate hate crimes.

Nonofficial juvenile justice institutions. Community resources and agencies that make up a component of the juvenile justice system. Procedures are usually informal and as welcoming as possible, depending on the circumstances of each case. Community-based agencies are responsible for rehabilitating and rescuing young people.

Nonsecure correctional facilities. Juvenile correctional facilities that typically require lockups only at specified times during the day or week such as at night.

Nonsecure residential facilities. Institutions that provide a more wholesome atmosphere than do locked detention facilities. They are often managed as homes or rescue facilities that provide different kinds of services ranging from shelter to intensive treatment. Examples of nonsecure residential facilities include shelter care facilities, group homes, and foster homes.

Notice in lieu of summons. Alternative procedural option available to the court that is delivered to juveniles and parents. It is an informal "softer" alternative to a formal summons and serves as a notification to appear before the court.

Notification. Notice that must be given to all parties concerned before a proceeding.

Offenses known to the police. Offenses reported to or observed by the police.

Official data. Information compiled by law enforcement agencies, courts, and correctional systems at the national, regional, and local levels. These data may be reported in the Uniform Crime Reports, juvenile court data, or National Crime Victimization Survey.

Official procedures. Clearly delineated guidelines that officers must follow when they formally process juveniles. These procedures are written into juvenile court acts and are reflected in internal procedures adopted by law enforcement agencies.

Office of Juvenile Justice and Delinquency Prevention. Organization that coordinates national policy for preventing juvenile delinquency and victimization. Established under the Juvenile Justice and Delinquency Prevention Act of 1974, it conducts independent research, assists experts and agencies in program evaluation, and recommends initiatives for delinquency prevention and treatment.

Original gangsters. Gang-bangers who are recognized by their reputation for affiliation with a particular gang. Sometimes referred to as "OGs," they receive a street-level respect for the things they have done to enhance their own status and the status of the gang. These activities may include drive-by shootings, imprisonment, killings, and other gang-related behavior. They are hardcore members who have never "sold out" their gang identification.

Outward Bound. Wilderness programs offered in school-like environments where participants learn about individual confidence, trust in others, and the meaning of personal character. Juveniles are generally housed on-site at schools and camps. Professional counselors manage the challenge programs, which usually last for 3 weeks and consist of four phases.

Overpolicing. Policing of a given area that is much more than would typically be necessary to maintain public order.

Parens patriae. Concept of the monarch as "father of the country." It is an ancient English doctrine allowing the state to intervene as a surrogate parent in the best interests of children whose parents have failed in their duties to protect, care for, and control their children.

"Parent figure" judge. Judge who is often genuinely concerned about the total well-being of juveniles who appear before the court.

Parole. Supervised early release from a residential correctional facility. Offenders' sentences begin in confinement and end in the community under supervision from a parole agency, assuming that they comply with the terms of parole. Participants are juvenile delinquents or criminals who have already served time in a correctional institution.

Parole boards. Independent parole or aftercare committees whose members are usually appointed by governors and approved by state legislatures. They are unaffiliated with state parole agencies or correctional facilities.

Part I offenses. Eight serious crimes reported in the Federal Bureau of Investigation's Uniform Crime Reports. These offenses are considered to be inherently serious crimes or are those that occur frequently. Part I crimes are punishable by long-term incarceration or capital punishment.

Part II offenses. A total of 21 lesser offenses reported in the Federal Bureau of Investigation's Uniform Crime Reports. These offenses are considered to be less serious than Part I crimes, do not occur frequently, or are considered to be victimless crimes.

Pedophile. Criminal offender who seeks out children for purposes of sexual gratification.

"People." One of two alliances of gangs, or "sets," formed in Chicago. Along with the "folks," the "people" is an amalgamation of gangs that cuts across racial and ethnic identities. These alliances are not quite mega-gangs but rather groupings that identify fellow members as allies.

Personality inventories. Psychological assessment tools examining people's personality characteristics.

Petition. Form process whereby a party initiates proceedings in a court of law.

Phrenology. Study of the shape of the skull.

Physical neglect. Abandonment, expulsion from the home, delay or failure to seek remedial health care, inadequate supervision, disregard for hazards in the home, and/or inadequate provision of food, clothing, or shelter.

Physiognomy. Ascribing moral and behavioral traits to physical appearance. In particular, facial characteristics are deemed to be indicators of moral character. It was popular from the Enlightenment through the 19th century.

PINS. Persons in need of supervision.

Police discretion. Exercise of judgment by individual officers on what type of action to take in a particular situation.

Police dispositions. Any contacts made by police officers regarding the welfare and safety of juveniles. Officers may exercise professional discretion to formally or informally resolve the dispositions. Many personal and legal factors affect the outcomes of police dispositions.

Police observational studies. Observations provided by researchers that help in better assessing the extent of unofficial or hidden delinquency, abuse, and neglect.

Police–school liaison officers. Officers located in schools who serve as sources of information and counselors for students.

Politicized gangs. Gangs that use violence for the purpose of promoting political change by instilling fear in innocent people.

Positive peer culture. Orientation used in some treatment programs in which the youthful offender is surrounded by a positive and prosocial set of peers in the program.

Positivist school of criminology. Deterministic approach to criminality and delinquency. This approach holds that biology, culture, and social experiences can be sources of deviant behavior.

Postadjudication detention. Confinement of juveniles in correctional facilities under court order after cases have been adjudicated. These include short-term facilities as well as long-term confinement facilities such as training schools, alternative detention programs (e.g., wilderness ranches, boot camps), and diagnostic centers.

Postadjudication intervention. Intervention that is provided after the adjudication phase.

Positivist school of criminology. School that emphasizes the need for empirical research in criminology, with some stressing the importance of environment as a causal factor in crime.

Postclassical theory. Theory involving the notion that people rationally consider the risks and rewards before they commit crimes. It is also called rational choice theory.

Preadjudication detention. Confinement of juveniles in youth facilities prior to the final dispositions of their cases by a juvenile judge. These are short-term confinement facilities such as nonsecure residential facilities, juvenile detention centers, and jails.

Preadjudication intervention. Intervention that is provided before the adjudication phase.

Predisposition report. Report filed by a probation officer with the juvenile judge after findings of fact are entered. It contains a social history of the juvenile and recommends a disposition to the court. This report is sometimes referred to as a presentence investigation.

Preliminary conference. Voluntary meeting arranged by a juvenile probation officer with the victim, the juvenile, and typically the juvenile's parents or guardian in an attempt to negotiate a settlement without taking further official action. Juvenile court acts clearly indicate those persons who are eligible to file a petition.

Preliminary hearing. Initial appearance in adult court.

Prenatal exposure to diseases and drugs. When children are born to mothers who either had a venereal disease or abused substances during pregnancy. Babies can contract the mother's disease or suffer mental retardation or addiction from the mother's prenatal alcohol or drug abuse.

Preponderance of the evidence. Standard of proof applied for the resolution of status offense cases. It technically refers to the greater weight of evidence that is more credible and convincing to the mind.

Preservice/Inservice training. Training given to law enforcement personnel. Preservice training is the initial academy-level training given to personnel when they begin their careers in law enforcement or corrections. Inservice training is additional training provided while the officers are employed with their respective agencies.

Prevention services. Community-based intervention services that anticipate potential problems and that are delivered as a means of reducing the probability that juveniles will return to the juvenile justice system.

Primary prevention. Intervention with juveniles who have not yet begun breaking the law or otherwise engaging in antisocial deviance.

Private attorneys. Attorneys who have private practices and are either personally retained or appointed as part of a court rotation system.

Private counsel. Attorneys who are sometimes retained or appointed to represent the interests of juveniles in court.

Private detention facilities. Detention facilities that tend to house fewer delinquents than do public facilities and that are less oriented toward strict custody than are facilities operated by the state department of corrections.

Probation. Sentence that is served in the community under supervision by a probation officer. The offender's sentence begins and ends in the community, assuming that he or she complies with the terms of probation.

Probation as conditional release. Community supervision that is purely conditional on offenders' compliance with any variety of court-mandated requirements set by the judge.

Probationary conditions (general). Conditions of probation that the juvenile court designs as "bottom-line" requirements for all probationers.

Probationary conditions (specific). Conditions of probation that the juvenile court uniquely designs for individual offenders.

Prosecutor/State's attorney. Official who makes the final decision about whether a juvenile will be dealt with in juvenile court. The prosecutor carries forth charges against the juvenile.

Prosecutorial discretion. Exercise of judgment by prosecutors on whether to refer juveniles to the court and, if so, on what type of charges to bring against them.

Prosecutorial waiver. System that permits prosecutors to exercise their discretion as to which court system (adult or juvenile) a case will be brought before.

Psychoanalytic approach. Theory proposed by Sigmund Freud arguing that humans have three personality components: the id, ego, and superego. Human personalities also progress through several phases of childhood development. Failure to progress through these phases in a healthy manner can lead to deviant behavior, as can failure to develop a healthy balance among the id, ego, and superego.

Psychological factors. Factors related to a juvenile's likelihood of future involvement in delinquency. These include dependency needs, family rejection, and impulse control.

Psychological theories. Theories that have their basis in the field of psychology as opposed to sociology.

Psychopathology theory. Concept of the psychopathic personality that describes criminals who behave cruelly and seemingly with no empathy for their victims. The fact that some criminals are apparently unable to appreciate the feelings of their victims led to a great deal of research on this behavior. Psychopaths, also known as sociopaths, are considered to be people who have no conscience—in Freudian terms, no superego. Aggressiveness and impulsiveness are typical manifestations of the psychopathic personality, and that is why many psychopaths become lawbreakers.

Psychopath. Aggressive criminal who acts impulsively for no apparent reason.

Public defenders. Court-appointed defense counsel who are full-time state employees and are essentially counterparts of prosecuting attorneys.

Public detention facilities. Detention facilities that are frequently located near large urban centers and often house large numbers of delinquents in a cottage- or dormitory-type setting.

Pure diversion. Preadjudicatory diversion. This is a process of immediate diversion, meaning that young offenders are sent directly into community-based programs before they are adjudicated by a juvenile court and processed into the formal juvenile justice system.

Purpose statement of a juvenile court act. Statement that spells out the intent or basic philosophy of the act.

Radical criminology. Critical approach to criminal causation first propounded during the 1960s and 1970s. Proponents argue that because power and wealth have been distributed unequally, those who have been shut out politically and economically understandably resort to criminal antagonism against the prevailing order.

Radical nonintervention. Approach that encourages law and policymaking organizations to be tolerant of the widest possible diversity of behaviors and attitudes. Such a process would then limit the amount of intervention necessary with juveniles.

Rational choice theory. Theory involving the notion that people rationally consider the risks and rewards before they commit crimes. It is also called postclassical theory.

Referees. Hearing officers who serve as assistants and advisers to the court. Refereed findings are advisory in nature, and judges have the authority to accept, modify, or reject recommendations.

Reform schools. Juvenile facilities established during the child savers era in the mid-19th-century United States. They were a programmatic counterpoint to houses of refuge in that the reform schools sought to create a nurturing environment rather than a harsh correctional one. Education and trade/craft were emphasized, as were home-like environments.

Rehabilitation. Treatment programs that try to salvage young offenders. This is often defined as a "liberal" approach to juvenile corrections.

Rehabilitation model. Original approach toward juvenile justice and still a fundamental concept. Rehabilitation refers to using institutions and programs to reclaim troubled juveniles. Under this model, methods and agencies are established to mold delinquents into productive adults and to mold victims into healthy adults.

Relative deprivation. Deprivation that is measured in comparison with the rest of society.

Remedial institutions. Child protective and health care agencies that essentially try to investigate and remediate the damaging effects of troubled and needy environments.

"Representing." How gang members present themselves as such. It involves the adoption of symbolic identifiers to set themselves apart and distinguish themselves from other gangs. Identifiers create a sense of self-importance and "specialness" for members.

Restitution. Conditions of probation, parole, or aftercare that require juvenile offenders to compensate victims and society for the harm they caused. Offenders are required to restore the victims to wholeness to the greatest extent possible.

Restorative justice. Type of justice in which the core concepts of accountability, competency, and public safety are used in mediation that includes the crime victim(s), the offender(s), and the community.

Restorative or balanced justice. Type of justice that uses mediation that includes the victim(s), the offender(s), and the community, as with restorative justice. The term *balanced justice* is used in some cases because this type of justice addresses all parties rather than focusing on only one party (i.e., the offender) to the exclusion of others.

Retribution. Outright punishment of youthful offenders that represents the theoretical separation of rehabilitation from penalties based on societal anger over the behavior of juvenile delinquents. Treatment is secondary to retaliation against juveniles who break the law.

Revitalized juvenile justice. Justice system that would include swift intervention with early offenders, an individualized comprehensive needs assessment, transfer of serious or chronic offenders, and intensive aftercare. Such a system would require the coordinated efforts of law enforcement, treatment, correctional, judicial, and social services personnel.

Revocation of probation. Cessation of probation with a corresponding imposition or execution of the sentence that could have been given originally by the judge.

Risk factors. Indicators of routes juveniles take toward juvenile delinquency. Risk factor environments include families, schools, peer associations, neighborhoods, and individual idiosyncrasies.

Role identity confusion. Confusion that occurs when probation officers are unclear about the expectations placed on them when they attempt to balance the competing interests of their "policing" role and their "reform"-oriented role.

Roper v. Simmons. A 2005 case in which the U.S. Supreme Court held that states cannot execute offenders who are under the age of 18 years.

Routine activities theory. Approach in which crime is simply a function of people's everyday behavior.

Runaways. A subcategory of missing children who voluntarily leave their households, usually under duress, and are forced to fend for themselves.

Scared Straight. First popularized in the 1978 film *Scared Straight,* program in which groups of young offenders spend a day in a maximum-security prison. They tour the facility and are placed in a graphic and intensive encounter session with hardened adult convicts.

Scientific theory. Set of two or more related, empirically testable assertions (statements of alleged facts or relationships among facts about a particular phenomenon), as defined by Fitzgerald and Cox.

Scope of a juvenile court act. Scope that is indicated by sections dealing with definitions, age, jurisdiction, and waiver.

Sealing of records. Preservation of files and records under strict confidentiality procedures. These files cannot be accessed without a court order, which is theoretically difficult to obtain.

Secondary diversion. Postadjudicatory diversion. This refers to the release of juveniles who have already been processed into the formal juvenile justice system. They are released into community-based programs prior to final disposition.

Secondary prevention. Intervention with juveniles who have only recently begun engaging in antisocial deviance and are in an early phase of committing relatively minor or status offenses. This is usually predelinquency prevention.

Secure correctional facilities. Juvenile correctional institutions that literally lock in juvenile offenders. Residents have no opportunity or discretion to leave the facilities.

Self-report studies. Studies constructed to resolve the false dichotomy between labeled and nonlabeled juveniles. They use self-reported data that inquires about respondents' own criminal behavior within a specified time period.

Self-reported data. Surveys of juveniles and adults asking respondents about their personal criminal behaviors within a specified time period.

Serious Habitual Offender Comprehensive Action Program. *See* SHOCAP.

Sexual abuse. Abuse that consists of the involvement of children in sexual activity to provide sexual gratification or financial benefit to the perpetrator, including contacts for sexual purposes, prostitution, pornography, or other sexually exploitative activities.

Sexual exploitation. Expanded statutory definition of sexual abuse. Such statutes typically include references to exploitation for pornographic purposes and to prostitution.

Shelter care. Nonsecure residential facilities that temporarily house juveniles. Shelters are typically used as temporary housing for status offenders and others while they await final placement to another facility or their family homes. Most shelters are not designed to treat or punish juveniles.

SHOCAP. Serious Habitual Offender Comprehensive Action Program, a multidisciplinary interagency case management and information-sharing system intended to help the agencies involved make informed decisions about juveniles who repeatedly engage in delinquent acts.

"Shock" programs. Programs that temporarily place juveniles under intensive, but guided, pressure with the goal of "shocking" them out of their antisocial behavior. Typical shock programs include boot camps and the Scared Straight model.

Shock intervention. Term used synonymously with "boot camp." The purpose is to "shock" residents into socially productive conformity.

Short-term temporary care facilities. One of two general types of juvenile correctional institutions. These facilities include jails, youth shelters, juvenile detention homes, and reception centers. The other general type of juvenile correctional institution is long-term correctional facilities.

Simon City Royals. Part of the "folks" alignment in gang membership. This group is a Caucasian gang that originated in Chicago during the early 1960s. Originally formed to stop the "invasion" of Hispanic gangs into their area of the city, the Royals became actively involved in burglaries and home invasions during the 1980s. As leaders of the gang were imprisoned, they formed alliances with the Black Gangster Disciples for protection, and the alliances continue today.

Sixth Amendment. Declaration that in all criminal prosecutions, the accused enjoys the right to a speedy and public trial. However, juvenile court acts prohibit public hearings on the grounds that opening such hearings would be detrimental to the children.

Skinhead gangs. "Lifestyle" gangs that are usually associated with white supremacist ideologies. Some are associated with neo-Nazi or other racial supremacist movements. As a counterpoint, some skinheads are antiracist groups that may have multiethnic memberships. All cut their hair very short and adopt distinctive dress such as boots, suspenders, and leather (or "bomber") jackets.

Social background investigations. Investigations that typically include information about the children, the children's parents, school, work, and general peer relations, as well as other environmental factors.

Social disorganization. Urban theory holding that poor urban communities are innately dysfunctional and give rise to criminal behavior.

Social ecology (structural) theory. Sociological theory of causation holding that the social structure of inner-city areas affects the quality of life of inhabitants. Structural elements include overcrowding, poor sanitation, unemployment, poverty, poor schools, transience, out-of-wedlock births, and low employment. These factors contribute to high crime and delinquency rates because of the resulting endemic social instability.

Social factors. Various factors that are not physiological in nature and that may increase the risk of problems such as abusing drugs and engaging in delinquent behavior.

Social promotions. When teachers pass troubled juveniles on to their colleagues by refusing to fail these "problem youth."

Socialization process. Process of learning moral and social norms of behavior, as defined within children's immediate environment.

Socioeconomic status. Individuals' social and economic position within society.

Sociological factors. Factors that may include a person's place of residence, school attended, age, race, nationality, and the like.

Sociological theories. Theories that look for causes of delinquency in society as well as in the individual.

Somatotypes. Theory of delinquent and criminal causation that classifies three body types as determinants of deviant behavior: *mesomorphs,* who are muscular, sinewy, narrow in waist and hips, and broad-shouldered; *ectomorphs,* who are fragile, thin, narrow, and delicate; and *endomorphs,* who are pudgy, round, soft, short-limbed, and smooth-skinned. Predominance of mesomorphic traits is theoretically found in delinquents and criminals.

Spergel model. Model that involves developing a coordinated team approach to delivering services and solving problems. Strategies involved include mobilization of community leaders and residents; use of outreach workers to engage juveniles in gangs; access to academic, economic, and social opportunities; and gang suppression activities.

Standard of preponderance of evidence. Standard of proof applied for the resolution of status offense cases. It technically refers to the greater weight of evidence that is more credible and convincing to the mind.

Stationhouse adjustment. Decision that occurs when a juvenile is brought into a police station house and his or her parents are called in for consultation. Officers can release the juvenile into the parents' custody, thereby ending the case prior to official processing into the juvenile justice system.

Status offenders. Offenders who violate laws that regulate the behavior of persons under the age of majority. These laws govern individuals who have not reached the age of majority and do not apply to adults.

Status offenses. Legal offenses applicable only to children and not to adults.

Stigma. Imprinting of disgrace or shame on a person so that he or she is thereafter judged by adults and peers in accordance with this impression.

Stoner gangs. "Lifestyle" gangs that are organized around drug use and hard rock music. They represent a countercultural alternative to mainstream society and may express themselves with piercings, tattoos, black clothing, and "Gothic" appearances.

Strain theory. Sociological theory of criminal causation that generally refers to a state of "normlessness" in society. Modern strain theory focuses on the availability of goals and means. When the greater society encourages its members to use acceptable means to achieve acceptable goals, and not all members have an equal availability of resources to achieve these goals, members may resort to illegitimate and illicit means.

Street corner adjustment. Informal response that officers may take to nonserious delinquency and/or vagrancy to resolve nuisance behavior committed by juveniles.

Street corner family. Based on the idea that the gang is the member's family and that gangs satisfy deep-seated needs of adolescents. It is associated with the work of William F. Whyte, who portrayed the gang as a "street corner family."

Street corner police decisions. Decisions that occur at the initiation of interaction between police officers and juveniles. These typically involve asking teens to move along, releasing a juvenile after questioning, or issuing a reprimand before release.

Street gangs. Gangs that routinely engage in criminal behavior such as drug trafficking, burglary, vehicular theft, and assault.

Tagger crews and posses. Graffiti gangs whose members vandalize property by marking their territories with graffiti messages. Some tagger crews can be quite artistic, albeit by vandalizing someone else's property.

Taking into custody. When an officer detains a juvenile and keeps him or her under supervision, watch, and/or guard. When delinquency is the alleged reason for taking into custody, law enforcement officers must adhere to appropriate constitutional guidelines.

Technical violation. Minor infraction committed by the probationer while on community supervision. Technical violations are generally worked out between the probationer and the probation officer, and they usually do not result in revocation action unless the probationer develops a complete disregard for the terms or conditions of probation.

Techniques of neutralization. Mental defenses to delinquency internalization that include (a) denial of responsibility (for the consequences of delinquent actions), (b) denial of injury (to the victim or larger society), (c) denial of a victim (the victim "had it coming"), (d) condemnation of the condemners (as hypocrites or spiteful), and (e) appeal to higher loyalties (e.g., to the gang), as defined by Sykes and Matza. Using these techniques, juveniles drift in an out of the delinquent subculture over time.

Teen courts. Courts made up of teens under 17 years of age who process cases by acting as prosecutor, defense counsel, bailiff, and clerk and who determine the punishment for the cases by acting as the jury. An adult attorney acts as the judge to ensure the fairness and legality of the sentencing. Offenders are required to complete the sentences handed down by the teen jury.

Territorial jealousy. Belief commonly held by agency personnel that attempts to coordinate efforts are actually attempts to invade the territory they have staked out for themselves.

Tertiary prevention. Intervention with juveniles who have engaged in serious and chronic deviance and who have already entered the juvenile justice system. These juveniles are technically in need of treatment rather than prevention because efforts to prevent the onset of delinquency have failed. These juveniles should be approached within the context of rehabilitation rather than prevention.

Theory of differential anticipation. Developed by Glaser, theory that combines differential association and control theory and is compatible with biological and personality theories. It assumes that a person will try to commit a crime wherever and whenever the expectations of gratification from doing so—as a result of social bonds, differential learning, and perceptions of opportunity—exceed the unfavorable anticipations from these sources.

Theory of differential association. Developed by Sutherland, theory that combines some of the principles of behaviorism (or learning theory) with the notion that learning takes place in interaction within social groups.

Therapeutic approach. Informal and unofficial approach, being grounded in approaches that emphasize treatment and/or casework.

Throwaways. Subcategory of missing children who have been ejected from their households and forced to fend for themselves.

Totality of circumstances. Approach that considers surrounding contextual facts in determining the validity of the waiver. Circumstances considered include the age, competency, and educational level of the juvenile; his or her ability to understand the nature of the charges; and the methods used in and length of the interrogation.

Training schools. Secure residential correctional facilities managed as either public or private institutions. Their underlying mission is to promote rehabilitation by providing intensive training and treatment. Many facilities are highly secure institutions designed in much the same manner as adult jails and prisons.

Treatment model. Approach that applies a therapeutic standard for evaluating the effectiveness of intervention. It is also referred to as the *medical model.* Psychological counseling, physical (health) regimens, and behavior modification are stressed as foundations for full rehabilitation. Punishment and detention are rejected as being counterproductive to successful treatment and rehabilitation. This approach is applicable for juvenile offenders and victims.

Treatment services. Community-based intervention services that manage existing problems and are delivered to rehabilitate juveniles prior to their reintegration into the community.

Trephining. Process that consists of drilling holes in the skulls of those perceived as deviants to allow the evil spirits to escape.

Turf wars. Fights over claimed urban territory by youth gangs.

Types of neglect. Consist of physical, emotional, and educational neglect.

UCRs. *See* Uniform Crime Reports.

Unconditional probation. Probation programs that typically permit some freedom of movement and require relatively low-intensive communications with probation officers via mail or telephone.

Underclass. Socioeconomic designation in which large numbers of inner-city poor people are caught in a chronic generational cycle of poverty, low educational achievement, teenage parenthood, chronic unemployment, and welfare dependence. Theorists argue that antisocial behaviors become norms within chronically impoverished inner-city environments.

Uniform Crime Reports. Reports published annually by the Federal Bureau of Investigation (FBI) providing a compilation of arrest data and categorizing offenses as Part I and Part II offenses. They are often referred to by the acronym UCRs. Police agencies submit information on clearances (arrest data) to the FBI annually. The FBI then constructs a statistical profile of crime in the United States based on these arrest data.

Uniform Juvenile Court Act. Act providing judicial procedures so that all parties are assured of fairness and recognition of legal rights.

Unofficial police procedures. Exercise of police discretion at certain decision points. At these decision points, police officers come to a decision on whether to begin formal processing of juveniles or to resolve the situation outside of officially mandated procedures.

Unofficial probation. Option that occurs when the prosecutor indicates that he or she has a prosecutable case but also indicates that prosecution will be withheld if the suspect in question agrees to behave according to certain guidelines.

Unofficial sources of data. Methods for determining just how much crime or delinquency remains hidden from official measures of such behavior.

Unruly children. Children who may be disposed of by the court in any authorized disposition allowable for the delinquent except commitment to the state correctional agency. However, if an unruly child is found to be not amenable to treatment under the disposition, the court, after another hearing, may make any disposition otherwise authorized for the delinquent.

Vice Lords. Also known as the Conservative Vice Lord Nation, gang whose members have aligned themselves with other groups such as the Latin Kings. These gangs are collectively referred to as "people" and are enemies of the Black Gangster Disciples.

Vice Queens. Female partner group of the Vice Lords. This female gang has broken the gender mold and engages in strong-arming, auto theft, and aggressive and violent behaviors.

Victims of crime impact panels. Community-based panels that strive to teach competency to offenders. They also attempt to develop offender empathy for understanding the impacts on the victims.

Victim–offender mediation programs. Programs that have a strong grounding in restorative justice principles by bringing the victim and the offender together in mediation. The desired result is closure and emotional healing on the part of the victim, with accountability and remorse being observed on the part of the offender.

Victim–offender reconciliation program. Program that brings the victim and the offender together to reconcile a criminal wrong that the offender has committed against the victim. It is a much more informal process than are standard court proceedings.

Victim survey research. Survey research that is derived from data provided by victims of crime.

Victimization surveys. Data derived from surveys of victims of crimes. Most victimization surveys compile household data. The best-known victimization survey is the National Crime Victimization Survey, an annual household survey conducted by the Bureau of Justice Statistics.

Violent Crime Index offenses. Consist of murder, forcible rape, robbery, and aggravated assault.

Waiver. Process whereby a juvenile court may relinquish jurisdiction and remand a case to the adult criminal court. In many states, waiver is mandatory for serious offenses. In other situations, the prosecutor determines whether waiver will be invoked.

Wannabes. Youngsters who emulate older gang members by imitating their dress, speech, and symbols. These young imitators want to be identified as gang members and frequently aspire to joining the group.

Ward of the state. Special status accorded to juveniles who have been removed from parental care and placed under the protection of the state.

White v. Illinois. U.S. Supreme Court ruling that affirmed the use of hearsay statements in child sexual abuse cases. In this case, the 4-year-old child victim did not testify, but others to whom she had talked about the assault (her mother, a doctor, a nurse, and a police officer) were allowed to testify.

Wilderness programs. Juvenile corrections programs centered on minimum-security residential correctional institutions that are located in rural settings. These programs are usually reserved for first-time offenders and/or juveniles who have committed minor offenses.

Examples include forestry camps, ranches, and well-established programs such as Outward Bound.

Wilding. Rampages by young people in public areas such as parks and shopping malls. Perpetrators randomly beat and/or sexually assault people, sometimes seriously injuring them.

***Winship* case.** U.S. Supreme Court ruling that the standard of proof for conviction of juveniles should be the same as that for adults in criminal court—proof beyond a reasonable doubt—in juvenile court proceedings involving delinquency.

XYY chromosome. Research suggesting that men with an XYY chromosomal pattern are more prevalent in prison populations than in society. These "supermales" are theoretically more aggressive than typical XY males.

Youth culture. Lifestyles that are peculiar to social associations of young people. There is no single youth culture but rather an assortment of lifestyles and selected identifications.

Youth-Focused Community Policing. Program providing information-sharing activities that promote proactive partnerships among the police, juveniles, and community agencies cooperating to identify and address juvenile problems in a manner consistent with a community policing philosophy.

Youth gangs. Term commonly used for gangs that routinely engage in criminal behavior such as drug trafficking, burglary, vehicular theft, and assault.

Youth Service Bureaus. Program originally recommended in 1967 by a presidential commission report titled *The Challenge of Crime in a Free Society.* Youth Service Bureaus were intended to be a primary center for dealing with juvenile delinquents and status offenders. The bureaus coordinated their activities with juvenile courts, the police, and probation agencies.

References

Abadinsky, H. (2003). *Organized crime* (7th ed.). Belmont, CA: Wadsworth/Thomson Learning.

Abadinsky, H., & Winfree, L. T., Jr. (1992). *Crime and justice* (2nd ed.). Chicago: Nelson–Hall.

Abundant Life Academy. (2006). Boot camps are proving a helping hand for troubled teens. Available: www.restoretroubledteens.com/Boot-Camps-for-Teens.html

Ackerman, W. V. (1998). Socioeconomic correlates of increasing crime rates in smaller communities. *Professional Geographer, 50,* 372–387.

Acoca, L. (1998). Outside/Inside: The violation of American girls at home, on the streets, and in the juvenile justice system. *Crime & Delinquency, 44,* 561–589.

Adler, A. (1931). *What life should mean to you.* London: Allen & Unwin.

Agnew, R. (1985). A revised strain theory of delinquency. *Social Forces, 64,* 151–167.

Agnew, R. (2001). Building on the foundation of general strain theory: Specifying the types of strain most likely to lead to crime and delinquency. *Journal of Research in Crime and Delinquency, 38,* 319–363.

Akers, R. L. (1964). Socioeconomic status and delinquent behavior: A retest. *Journal of Research in Crime and Delinquency, 10,* 38–46.

Akers, R. L. (1985). *Deviant behavior: A social learning approach* (3rd ed.). Belmont, CA: Wadsworth.

Akers, R. L. (1992). Linking sociology and its specialties: The case of criminology. *Social Forces, 71,* 1–16.

Akers, R. L. (1994). *Criminological theories: Introduction and evaluation.* Los Angeles: Roxbury.

Akers, R. L. (1998). *Social learning and social structure: A general theory of crime and deviance.* Boston: Northeastern University Press.

Akers, R. L., & Sellers, C. S. (2004). *Criminological theories: Introduction, evaluation, and application* (4th ed.). Los Angeles: Roxbury.

Alabama Code. (1995).

Allen, T. T. (2005). Taking a juvenile into custody: Situational factors that influence police officers' decisions. *Journal of Sociology and Social Welfare, 32,* 121–129.

Alwin, D. F., & Thornton, A. (1984). Family origins and the schooling process: Early versus late influence of parental characteristics. *American Sociological Review, 49,* 784–802.

American Bar Association. (1977). *Standards relating to counsel for private parties.* Cambridge, MA: Ballinger.

American Correctional Association. (2005). *Public correctional policy on private sector involvement in corrections.* Available: www.aca.org/government/policyresolution/view.asp?ID=33

Anderson, E. (1990). *Streetwise.* Chicago: University of Chicago Press.

Aniskievicz, R., & Wysong, E. (1990). Evaluating DARE: Drug education and the multiple meanings of success, *Policy Studies Review, 9,* 727–747.

Ariessohn, R. M. (1981). Recidivism revisited. *Juvenile Family Court Journal, 32*(4), 59–68.

Arizona Revised Statutes Annotated. (1999).

Arkansas Code Annotated. (1999).

Armagh, D. (1998). A safety net for the Internet: Protecting our children. *Juvenile Justice Journal, 5*(1), 9–15. Available: http://ojjdp.ncjrs.org/jjjournal/jjjournal598/net.html

Atkins, B., & Pogrebin, M. (1978). *The invisible justice system: Discretion and the law.* Cincinnati, OH: Anderson.

Attorney General of Texas. (2002). *Gangs in Texas 2001: An overview.* Austin: Author. Available: www.oag.state.tx.us/AG_Publications/pdfs/2001gangrept.pdf

Babicky, T. (1993a, July). *Gangs fact sheets: A reference guide.* Springfield: Illinois Department of Corrections.

Babicky, T. (1993b, September). *Gangs and gang activity.* Springfield: Illinois Department of Corrections.

Barlow, H. D. (2000). *Criminal justice in America.* Upper Saddle River, NJ: Prentice Hall.

Baron, S. W. (2004). General strain, street youth, and crime: A test of Agnew's revised theory. *Criminology, 42,* 457–483.

Bartollas, C. (1993). *Juvenile delinquency* (3rd ed.). New York: Macmillan.

Battistich, V., & Hom, A. (1997). The relationship between students' sense of their school as a community and their involvement in problem behaviors. *American Journal of Public Health, 87,* 1997–2001.

Bazemore, G., & Day, S. E. (1996). Restoring the balance: Juvenile and community justice. *Juvenile Justice, 3*(1), 3–14.

Bazemore, G., & Senjo, S. (1997). Police encounters with juveniles revisited: An exploratory study of themes and styles in community policing. *Policing: An International Journal of Police Strategy and Management, 20,* 60–82.

Bazemore, G., & Washington, C. (1995). Charting the future of the juvenile justice system: Reinventing mission and management. *Spectrum: The Journal of State Government, 68*(2), 51–66.

Becker, H. S. (1963). *The outsiders.* New York: Free Press.

Beirne, P., & Quinney, R. (1982). *Marxism and the law.* New York: John Wiley.

Belknap, J., Morash, M., & Trojanowicz, R. (1987). Implementing a community policing model for work with juveniles. *Criminal Justice and Behavior, 14,* 211–245.

Bell, D. J., & Bell, S. (1991). The victim–offender relationship as a determinant factor in police dispositions of family violence incidents: A replication study. *Policing and Society, 1,* 225–234.

Bellair, P. E., & McNulty, T. L. (2005). Beyond the bell curve: Community disadvantage and the explanation of black–white differences in adolescent violence. *Criminology, 43,* 1135–1169.

Benekos, P. J., & Merlo, A. V. (Eds.). (2004). *Controversies in juvenile justice and delinquency.* Cincinnati, OH: Anderson.

Bernard, T. (1990). Twenty years of testing theories: What have we learned and why? *Journal of Research in Crime and Delinquency, 27,* 325–347.

Bernard, T. J. (1992). *The cycle of juvenile justice.* New York: Oxford University Press.

Bilchik, S. (1998a). A juvenile justice system for the 21st century. *Crime & Delinquency, 44,* 89–101.

Bilchik, S. (1998b). *Juvenile Mentoring Program: 1998 report to Congress.* Washington, DC: U.S. Department of Justice.

Bilchik, S. (1999a, December). Juvenile justice: A century of change. *Juvenile Justice Bulletin.* (Washington, DC: U.S. Department of Justice)

Bilchik, S. (1999b, August). *OJJDP research: Making a difference for juveniles.* Washington, DC: U.S. Department of Justice.

Bilchik, S. (1999c, December). *1997 National Youth Gang Survey.* Washington, DC: U. S. Department of Justice.

Bishop, D. (2004). Injustice and irrationality in contemporary youth policy. *Criminology and Public Policy, 3,* 633–644.

Bishop, S. J., Murphy, J. M., & Hicks, R. (2000). What progress has been made in meeting the needs of seriously maltreated children? The course of 200 cases through the Boston Juvenile Court. *Child Abuse & Neglect, 24,* 599–610.

Bjerregaard, B., & Smith, C. (1993). Gender differences in gang participation, delinquency, and substance abuse. *Journal of Quantitative Criminology, 4,* 329–355.

Black, D. J., & Reiss, A. J., Jr. (1970). Police control of juveniles. *American Sociological Review, 35,* 63–77.

Blackstone, W. (1803). *Commentaries on the laws of England* (12th ed., Vol. 4). London: Strahan.

Blair, S. L., Blair, M. C. L., & Madamba, A. B. (1999). Racial/Ethnic differences in high school students' academic performance: Understanding the interweave of social class and ethnicity in family context. *Journal of Comparative Family Studies, 30,* 539–555.

Blankenship, R. L., & Singh, B. K. (1976). Differential labeling of juveniles: A multivariate analysis. *Criminology, 13,* 471–490.

Bloch, H., & Neiderhoffer, A. (1958). *The gang: A study in adolescent behavior.* New York: Philosophical Library.

Bloom, B., Owen, B., & Covington, S. (2003). *Gender responsive strategies: Research, practice, and guiding principles for women offenders.* Washington, DC: National Institute of Corrections.

Blumberg, A. S. (1967). *Criminal justice.* Chicago: Quadrangle.

Bohm, R. M. (2001). *A primer on crime and delinquency theory* (2nd ed.). Belmont, CA: Wadsworth.

Bookin, H., & Horowitz, R. (1983). The end of the youth gang: Fad or fiction? *Criminology, 21,* 585–602.

Boot Camps for Teens. (n.d.). *Welcome to boot camps for teens.* Available: www.bootcampsforteens.com

Booth, A., & Osgood, D. W. (1993). The influence of testosterone on deviance in adulthood: Assessing and explaining the relationship. *Criminology, 31,* 93–117.

Botch, D. (2006). Reforming the American justice system. *Public Administration Review, 66,* 640–643.

Bradley, C. M. (2006, March/April). The right decision on the juvenile death penalty. *Judicature, 89,* 302–305. Available: http://proquest.umi.com/pqdweb?did=1039951891&sid=3&Fmt=3&clientId=59796&RQT=309&VName=PQD

Bradshaw, W., & Roseborough, D. (2005). Restorative justice dialogue: The impact of mediation and conferencing on juvenile recidivism. *Federal Probation, 69*(2), 15–21.

Braithwaite, J. (1989). *Crime, shame, and reintegration.* New York: Cambridge University Press.

Braun, C. (1976). Teacher expectations: Sociopsychological dynamics. *Review of Educational Research, 46,* 185–213.

Breed v. Jones, 421 U.S. 519, 95 S.Ct. 1779 (1975).

Brown, B. (2006). Understanding and assessing school police officers: A conceptual and methodological comment. *Journal of Criminal Justice, 34,* 591–598.

Brown, J., Cohen, P., Johnson, J., & Salzinger, S. (1998). A longitudinal analysis of risk factors for child maltreatment: Findings of a 17-year prospective study of officially recorded and self-reported child abuse and neglect. *Child Abuse & Neglect, 22,* 1065–1078.

Brownfield, D. (1990). Adolescent male status and delinquent behavior. *Sociological Spectrum, 10,* 227–248.

Browning, K., & Loeber, R. (1999, February). *Highlights of findings from the Pittsburgh Youth Study* (OJJDP Fact Sheet No. 95). Washington, DC: U.S. Department of Justice.

Buerger, M. E., Cohn, E. G., & Petrosino, A. J. (2000). Defining the hot spots of crime. In R. W. Glensor, M. E. Corriea, & K. J. Peak (Eds.), *Policing communities: Understanding crime and solving problems* (pp. 138–150). Los Angeles: Roxbury.

Buikhuisen, W., & Jongman, R. W. (1970). A legislative classification of juvenile delinquents. *British Journal of Criminology, 10,* 109–123.

Bumphus, V. W., & Anderson, J. F. (1999). Family structure and race in a sample of criminal offenders. *Journal of Criminal Justice, 27,* 309–320.

Burch, J., & Kane, C. (1999, July). *Implementing the OJJDP comprehensive gang model* (OJJDP Fact Sheet No. 112). Washington, DC: U.S. Department of Justice.

Bureau of Justice Assistance. (2003). *Drug court monitoring, evaluation, and management information systems: National scope needs assessment* (U.S. Department of Justice Monograph). Washington, DC: Government Printing Office.

Bureau of Justice Statistics. (2005a). *Family violence statistics.* Available: www.ojp.gov/bjs/pub/pdf/fvs02.pdf

Bureau of Justice Statistics. (2005b). *Incident-based statistics.* Available: http://ojp.usdoj.gov/bjs

Burgess, E. W. (1952). The economic factor in juvenile delinquency. *Journal of Criminal Law, 43,* 29–42.

Burgess, R. L., & Akers, R. L. (1968). A differential association-reinforcement theory of criminal behavior. *Social Problems, 14,* 128–147.

Bursik, R. J., & Grasmick, H. G. (1995). The effect of neighborhood dynamics on gang behavior. In M. Klein, C. L. Maxson, & J. Miller (Eds.), *The modern gang reader* (pp. 114–123). Los Angeles: Roxbury.

Butts, J. A. (1997). Necessarily relative: Is juvenile justice speedy enough? *Crime & Delinquency, 43,* 3–23.

Bynum, J. E., & Thompson, W. E. (1999). *Juvenile delinquency* (4th ed.). Boston: Allyn & Bacon.

Bynum, J. E., & Thompson, W. E. (1992). *Juvenile delinquency* (2nd ed.). Boston: Allyn & Bacon.

Cadzow, S. P., Armstrong, K. L., & Fraser, J. A. (1999). Stressed parents with infants: Reassessing physical abuse risk factors. *Child Abuse & Neglect, 23,* 845–853.

Califano, J. A., Jr., and Colson, C. W. (2005, January). Criminal neglect. *USA Today Magazine,* pp. 34–35.

California Welfare and Institutional Code. (1998).

Campbell, A. (1995). Female participation in gangs. In M. W. Klein, C. L. Maxson, & J. Miller (Eds.), *The modern gang reader* (pp. 70–77). Los Angeles: Roxbury.

Campbell, A. (1997). Self definition by rejection: The case of gang girls. In G. L. Mays (Ed.), *Gangs and gang behavior* (pp. 129–149). Chicago: Nelson–Hall.

Canter, R. (1982). Family correlates of male and female delinquency. *Criminology, 20,* 149–167.

Carelli, R. (1990, June 28). Court backs sparing children in abuse cases. *Peoria Journal Star,* p. A2.

Carnevale Associates. (2006). *A longitudinal evaluation of the new curricula for the D.A.R.E. middle (7th grade) and high school (9th grade) programs: Take Charge of Your Life—Year Four progress report.* Available: www.dare.com/home/Resources/documents/DAREMarch06ProgressReport.pdf

Cauchon, D. (1993, October 11). Studies find drug program not effective. *USA Today,* pp. 1A–2A.

Cavan, R. S. (1969). *Juvenile delinquency: Development, treatment, control* (2nd ed.). Philadelphia: J. B. Lippincott.

Center for Restorative Justice and Mediation. (1985). *Principles of restorative justice.* St. Paul: University of Minnesota, School of Social Work.

Center for Restorative Justice and Mediation. (1996a). *Restorative justice: For victims, communities and offenders—What is restorative justice?* St. Paul: University of Minnesota, School of Social Work.

Center for Restorative Justice and Mediation. (1996b). *Restorative justice: For victims, communities, and offenders—What is the community's part in restorative justice?* St. Paul: University of Minnesota, School of Social Work.

Centerwood, B. S. (1992). Television and violence: The scale of the problem and where to go from here. *Journal of the American Medical Association, 267,* 3059–3063.

Chambliss, W. (1973). The saints and the roughnecks. *Society, 11,* 24–31.

Chambliss, W. J. (1984). *Criminal law in action* (2nd ed.). New York: John Wiley.

Chambliss, W. J., & Mandoff, M. (1976). *Whose law, what order?* New York: John Wiley.

Champion, D. (2002). *Probation, parole, and community corrections* (4th ed.). Upper Saddle River, NJ: Prentice Hall.

Charton, S. (2001, July 16). Sheriff says making kids shovel manure stinks. *Chicago Sun-Times,* p. 28.

Chesney-Lind, M. (1999). Challenging girls' invisibility in juvenile court. *Annals of the American Academy of Political & Social Science, 564,* 185–202.

Child Abuse Prevention Program. (2006). *Programs.* Available: www.childabusepreventionprogramorg/programs.htm

Child Trends DataBank. (2006). *Youth who feel unsafe at school.* Available: www.childtrendsdatabank.org/indicators/38UnsafeatSchool.cfm

Child Welfare Information Gateway. (2006). *A new way to stay connected.* Available: www.childwelfare.gov

Children's Bureau. (1969). *Legislative guide for drafting family and juvenile court acts.* Washington, DC: U.S. Government Printing Office.

Children's Defense Fund. (2006). *Poverty.* Available: www.childrensdefense.org/site/PageNavigator/c2pp_poverty

Children's Defense Fund. (2007). *CDF gun report: 2,827 child, teen deaths by firearms in one year exceed total U.S. combat fatalities during three years in Iraq.* Available: www.childrensdefense.com/site/News2?page=NewsArticle&id=8491

City of Phoenix. (2006). *Latchkey children.* Available: www.ci.phoenix.az.us/FIRE/keykids.html

Clark, J. P., & Tifft, L. L. (1966). Polygraph and interview validation of self-reported deviant behavior. *American Sociological Review, 4,* 516–523.

Clear, T. R., & Cole, G. F. (2003). *American corrections* (6th ed.). Belmont, CA: Wadsworth/Thomson Learning.

Clemson University. (2002). *The Olweus Bullying Prevention Program: Background and program overview.* Clemson, SC: Author.

Clifford, R. D., Bryant, D., & Early, D. M. (2005, October). A pre-K programs status report. *Principal,* pp. 50–53.

Cloward, R. A., & Ohlin, L. E. (1960). *Delinquency and opportunity.* Glencoe, IL: Free Press.

Cohen, A. K. (1955). *Delinquent boys: The culture of the gang.* Glencoe, IL: Free Press.

Cohn, A. W. (1999). Juvenile focus. *Federal Probation, 58,* 87–91.

Cohn, A. W. (2004a). Juvenile justice in transition: Is there a future? *Federal Probation, 63*(2), 61–67.

Cohn, A. W. (2004b). Planning for the future of juvenile justice. *Federal Probation, 68*(3), 39–44.

Colorado Revised Statutes Annotated. (1999).

Conklin, J. E. (1998). *Criminology* (6th ed.). Boston: Allyn & Bacon.

Connecticut General Statutes Annotated. (1991).

Connecticut General Statutes Annotated. (1999).

Consolidated Laws of New York Annotated. (1975).

Coohey, C. (1998). Home alone and other inadequately supervised children. *Child Welfare, 77,* 291–310.

Corbitt, W. A. (2000). Violent crimes among juveniles: Behavioral aspects. *FBI Law Enforcement Bulletin, 69*(6), 18–21.

Costello, B. J., & Vowell, P. R. (1999). Testing control theory and differential association: A reanalysis of the Richmond Youth Project data. *Criminology, 37,* 815–842.

Cote, S. (2002). *Criminological theories: Bridging the past to the future.* Thousand Oaks, CA: Sage.

Cothern, L. (2000, November). *Juveniles and the death penalty.* Washington, DC: U.S. Department of Justice, Coordinating Council on Juvenile Justice and Delinquency Prevention.

Covington, J. (1984). Insulation from labeling: Deviant defenses in treatment. *Criminology, 22,* 619–643.

Cox, S. M. (1975). Review of "Critique of Legal Order." *Teaching Sociology, 3*(1), 97–99.

Cox, S. M., & Wade, J. E. (1996). *The criminal justice network: An introduction.* Boston: McGraw–Hill.

Crack, hour by hour. (1988, November 28). *Newsweek,* pp. 20–29.

Cromwell, P. F., Jr., Killinger, G. G., Kerper, H. B., & Walker, C. (1985). *Probation and parole in the criminal justice system* (2nd ed.). St. Paul, MN: West.

Crosson-Tower, C. (1999). *Understanding child abuse and neglect* (4th ed.). Boston: Allyn & Bacon.

Cunningham, S. M. (2003). The joint contribution of experiencing and witnessing violence during childhood on child abuse in the parent role. *Violence and Victims, 18,* 619–639.

Curran, D. J., & Renzetti, C. M. (1994). *Theories of crime.* Boston: Allyn & Bacon.

Curry, G. D. (1998). Female gang involvement. *Journal of Research in Crime and Delinquency, 35,* 100–118.

Curry, G. D., & Spergel, I. A. (1988). Gang homicide, delinquency, and community. *Criminology, 26,* 381–405.

Cyr, M., McDuff, P., Wright, J., Theriault, C., & Cinq-Mars, C. (2005). Clinical correlates and repetition of self-harming behaviors among female adolescent victims of sexual abuse. *Journal of Child Sexual Abuse, 14*(2), 49–68.

Daly, K., & Chesney-Lind, M. (1988), Feminism and criminology. *Justice Quarterly, 5,* 497–538.

Dart, R. W. (1992). *Street gangs.* Chicago: Chicago Police Department.

Davidson, H. (1981). The guardian ad litem: An important approach to the protection of children. *Children Today, 10*(2), 20–23.

Davidson, N. (1990). Life without father. *Policy Review, 51,* 40–44.

Davidson, W. S., Redner, R., & Amdur, R. L. (1990). *Alternative treatments for troubled youth: The case of diversion from the justice system.* New York: Plenum.

Davis, K. C. (1975). *Police discretion.* St. Paul, MN: West.

Davis, N. J. (1999). *Youth crisis: Growing up in a high-risk society.* Westport, CT: Praeger.

Davis, S. M. (2001). *Rights of juveniles.* New York: Clark Boardman/West.

Dawkins, M. P. (1997). Drug use and violent crime among juveniles. *Adolescence, 32,* 395–405.

Dawley, D. (1973). *A nation of lords.* New York: Doubleday.

Death Penalty Information Center. (n.d.). U. S. Supreme Court: *Roper v. Simmons.* Available: www.deathpenaltyinfo.org/article.php?scid=38&did=885

Demuth, S., & Brown, S. L. (2004). Family structure, family processes, and adolescent delinquency: The significance of parental absence versus parental gender. *Journal of Research in Crime and Delinquency, 41,* 58–81.

Denno, D. W. (1994). Gender, crime, and the criminal law defenses. *Journal of Criminal Law and Criminology, 85,* 80–180.

Dentler, R. A., & Monroe, L. J. (1961). Early adolescent theft. *American Sociological Review, 26,* 733–743.

DePaul, J., & Domenech, L. (2000). Childhood history of abuse and child abuse potential in adolescent mothers: A longitudinal study. *Child Abuse & Neglect, 24,* 701–713.

Dighton, D. (2003, Summer). Minority overrepresentation in the criminal justice and juvenile justice systems. *The Compiler, 22,* 1–6.

DiLillo, D., Tremblay, G. C., & Peterson, L. (2000). Linking childhood sexual abuse and abusive parenting: The mediating role of maternal anger. *Child Abuse & Neglect, 24,* 767–779.

District of Columbia Code. (1997).

Dorne, C., & Gewerth, K. (1998). *American juvenile justice: Cases, legislation, and comments.* San Francisco: Austin & Winfield.

Drowns, R. W., & Hess, K. M. (1990). *Juvenile justice.* St. Paul, MN: West.

Dugdale, R. L. (1888). *The Jukes: A study in crime, pauperism, disease, and heredity.* New York: Putnam.

Dukes, R. L., Stein, J. A., & Ullman, J. B. (1997). Long term impact of Drug Abuse Resistance Education (D.A.R.E.). *Evaluation Review, 21,* 483–500.

Dunn, M. J. (1999). Focus on community policing: Police liaison for schools. *FBI Law Enforcement Bulletin, 68*(9), 7–9.

Echeburua, E., Fernandez-Montalvo, J., & Baez, C. (2000). Relapse prevention in the treatment of slot-machine pathological gambling. *Behavior Therapy, 31,* 351–364.

Eddings v. Oklahoma, 455 U.S. 104 (1982).

Edwards, L. P. (2005). The role of the juvenile court judge revisited. *Juvenile and Family Court Journal, 56*(1), 33–45.

Ehrenreich, B. (1990). The hourglass society. *New Perspectives Quarterly, 7,* 44–46.

Eitzen, D. S., & Zinn, M. B. (1992). *Social problems* (5th ed.). Boston: Allyn & Bacon.

Elliott, D., & Ageton, S. (1980). Reconciling race and sex differences in self-reported and official estimates of delinquency. *American Sociological Review, 45,* 95–110.

Ellis, R. A., O'Hara, M., & Sowers, K. (1999). Treatment profiles of troubled female adolescents: Implications for judicial disposition. *Juvenile and Family Court Journal, 50*(3), 25–40.

Ellis, R. A., & Sowers, K. M. (2001). *Juvenile justice practice: A cross-disciplinary approach to intervention.* Belmont, CA: Wadsworth.

Elrod, P., & Ryder, R. S. (2005). *Juvenile justice: A social, historical, and legal perspective* (2nd ed.). Sudbury, MA: Jones & Bartlett.

Emery, R. E. (1982). Interparental conflict and the children of discord and divorce. *Psychological Bulletin, 92,* 310–330.

Empey, L. T., & Stafford, M. C. (1991). *American delinquency: Its meaning and construction* (3rd ed.). Belmont, CA: Wadsworth.

Empey, L. T., Stafford, M. C., & Hay, C. H. (1999). *American delinquency: Its meaning and construction* (4th ed.). Belmont, CA: Wadsworth.

Engel, R. S., Sobol, J. J., & Worden, R. E. (2000). Further exploration of the demeanor hypothesis: The interaction effects of suspects' characteristics and demeanor on police behavior. *Justice Quarterly, 17,* 235–258.

Ennett, S., Tobler, N. S., Ringwalt, C. L., & Flewelling, R. L. (1994). How effective is drug abuse resistance education? A meta-analysis of project DARE outcome evaluations. *American Journal of Public Health, 84,* 1394–1401.

Ericson, N. (2001). *Public/Private ventures' evaluation of faith based programs* (OJJDP Fact Sheet No. 38). Washington, DC: Office of Juvenile Justice and Delinquency Prevention.

Erikson, K. (1962). Notes on the sociology of deviance. *Social Problems, 9,* 301–314.

Ervin, J. D., & Swilaski, M. (2004). Community outreach through children's programs. *Police Chief, 71*(9), 42–45.

Esbensen, F-A. (2004). *Evaluating G.R.E.A.T.: A school-based gang prevention program* (Research for Policy). Washington, DC: National Institute of Justice.

Esbensen, F., Deschenes, E. P., & Winfree, L. T. (1999). Differences between gang girls and gang boys: Results from a multisite survey. *Youth & Society, 31,* 27–53.

Espiritu, R. C., Huizinga, D., Crawford, A., & Loeber, R. (2001). Epidemiology of self-reported delinquency. In R. Loeber & D. P. Farrington (Eds.), *Child delinquents: Development, intervention, and service needs* (pp. 47–66). Thousand Oaks, CA: Sage.

Evans, D. G., & Sawdon, J. (2004, October). The development of a gang exit strategy: The Youth Ambassador's Leadership and Employment Project. *Corrections Today.* Available: www.cantraining.org/BTC/docs/Sawdon%20 Evans%20CT%20Article.pdf

Evans, W. P., Fitzgerald, C., & Weigel, D. (1999). Are rural gang members similar to their urban peers? Implications for rural communities. *Youth & Society, 30,* 267–282.

Fader, D. P., Harris, P. W., Jones, P. R., & Poulin, M. E. (2001). Factors involved in decisions on commitment to delinquency programs for first-time juvenile offenders. *Justice Quarterly, 18,* 323–342.

Fagan, J., & Pabon, E. (1990). Contributions of delinquency and substance use to school dropout among inner-city youths. *Youth & Society, 21,* 306–354.

Fanton, J. (2006, May 14). Illinois a national leader in juvenile justice reforms. *Peoria Journal Star,* p. A5.

Farrington, D. P. (2003). Developmental and life-course criminology: Key theoretical and empirical issues—The 2002 Sutherland Address. *Criminology, 41,* 221–255.

Farrington, D. P., Jollife, D., Hawkins, J. D., Catalano, R. F., Hill, K. G., & Kosterman, R. (2003). Comparing delinquency careers in court records and self-reports. *Criminology, 41,* 933–959.

Farrington, D. P., Loeber, R., Stouthamer-Loeber, M., Van Kammen, W. B., & Schmidt, L. (1996). Self-reported delinquency and a combined delinquency scale based on boys, mothers, and teachers: Concurrent and predictive validity for African Americans and Caucasians. *Criminology, 34,* 493–517.

Faust, F. L., & Brantingham, P. J. (1974). *Juvenile justice philosophy.* St. Paul, MN: West.

Federal Bureau of Investigation. (2005). *Crime in the United States, 2004.* Available: www.fbi.gov/ucr/cius_04

Federal Bureau of Investigation. (2006). *Crime in the United States, 2005.* Available: www.fbi.gov/ucr/cius_05

Feiler, S. M., & Sheley, J. F. (1999). Legal and racial elements of public willingness to transfer juvenile offenders to adult court. *Journal of Criminal Justice, 27*(1), 55–64.

Fenwick, C. R. (1982). Juvenile court intake decision making: The importance of family affiliation. *Journal of Criminal Justice, 10,* 443–453.

Finkelhor, D., & Ormrod, R. (2001, October). Homicides of children and youth. *Juvenile Justice Bulletin.* (Washington, DC: Office of Juvenile Justice and Delinquency Prevention)

Finkenauer, J. O. (1982). *Scared straight.* Englewood Cliffs, NJ: Prentice Hall.

Fishbein, D. H. (1990). Biological perspectives in criminology. *Criminology, 28,* 27–72.

Fishman, L. T. (1995). The Vice Queens: An ethnographic study of black female gang behavior. In M. W. Klein, C. L. Maxson, & J. Miller (Eds.), *The modern gang reader* (pp. 83–92). Los Angeles: Roxbury.

Fitzgerald, J. D., & Cox, S. M. (2002). *Research methods and statistics in criminal justice: An introduction* (3rd ed.). Belmont, CA: Wadsworth/Thomson.

Flannery, D. J., Williams, L. L., & Vazsonyi, A. T. (1999). Who are they and what are they doing? Delinquent behavior, substance abuse, and early adolescents' after-school time. *American Journal of Orthopsychiatry, 69,* 247–253.

Fleener, F. T. (1999). Family as a factor in delinquency. *Psychological Reports, 85*(1), 80–81.

Florida Statutes Annotated. (1999).

Florida Statutes Annotated. (2005).

Florida Statutes Annotated. (2006).

Florsheim, P., Shotorbani, S., & Guest-Warnick, G. (2000). Role of the working alliance in the treatment of delinquent boys in community-based programs. *Journal of Clinical Child Psychology, 29,* 94–107.

Ford, J. D., Chapman, J., Mack, M., & Pearson, G. (2006). Pathways from traumatic child victimization to delinquency: Implications for juvenile and permanency court proceedings and decisions. *Juvenile and Family Court Journal, 57*(1), 13–28.

Forum on Child and Family Statistics. (2006). *Population and family characteristics* (America's Children in Brief). Available: http://childstats.gov/americaschildren/pop.asp

Fox, J. A., & Levin, J. (1994). Firing back: The growing threat of workplace homicide. *Journal of Criminal Law, Criminology, and Police Science, 563,* 16–30.

Fox, S. (1984). *Juvenile courts in a nutshell* (2nd ed.). Saint Paul, MN: West Publishing Co.

Frazier, C. E., Bishop, D. M., & Henretta, J. C. (1992). The social context of race differentials in juvenile justice dispositions. *Sociological Quarterly, 33,* 447–458.

Friedenberg, E. Z. (1965). *Coming of age in America.* New York: Random House.

Friend, T. (2000, June 27). Genetic map is hailed as new power. *USA Today,* p. 1A.

Gagnon v. Scarpelli, 411 U. S. 778 (1973).

Gallegos v. Colorado, 370 U.S. 49 (1969).

Gandhi, A. G., Murphy-Graham, E., Anthony, P., Chrismer, S. S., & Weiss, C. H. (2007). The devil is in the details: Examining the evidence for "proven" school-based drug abuse prevention programs. *Evaluation Review, 31,* 43–74.

Gang Resistance Education and Training. (2006). *Welcome to the G.R.E.A.T. Web site.* Available: www.greatonline.org

Gardner, E., Rodriguez, N., & Zatz, M. S. (2004). Criers, liars, and manipulators: Probation officers' views of girls. *Justice Quarterly, 21,* 547–579.

Garrett, M., & Short, J. F. (1975). Social class and delinquency: Predictions and outcomes of police–juvenile encounters. *Social Problems, 22,* 368–383.

Garry, E. (1996). *Truancy: First step to a lifetime of problems.* Washington, DC: U.S. Department of Justice, Office of Juvenile Justice and Delinquency Prevention.

Georgia Code Annotated. (1999).

Glaser, D. (1960). Differential association and criminological prediction. *Social Problems, 8,* 6–14.

Glaser, D. (1978). *Crime in our changing society.* New York: Holt, Rinehart & Winston.

Glensor, R. W., Correia, M. E., & Peak, K. J. (Eds.). (2000). *Policing communities: Understandinig crime and solving problems.* Los Angeles: Roxbury.

Gluck, S. (1997, June). Wayward youth, super predataor: An evolutionary tale of juvenile delinquency froms the 1950s to the present. *Corrections Today, 59,* 62–64.

Glueck, S., & Glueck, E. (1950). *Unraveling juvenile delinquency.* Cambridge, MA: Harvard University Press.

Goddard, H. H. (1914). *Feeblemindedness: Its causes and consequences.* New York: Macmillan.

Goldberg, R. (2003). *Drugs across the spectrum* (4th ed.). Belmont, CA: Wadsworth/Thomson Learning.

Goldman, J., Salus, M. K., Wolcott, D., & Kennedy, K. Y. (2003). *A coordinated response to child abuse and neglect: The foundation for practice.* Washington, DC: U.S. Department of Health and Human Services. Available: www.childwelfare.gov/pubs/usermanuals/foundation/foundatione.cfm

Goldstein, S. L. & Tyler, R. P. (1998). Frustrations of inquiry: Child sexual abuse allegations in divorce and custody cases. *FBI Law Enforcement Bulletin, 67*(7), 1–6.

Goodkind, S., Ng, I., & Sarri, R. C. (2006). The impact of sexual abuse in the lives of young women involved or at risk of involvement with the juvenile justice system. *Violence Against Women, 12,* 456–477.

Goring, C. (1913). The *English convict.* London: Her Majesty's Stationery Office.

Gorman-Smith, D., Tolan, P. H., & Loeber, R. (1998). Relation of family problems to patterns of delinquent involvement among urban youth. *Journal of Abnormal Child Psychology, 26,* 319–333.

Gottfredson, M. R., & Hirschi, T. (1990). *A general theory of crime.* Stanford, CA: Stanford University Press.

Gough, H. G. (1948). A sociological theory of psychopathy. *American Journal of Sociology, 53,* 359–366.

Gough, H. G. (1960). Theory and measurement of socialization. *Journal of Consulting Psychology, 24,* 23–30.

Greenberg, D. F. (1999). The weak strength of social control theory. *Crime & Delinquency, 45,* 66–81.

Gresham, F. M., MacMillan, D. L., & Bocian, K. M. (1998). Comorbidity of hyperactivity–impulsivity–inattention and conduct problems: Risk factors in social, affective, and academic domains. *Journal of Abnormal Child Psychology, 26,* 393–406.

Griffin, B. S., & Griffin, C. T. (1978). *Juvenile delinquency in perspective.* New York: Harper & Row.

Haley v. Ohio, 332 U.S. 596 (1948).

Halleck, S. (1971). *Psychiatry and the dilemmas of crime.* Berkeley: University of California Press.

Hamarman, S., & Bernet, W. (2000). Evaluating and reporting emotional abuse in children: Parent-based, action-based focus aids in clinical decision-making. *Journal of the American Academy of Child & Adolescent Psychiatry, 39,* 928–930.

Hamm, M. S. (1993). *American skinheads: The criminology and control of hate crime.* Westport, CT: Praeger.
Hanser, R. D. (2007a, forthcoming). Gang crimes: Latino gangs in America. In F. Shanty (Ed.), *Organized crime: An international encyclopedia.* Santa Barbara, CA: ABC–CLIO.
Hanser, R. D. (2007b). *Special needs offenders in the community.* Upper Saddle River, NJ: Prentice Hall.
Hanson, R. F., Resnick, H. S., & Saunders, B. E. (1999). Factors related to the reporting of childhood rape. *Child Abuse & Neglect, 23,* 559–569.
Harcourt, B. E., & Ludwig, J. (2006). Broken windows: New evidence from New York City and a five-city social experiment. *University of Chicago Law Review, 73,* 271–320.
Harper, G. W., & Robinson, W. L. (1999). Pathways to risk among inner-city African-American adolescent females: The influence of gang membership. *American Journal of Community Psychology, 27,* 383–404.
Harris, P. W., & Jones, P. R. (1999). Differentiating delinquent youths for program planning and evaluation. *Criminal Justice & Behavior, 26,* 403–434.
Hart, T. C., & Rennison, C. (2003, March). *Reporting crime to the police: 1992–2000.* Washington, DC: U.S. Department of Justice.
Hatchette, G. (1998). Why we can't wait: The juvenile court in the new millennium. *Crime & Delinquency, 44,* 83–88.
Hawaii Revised Statutes. (1993).
Hawkins, J. D., & Lishner, D. M. (1987). Schooling and delinquency. In E. Johnson (Ed.), *Handbook on crime and delinquency prevention* pp. 179–222. Westport, CT: Greenwood.
Hay, C., & Evans, M. (2006). Violent victimization and involvement in delinquency: Examining predictions from general strain theory. *Journal of Criminal Justice, 34,* 261–274.
Hayes, H., & Daly, K. (2003). Youth justice conferencing and reoffending. *Justice Quarterly, 20,* 725–764.
Healy, W., & Bronner, A. (1936). *New light on delinquency and its treatment.* New Haven, CT: Yale University Press.
Heck, W. P. (1999). Basic investigative protocol for child sexual abuse. *FBI Law Enforcement Bulletin, 68*(10), 19–25.
Helping America's Youth. (n.d.). *Introduction to risk factors and protective factors.* Available: http://guide.helpin gamericasyouth.gov/programtool-factors.cfm
Hibbler, W. J. (1999). A message from the 14th annual National Juvenile Corrections and Detention Forum. *Corrections Today, 61*(4), 28–31.
Hieb, C. F. (1992). *Gang task force and Lakewood, CO, Police Department.* Lakewood, CO: Access Publishing.
High/Scope Educational Research Foundation. (2002). *High-quality preschool program found to improve adult status.* Available: www.highscope.org/Research/PerryProject/perrymain.htm
Hindelang, M. J., Hirschi, T., & Weis, J. G. (1981). *Measuring delinquency.* Beverly Hills, CA: Sage.
Hinzman, G., & Blome, D. (1991). Cooperation key to success of child protection center. *Police Chief, 58*(2), 24–27.
Hirschi, T. (1969). *Causes of delinquency.* Berkeley: University of California Press.
Hirschi, T., & Gottfredson, M. (1993). Rethinking the juvenile justice system. *Crime & Delinquency, 39,* 262–271.
Holsinger, K. (2000). Feminist perspectives on female offending: Examining real girls' lives. *Women & Criminal Justice, 12*(1), 23–51.
Holzman, H. R. (1996). Criminological research on public housing: Toward a better understanding of people, places, and spaces. *Crime & Delinquency, 42,* 361–378.
Hooton, E. (1939). *Crime and the man.* Cambridge, MA: Harvard University Press.
Howell, J. C., & Decker, S. H. (1999). *The youth gangs, drugs, and violence connection.* Washington, DC: Office of Juvenile Justice and Delinquency Prevention.
Huff, C. R. (1996). The criminal behavior of gang members and nongang at-risk youth. In C. Huff (Ed.), *Gangs in America* (2nd ed., pp. 75–102). Thousand Oaks, CA: Sage.
Hughes, L. A. (2005). Studying youth gangs: Alternative methods and conclusions. *Journal of Contemporary Criminal Justice, 21,* 98–117.
Hurst, Y. G., McDermott, M. J., & Thomas, D. L. (2005). The attitudes of girls toward the police: Differences by race. *Policing: An International Journal of Police Strategies and Management, 28,* 578-594.
Illinois Compiled Statutes. (1998).
Illinois Compiled Statutes. (1999).
Illinois Compiled Statutes. (2006).
Illinois Department of Corrections. (1985). *Training manual.* Springfield: Author.
Inciardi, J. A., Horowitz, R., & Pottieger, A. E. (1993). *Street kids, street drugs, street crime: An examination of drug use and serious delinquency in Miami.* Belmont, CA: Wadsworth.
Indiana Code Annotated. (1997).

Indystar.com. (2006, October 2). School violence around the world. Available: www2.indystar.com/library/factfiles/crime/school_violence/school_shootings.html

In re Gault, 387 U.S. 1, 87 S.Ct. 1428 (1967).

In re George T., 2002 N.Y. Int. 0161 (2002).

In re Holmes, 379 Pa. 599, 109 A 2d. 523 (1954), cert. denied, 348 U.S. 973, 75 S.Ct. 535 (1955).

In re Register, 84 N.C. App. 336, 352 S. E. 2d 889 (1987) [dictum].

In re William A., 393 Md. 690, 698–699, 548 A.2d 130, 134 (1988).

In re Winship, 397 U.S. 358, 90 S.Ct. 1068 (1970).

In this church, the little children suffer. (2001, March 23). *Omaha World Herald* [editorial].

Institute of Education Sciences. (2005). *Indicators of school crime and safety: 2005.* Available: http://nces.ed.gov/pubsearch/pubsinfo.asp?pubid=2006001

Institute of Judicial Administration and the American Bar Association. (1980). *Standards for the administration of juvenile justice.* Cambridge, MA: Ballinger.

Iowa Code Annotated. (1999).

Jacobs, J. (1977). *Stateville: The penitentiary in mass society.* Chicago: University of Chicago Press.

Jarjoura, G. R. (1996). The conditional effect of social class on the dropout-delinquency relationship. *Journal of Research in Crime and Delinquency, 33,* 232–255.

Jarjoura, G. R., Triplett, R. A., & Brinker, G. P. (2002). Growing up poor: Examining the link between persistent childhood poverty and delinquency. *Journal of Quantitative Criminology, 18,* 159–187.

Jeffery, C. R. (1978). Criminology as an interdisciplinary behavioral science. *Criminology, 16,* 149–169.

Jeffery, C. R. (1996). The genetics and crime conference revisited. *The Criminologist, 21*(2), 1–3.

Johnson, K. (2006, July 13). Police tie jump in crime to juveniles. *USA Today,* p. 1A.

Johnson, R. E. (1980). Social class and delinquent behavior. *Criminology, 18,* 86–93.

Johnson, S. (1998). Girls in trouble: Do we care? The number of delinquent girls is on the rise; only a coordinated, multiagency approach can turn the tide. *Corrections Today, 60*(7), 136–138.

Jones v. Commonwealth, 185 Va. 335, 38 S.E.2d 444 (1946).

Jones, M. B., & Jones, D. R. (2000). The contagious nature of antisocial behavior. *Criminology, 39,* 25–46.

Kanof, M. (2003). *Youth illicit drug use prevention: DARE long term evaluations and federal efforts to identify effective programs.* Washington, DC: U.S. General Accounting Office.

Kaser-Boyd, N. (2002). Children who kill. In N. G. Ribner (Ed.), *Handbook of juvenile forensic psychology* (pp. 159–229). San Francisco: Jossey–Bass.

Katkin, D., Hyman, D., & Kramer, J. (1976). *Juvenile delinquency and the juvenile justice system.* North Scituate, MA: Duxbury.

Kaufman, J., & Zigler, E. F. (1987). Do abused children become abusive parents? *American Journal of Orthopsychiatry, 40,* 953–959.

Kelley, D. H. (1977). Labeling and the consequences of wearing a delinquent label in a school setting. *Education, 97,* 371–380.

Kempf, K. L. (1992). *The role of race in juvenile justice processing in Pennsylvania.* Shippensburg, PA: Center for Juvenile Justice Training and Research.

Kent v. United States, 383 U.S. 541, 86 S.Ct. 1045 (1966).

Klein, M. W. (1967). *Juvenile gangs in context.* Englewood Cliffs, NJ: Prentice Hall.

Klein, M. W. (2005). The value of comparisons in street gang research. *Journal of Contemporary Criminal Justice, 21,* 135–151.

Klein-Saffran, J., Chapman, D. A., & Jeffers, J. L. (1993). Boot camp for prisoners. *Law Enforcement Bulletin, 62*(10), 13–16.

Klinger, D. A. (1996). More on demeanor and arrest in Dade County. *Criminology, 34,* 61–82.

Klockars, C. B. (1979). The contemporary crises of Marxist criminology. *Criminology, 16,* 477–515.

Klofas, J., & Stojkovic, S. (Eds.). (1995). *Crime and justice in the year 2010.* Belmont, CA: Wadsworth.

Knudsen, D. D. (1992). *Child maltreatment: Emerging perspectives.* Dix Hills, NY: General Hall.

Kowaleski-Jones, L. (2000). Staying out of trouble: Community resources and problem behavior among high-risk adolescents. *Journal of Marriage and the Family, 62,* 449–464.

Krasner, L., & Ullman, L. P. (1965). *Research in behavior modification.* New York: Holt, Rinehart & Winston.

Kratcoski, P. C., & Kratcoski, L. D. (1995). *Juvenile delinquency* (4th ed.). New York: McGraw–Hill.

Kretschmer, E. (1925). *Physique and character* (W. Sprott, Trans.). New York: Harcourt, Brace, and World.

Krisberg, B. (2005). *Juvenile justice: Redeeming our children.* Thousand Oaks, CA: Sage.

Krisberg, B., Austin, J., & Steele, P. A. (1989). *Unlocking juvenile corrections: Evaluating the Massachusetts Department of Youth Services.* San Francisco: National Council on Crime and Delinquency.

Kvaraceus, W. C. (1945). *Juvenile delinquency and the school.* New York: World Book.

Lander, B. (1970). An ecological analysis of Baltimore. In M. E. Wolfgang, L. Savitz, & N. Johnston (Eds.), *Sociology of crime and delinquency* (2nd ed., pp. 247–265). New York: John Wiley.

Lane, J., & Turner, S. (1999). Interagency collaboration in juvenile justice: Learning from experience. *Federal Probation, 63*(2), 33–39.

Lanier, M., & Henry, S. (1998). *Essential criminology.* Boulder, CO: Westview.

Lardiero, C. J. (1997). Of disproportionate minority confinement. *Corrections Today, 59*(3), 14–16.

Laub, J. H., & MacMurray, B. K. (1987). Increasing the prosecutor's role in juvenile court: Expectations and realities. *Justice System Journal, 12,* 196–209.

Leiber, M., & Fox, K. (2005). Race and the impact of detention on juvenile justice decision making. *Criminology, 51,* 470–497.

Leiber, M. J., & Stairs, J. M. (1999). Race: Contexts and the use of intake diversion. *Journal of Research in Crime and Delinquency, 36,* 56–86.

Levinson, R. B., & Chase, R. (2000). Private sector involvement in juvenile justice. *Corrections Today, 62*(2), 156–159.

Liddle, H. A., & Hogue, A. (2000). A family-based, developmental-ecological preventive intervention for high-risk adolescents. *Journal of Marital and Family Therapy, 26,* 265–279.

Lindner, C. (2004, Spring). A century of revolutionary changes in the United States court systems. *Perspectives,* pp. 24–29.

Lizotte, A. J., Tesorio, J. M., Thornberry, T. P., & Krohn, M. D. (1994). Patterns of adolescent firearms ownership and use. *Justice Quarterly, 11,* 51–74.

Lotz, R., & Lee, L. (1999). Sociability, school experience, and delinquency. *Youth & Society, 31,* 199–223.

Louisiana Children's Code Annotated. (1999).

Lowencamp, C. T., Cullen, F. T., & Pratt, T. C. (2003). Replicating Sampson and Groves's test of social disorganization theory: Revisiting a criminological classic. *Journal of Research in Crime and Delinquency, 40,* 351–373.

Ludwig, F. J. (1955). *Youth and the law: Handbook on laws affecting youth.* Brooklyn, NY: Foundation Press.

Lundman, R. J. (1993). *Prevention and control of juvenile delinquency* (2nd ed.). New York: Oxford University Press.

Lundman, R. L., Sykes, R. E., & Clark, J. P. (1978). Police control of juveniles: A replication. *Journal of Research in Crime and Delinquency, 15,* 74–91.

Lyerly, R. R., & Skipper, J. K. (1981). Differential rate of rural–urban delinquency. *Criminology, 19,* 385–399.

Lynam, D. R. (1998). Early identification of the fledgling psychopath: Locating the psychopathic child in the current nomenclature. *Journal of Abnormal Psychology, 107,* 566–575.

MacDonald, S. S., & Baroody-Hard, C. (1999). Communication between probation officers and judges: An innovative model. *Federal Probation, 63*(1), 42–50.

MacKenzie, D. L., & Souryal, C. C. (1991, October). Boot camp survey. *Corrections Today,* pp. 90–96.

Madigan, C. (1989, February 5). Minnesota shows the way in childhood assistance. *Chicago Tribune,* sec.1, p. 7.

Main, F. (2001, January 16). Gangs go global. *Chicago Sun–Times,* p. 3.

Main, F., & Spielman, F. (2000, October 5). Gang battles terrorize schools: Students locked inside after series of shootings near campuses. *Chicago Sun–Times,* p. 1.

Marshall, W. L., Cripps, E., Anderson, D., & Cortoni, F. A., (1999). Self-esteem and coping strategies in child molesters. *Journal of Interpersonal Violence, 14*(9), 955–963.

Martens, W. H. J. (1999). Marcel: A case report of a violent sexual psychopath in remission. *International Journal of Offender Therapy and Comparative Criminology, 43,* 391–399.

Martin, G., & Peas, J. (1978). *Behavior modification: What it is and how to do it.* Englewood Cliffs, NJ: Prentice Hall.

Martin, J. R., Schulze, A. D., & Valdez, M. (1988). Taking aim at truancy. *FBI Law Enforcement Bulletin, 57*(5), 8–12.

Martinson, T. M. (1974). What works? Questions and answers about prison reform. *Public Interest, 35*(2), 22–54.

Maryland Annotated Code. (2001).

Maryland Courts and Judicial Procedures Code Annotated. (1998).

Massachusetts General Laws Annotated. (1999).

May, D. C. (1999). Scared kids, unattached kids, or peer pressure: Why do students carry firearms to school? *Youth & Society, 31,* 100–127.

Mays, G. L. (1997). *Gangs and gang behavior.* Chicago: Nelson–Hall.

Mays, G. L., & Winfree, L. T., Jr. (2000). *Juvenile justice.* Boston: McGraw–Hill.

McGloin, J. M. (2005). Policy and intervention considerations of a network analysis of street gangs. *Criminology and Public Policy, 4,* 607–636.

McKeiver v. Pennsylvania, 403 U.S. 528, 91 S. Ct. (1971).

McKinney. (1975). *Consolidated laws of New York, Annotated.* New York: West Group.

McMahon, P. (2002, March 7). Mother convicted of Munchausen child abuse loses appeal. *Sun–Sentinel* (South Florida). Available: www.vachss.com/help_text/archive/kathy_bush.html

McPhee, M. (2000, November 26). Gangs waging war on street: Crips and Bloods take their deadly battle to Brooklyn. *Daily News* (New York), p. 28.

Meddis, S. V. (1993, October 29). In a dark alley, most-feared face is a teen's. *USA Today,* p. 6A.

Mednick, S. A., & Christiansen, K. O. (1977). *Biological bases of criminal behavior.* New York: Gardner.

Meehan, A. J. (1992). I don't prevent crime, I prevent calls: Policing a negotiated order. *Symbolic Interaction, 15,* 455–480.

Memory loss as a survival method. (1991, November). *Healthfacts,* pp. 1–6. (New York: Center for Medical Consumers)

Mempa v. Rhay, 389 U.S. 128 (1967).

Mercer, R., Brooks, M., & Bryant, P. T. (2000). Global positioning satellite system: Tracking offenders in real time (Florida). *Corrections Today, 62*(4), 76–80.

Merton, R. K. (1938). Social structure and anomie. *American Sociological Review, 3,* 672–682.

Merton, R. K. (1955). *Social theory and social structure.* New York: Free Press.

Michie's Legal Resources. Available: www.michie.com.

Michigan Compiled Laws Annotated. (1999).

Mihalic, S., Fagan, A., Irwin, K., Ballard, D., & Elliot, D. (2004). *Blueprints for violence prevention.* Washington, DC: U.S. Department of Justice, Office of Juvenile Justice and Delinquency Prevention.

Mihalic, S., Irwin, K., Elliot, D., Fagan, A., & Hansen, D. (2001). Blueprints for violence prevention. *Juvenile Justice Bulletin.* (Washington, DC: U.S. Department of Justice) Available: www.ncjrs.gov/html/ojjdp/jjbul2001_7_3/contents.html

Miller, W. B. (1958). Lower class culture as a generating milieu of gang delinquency. *Journal of Social Issues, 14*(3), 5–19.

Miller, W. B. (1974). American youth gangs: Past and present. In A. Blumberg (Ed.), *Current perspectives on criminal behavior* (pp. 410–420). New York: Knopf.

Miller, W. B. (2001, April). *The growth of gangs in the United States: 1970–1998* (OJJDP Report). Washington, DC: U.S. Department of Justice.

Minnesota Department of Corrections. (n.d.). *Using crime to reweave the social fabric: Developing social capital/informal social control through restorative processes—Stories from the street.* Available: www.corr.state.mn.us/rj/publications/usingcrime.htm

Minnesota Statutes Annotated. (1998).

Mississippi Code Annotated. (1999).

Mitchell, D. B., & Kropf, S. E. (2002). Youth violence: Response of the judiciary. In G. S. Katzman (Ed.), *Securing our children's future: New approaches to juvenile justice and youth violence* (pp. 432–444). Washington, DC: Brookings Institution Press and Governance Institute.

Moffitt, T. E. (1993). Adolescence-limited and life-course persistent antisocial behavior: A developmental taxonomy. *Psychological Review, 100,* 674–701.

Moffitt, T., & Caspi, A. (2001). Childhood predictors differentiate life-course persistent and adolescent-limited antisocial pathways among males and females. *Development and Psychopathology, 13,* 355–375.

Monk-Turner, E. (1990). The occupational achievements of community and four-year college entrants. *American Sociological Review, 55,* 719–725.

Montana Code Annotated. (1999).

Moon, M. M., Sundt, J. L., Cullen, F. T., & Wright, J. P. (2000). Is child saving dead? Public support for juvenile rehabilitation. *Crime & Delinquency, 46,* 38–60.

Moore, B. M. (2001/2002). Blended sentencing for juveniles: The creation of a third criminal justice system? *Journal of Juvenile Law, 22,* 126–138.

Morash, M. (1984). Establishment of a juvenile police record: The influence of individual and peer group characteristics. *Criminology, 22,* 97–111.

Morash, M., & Chesney-Lind, M. (1991). A reformulation and partial test of the power control theory of delinquency. *Justice Quarterly, 8,* 347–379.

Morrissey v. Brewer, 408 U.S. 471 (1972).

Moyer, I. (1981). Demeanor, sex, and race in police processing. *Journal of Criminal Justice, 9,* 235–246.

Moyer, I. L. (2001). *Criminological theories: Traditional and nontraditional voices and themes.* Thousand Oaks, CA: Sage.

Naffine, N. (1996). *Feminism and criminology.* Philadelphia: Temple University Press.

National Advisory Commission on Criminal Justice Standards and Goals. (1973). *Courts.* Washington, DC: U.S. Department of Justice.

National Conference of Commissioners on Uniform State Laws. (1968). *Uniform Juvenile Court Act.* Philadelphia: Author.

National Criminal Justice Reference Service. (2005). *Gangs: Facts and figures.* Washington, DC: U.S. Department of Justice. Available: www.ncjrs.org/spotlight/gangs/facts.html

National Institute of Justice. (1988, September/October). Targeting serious juvenile offenders for prosecution can make a difference. *NIJ Reports,* pp. 9–12. (Washington, DC: U.S. Department of Justice)

National Institute of Justice. (1994, October). *The DARE program: A review of prevalence, user satisfaction, and effectiveness.* Washington, DC: Author.

National Institute on Drug Abuse. (n.d.). *Assessing drug abuse within and across communities: Community epidemiology surveillance networks on drug abuse* (2nd ed.). Washington, DC: U.S. Department of Health and Human Services.

National Victims Center. (1998). *Promising practices and strategies for victim services in corrections.* Washington, DC: U.S. Department of Justice. Available: www.ojp.usdoj.gov/ovc/publications/infores/victims/victserv.pdf

National Youth Gang Center. (1997). *1995 National Youth Gang Survey.* Washington, DC: U.S. Department of Justice, Office of Juvenile Justice and Delinquency Prevention.

Nebraska Revised Statutes. (1998).

Neubauer, D. W. (2002). *America's courts and the criminal justice system* (7th ed.). Belmont, CA: Wadsworth/Thomson Learning.

New Jersey Statutes Annotated. (1999).

New Mexico Statutes Annotated. (1995).

New York Family Court Act. (1999).

New York Sessions Laws. (1962).

North Carolina General Statutes. (1999).

North Dakota Century Code. (1991).

North Dakota Century Code. (1999).

Novotney, L., Mertinko, E., Lange, J., & Baker, T. K. (2000). Juvenile Mentoring Program: A progress report. *Juvenile Justice Bulletin.* (Washington, DC: Office of Juvenile Justice and Delinquency Prevention)

NYC mom arrested in son's death. (2006, November 16). *Journal Star,* p. A8. (Peoria, IL)

Nyquist, O. (1960). *Juvenile justice: A comparative study with special reference to the Swedish Welfare Board and the California juvenile court system.* London: Macmillan.

Oates, K., Jones, D., Denson, D., Sirotnak, A., Gary, N., & Krugman, R. (2000). Erroneous concerns about sexual abuse. *Child Abuse & Neglect, 24,* 149–157.

Office of Juvenile Justice and Delinquency Prevention. (1979). *Delinquency prevention: Theories and strategies.* Washington, DC: U.S. Government Printing Office.

Office of Juvenile Justice and Delinquency Prevention. (1980). *Juvenile justice: Before and after the onset of delinquency.* Washington, DC: U.S. Government Printing Office.

Office of Juvenile Justice and Delinquency Prevention. (1992). *Arrests of youth, 1990.* Washington, DC: U.S. Government Printing Office.

Office of Juvenile Justice and Delinquency Prevention. (1996, March). *Juvenile probation: The workhorse of the juvenile justice system.* Washington, DC: U.S. Government Printing Office.

Office of Juvenile Justice and Delinquency Prevention. (2000, September). Preventing adolescent gang involvement. *Juvenile Justice Bulletin.* (Washington, DC: U.S. Department of Justice)

Office of Juvenile Justice and Delinquency Prevention. (2001, January). The decline in child sexual abuse cases. *Juvenile Justice Bulletin.* Available: www.ncjrs.gov/html/ojjdp/jjbu12001_1_1/page4.html

Office of Juvenile Justice and Delinquency Prevention. (2003). *State statutes define who is under juvenile court jurisdiction* (Juveniles in Court, OJJDP National Report Series Bulletin). Washington, DC: U.S. Department of Justice.

Office of Juvenile Justice and Delinquency Prevention. (2006). *Juvenile offenders and victims: 2006 national report.* Washington, DC: U.S. Department of Justice. Available: www.ojjdp.ncjrs.gov/ojstatbb/nr2006/index.html

Ohlin, L. E. (1998). The future of juvenile justice policy and research. *Crime & Delinquency, 44,* 143–153.

Oklahoma Statutes Annotated. (1998).

Oregon Revised Statutes. (1995).

Osborne, D., & Gaebler, T. (1992). *Reinventing government: How the entrepreneurial spirit is transforming the public sector.* Reading, MA: Addison–Wesley.

Osgood, D. W., & Chambers, J. M. (2000). Social disorganization outside the metropolis: An analysis of rural youth violence. *Criminology, 38,* 81–115.

Pagani, L., Boulerice, B., & Vitaro, F. (1999). Effects of poverty on academic failure and delinquency in boys: A change and process model approach. *Journal of Child Psychology and Psychiatry and Allied Disciplines, 40,* 1209–1219.

Palumbo, M. G., & Ferguson, J. (1995). Evaluating Gang Resistance Education and Training: Is the impact the same as Drug Abuse Resistance Education (DARE)? *Evaluation Review, 19,* 597–619.

Papachristos, A. (2005, March/April). Gang world. *Foreign Policy,* pp. 48–55. Available: www.foreignpolicy.com/story/files/story2798.php

Parker-Jimenez, J. (1997). An offender's experience with the criminal justice system. *Federal Probation, 61*(1), 47–52.

Parsons, A. (1998, August). Meth and cocaine: Addictive drugs alike but different. *Southeast Missourian,* pp. 1–4.

Pattillo, M. E. (1998). Sweet mothers and gang bangers: Managing crime in a black middle-class neighborhood. *Social Forces, 76,* 747–774.

Paul, R. H., Marx, B. P., & Orsillo, S. M. (1999). Acceptance-based psychotherapy in the treatment of an adjudicated exhibitionist: A case example. *Behavior Therapy, 30,* 149–162.

Paxson, C. H., & Waldfogel, J. (1999). Parental resources and child abuse and neglect. *American Economic Review, 89,* 239–244.

Peak, K. J. (1999). Gangs: Origins, status, community responses, and policy implications. In R. Muraskin & A. R. Roberts (Eds.), *Visions for change: Crime and justice in the twenty-first century* (pp. 51–63). Upper Saddle River, NJ: Prentice Hall.

Pears, K. C., & Capaldi, D. M. (2001). Intergenerational transmission of abuse: A two-generational prospective study of an at-risk sample. *Child Abuse & Neglect, 25,* 1439–1461.

People ex rel. O'Connor v. Turner, Ill. 280, 286 (1870).

People v. Dominquez, 256 Cal.App. 2d 623 (1967).

Perkiss, M. (1989, January 22). Program aims to help mildly abused children without foster care. *Macomb Journal,* p. 2C.

Perlmutter, B. F. (1987). Delinquency and learning disabilities: Evidence for compensatory behaviors and adaptation. *Journal of Youth and Adolescence, 16,* 89–95.

Peters, J. M. (1991). Specialists a definite advantage in child sexual abuse cases. *Police Chief, 58*(2), 21–23.

Peters, S. R., & Peters, S. D. (1998). Violent adolescent females. *Corrections Today, 60*(3), 28–29.

Petition of Ferrier, 103 Ill. 367, 371 (1882).

Piliavin, I., & Briar, S. (1964). Police encounters with juveniles. *American Journal of Sociology, 70,* 206–214.

Piquero, A. R., Gomez-Smith, Z., & Langton, L. (2004). Discerning fairness where others may not: Low self-control and unfair sanctions perceptions. *Criminology, 42,* 699–734.

Piscotta, A. W. (1982). Saving the children: The promise and practice of *parens patriae,* 1838–98. *Crime & Delinquency, 28,* 424–425.

Plass, P. S., & Carmody, D. C. (2005). Routine activities of delinquent and non-delinquent victims of violent crime. *American Journal of Criminal Justice, 29,* 235–246.

Platt, A. (1977). *The child savers* (2nd ed.). Chicago: University of Chicago Press.

Polk, K. (1984). The new marginal youth. *Crime & Delinquency, 30,* 462–480.

Polk, K., & Schafer, W. B. (1972). *School and delinquency.* Englewood Cliffs, NJ: Prentice Hall.

Pollock, J. M. (1994). *Ethics in crime and justice: Dilemmas and decisions* (2nd ed.). Belmont, CA: Wadsworth.

Porterfield, A. L. (1946). *Youth in trouble.* Fort Worth, TX: Leo Potishman Foundation.

Portwood, S. G., Grady, M. T., & Dutton, S. E. (2000). Enhancing law enforcement identification and investigation of child maltreatment. *Child Abuse & Neglect, 24,* 195–207.

Postman, N. (1991). Quoted in D. Osborne & T. Gaebler. *Reinventing government: How the entrepreneurial spirit is transforming the public sector,* p. 19. Reading, MA: Addison Wesley.

Poythrees, N. G., Edens, J. F., & Lilienfeld, S. O. (1998). Criterion-related validity of the Psychopathic Personality Inventory in a prison sample. *Psychological Assessment, 10,* 426–430.

President's Commission on Law Enforcement and Administration of Justice. (1967). *Task force report: Juvenile delinquency and youth crime.* Washington, DC: U.S. Government Printing Office.

Puzzanchera, C. H. (2003). *Delinquency cases waived to criminal court, 1990–1999* (OJJDP Fact Sheet No. 35). Washington, DC: U.S. Department of Justice, Office of Juvenile Justice and Delinquency Prevention.

Quinney, R. (1970). *The social reality of crime.* Boston: Little, Brown.

Quinney, R. (1974). *Critique of legal order: Crime control in capitalist society.* Boston: Little, Brown.

Quinney, R. (1975). *Criminology.* Boston: Little, Brown.

Rafter, N. (2004). Earnest Hooton and the biological tradition in American criminology. *Criminology, 42,* 735–772.

Rebellon, C. J. (2002). Reconsidering the broken homes/delinquency relationship and exploring its mediating mechanism(s). *Criminology, 40,* 103–136.

Reckless, W. C. (1961). A new theory of delinquency and crime. *Federal Probation, 25,* 42–46.

Reckless, W. C. (1967). *The crime problem.* New York: Appleton–Century–Crofts.

Regoli, R. M., & Hewitt, J. D. (1994). *Delinquency in society: A child-centered approach.* New York: McGraw–Hill.

Reid, S. T. (2006). *Crime and criminology* (11th ed.). New York: McGraw–Hill.

Rendleman, D. R. (1974). *Parens patriae:* From chancery to the juvenile court. In F. L. Faust & P. J. Brantingham (Eds.), *Juvenile justice* (pp. 72–117). St. Paul, MN: West.

Restorative Justice for Illinois. (1999). *What is restorative justice?* Des Plaines, IL: LSSI/Prison and Family Ministry.

Rinehart, W. (1991). *Convicted child molesters.* Unpublished M.A. thesis, Western Illinois University.

Roberts, A. R. (1989). *Juvenile justice: Policies, programs, services.* Chicago: Dorsey.

Robinson, S. (1999). Juvenile offenders are worth the effort. *Corrections Today, 61*(2), 8.

Rodney, E. H., & Mupier, R. (1999). Comparing the behaviors and social environments of offending and non-offending African-American adolescents. *Journal of Offender Rehabilitation, 30*(1/2), 65–80.

Rosenberg, M. S., & Repucci, N. D. (1983). Abusive mothers: Perceptions of their own and their children's behaviors. *Journal of Consulting and Clinical Psychology, 51,* 674–682.

Ross, R. R., & McKay, H. B. (1978). Behavioral approaches to treatment in corrections: Requiem for a panacea. *Canadian Journal of Criminology, 20,* 279–298.

Russi, K. (1984). Operation K.I.D.: A community approach to child protection. *Police Chief, 46*(2), 35–36.

Sanders, W. B. (1974). Some early beginnings of the children's court movement in England. In F. L. Faust & P. J. Brantingham (Eds.), *Juvenile justice philosophy* (pp. 46–51). St. Paul, MN: West.

Satterfield, J. H. (1987). Childhood diagnostic and neurophysiological predictors of teenage arrest rates: An eight-year prospective study. In S. A. Mednick, T. E. Moffit, & S. S. Stack (Eds.), *The causes of crime: New biological approaches* (pp. 146–167). Cambridge, UK: Cambridge University Press.

Scaramella, G. L. (2000). Methamphetamines: A blast from the past. *Crime & Justice International, 16,* 7–8.

Schafer, W. E., & Polk, K. (Eds.). (1967). Delinquency and the schools. In *Task force report: Juvenile delinquency and youth crime* (President's Commission on Law Enforcement and the Administration of Justice). Washington, DC: U.S. Government Printing Office.

Scherer, R. (2005, August 18). Child porn rising on the Web. *Christian Science Monitor.* Available: www.csmonitor .com/2005/0818/p01s01-stct.html

Schinke, S. P., & Gilchrist, L. D. (1984). *Life counseling skills with adolescents.* Baltimore, MD: University Park Press.

Schur, E. M. (1973). *Radical non-intervention: Rethinking the delinquency problem.* Englewood Cliffs, NJ: Prentice Hall.

Schwartz, I. M., Weiner, N. A., & Enosh, G. (1998). Nine lives and then some: Why the juvenile court does not roll over and die. *Wake Forest Law Review, 33,* 533–552.

Schwartz, I. M., Weiner, N. A., & Enosh, G. (1999). Myopic justice? The juvenile court and child welfare systems. *Annals of the American Academy of Political & Social Science, 564,* 126–141.

Scott, J. W., & Vaz, E. W. (1963). A perspective on middle-class delinquency. *Canadian Journal of Economics and Political Science, 29,* 324–335.

Scudder, R. G., Blount, W. R., Heide, K. M., & Silverman, I. J. (1993). Important links between child abuse, neglect, and delinquency. *International Journal of Offender Therapy and Comparative Criminology, 37,* 310–323.

Sealock, M. D., & Simpson, S. (1998). Unraveling bias in arrest decisions: The role of juvenile offender type-scripts. *Justice Quarterly, 15,* 427–457.

Sedlak, A. J., Doueck, H. J., Lyons, P., & Wells, S. J. (2005). Child maltreatment and the justice system: Predictors of court involvement. *Research on Social Work Practice, 15,* 389–407.

Shaw, C. R., & McKay, H. D. (1942). *Juvenile delinquency and urban areas.* Chicago: University of Chicago Press.

Shaw, C. R., & McKay, H. D. (1969). *Juvenile delinquency and urban areas* (Rev. ed.). Chicago: University of Chicago Press.

Sheldon, W. H. (1949). *Varieties of delinquent youth: An introduction to constitutional psychiatry.* New York: Harper & Row.

Shelton, T. L., Barkley, R. A., & Crosswait, C. (2000). Multimethod psychoeducational intervention for preschool children with disruptive behavior: Two-year post-treatment follow-up. *Journal of Abnormal Child Psychology, 28,* 253–266.

Sherman, L. W., & Weisburd, D. (1995). General deterrent effects of police patrol in crime "hot spots": A randomized study. *Justice Quarterly, 12,* 625–640.

Sherman, L. W., Gottfredson, D., MacKenzie, D., Eck, J., Reuter, P., & Bushaway, S. (1997). *Preventing crime: What works, what doesn't, what's promising—A report to the United States Congress.* Washington, DC: U.S. Government Printing Office. Available: www.ncjrs.org

Shorkey, C. T., & Armendariz, J. (1985). Personal worth, self-esteem, anomia, hostility, and irrational thinking of abusing mothers: A multivariate approach. *Journal of Clinical Psychology, 41,* 414–421.

Short, J. F., & Nye, F. I. (1958). Extent of unrecorded juvenile delinquency: Some tentative conclusions. *Journal of Criminal Law, Criminology, and Police Science, 49,* 296–302.

Short, J. F., & Strodtbeck, F. (1965). *Group process and gang delinquency.* Chicago: University of Chicago Press.

Shusta, R. M., Levine, D. R., Wong, H. Z., & Harris, P. R. (2005). *Multicultural law enforcement: Strategies for peace-keeping in a diverse society* (3rd ed.). Upper Saddle River, NJ: Prentice Hall.

Sickmund, M. (2003). *Juvenile offenders and victims* (National Report Series). Washington, DC: U.S. Department of Justice, Office of Juvenile Justice and Delinquency Prevention.

Sickmund, M., Snyder, H. N., & Poe-Yamagata, E. (1997). *Juvenile offenders and victims: 1997 update on violence.* Washington, DC: U.S. Department of Justice, Office of Juvenile Justice and Delinquency Prevention.

Siegel, L. J., & Senna, J. J. (1994). *Juvenile justice: Theory, practice, and law* (5th ed.). St. Paul, MN: West.

Siegel, J. A., & Williams, L. M. (2003). The relationship between child sexual abuse and female delinquency and crime: A prospective study. *Journal of Research in Crime and Delinquency, 40,* 71–95.

Siegel, L. J., Welsh, B. C., & Senna, J. J. (2003). *Juvenile delinquency: Theory, practice, and law* (8th ed.). Belmont, CA: Wadsworth/Thomson Learning.

Simms, S. O. (1997, December). Restorative juvenile justice: Maryland's legislature reaffirms commitment to juvenile justice reform. *Corrections Today, 59,* 94–97.

Simons, R. L., Simons, L. G., Burt, C. H., Brody, G. H., & Cutrona, C. (2005). Collective efficacy, authoritative parenting, and delinquency: A longitudinal test of a model integrating community- and family-level process. *Criminology, 43,* 989–1030.

Simonsen, C. E. (1991). *Juvenile justice in America* (2nd ed.). New York: Macmillan.

Simonsen, C. E., & Gordon, M. S. (1982). *Juvenile justice in America* (2nd ed.). New York: Macmillan.

Skinner, B. F. (1953). *Science and human behavior.* New York: Macmillan.

Smith, C. A., & Stern, S. B. (1997). Delinquency and antisocial behavior: A review of family processes and intervention research. *Social Service Review, 71,* 382–420.

Smykla, J. O., & Willis, T. W. (1981). The incidence of learning disabilities and mental retardation in youth under the jurisdiction of the juvenile court. *Journal of Criminal Justice, 9,* 219–225.

Snyder, H. N. (2000, December). Juvenile arrests 1999. *Juvenile Justice Bulletin.* (Washington, DC: U.S. Department of Justice)

Snyder, H. N., & Sickmund, M. (1999, November). *Juvenile offenders and victims: 1999 national report.* Washington, DC: U.S. Department of Justice.

Snyder, H. N., & Sickmund, M. (2006, March). *Juvenile offenders and victims: 2006 national report.* Washington, DC: U.S. Department of Justice. Available: www.ojjdp.ncjrs.gov/ojstatbb/nr2006/index.html

Sorrells, J. (1980). What can be done about juvenile homicide? *Crime & Delinquency, 26,* 152–161.

South Dakota Codified Laws Annotated. (1999).

Stanford v. Kentucky, 492 U.S. 361 (1989).

Stark, R. (1987). Deviant places: A theory of the ecology of crime. *Criminology, 25,* 893–909.

Statistical base still unfolding. (1992). *NRCCSA News,* 1(1), 5–8. (Huntsville, AL: National Resource Center on Child Sexual Abuse)

Stearns, M., & Garcia, M. (2001, July 29). Law challenges preacher's principles: Religious development's founder defends the way he ministers to children. *Kansas City Star,* p. A1.

Stephens, G. (1998, May). Saving the nation's most precious resources: Our children. *USA Today Magazine,* pp. 54–57.

Stephens, R. D., & Arnette, J. L. (2000). *From the schoolhouse to the courthouse: Making successful transitions.* Washington, DC: U.S. Department of Justice, Office of Juvenile Justice and Delinquency Prevention.

Stern, R. S. (1964). *Delinquent conduct and broken homes.* New Haven, CT: College and University Press.

Streib, V. L. (2000). *The juvenile death penalty today: Death sentences and executions for juvenile crimes, January 1, 1973–June 30, 2000.* Ada, OH: Northern University Clause W. Pettit College of Law.

Stroud, D. D., Martens, S. L., & Barker, J. (2000). Criminal investigations of child sexual abuse: A comparison of cases referred to the prosecutor to those not referred. *Child Abuse & Neglect, 24,* 689–700.

Stuckey, G., Roberson, C., & Wallace, H. (2004). *Procedures in the justice system* (7th ed.). Upper Saddle River, NJ: Prentice Hall.

Sudnow, D. (1965). Normal crimes: Sociological features of the penal code in a public defender office. *Social Problems, 12,* 255–276.

Susman, T. (2002, August 21). Doubting the system: Laws on juveniles stir debate over punishment and racism. *Los Angeles Times.*

Sutherland, E. H. (1939). *Principles of criminology* (3rd ed.). Philadelphia: J. B. Lippincott.

Sutherland, E. H., & Cressey, D. R. (1978). *Criminology* (10th ed.). New York: J. B. Lippincott.

Sutherland, E. H., Cressey, D. R., & Luckenbill, D. F. (1992). *Criminology* (11th ed.). Dix Hills, NJ: General Hall.

Sutphen, R., Kurtz, D., & Giddings, M. (1993). The influence of juveniles' race on police decision-making: An exploratory study. *Juvenile and Family Court Journal, 44*(2), 69–76.

Sykes, G. M., & Matza, D. (1957). Techniques of neutralization: A theory of delinquency. *American Sociological Review, 22*, 664–670.

Szymanski, L. A. (2005). Clear and convincing evidence as burden of proof for pre-adjudication detention. *Juvenile and Family Law Digest, 37*(1), 3963–3982.

Tappan, P. (1949). *Juvenile delinquency.* New York: McGraw–Hill.

Taylor, J., McGue, M., & Iacono, W. G. (2000). A behavioral genetic analysis of the relationship between the Socialization Scale and self-reported delinquency. *Journal of Personality, 69*, 29–50.

Taylor, R. L. (1994). Black males and social policy: Breaking the cycle of disadvantage. In R. G. Majors & J. U. Gordon (Eds.), *The American black male: His present status and his future* (pp. 148–166). Chicago: Nelson–Hall.

Tennessee Code Annotated. (1996).

Terry, R. M. (1967). The screening of juvenile offenders. *Journal of Criminal Law, Criminology, and Police Science, 58*, 173–181.

Texas Family Code Annotated. (1996).

Texas Family Code Annotated. (2005).

Thompson v. Oklahoma, 487 U.S. 815 (1988).

Thompson, G. (2004, September 26). Shuttling between nations: Latino gangs confound the law. *The New York Times*, p. 1. Available: www.sawers.com/deb/nytimes%20gang%20article.htm

Thornberry, T. P. (1987). Toward an interactional theory of delinquency. *Criminology, 25*, 863–891.

Thornberry, T. P., Huizinga, D., & Loeber, R. (2004). The Causes and Correlates studies: Findings and policy implications. *Juvenile Justice, 9*(1). Available: www.ncjrs.gov/html/ojjdp/203555/jj2.html

Thornberry, T., Moore, M., & Christenson, R. L. (1985). The effects of dropping out of high school on subsequent criminal behavior. *Criminology, 23*, 3–18.

Thornton, W. E., Voight, L., & Doerner, W. G. (1987). *Delinquency and justice* (2nd ed.). New York: Random House.

Thrasher, F. M. (1927). *The gang.* Chicago: University of Chicago Press.

Tittle, C., Villemez, W., & Smith, D. (1978). The myth of social class and criminality. *American Sociological Review, 43*, 643–656.

Too poor to be defended [editorial]. (1998, April 9). *The Economist*, pp. 21–22.

Torbet, P. M. (1996, March). Juvenile probation: The workhorse of the juvenile justice system. *Juvenile Justice Bulletin.* (Washington, DC: U.S. Department of Justice)

Tower, C. C. (1993). *Understanding child abuse and neglect.* Boston: Allyn & Bacon.

Turk, A. (1969). *Criminality and legal order.* Chicago: Rand McNally.

Turkheimer, E. (1998). Heritability and psychological explanations. *Psychological Review, 105*, 782–791.

Twentieth Century Fund Task Force on Sentencing Policy Toward Young Offenders. (1987). *Confronting youth crime: Report of the Twentieth Century Fund Task Force on Sentencing Policy Toward Young Offenders* (Background Paper by Franklin E. Zimring). New York: Holmes & Meier.

Umbreit, M., Vos, B., & Coates, R. B. (2006). Victim offender mediation: An evolving evidence-based practice. In D. Sullivan & L. Tifft (Eds.), *The handbook of restorative justice: A global perspective* (pp. 52–61). New York: Routledge.

Umbreit, M., Vos, B., Coates, R. B., & Lightfoot, E. (2005). Restorative justice in the twenty-first century: A social movement full of opportunities and pitfalls. *Marquette Law Review, 89*, 251–304.

Unnever, J. D., Cullen, F. T., & Pratt, T. C. (2003). Parental management, ADHD, and delinquency involvement: Reassessing Gottfredson and Hirschi's general theory. *Justice Quarterly, 20*, 471–500.

U.S. Department of Health and Human Services. (2005). *Child maltreatment, 2003.* Available: www.acf.hhs.gov/programs/cb

U.S. Department of Health and Human Services. (2006). *Child maltreatment, 2004.* Available: www.acf.dhhs.gov/programs/cb/pubs/cm04/index.htm

U.S. Department of Justice. (1973). *Prosecution in juvenile courts: Guidelines for the future.* Washington, DC: Author.

U.S. Department of Justice. (1998). *1998 annual report on school safety.* Washington, DC: Author.

U.S. Department of Justice. (2006). *Attorney General Gonzales announces implementation of Project Safe Childhood.* Washington, DC: Author. Available: www.ojjdp.ncjrs.gov/enews/06juvjust/060524.html

U.S. offers funds, training to combat predators using Web to exploit youth. (2006). *Juvenile Justice Digest, 34*(10), 1–2.

Utah Code Annotated. (2006).

Valdez, A. (2005). *Gangs: A guide to understanding street gangs.* San Clemente, CA: LawTech Publishing.

Vander Ven, T. M., Cullen, F. T., Carrozza, M. A., & Wright, J. P. (2001). Home alone: The impact of maternal employment on delinquency. *Social Problems, 48*, 236–257.

Index

Abadinsky, H., 7, 288, 292, 297
Abandonment, 125–126, 262, 267–268
Abbott, R. D., 60
Abuse/neglect, 262–268, 264f, 275, 324
  adjudicatory hearings and, 156, 157
  CASAs and, 204
  correctional facilities and, 258
  custody and, 144
  delinquency and, 82
  differential association theory and, 102
  family services and, 203
  foster homes and, 245
  judges and, 196, 197, 199
  juvenile court acts and, 112, 114, 116, 117–119, 121, 125, 126, 140
  juvenile court proceedings and, 193
  juvenile law and, 309
  legal definitions and, 22–24
  police and, 170, 177, 178f, 179, 181, 182, 186
  prevention programs and, 212–213, 214, 231–232, 233
  probation and, 201, 202, 245
  prosecutors and, 192
  reporting, 28–29, 30t, 35
  rights of juveniles and, 141
  spanking and, 23, 266
  statistics and, 27
  treatment centers and, 248
  violent crime and, 25
  See also Emotional abuse; Sexual abuse
Abused and Neglected Child Reporting Act, 263, 266, 321, 363, 366
Accountability:
  juvenile court acts and, 114, 120, 136
  juvenile courts and, 308
  juvenile justice system and, 317
  police and, 171
  prevention programs and, 228
  probation and, 243
  restorative justice and, 224
Ackerman, W. V., 56
Acoca, L., 283
Adjudicatory hearings, 112, 113t, 120, 125, 156–160, 159f, 321, 337
  judges and, 199, 238
  petition and, 154
  probation officers and, 201
  social background investigation and, 160
Adler, A., 94
African Americans, 36
  arrests and, 23c, 310–311
  crime and, 79
  custody and, 25
  discriminatory practices and, 73
  education and, 51, 219
  family and, 48
  fear of attack at school and, 53–54, 55f
  gangs and, 285, 292
  harsher dispositions and, 34
  police and, 172, 173, 175
Age:
  crime and, 61–67, 62–66t
  gangs and, 278, 294
  juvenile court acts and, 120–122, 121t, 128
  police and, 172
  waivers and, 130, 133–134
Age ambiguity, 20, 22, 321
Age of responsibility, 4, 17, 321
Ageton, S., 56
Agnew, R., 99
Akers, R. L., 17, 32, 93, 94, 95, 102, 105, 107
Alabama Code, 126

Alcohol:
education programs and, 218
middle-class youths and, 56, 57
neglect and, 268
police and, 172
prevention programs and, 228
probation and, 241
probation officers and, 202
schools and, 58–59
See also Drugs
Allen, T. T., 171, 172
Alternatives to incarceration, 252, 255, 259, 298–300, 322
Alwin, D. F., 50
Amdur, R. L., 228
American Bar Association, 113, 136, 192, 205
American Civil Liberties Union, 145
American Correctional Association, 242
American Humane Association, 27
American Indians/Alaskan Natives, 74, 75t, 76t, 77t
Anderson, D., 105
Anderson, E., 173
Anderson, J. F., 48
Aniskievicz, 182
Anomalies, 14, 88, 322
Anomie theory, 98–99, 322
Anthony, P., 182
Appeals, 322
Ariessohn, R. M., 299
Arizona Revised Statutes Annotated, 134
Arkansas Code Annotated, 128
Armagh, D., 272
Armendariz, J., 105
Armstrong, K. L., 267
Arnette, J. L., 216
Arrests, 143–144
adults/juvenile crime and, 42
custody and, 144
gangs and, 320
judges and, 197
minorities and, 175
police and, 171, 172
statistics and, 25–26
Asians, 74, 75t, 285, 292
At-risk youths, 4, 16, 219
Atavists, 88, 322
Atkins, B., 191
Attorney General of Texas, 292
Attorneys, 206
juvenile codes and, 205
juvenile court proceedings and, 192–194
police and, 180
Augustus, J., 239, 335
Austin, J., 253
Automatic waiver, 16, 130, 322
Away syndrome, 252, 322
Ayers, C. D., 60

Babicky, T., 291, 294
Baez, C., 97
Bail, 142–143, 153, 322
Baker, T. K., 227
Balance, G. S., 309
Ballard, D., 214
Barker, J., 271
Barkley, R. A., 97
Barlow, H. D., 194, 195
Baron, S. W., 99
Baroody-Hard, C., 199
Bartollas, C., 61, 185
Battistich, V., 51
Bazemore, G., 172, 229, 243, 244, 309
Beaver, K. M., 89
Beccaria, C., 85
Becker, H. S., 98, 103
Behavioral definitions, 20, 23, 24–25, 36, 322
Behaviorism, 95, 97
Beirne, P., 104
Belknap, J., 229
Bell, D. J., 35
Bell, S., 35
Bellair, P. E., 74
Benekos, P. J., 73
Bentham, J., 85
Bernard, T. J., 4, 5, 83, 251
Bernet, W., 268
Beyond a reasonable doubt, 11, 12, 120, 123, 124, 156–157, 199, 323
Big Brothers/Big Sisters of America (BBBSA), 219, 229
Bilchik, S., 2, 16, 130, 227, 278, 283, 285, 287, 290, 309
Biological theories, 88–90, 108, 323
Bishop, D., 308
Bishop, D. M., 34, 73
Bishop, S. J., 199, 272
Bjerregaard, B., 295, 296
Black, D. J., 26, 34, 171, 179
Black Disciples, 291, 323

Blacks. *See* African Americans
Blackstone, W., 5
Blackstone Rangers, 286–287, 287, 323
Blair, M. C. L., 50
Blair, S. L., 50
Blankenship, R. L., 103
Blazak, R., 56, 57, 173, 278, 282, 283
Blended sentencing, 128, 317
Bloch, H., 294
Blome, D., 232
Bloods, 286, 290, 291, 323
Bloom, B., 283
Blount, W. R., 82
Blumberg, A. S., 195
Bocian, K. M., 95
Bohm, R. M., 85, 89, 94, 95, 101, 105
Bookin, H., 278
Boot camp, 255–256, 323, 335, 338, 342, 346
Booth, A., 90
"Born criminal" theory, 88
Botch, D., 194
Boulerice, B., 32
Boy. *See* Males
Bradley, C. M., 11
Bradshaw, W., 220
Braithwaite, J., 104
Brantingham, P. J., 9, 10, 123
Braun, C., 50
Breed v. Jones, 11, 135, 324, 329
Briar, S., 34, 171, 175
Brinker, G. P., 58
Broder, P. K., 53
Brody, G. H., 47
Broken homes, 47–48, 49, 50, 78, 83, 96, 99, 324
Bronner, A., 94
Brooks, M., 203
Brown, B., 181, 182, 185, 297
Brown, J., 267, 268
Brown, L., 297
Brown, S. L., 48
Brownfield, D., 53
Browning, K., 48, 53
Bruinius, H., 91, 92
Bryant, D., 226
Bryant, P. T., 203
Buck v. Bell, 91
Buerger, M. E., 86
Buikhuisen, W., 299
Bullying, 54, 214, 282
Bullying Prevention Program, 219
Bumphus, V. W., 48
Burch, J., 298
Bureau of Justice Assistance, 228
Bureau of Justice Statistics, 26, 27, 262, 339, 351
Burgess, E. W., 326
Burgess, R. L., 99, 102
Bursik, R. J., 58, 278
Burt, C. H., 47
Bush, G. W., 221
Butts, J. A., 157
Bynum, J. E., 5, 32, 50

Cadzow, S. P., 267
Califano, J. A., Jr., 317
California Personality Inventory (CPI), 94
California Welfare and Institutional Code, 144, 160
Campbell, A., 295, 296
Canter, R., 47, 48
Capaldi, D. M., 266
Capital punishment, 4, 11, 108, 254, 324, 341
Capizzano, J., 49
Carelli, R., 272
Carmody, D. C., 86
Carrozza, M. A., 49
Caspi, A., 298, 299, 300
Caucasian Americans *See* Whites
Cauchon, D., 217
Cavan, R. S., 8
Center for Restorative Justice and Mediation, 243, 244
Center for the Study and Prevention of Violence, 219
Centerwood, B. S., 283
Chambers, J. M., 293
Chambliss, W. J., 58, 104
Champion, D., 202, 203
Chancery courts, 5, 12, 324
Chapman, D. A., 256
Chapman, J., 197
Charton, S., 84
Chase, R., 242
Chesney-Lind, M., 56, 67, 105
Child abuse. *See* Abuse/neglect
Child Abuse Prevention Program, 232
Child death review teams, 273, 324
Child Development–Community Policing (CDCP), 273
Child neglect, 117, 267–268, 324
Child Trends DataBank, 53, 54

Child Welfare Department, 161
Child Welfare Information Gateway, 262
Children and family services, 190, 199, 203–204, 206, 224–225, 231, 233, 324
Children's Bureau, 142, 227
Children's Defense Fund, 25, 267
Chrismer, S. S., 182
Christenson, R. L., 51
Christiansen, K. O., 90
Churches, 211, 213, 240
Cinq-Mars, C., 271
City Custom of Apprentices, 6, 12
Clark, J. P., 26, 32, 33
Classical theory, 83, 85, 325
Clear, T. R., 298, 300
Clear and convincing evidence, 157, 158, 325
Clifford, R. D., 226
Clinton, B., 297
Cloward, R. A., 56, 57, 98
Coates, R. B., 220, 244
Cocaine, 59–60, 290
  See also Drugs
Cohen, A. K., 56, 98, 278, 285
Cohen, P., 267
Cohn, A. W., 13, 59, 60, 316
Cohn, E. G., 86
Cole, G. F., 298, 300
Colorado Revised Statutes Annotated, 128, 142, 144
Colson, C. W., 317
Comerci, G., 271
Common law, 5, 122, 127, 325
Community Action Team (CAT), 298
Community-based programs, 15, 239
Community Oriented Policing Services (COPS), 27
Community Response to Crime program, 243, 326
Conceptual schemes, 83, 84, 106, 326
Concurrent jurisdiction, 128–129, 130, 326
Conditioning, 95, 97, 99, 326
Conflict/radical/critical/marxist theories, 104–106
Conklin, J. E., 32, 85, 86, 88
Connecticut General Statutes Annotated, 130, 142, 144
Consolidated Laws of New York Annotated, 123
Constitutional rights, 116, 119–120, 156
Continuance under supervision, 158, 159f, 192, 326
Coohey, C., 49
Corbitt,W. A., 303
Corporal punishment, 23, 126, 231, 266
Correctional facilities, 251–254, 255–258, 259
Correia, M. E., 185
Cortoni, F. A., 105
Costello, B. J., 106
Cote, S., 86
Cothern, L., 254
Counseling, 108
  correctional facilities and, 255, 256
  education programs and, 219
  family services and, 225
  females and, 73, 173
  gangs and, 301
  police and, 181–182
  prevention programs and, 229
  probation and, 202, 240
  treatment centers and, 248
Court-appointed counsel, 193, 327
Court Appointed Special Advocates (CASAs), 204, 227, 327
Covington, J., 103
Covington, S., 283
Cox, S. M., 83, 105, 220, 345
Crack, hour by hour, 288
Crack cocaine, 22, 59, 60, 290, 291, 311, 323, 327
  See also Drugs
Crawford, A., 33
Cressey, D. R., 94, 101, 197
Crime:
  age and, 61–67, 62–66t
  black juveniles and, 79
  controlling, 85
  drugs and, 60
  gangs and, 290, 292–293, 297–298
  gender and, 67–73, 68–72t
  increase of, 280–281
  male teenagers and, 283
  reporting, 28–29, 30t, 35
Criminal sexual abuse, 269, 327
  See also Rape; Sexual abuse
Criminal subculture, 57, 327
Cripps, E., 105
Crips, 286, 290, 291, 323
Cromwell, P. F., Jr., 239
Crook, S., 11
Crosson-Tower, C., 266, 272
Crosswait, C., 97
Cullen, F. T., 3, 47, 49, 89, 101
Cunningham, S. M., 266

Curran, D. J., 102
Curry, G. D., 294, 296
Cutrona, C., 47
Cyr, M., 271

Daly, K., 105, 244
Dart, R. W., 291
Data, 26, 27, 30, 36
  abuse and, 267
  crime and, 280–281
  domestic violence and, 262–263
  gangs and, 288, 290
  unofficial sources of, 31–35
  See also Statistics
Davidson, H., 193
Davidson, N., 47
Davidson, W. S., 228
Davis, K. C., 171
Davis, N. J., 57
Davis, S. M., 112, 141
Dawkins, M. P., 60
Dawley, D., 304
Day, S. E., 243, 244
Death Penalty Information Center, 11, 254
Decker, S. H., 61
Defense counsel, 190, 191, 192–196, 328
Deficit Hyperactivity Disorder (ADHD), 89–90
Delinquency, neglect, abuse, and dependency cases, 26, 112, 328, 338
Delinquent subculture, 57, 98, 253, 254, 259, 278, 328
Delone, M., 172, 297
Demeanor, 171–172, 173, 328
Demonology, 84, 108, 328
Demuth, S., 48
Denno, D. W., 90
Dentler, R. A., 56
Department of Health, Education, and Welfare, 213
DePaul, J., 266
Deschenes, E. P., 295
Detention, 152–153, 156, 180, 248, 275, 328, 339, 346
Detention hearings, 113t, 147–152, 150–151f, 152, 328
Detention/shelter care, 152–158, 160, 161–162
  See also Shelter care
Deterrence theory, 85–86, 328
Differential association theory, 101–102
Dighton, D., 229
DiLillo, D., 266
Discretionary waiver, 130, 329
Discrimination, 73, 79, 108
  gangs and, 297
  juvenile justice system and, 309
  prevention programs and, 231
  See also Racism
Dispositional hearing, 112, 113t, 158, 160, 161–162, 238–239, 329
Dispositional order, 162f
District of Columbia Code, 124, 125, 126, 161
Diversion programs, 214, 216, 316
  criticisms of, 230–231
Divorce/separation, 47–48, 262–263
Doerner, W. G., 60
Domenech, L., 266
Domestic violence, 35, 96, 175, 222, 262–263, 263, 274, 329
  See also Abuse/neglect; Violence
Dorne, C., 6, 180, 254
Double jeopardy, 11, 135, 136, 324, 329
Doueck, H. J., 193
Dropouts, 42, 59, 216, 329
  African Americans and, 73
  crime and, 42
  delinquency and, 51
  drugs and, 59
  earlier intervention and, 211
  economic needs and, 58
  federal programs and, 226, 227
  neglect and, 268
  prevention programs and, 229
Drowns, R. W., 6, 170
Drug Abuse Resistance Education (DARE), 182, 217, 218, 302, 328
Drug courts, 222, 228, 229, 329
Drugs, 58–61
  African Americans and, 73
  amphetamines and, 60
  cocaine and, 59–60, 290
  correctional facilities and, 253
  crack cocaine and, 22, 59, 60, 290, 291, 311, 323, 327
  DARE and, 217
  education programs and, 217, 219
  educational systems and, 53, 54
  emotional abuse and, 268
  females and, 73, 283
  gangs and, 284, 288, 290–291, 293, 294, 297. *See also* Gangs
  juvenile justice system and, 316

male adolescent delinquency and, 41
marijuana and. *See* Marijuana
methamphetamines and, 22, 58, 60, 61, 119, 338
middle-class youths and, 57
neglect and, 268
police and, 172, 182
prevention programs and, 231
probation and, 240, 241
probation officers and, 202
sexual abuse and, 271
social factors and, 43
tobacco and, 217, 218
trafficking and, 290, 338
treatment centers and, 248
violence and, 283–284
waivers and, 134
See also Alcohol
Due process, 119, 120, 122, 123, 124, 134, 156, 157, 329
Dugdale, R. L., 89
Dukes, R. L., 217
Dunlop, L. C., 32
Dunn, M. J., 217
Dutton, S. E., 179

Early, D. M., 226
Echeburua, E., 97
Eddings v. Oklahoma, 254
Edens, J. F., 94
Education, xi, 50–51, 78–79
abuse and, 267
correctional facilities and, 255
employment and, 58
federal programs and, 225–228
history and, 6, 9
juvenile court acts and, 125–126
juvenile justice system and, 309
neglect and, 268
prevention programs and, 213
See also Schools; Training
Educational neglect, 266
Edwards, L. P., 195, 196, 197
Ehrenreich, B., 58
Eighth Amendment, 254
Eitzen, D. S., 226
El Rukn, 286, 296, 323, 329
Electronic monitoring, 21, 203, 242, 330
Elliott, D., 56, 214, 218
Ellis, R. A., 13, 67, 191
Elrod, P., 57
Emery, R. E., 48
Emotional abuse, 266, 268, 330
Emotional problems, 256, 299
Empey, L. T., 26, 31, 33
Employment:
correctional facilities and, 255, 258
educational systems and, 51
federal programs and, 225–228
prevention programs and, 229
probation officers and, 202
social background investigation and, 160
See also Unemployment
Engel, R. S., 34, 172
Ennett, S., 182
Enosh, G., 13, 157
Era of socialized juvenile justice, 9, 330
Ericson, N., 221
Erikson, K., 103
Ervin, J. D., 181, 182, 297, 298, 300
Esbensen, F., 218, 295, 296
Espiritu, R. C., 33
Eugenics, 91–92
Evans, D. G., 300, 301
Evans, M., 99
Evans, W. P., 293
Exclusive jurisdiction, 127, 128–129, 130, 330

Fader, D. P., 50
Fagan, A., 214, 218
Fagan, J., 59
Faith-based initiatives, 221, 317, 330
Faith-based mentors, 15
Families, 78, 79
correctional facilities and, 259
delinquent behavior and, 43, 44c, 45, 46–50
domestic violence and, 35, 96, 175, 222, 262–263, 274, 329
educational systems and, 50
family services and, 203–204
foster homes and, 245–247
gangs and, 284, 285, 294, 303
juvenile court acts and, 114, 116, 120, 127, 136
juvenile justice system and, 309
prevention programs and, 213–214, 230, 231, 233
prisons and, 315
probation and, 240, 241–242, 245
probation officers and, 203

sexual abuse and, 271
violent juveniles and, 299
Family court, 192
Family services, 203–204, 206
Fanton, J., 4, 20
Farrington, D. P., 32, 33, 106
Faust, F. L., 9, 10, 123
Federal Bureau of Investigation (FBI), 61, 280, 322, 350
female violence and, 283
gangs and, 296
incident-based reporting system and, 26
statistics and, 25
violent crime arrest rates and, 311
Federal programs, 225–228
Feiler, S. M., 73
Females, 34, 36
abuse and, 267. *See also* Abuse/neglect
broken homes and, 48
crime and, 67–73, 68–72t
custody and, 25
domestic violence and, 262
feminism and, 105
gangs and, 283, 295, 301
juvenile court acts and, 115–116
police and, 172, 173
violence and, 283
See also Gender
Fenwick, C. R., 49
Ferguson, J., 182
Fernandez-Montalvo, J., 97
Ferri, E., 87
Fifth Amendment, 11, 135, 141
Finkelhor, D., 263, 273, 274, 297
Finkenauer, J. O., 229
Firearms, 191, 281–284
See also Weapons
Fishbein, D. H., 88, 90
Fishman, L. T., 295
Fitzgerald, J. D., 83, 293, 345
Flannery, D. J., 49
Fleener, F. T., 47
Flewelling, R. L., 182
Florida Statutes Annotated, 128, 268, 269
Florsheim, P., 97
Folks, 291, 331
Follow Through, 226, 331
Ford, J. D., 197, 201
Fort, J., 296
Forum on Child and Family Statistics, 48
Foster care, 73, 272
Foster homes, 78, 245–247, 246, 247, 248, 275, 331, 339, 352
dispositional hearings and, 238
prevention programs and, 231
Fourteenth Amendment, 10, 11, 134, 141, 254
Fourth Amendment, 141, 331
Fox, J. A., 94
Fox, K., 196
Fox, S., 126
Fraser, J. A., 267
Frazier, C. E., 34, 73
Free will approach, 85, 331
Freud, S., 93, 94, 95, 343
Friedenberg, E. Z., 210
Friend, T., 90
Functionally related agencies, 214, 332

Gaebler, T., 320
Gagnon v. Scarpelli, 241
Gallegos v. Colorado, 144
Gandhi, A. G., 182
Gang nation, 292
Gang Resistance Education and Awareness Training (GREAT), 182, 218, 299, 333
Gangs, 278, 284–285, 290, 291, 295, 302–303, 320, 348
characteristics of, 287–288, 294
crime and, 297–298
drugs and, 60–61, 79. *See also* Gangs
ecological/social disorganization approach and, 101
economic needs and, 58
educational systems and, 51, 53
federal programs and, 227
females and, 283
GREAT and, 218
history and, 285–287
interactional theory and, 107
intervention programs and, 301
labeling theory and, 103
membership and, 293–294
NYGC and, 2, 300
programs and, 182–183, 211
prosecutors and, 191
recruitment and, 296–297, 320
strain theory and, 98
violence and, 279–281, 297–298
wilding and, 288
Garcia, M., 84

Gardner, E., 202
Garofalo, R., 87
Garrett, M., 31
Garry, E., 217
Gault case, 10, 12, 17, 120, 123, 124, 141, 144
  adjudicatory hearings and, 157
  dispositional hearings and, 238
  notification and, 155, 156
  prosecutors and, 191
Gender:
  crime and, 67–73, 68–72t
  gangs and, 294–295
  juvenile law and, 309
  police and, 171, 172, 182
  See also Females; Males
Georgia Code Annotated, 125, 126, 141, 143, 157, 160, 161
"Get tough" approach, 13, 83, 108, 308, 332
  deterrence theory and, 86
  gangs and, 303
  juvenile court acts and, 112
  rational choice theory and, 85
  violence and, 3, 279
Gewerth, K., 6, 180, 254
Giddings, M., 73, 172
Gilchrist, L. D., 60
Girls. *See* Females
Glaser, D., 102, 349
Glensor, R. W., 185
Gluck, S., 252
Glueck, E., 45, 88, 89
Glueck, S., 45, 88, 89
Goddard, H. H., 88, 92
Goldberg, R., 182
Goldman, J., 266
Goldstein, S. L., 179
Gomez-Smith, Z., 106
Gonzales, A., 271
Goodkind, S., 201, 269
Gordon, M. S., 7
Goring, C., 92
Gorman-Smith, D., 47
Gottfredson, M. R., 17, 106
Gough, H. G., 94
Grady, M. T., 179
Graffiti, 290, 293, 296, 332
Grasmick, H. G., 58, 278
Greenberg, D. F., 106
Gresham, F. M., 95
Griffin, B. S., 6
Griffin, C. T., 6
Group homes, 248
Guardian ad litem, 141, 193, 333
Guest-Warnick, G., 97
Guns *See* Weapons

Haley v. Ohio, 144
Halfway houses, 255
Halleck, S., 94
Hamarman, S., 268
Hamm, M. S., 283
Hansen, D., 218
Hanser, R. D., 202, 203, 300, 301
  gangs and, 278, 290, 292, 295, 297, 298
  juvenile court and, 196
  parents and, 299
  probation and, 196, 203
  violence and, 282, 283
Hanson, R. F., 271
Harcourt, B. E., 40
Harper, G. W., 295
Harris, P. R., 172
Harris, P. W., 50, 229, 257
Hart, T. C., 26, 27
Haskell, M., 51, 226
Hatchette, G., 308
Hawaii Revised Statutes, 142
Hawkins, J. D., 51
Hay, C., 26, 99
Hayes, H., 244
Head Start, 226, 331, 333
Healy, W., 94
Hearing officers, 196–197
Heck, W. P., 273
Heide, K. M., 82
Helping America's Youth, 43
Henretta, J. C., 34, 73
Henry, S., 85, 86, 92, 93, 278
Hess, K. M., 6, 170
Hewitt, J. D., 127, 171
Hibbler, W. J., 258
Hicks, R., 199, 272
Hieb, C. F., 291, 297, 303
High/Scope Perry Preschool Project, 219, 226
Hindelang, M. J., 32
Hinzman, G., 232
Hirschi, T., 17, 32, 105, 106
Hispanics:
  custody and, 25
  fear of attack at school and, 53–54, 55f

females and, 295
gangs and, 285, 290, 291, 292
police and, 172
Hogue, A., 213
Holmes case, 10, 333
Holsinger, K., 67
Holzman, H. R., 41
Hom, A., 51
Home monitoring, 15
Homicides:
children and, 263
gangs and, 290, 291
juvenile court acts and, 298
Hooton, E., 88, 90
Horowitz, R., 59, 256, 278
Houses of refuge, 7, 333, 335, 344
Howell, J. C., 61
Huff, C. R., 288, 290, 297, 298
Hughes, L. A., 290, 296, 297, 298
Huizinga, D., 33, 41
Hurst, Y. G., 173
Hybrid sentencing, 308, 317, 333
Hyman, D., 213
Hyperactivity, 89–90, 95

Iacono, W. G., 32
Id, ego, and superego, 93, 334, 343
Illinois Compiled Statutes, 20, 125, 140, 159, 176, 191, 266
Illinois Industrial School Act, 8
Illinois Juvenile Court Act, 20, 22, 112
In loco parentis, 5, 7, 8–9, 17, 158, 334
In need of supervision, 122, 124–125, 127, 140, 142, 144, 160, 161, 334
In re Gault, 10, 120, 141
In re George T., 157
In re Holmes, 10
In re Register, 127
In re William, A., 127
In re Winship, 11, 157
Inciardi, J. A., 59, 256
Indiana Code Annotated, 122, 130
Indystar.com, 54
Institute of Education Sciences, 54
Integrated theories, 106–107
Intensive supervision, 203, 242, 334
Internet Crimes Against Children, 273
Internet exploitation, 271–272, 334
Interrogation, 141, 144–147, 165, 180, 197, 335
Intervention, 272–274, 294, 335
Iowa Code Annotated, 129, 160
Irwin, K., 214, 218

Jacobs, 286, 294
Jails, 153, 180
See also Prisons
Jamieson, E., 367
Jargon, 296, 335
Jarjoura, G. R., 51, 58
Jeffers, J. L., 256
Jeffery, C. R., 90
Jenkins, P. H., 229
John Augustus, 239, 335
Johnson, J., 267
Johnson, K., 2, 3
Johnson, R. E., 57
Johnson, S., 67, 213
Jones, D. R., 89
Jones, M. B., 89
Jones, P. R., 50, 51, 257
Jones v. Commonwealth, 240
Jongman, R. W., 299
Judges, 206
correctional facilities and, 253
juvenile codes and, 205
juvenile courts and, 196–197, 199
leadership and, 316
probation and, 240
violent juveniles and, 300
Jurisdiction, 127–129, 130
Juvenile Causes Act, 4
Juvenile court judge, 194, 196–197, 199, 202, 204, 205, 336
Juvenile institutions, 251–254
Juvenile Justice and Delinquency Act, 12
Juvenile Mentoring Program (JUMP), 226–227, 335
Juvenile probation officer, 190, 199, 201–203, 205, 206, 337, 377

Kane, C., 298
Kanof, M., 217
Kaser-Boyd, N., 282, 283, 299
Katkin, D., 213
Kaufman, J., 102
Keilitz, I., 53
Kelley, D. H., 51
Kelly, J. B., 47
Kempf, K. L., 73
Kennedy, K. Y., 266

Kent v. United States, 10, 17, 123, 134, 141, 160, 191, 337
Kerper, H. B., 239
Killinger, G. G., 239
Klaver, J., 191
Klein, M. W., 278, 284, 287, 290
Klein-Saffran, J., 256
Klinger, D. A., 172
Klockars, C. B., 105
Klofas, J., 83, 84
Knudsen, D. D., 31, 87, 95, 102
Kowaleski-Jones, L., 51, 213, 214
Kramer, J., 213
Krasner, L., 97
Kratcoski, L. D., 170
Kratcoski, P. C., 170
Kretschmer, E., 89
Krisberg, B., 50, 73, 74, 104, 253, 273
Krohn, M. D., 281
Kropf, S. E., 22
Kurtz, D., 73, 172
Kvaraceus, W. C., 51

Labeling, 103–104, 108, 241–242, 337
  diversion programs and, 214
  educational systems and, 51
  gangs and, 286, 298
  juvenile court acts and, 114
  prevention and, 212
  probation and, 242
  self-concept and, 106
Lander, B., 101
Lane, J., 213
Lange, J., 227
Langton, L., 106
Lanier, M., 85, 86, 92, 93
Lardiero, C. J., 175
Latchkey children, 49, 268, 337
Latin Kings, 291, 337, 351
Latinos. *See* Hispanics
Laub, J. H., 191
Lawgiver judge, 199, 338
Lawyers. *See* Attorneys
Learning disabilities, 50, 53, 89, 90, 338
Learning theory, 95, 97, 101, 102, 212, 338, 349
Lee, L., 51
Legal definitions, 20–24, 36, 113, 114, 123, 266, 338
Legalistic approach, 12, 338
Leiber, M. J., 73, 74, 196
Levin, J., 94
Levine, D. R., 172
Levinson, R. B., 242
Liddle, H. A., 213
Lightfoot, E., 220
Likert scale, 173
Lilienfeld, S. O., 94
Lindner, C., 308
Lishner, D. M., 51
Lizotte, A. J., 281
Loeber, R., 33, 41, 47, 48, 53
Lombroso, C., 87, 88, 93
Lotz, R., 51
Louisiana Children's Code Annotated, 130, 134
Lowencamp, C. T., 101
Lower social class, 57, 58, 79, 295, 350
  delinquency and, 56
  educational systems and, 51, 53
  gangs and, 286, 293
  legal definitions and, 23, 24
  peer recognition and, 57
  self-report studies and, 32
  See also Poverty
Luckenbill, D. F., 101, 197
Ludwig, F. J., 4
Ludwig, J., 40
Lundman, R. L., 26, 211, 229
Lyerly, R. R., 101
Lynam, D. R., 95
Lyons, P., 193

MacArthur Foundation, 21
MacDonald, S. S., 199
Mack, M., 197
MacKenzie, D. L., 256
MacMillan, D. L., 95
MacMillan, H., 367
MacMurray, B. K., 191
Madamba, A. B., 50
Madigan, C., 232
Main, F., 291
Males:
  broken homes and, 48
  crime and, 67–73, 68–72t, 283
  domestic violence and, 262
  peer recognition and, 57
  police and, 171, 172, 173
  See also Gender
Mandated reporters, 177, 179, 338
Mandatory sentencing laws, 298

Mandoff, M., 104
Marijuana:
DARE and, 217
education programs and, 218
experimentation with, 36
gangs and, 290
schools and, 58–59
See also Drugs
Marshall, W. L., 105
Martens, S. L., 271
Martens, W. H. J., 94
Martin, G., 97
Martin, J. R., 182
Martinson, R. M., 251
Martinson, T. M., 251
Marx, B. P., 97
Maryland Annotated Code, 269
Maryland Courts and Judicial Procedures Code Annotated, 129
Massachusetts General Laws Annotated, 142
Matza, D., 98, 328, 349
May, D. C., 106
Mays, G. L., 170, 171, 179, 185, 191, 210, 211, 230, 278, 294, 297
McDermott, M. J., 173
McDuff, P., 271
McGloin, J. M., 278, 284, 285, 287, 290, 297, 298
McGue, M., 32
McKay, H., 278
McKay, H. B., 97
McKay, H. D., 98, 99, 101, 278, 285, 329
McKeiver v. Pennsylvania, 11, 142, 156, 338
McKinney, 122, 123, 124, 127, 157
McMahon, P., 273
McNulty, T. L., 74
McPhee, M., 291
Meddis, S. V., 283
Media:
boot camps and, 256
gangs and, 278, 286, 290, 297, 303
police and, 183
politics and, 279–280
probation and, 239–240
school violence and, 55
statistics and, 27
violence and, 283, 302
Medical care, 267, 268
Mednick, S. A., 90
Meehan, A. J., 170, 176
Mempa v. Rhay, 241
Mens rea, 5, 122, 338
Mental health issues:
African Americans and, 73
correctional facilities and, 252, 258
delinquency and, 53
dispositional hearings and, 162
females and, 73
juvenile court acts and, 127, 128
labeling theory and, 103
male adolescent delinquency and, 41
probation officers and, 202
sexual abuse and, 269
Mentoring, 226–227
Mercer, R., 203
Merlo, A.V., 73
Mertinko, E., 227
Merton, R. K., 56, 98
Methamphetamines, 22, 58, 60, 61, 119, 338
See also Drugs
Michie, 125, 130, 141
Michigan Compiled Laws Annotated, 123, 128, 130, 142
Middle-class, 31, 40, 56–57, 79
educational systems and, 50, 51
females and, 295
gangs and, 286, 293, 297
Mihalic, S., 214, 218, 219
Miller,W. B., 56, 57, 98, 286, 287
Minnesota Multiphasic Personality Inventory (MMPI), 94
Minnesota Statutes Annotated, 142
Minor Requiring Authoritative Intervention (MRAI), 205
Minorities:
educational systems and, 50
gangs and, 278
juvenile law and, 309
police and, 172
violence and, 283
Miranda, 141, 144, 145, 146, 165
Mississippi Code Annotated, 129
Mitchell, D. B., 22
Moffitt, T., 298, 299, 300
Monikers, 296, 303, 338
Monk-Turner, E., 51
Monroe, L. J., 56
Montana Code Annotated, 141, 142
Moon, M. M., 3
Moore, B. M., 317
Moore, M., 51

Morash, M., 34, 56, 171, 229
Morrissey v. Brewer, 241
Mothers Advocating Juvenile Justice, 315
Mothers Against Drunk Driving (MADD), 244
Moyer, I. L., 84, 85, 101, 105, 171
Munchausen's Syndrome By Proxy (MSBP), 273, 338
Mupier, R., 51
Murders. *See* Homicides
Murphy, J. M., 182, 199, 272
Murphy-Graham, E., 182

Naffine, N., 105
Narcotics. *See* Drugs
National Center for Juvenile Justice, 26, 336, 338
National Center for Missing & Exploited Children (NCMEC), 271
National Center on Child Abuse and Neglect, 27, 338
National Children's Advocacy Center, 27, 339
National Crime Victimization Survey (NCVS), 27, 30, 32, 339, 340, 351
National Criminal Justice Reference Service (NCJRS), 37, 288, 290, 291
National Incident-Based Reporting System (NIBRS), 26, 339
National Institute of Justice, 182, 278, 300
National Institute on Drug Abuse, 228
National Probation Act, 239, 339
National Youth Gang Center (NYGC), 2, 300
Nebraska Revised Statutes, 123, 142
Neglect, 267–268
  See also Abuse/neglect
Neiderhoffer, A., 294
Neighborhood Advisory Groups (NAGs), 298
Neo-Nazism, 283, 339
Neoclassical approach, 85, 339
Neubauer, D. W., 191, 193, 194, 195, 196, 201
New Jersey Statutes Annotated, 130
New Mexico Statutes Annotated, 141
New York Family Court Act, 124, 127, 157
New York Sessions Laws, 122
Ng, I., 201, 269
North Carolina General Statutes, 144
North Dakota Century Code, 126, 157
Notification, 147, 155–156, 177, 340
Novotney, L., 227
Nye, F. I., 32, 56
Nyquist, O., 4

Oates, K., 274
Offenses, 280, 283, 285, 290, 294, 297, 300, 302, 340
Offenses known to the police, 31t, 340
Office of Juvenile Justice and Delinquency Prevention (OJJDP), 22, 26, 121, 122, 203, 312, 340
  custody and, 25
  data and, 27
  earlier intervention and, 211
  grants and, 297–298
  habitual juvenile offenders and, 300
  juvenile offenders as adults and, 312
  police and, 171
  probation and, 242
  programs and, 231
  sexual abuse and, 269
  violent crimes and, 280
Official procedures, 179–180, 340
O'Hara, M., 67
Ohlin, L. E., 56, 57, 98, 308, 309
OJJDP Fact Sheet, 291
Oklahoma Statutes Annotated, 143
One-parent families, 47–48, 78
  See also Families
Oregon Revised Statutes, 142
Ormrod, R., 263, 273, 274, 297
Orsillo, S. M., 97
Osborne, D., 320
Osgood, D. W., 90, 293
Owen, B., 283

Pabon, E., 59
Pacific Islanders, 74, 75t
Pagani, L., 32
Palumbo, M. G., 182
Papachristos, A., 320
Parens patriae, 5, 8, 10, 17, 130, 144, 274, 324, 336, 340
"Parent figure" judge, 197, 340
Parents/guardians:
  abuse and, 266–267
  adjudicatory hearings and, 158
  CASAs and, 204
  detention and, 147
  detention/shelter care and, 152, 153
  dispositional hearings and, 161
  drug abuse and, 60
  education programs and, 219

families and, 47, 48. *See also* Families
federal programs and, 226
firearms and, 282
foster homes and, 245–247
gangs and, 297, 303
interrogation and, 144, 147
judges and, 197, 199
juvenile court acts and, 114
juvenile court proceedings and, 192–194
plea bargaining and, 195
police and, 171, 176, 177, 179–180
preliminary conferences and, 153
prevention programs and, 213
probation officers and, 201, 202
prosecutors and, 192
sexual abuse and, 269, 274
social background investigation and, 160
violent juveniles and, 299
work and, 49
Parker-Jimenez, J., 202
Parsons, A., 60
Pattillo, M. E., 293
Paul, R. H., 97
Paxson, C. H., 267
Peak, K. J., 185, 284, 292, 294
Pears, K. C., 266
Pearson, G., 197
Peas, J., 97
Pedophiles, 269, 270, 341
Peer pressure, 56, 57, 217, 253
People, 291, 331, 337, 341
People v. Dominquez, 240
Perkiss, M., 231
Perlmutter, B. F., 90
Personality inventories, 93, 94, 341
Peters, J. M., 34
Peters, S. D., 67, 283
Peters, S. R., 67, 283
Peterson, L., 266
Petition of Ferrier, 8, 365
Petitions, 154, 155f
adjudicatory hearings and, 156
police and, 171
preliminary conferences and, 153–154
Petrosino, A. J., 86
Pevalin, D. J., 59
Phrenology, 88, 341
Physical abuse, 263–268, 264f
Piliavin, I., 34, 171, 175
Piquero, A. R., 106
Piscotta, A. W., 8
Plass, P. S., 86
Platt, A., 104
Plea bargaining, 192, 194, 195–196, 197
Poe-Yamagata, E., 281
Pogrebin, M., 191
Police:
community-oriented policing and, 185
family services and, 203
gangs and, 320
informal adjustments and, 175–177, 179, 186, 192
juvenile courts and, 185
overpolicing and, 173, 340
probation officers and, 202
programs and, 181–184, 185
promotion/recognition and, 180, 181
training and, 173, 175, 177, 179–181, 182, 186, 187
violent juveniles and, 300
Police discretion, 170–173, 175, 341, 350
Police observational studies, 36, 341
Police–school liaison officers, 182, 341
Politicized gangs, 241, 283
Politics, 16, 108
community-oriented policing and, 185
conflict theories and, 104
family services and, 225
gangs and, 278, 283, 286, 287, 290
juvenile justice system and, 316
prosecutors and, 191
violence and, 279
Polk, K., 50, 51, 53
Pollock, J. M., 194
Porterfield, A. L., 32
Portwood, S. G., 179
Positive peer culture, 248, 257, 341
Positivist school, 87, 341, 342
Postadjudication intervention, 211, 342
Postclassical theory, 85, 342, 344
Postman, 320
Pottieger, A. E., 59, 256
Poulin, M. E., 50
Poverty, 40
abuse and, 267
delinquency and, 56
ecological/social disorganization approach and, 100, 101

emotional abuse and, 268
females and, 295–296
gangs and, 288
juvenile justice system and, 309
prevention programs and, 213
See also Lower social class
Poythrees, N. G., 94
Pratt, T. C., 89, 101
Preadjudication intervention, 211, 342
Pregnancy, 43, 219, 240
Preliminary conference, 113t, 140, 153–154, 201, 337, 342
Preliminary inquiry, 154
Preservice/Inservice training, 205, 342
Prevention, 210
educational systems and, 53
history and, 4, 6, 9, 10
police and, 172–173, 186
Prevention programs, 36, 211–214, 228–229, 233
child abuse/neglect and, 231–232
criticisms of, 230–231
Primary prevention, 211–214, 342
Prisons:
families and, 315
gangs and, 303, 320
medical treatment and, 312
racism and, 310–311
Private counsel, 31, 114, 193, 194, 343
Private detention facilities, 252, 343
Probable cause, 147
Probation, 239–245, 299
conditional release and, 241, 343
unofficial, 160, 166, 192, 199, 350
Probation officers, 190, 199, 201–203, 205, 206, 377
Promoting Alternative Thinking Strategies (PATHS), 219
Prosecutors, 190–192, 343
correctional facilities and, 253
defense counsels and, 194–196
violent juveniles and, 300
Prostitution, 60, 63t, 66t, 67, 69t, 71t
African Americans and, 74
drugs and, 60
gangs and, 287, 288
sexual abuse and, 268–269
Psychoanalytic approach, 93–94, 97, 343
Psychological abuse, 268
Psychological disorders, 256
Psychological factors, 278, 343
Psychological theories, 10, 92–97, 343
Psychopathic Personality Inventory (PPI), 94–95
Psychopaths, 94–95, 343
Psychotherapy, 108, 256
Public defenders, 193–194, 253
Public detention facilities, 252, 344
Punishment, 108
anger and, 106
classical theory and, 85
correctional facilities and, 252, 255
deterrence theory and, 85–86
gangs and, 290
judges and, 199
juvenile court acts and, 114, 120
juvenile justice system and, 317
plea bargaining and, 195
police and, 177, 179, 187
politics and, 279, 286
prevention programs and, 228, 230
prosecutors and, 191, 192
routine activities theory and, 86–87
schools and, 318
treatment and, 308
waivers and, 130
Pure diversion, 214, 344
Purpose statement of a juvenile court act, 112, 344
Puzzanchera, C. H., 103
Puzzanchera, C. M., 37

Quinney, R., 98, 104

Race, 73–77, 75–77t
gangs and, 278
juvenile justice system and, 312
juvenile law and, 309
police and, 171, 172, 173, 182
Racism:
gangs and, 283, 301
juvenile law and, 309
White legislatures and, 310
See also Discrimination
Radical nonintervention, 230, 344
Rafter, N., 88, 90
Rankin, J., 48
Rape:
gangs and, 288
increase of, 280
juvenile court acts and, 298
treatment centers and, 248

Rational choice theory, 85, 342, 344
Rebellon, C. J., 48
Reckless,W. C., 105
Redner, R., 228
Reform schools, 7, 8–9, 12, 344
Regoli, R. M., 127, 171
Rehabilitation:
  adjudicatory hearings and, 158
  capital punishment and, 254
  correctional facilities and, 251–254, 255, 256
  dispositional hearings and, 162
  history and, 3, 8, 10
  juvenile court acts and, 112, 114, 120, 123, 124
  juvenile justice system and, 317
  petition and, 154
  police and, 179
  probation officers and, 202
  programs and, 35, 114
  waivers and, 130, 134
Reid, S. T., 85, 88, 97, 106
Reiss, A. J., Jr., 26, 34, 171, 179
Religion, 228–229, 240, 255
Rendleman, D. R., 5
Rennison, C., 26, 27
Renzetti, C. M., 102
Reporting crimes, 28–29, 30t, 35
Repucci, N. D., 105
Resnick, H. S., 271
Restorative justice, 220–224, 229, 308, 317, 318, 345
Restorative or balanced justice, 308, 345
Revitalized juvenile justice, 309, 345
Revocation of probation, 162, 201, 240–241, 345
Rich, W. D., 53
Rinehart, W., 33
Ringwalt, C. L., 182
Roberson, C., 190
Roberts, A. R., 54, 197
Robinson, S., 255
Robinson, W. L., 295
Rodney, E. H., 51
Rodriguez, N., 202
Roesch, R., 191
Role identity confusion, 202, 345
Roper v. Simmons, 11, 12–13, 254, 345
Roseborough, D., 220
Rosenburg, 105
Ross, R. R., 97
Routine activities theory, 86–87, 345
Russi, K., 232
Ryder, R. S., 57

Salus, M. K., 266
Salzinger, S., 267
Sanders, W. B., 6
Sarri, R. C., 201, 269
Satterfield, J. H., 90
Saunders, B. E., 271
Sawdon, J., 300, 301
Scaramella, G. L., 58, 61
Scared Straight, 228–229, 345, 346
Schafer, W. E., 50, 51
Scherer, R., 271
Schinke, S. P., 60
Schmidt, L., 33
Schools, 50–51, 53–55, 79, 216
  community-oriented policing and, 185
  delinquency and, 51, 53
  dropouts and. *See* Dropouts
  drugs and, 58–59, 60–61
  early detection and, 233
  education programs and, 219
  fear and, 53–54, 55f
  gangs and, 278, 284–285, 297, 318–319
  male adolescent delinquency and, 41
  police and, 182, 217
  prevention programs and, 213, 214, 228–229, 230, 231
  probation and, 240, 242
  recruiting gang members and, 297
  safety and, 53, 54
  social background investigation and, 160
  truancy and. *See* Truancy
  violence and, 74, 299
  See also Education; Teachers
Schulze, A. D., 182
Schur, E. M., 230, 231
Schwartz, I. M., 13, 157
Scientific theory, 83–84, 345
Scope of a juvenile court act, 112, 345
Scott, J. W., 56
Scudder, R. G., 82, 102
Sealock, M. D., 172
Secondary diversion, 214, 346
Secondary prevention, 211, 346
Sedlak, A. J., 193, 203, 204
Seigel, 300
Seizures, 180, 197

Self-esteem:
abuse and, 266
control theories and, 105
education programs and, 219
educational systems and, 51
female gang members and, 296
gangs and, 293
punishment and, 87
Self-report studies, 32–33, 346
Sellers, C. S., 17, 94, 107
Senjo, S., 172, 229
Senna, J. J., 191, 192, 194, 197, 288
Sentencing order, 163–165f
Serious Habitual Offender Comprehensive Action Program (SHOCAP), 185, 346
Sexual abuse, 262, 266, 268–274, 327, 346
correctional facilities and, 253
females and, 73, 283. *See also* Females
parents and, 274
prevention programs and, 232
resource center and, 27
Sexual assault, 269, 327
See also Rape
Sexual exploitation, 269, 271, 346
Shaw, C. R., 98, 99, 101, 278, 285, 329
Sheldon, W. H., 88, 89
Sheley, J. F., 73
Shelter care, 127, 147, 152–153, 248
See also Detention/shelter care
Shelton, T. L., 97
Sherman, L. W., 86, 231
Shock intervention, 255–256, 346
Shorkey, 105
Short, J. F., 31, 32, 56, 288
Shotorbani, S., 97
Shusta, R. M., 172, 173, 175, 180, 181, 185
Sickmund, M., 4, 6, 9, 12, 14, 129, 133, 266, 268, 269, 281, 308
Siegal, L. J., 82, 102, 192, 194, 197
Siegel, J. A., 288, 297, 298, 299, 300, 367
Silverman, I. J., 82
Simms, S. O., 4
Simon City Royals, 291, 337, 346
Simons, L. G., 47
Simons, R. L., 47
Simonsen, C. E., 7, 49
Simpson, O. J., 195
Simpson, S., 172
Singh, B. K., 103
Sixth Amendment, 141, 156, 157, 347
Skinner, B. F., 95
Skipper, J. K., 101
Smith, C., 295, 296
Smith, C. A., 43
Smith, D., 56
Smykla, J. O., 53
Snyder, H. N., 4, 6, 9, 12, 266, 268, 269, 280, 281, 308
Sobol, J. J., 34, 172
Social background investigations, 160, 347
Social disorganization, 278, 347
Social factors, 43, 45–55, 347
Social promotions, 216, 347
Social service agencies, 140
Social workers, 202
Socialization process, 43, 45, 347
Socioeconomic status, 50, 53, 56, 57, 172, 347
Sociological factors, 173, 278, 347
Sociological theories, 83, 98–104, 347
Somatotypes, 89, 347
Sorrells, J., 299
Souryal, C. C., 256
South Dakota Codified Laws Annotated, 124
Sowers, K. M., 13, 67, 191
Spanking, 23, 231, 266
Spergel, I. A., 294
Spergel model, 297–298, 347–349
Spielman, F., 291
Spohn, C., 172, 297
Stafford, M. C., 26, 31
Stairs, J. M., 73, 74
Standard of preponderance of evidence, 125, 157, 348
Stanford v. Kentucky, 254
Stark, R., 101
Stationhouse adjustment, 140, 176–177, 186, 348
Statistical base, 273, 373
Statistics, 25–27, 30, 31, 36, 79
arrests and, 73–74, 75t
errors and, 30–31, 31t
See also Data
Status offenses, 12, 48, 122, 124, 136, 348
Statute of Artificers, 5–6
Stearns, M., 84
Steele, P. A., 253
Stein, J. A., 217
Stephens, G., 219
Stephens, R. D., 216

Sterilization, 90, 91–92
Stern, R. S., 48
Stern, S. B., 43, 48
Stojkovic, S., 83, 84
Stouthamer-Loeber, M., 33
Strain theory, 98–99
Street corner adjustment, 176, 348
Street corner family, 294, 348
Street gangs. *See* Gangs
Streib, V. L., 254
Strodtbeck, F., 288
Stroud, D. D., 271
Stuckey, G., 190, 194, 195, 196, 201
Substance abuse. *See* Drugs
Substance Abuse and Narcotics Education
Gang (SANE Gang), 183
Sudnow, D., 195
Summons, 144, 156
Sundt, J. L., 3
Surveys, 27, 30, 32, 36, 40
Susman, T., 309
Sutherland, E. H., 94, 101, 102, 197, 349
Sutpen, R., 73, 172
Swilaski, M., 181, 182, 297, 298, 300
Sykes, G. M., 98, 328, 349
Sykes, R. E., 26
Szymanski, L. A., 297, 298, 300

Taking into custody, 113t, 143–144, 348
Tannenbaum, F., 278
Tappan, P., 112, 154
Taylor, J., 32, 73
Taylor, R. L., 73
Teachers:
federal programs and, 226
firearms and, 282
gangs and, 303
prevention programs and, 213
sharing information and, 233
violence and, 299
See also Schools
Technical violations, 240–241, 349
Technology, 203, 242, 317
Teen courts, 145, 200, 228, 349
Temporary custody hearing order, 150–151f
Territorial jealousy, 214, 216, 233, 349
Terry, R. M., 34
Tertiary prevention, 211, 349
Tesorio, J. M., 281
Texas Family Code Annotated, 127, 128, 157
Texas Youth Commission., 255–256
Theories, 107
biological, 88–90
conflict/radical/critical/marxist, 104–106
integrated, 106–107
psychological, 92–97
scientific, 83–87
sociological, 98–104
Therapeutic approach, 12, 349
Theriault, C., 271
Thomas, D. L., 173
Thompson, G., 290, 292
Thompson v. Oklahoma, 254
Thompson,W. E., 5, 32, 50
Thornberry, T. P., 41, 51, 107, 281
Thornton, A., 50
Thornton, W. E., 60
Thrasher, F. M., 98, 278, 285, 295
Tifft, L. L., 32, 33
Tittle, C., 56
Tobacco, 217, 218
See also Drugs
Tobler, N. S., 182
Tolan, P. H., 47
Too poor to be defended, 207
Torbet, P.M., 201, 203, 205
Totality of circumstances, 141, 142, 144, 349
Tout, K., 49
Tower, C. C., 179, 180, 199
Training, 206
correctional facilities and, 255, 257, 258
education programs and, 218.
See also Education
federal programs and, 226
gang intervention programs and, 301
juvenile court acts and, 114, 120
juvenile court personnel and, 204–205
police and, 173, 175, 177, 179–181,
182, 186, 187
prevention programs and, 228
Treatment:
ABA and, 192–193
adjudicatory hearings and, 158
blacks and, 311
correctional facilities and, 253–254, 255, 256
dispositional hearings and, 162
family services and, 204
gangs and, 301
girls and, 116
judges and, 199

juvenile court acts and, 120, 123, 124
juvenile justice system and, 309, 317
petition and, 154
police and, 177, 179, 185
prevention programs and, 230
probation and, 241
probation officers and, 202
violent juveniles and, 300
waivers and, 130, 134
Treatment centers, 247–248
Treatment programs, 35, 252, 255, 300
Tremblay, G. C., 266
Trephining, 84, 350
Triplett, R. A., 58
Trojanowicz, R., 229
Truancy, 100
custody and, 144
females and, 73
juvenile court acts and, 122, 124
juvenile crime and, 13, 42, 43
neglect and, 268
prevention programs and, 229
Turk, A., 104
Turkheimer, E., 90
Turner, S., 213
Two-parent families, 48
See also Families
Tyler, R. P., 179

Ullman, J. B., 217
Ullman, L. P., 97
Umbreit, M., 220, 244
Underclass, 57, 58, 295, 350
See also Lower social class
Unemployment, 108
federal programs and, 226
juvenile justice system and, 309
prevention programs and, 213, 231
violent juveniles and, 299
See also Employment
Uniform Crime Reports (UCRs), 25, 26, 27, 350
Uniform Juvenile Court Act, 112, 113, 114, 116, 120, 122–123, 350
Unnever, J. D., 89
Unofficial probation, 160, 166, 192, 199, 350
See also Probation
Unofficial sources of data, 31, 350
See also Data
Unruly behavior, 350
adjudicatory hearings and, 157
detention and, 147
detention hearings and, 152
dispositional hearings and, 161
juvenile court acts and, 124–125, 127
notification and, 156
petition and, 154
Upper-class, 79
U.S. Bureau of the Census, 27
U.S. Department of Health and Human Services (DHHS), 25, 27, 142, 262–263, 267
U.S. Department of Justice, 27, 191, 271, 282
Utah Code Annotated, 123

Valdez, A., 182, 183, 278, 283, 284, 288, 290, 292, 295, 297, 300
Values:
education and, 50, 219
family and, 43, 45
middle-class youths and, 56, 57
Van Kammen, W. B., 33
Vander Ven, T. M., 49
Vandivere, S., 49
Vaz, E. W., 56
Vazsonyi, A., 49
Venkatesh, S. A., 41
Vermont Statutes Annotated, 142
Vice Lords, 286, 291, 323, 337, 351
Vice Queens, 295, 351
Victim survey research, 26, 32, 351
Victim–offender mediation programs, 242, 243, 244, 351
Victims Of Crime Impact Panels (VCIPs), 244, 351
Viljoen, J. L., 191, 195, 196, 197
Villemez, W., 56
Violence, 274, 275, 302
abuse/neglect and, 25. *See also* Abuse/neglect
African Americans and, 73–74
arrest rates and, 42
correctional facilities and, 253
decreasing of, 308
domestic, 35, 96, 175, 222, 262–263, 274, 329
drugs and, 60, 61
education programs and, 219
educational systems and, 53–54, 54–55

females and, 67, 73, 295. *See also* Females
firearms and, 281–284
gangs and, 2–3, 278, 279–281, 288, 290, 297–298. *See also* Gangs
incarceration and, 298–300
juvenile court acts and, 123
juvenile justice system and, 309, 312
Latino gangs and, 292
peer recognition and, 57
prevention programs and, 231
prosecutors and, 191
schools and, 53–54, 55f
waivers and, 135
See also Abuse/neglect
Violent Crime Index, 280, 351
Vitaro, F., 32
Vocational programs, 255
Voight, L., 60
Vold, G. B., 104
Vos, B., 220, 244
Voss, H. L., 32
Vowell, P. R., 106

Wade, J. E., 220
Wade, T. J., 59
Waivers, 129–135, 131–133t
Waldfogel, J., 267
Walker, C., 217, 220, 231, 239
Walker, S., 172, 173, 175, 297, 298
Wallace, H., 190
Wallerstein, J., 47
Walsh, A., 90
Walsh, C., 367
Walters, P. M., 185
Wannabes, 293, 298, 351
Wards of the court, 144, 158, 160, 245
Warrants, 144, 152
Washington, C., 309
Watson, D. W., 59
Weapons, 281–284
correctional facilities and, 253
educational systems and, 53, 54
firearms and, 191, 281–284
gangs and, 287, 288
juvenile justice system and, 316
waivers and, 134
Webber, A. M., 185
Weigel, D., 293
Weiner, N. A., 13, 157
Weis, J. G., 32
Weisburd, D., 86
Weiss, C. H., 182
Welfare, 144, 161, 213, 262
Welles, R. H., 179
Wells, E., 48
Wells, K., 22
Wells, S. J., 193
Welsh, B. C., 191, 288
Welsh, W. N., 191, 229, 288
Werthman, C., 34, 171
West, 142, 144, 157, 160
Weston, J., 298
White v. Illinois, 272–273, 351
Whites:
educational systems and, 50, 51
family and, 48
fear of attack at school and, 53–54, 55f
gangs and, 285, 291
police and, 172, 173
violent crimes and, 311
See also Middle-class
Whyte, W. F., 278, 294, 348
Wice, P. B., 194
Wilderness programs, 219–221, 242, 325, 340, 351
Wilding, 288, 352
Wilkins v. Missouri, 254
Wilkinson, R. A., 220
Wilks, J. A., 100
Williams, J. H., 60
Williams, J. M., 32
Williams, L. L., 49, 60
Williams, L. M., 82, 102
Willis, C. L., 53, 179
Willwerth, J., 49
Wilson, J. J., 285
Winfree, L. T., Jr., 7
gangs and, 294, 295, 297
police and, 170, 171, 179, 185
programs and, 210, 211, 230
prosecutors and, 191
Winship case, 11, 17, 157, 191, 352
Winters, C. A., 53
Wolcott, D., 266
Wolfe, D. A., 95
Women. *See* Females
Wong, H. Z., 172
Wooden, W. S., 56, 57, 173, 278, 282, 283

Worden, R. E., 34, 172
Wright, J. P., 3, 47, 49, 89, 271
Wyoming Statutes Annotated, 125n 2, 126, 133, 268
Wysong, 182

XYY chromosome, 90, 352

Yablonsky, L., 51, 226, 285
Yates, A., 271
Yogan, L. J., 51
Yoshikawa, H., 213
Youth culture, 56, 57, 352
Youth Development and Delinquency Prevention Administration, 213
Youth-Focused Community Policing (YFCP), 185, 187, 352

Zaslaw, J. G., 309
Zaslow, M. J., 49
Zatz, M., 202
Zehr, H., 53, 220
Zigler, E. F., 102
Zimmerman, J., 53
Zinn, M. B., 226

About the Authors

Steven M. Cox has been a member of the Law Enforcement and Justice Administration faculty at Western Illinois University since 1975. He earned his B.S. in psychology, M.A. in sociology, and Ph.D. in sociology at the University of Illinois in Urbana–Champaign. For the past 35 years, he has served as trainer and consultant to numerous criminal justice agencies in the United States and abroad. In addition, he has authored or coauthored numerous books and articles.

Rob D. Hanser is a full-time faculty member in the Department of Criminal Justice at the University of Louisiana at Monroe. He has dual licensure as a professional counselor in the states of Texas and Louisiana, is a certified anger resolution therapist, and has a specialty license in addictions counseling. He has worked as a child and adolescent therapist in an urban domestic violence shelter and has also worked as a secondary educator at an alternative school for troubled youth experiencing emotional and legal challenges.

Jennifer M. Allen is a full-time professor at Western Illinois University in the Department of Law Enforcement and Justice Administration. She has worked with juveniles in detention and on probation as well as with those victimized by abuse or neglect. She has published in the areas of ethics, restorative justice, juvenile delinquency and justice, and youth programming.

John J. Conrad served as chair of the Department of Law Enforcement and Justice Administration at Western Illinois University and was very active in the department, university, and surrounding community. After teaching for more than 30 years, he is now retired and enjoying his time traveling throughout the United States.